MOMENTUM: ART AND ECOLOGY IN CONTEMPORARY LATIN AMERICA

This publication is a project of the Patricia Phelps de Cisneros Institute for the Study of Art from Latin America, The Museum of Modern Art, New York

MOMENTUM
Art & Ecology in Contemporary Latin America

Edited by
INÉS KATZENSTEIN
MARÍA DEL CARMEN CARRIÓN
MADELINE MURPHY TURNER

THE MUSEUM OF MODERN ART, NEW YORK

CONTENTS

CONTENTS

Part II
Land Disputes and Colonial Legacies

Part III
Proposals for the Future

CONTENTS

CONTENTS

FOREWORD

There are many reasons to celebrate *Momentum: Art and Ecology in Contemporary Latin America*. It is the first publication produced by the Patricia Phelps de Cisneros Research Institute for the Study of Art from Latin America at The Museum of Modern Art, and it achieves one of the main goals set during the foundation of the Institute: to expand our knowledge of Latin American art through rigorous and experimental research, and deploying a strong network of collaborators, including artists, curators, scholars, and scientists from the region.

Momentum is also notable in its timing, since it focuses on an urgent discussion uniting the arts, the culture, the social policy, and the science of today: the ecological crisis. In this sense the Cisneros Institute research project that laid the ground for this book—one involving a multiplicity of platforms, including conferences, field trips, studio visits, and study groups—operated in virtuous synchrony with the museum's institutional commitment to sustainability and our growing acknowledgment that our intellectual and cultural work is inseparable from the biological and climatic systems that govern the planet. It is important to highlight that MoMA's curatorial and research programs extend to studying the ways in which artists, designers, and thinkers are opening new territory in imagining more sustainable ways of living, and that at the institutional level we are creating systems to reduce energy consumption and emissions by transforming the way we operate our buildings, striving toward achieving zero waste.

The Cisneros Institute was created thanks to the vision and generosity of our Trustee Patricia Phelps de Cisneros and the late Gustavo A. Cisneros with the aim of establishing within the museum a space permanently committed to Latin America. The Institute focuses on the study of the Cisneros's transformative gift of modern and contemporary Latin American art to MoMA, but as this book attests, it also goes beyond the gift, exploring artists and ideas new to the museum and its collection and connecting us with debates that challenge the ways we think about Latin America. The Cisneros Institute started operating in 2018, when we hired its inaugural director, Inés Katzenstein, who designed innovative research and fellowship programs that have placed Latin America at the center of many of the museum's conversations. We are honored to be housing the Institute and grateful to the Cisneros family for their trust.

Another feature of *Momentum* to celebrate is that while it was produced in collaboration with experts in the field and in both North and Latin America, it was conceived entirely in-house by the staff of the Institute: Inés Katzenstein, Director of the Cisneros Institute and Curator of Latin American Art; María del Carmen Carrión, Project Manager; and Madeline Murphy Turner, former Research Fellow of the Institute. I am grateful for their dedication to this project. I also want to recognize the inspiring work of the artists represented in these pages as well as of the authors of the book's essays, both original and reprinted, who prove the impressive leadership of the region in terms of ecological imagination. Their care for the planet is a guiding force for all of us invested in creating a sustainable environment now and for future generations.

—Glenn D. Lowry
The David Rockefeller Director
The Museum of Modern Art

ACKNOWLEDGMENTS

This volume involved the collaboration of many people who participated in different but fundamental ways.

Our profound gratitude goes first and foremost to Patricia Phelps de Cisneros for her forceful championing of Latin American art. Inscribed in the context of a museum such as The Museum of Modern Art, her vision of what has become the Patricia Phelps de Cisneros Research Institute for the Study of Art from Latin America created the conditions and the resources to make this project possible. Her absolute understanding and support of the academic independence of the Institute and of the value of our research is exemplary; it has allowed us to expand both the museum's knowledge of an underexplored field and its long-standing relationship to a vibrant field of art, broadening its scope in terms of artists, artistic vocabularies, and political challenges in times of profound social division and transformation.

In this, our first long-term research project, we are fortunate to have been able to rely on the feedback, intellectual guidance, and sustained support of the brilliant members of the Cisneros Institute Advisory Board: Adriana Cisneros, Stuart Comer, Tom Cummins, Mimi Haas, Jay Levenson, Harper Montgomery, and Gabriel Pérez-Barreiro. In them we have not only a sounding board but a group of active collaborators.

We also benefited from the input of editorial advisors Camila Marambio, Julieta González, and Graciela Speranza, whose profound commitment and collaboration are crucial to the book.

We are grateful for the effort and enthusiasm of the authors of the commissioned essays: Jens Andermann, Carla Acevedo-Yates, Lisa Blackmore, Helena Chávez Mac Gregor, Patricio del Real, Jose Falconi, Arnaud Gerspacher, Julieta González, Miguel A. López, Carla Macchiavello Cornejo, Camila Marambio, Joanna Page, Mara Polgovsky Ezcurra, Victor Manuel Rodríguez-Sarmiento, Irene V. Small, Graciela Speranza, and Patricia Vieira. We are also indebted to key writers whose brilliant ideas have proved fundamental to the broad field of art and the environment in Latin America, and whose work we are reprinting in this publication: Arturo Escobar, Gastón Gordillo, Eduardo Gudynas, Eduardo Kohn, Davi Kopenawa, Ailton Krenak, Floresmilo Simbaña, Daniel Steegman Mangrané, Maristella Svampa, Paulo Tavares, Gladys Tzul Tzul, and Eduardo Viveiros de Castro.

At MoMA, Glenn D. Lowry, the David Rockefeller Director, has trusted our vision and extended his unwavering support. Peter Reed, former Senior Deputy Director of Curatorial Affairs, and Sarah Suzuki, Associate Director, have guided us with care and goodwill.

From the beginning, this study was a concerted effort. Our sincere appreciation goes to Pablo Guardiola, Sofía Gallisá-Muriente, and Michael Linares, codirectors of Beta-Local in 2019, who hosted our initial think tank in San Juan, Puerto Rico. Our heartfelt gratitude extends to the participants of that convening for their abundant generosity, lucid thinking, and passion: Pablo Helguera, Sofía Hernández Chong-Cuy, Camila Marambio, Michy Marxuach, Rodrigo Moura, and the late Tamara Díaz Bringas. Our gathering in a post–Hurricane Maria archipelago set us to collectively decide that ecology and the potential of art to create awareness and the possibility of change were the right topics to kick off the work of the Cisneros Institute. Our MoMA colleagues Ruth Halvey, Sara Lookofsky, Iberia Pérez González, and Yudi Rafael accompanied us, thinking together, discovering artists and writers, and enriching our knowledge of the topic through weekly reading sessions that were mostly conducted online due to the COVID-19 pandemic. Immense respect goes to the many artists with whom we made online studio visits during the pandemic: Jennifer Allora and Guillermo Calzadilla, Hellen Ascoli, Firelei Baez, Adrián Balseca, Mabe Bethônico, Ursula

Biemann, Tania Candiani, Alia Farid, Ximena Garrido-Lecca, Jorge González, Iconoclasistas, Irene Kopelman, Dan Lie, Gilda Mantilla and Raymond Chávez, Alejandro Meitin, Jorgge Menna Barreto, Mónica Millán, Joiri Minaya, Delcy Morelos, Eduardo Navarro, Paulo Nazareth, Fernando Palma Rodríguez, Postcommodity, Naomi Rincón Gallardo, Thiago Rocha Pitta, Barbara Santos, Daniel Steegman Mangrané, Clarissa Tossin, Emerson Uyra, Erika Verzutti, and Castiel Vitorino Brasileiro. We are also grateful to our MoMA colleagues from the C-MAP Latin America study group who volunteered their time to lead virtual studio visits: Beverly Adams, Laura Braverman, Anna Burckhardt, Ruba Katrib, Inga Lāce, Lilia Taboada, Ana Torok, and Wong Binghao. Additionally, the internal research series *Pluriversal Perspectives* counted on the crucial participation of Laura Anderson Barbata, Sheroanawe Hakihiiwe, Max Jorge Hinderer Cruz, Sharon Lerner, Freddy Mamani, and Antonio Paucar.

Our research took public form first through online conferences and seminars and later through in-person gatherings and conferences. We want to thank our audiences in the United States and Latin America who joined our programs with enthusiasm, criticality, and joy. Our recognition also goes to the professional work of MoMA's AV team for their support and patience in helping us to navigate the then new territory of online programming. Deep thanks go to Camila Marambio, who conceived the first session, "*Cumbre Aconcagua*," as well as to the artists, theorists, activists, and lawyers who joined us for it: Maria Thereza Alves, Carolina Caycedo, Marisol de la Cadena, Francisca Fernández Droguett, Denise Ferreira da Silva, Bárbara Saavedra, Maristella Svampa, Catalina Valdés, Ignacio Valero, Cecilia Vicuña, and Nancy Yañez. Our second online conference, "Reweaving Ourselves: Contemporary Ecology through the Ideas of Juan Downey," was brilliantly designed by Julieta González, who invited María Belén Sáez de Ibarra, Minerva Cuevas, Arturo Escobar, Eduardo Kac, Javier Rivero Ramos, Beatriz Santiago Muñoz, Felicity Scott, Paulo Tavares, and Franco Viteri Gualinga to be participants. We send our appreciation to Tom Cummins and the Harvard team of Bruno Carvalho, Patricio del Real, and Thaïsa Way, who collaboratively conceived the online seminar "Recreating Territories," a program that brought together the ideas and practices of artists, architects, and theorists: Barreto, Caycedo, Garrido-Lecca, Gordillo, Leann Andrews, C. J. Alvarez, Felipe Correa, Ana María Durán, Guadalupe Maravilla, Esperanza Martínez, David de Rozas, Guillerme Wisnik, and Héctor Zamora. In June of 2022, we were at last able to convene in person in Valle de Bravo, Mexico, for the workshop "*El canto de la yerba bruja*," curated by Mauricio Marcín and Michy Marxuach, who gathered a thoughtful group of participants that included Colectivo Prácticas Narrativas, Alejandra Domínguez, Alia Farid, Jorge González, David Gutiérrez Castañeda, Ariel Guzik, Catalina Juárez, and Bernardo Zabalaga. The Harvard team continued to collaborate with us on a conference to close our research on this topic; our gratitude goes to Tom, Thaïsa, Bruno, and Patricio—along with their respective institutions and departments, Dumbarton Oaks and, at Harvard, the David Rockefeller Center for Latin American Studies Art, Film, and Culture program, the Mellon Urban Initiative, and the Department of History of Art and Architecture—who conceived with us "Energies and Imaginations: Hypotheses for a Present in Transition," presented at Harvard in October 2023. Our thanks to the artists, architects, and scholars who traveled long distances to join us there: Pedro Ignacio Alonso, Cara Daggett, Jota Mombaça, Marina Otero Verzier, Naufus Ramírez-Figueroa, and Elizabeth Wagemann. Also thanks to Marcela Ramos for her coordination work.

The support of MoMA's Department of Publications has been fundamental in the development of this book. Our gratitude goes to our colleagues Michelle Kuo, Hannah Kim, Anne Levine, Joseph Mohan, Matthew Pimm, Rebecca Roberts, and Curtis R. Scott, and to former staff Naomi Falk, Don McMahon, and Marc Sapir. We want to thank Sarah Demeuse, who offered insightful feedback on the new essays. On behalf of Camila Marambio, we want to extend a special thanks to Deborah Meacham. Graciela Speranza provided her expertise in editing texts originally written in Spanish, and David Frankel was our key collaborator in editing this ambitious publication. Christopher Winks and Jen/Eleana Hofer were our dedicated translators. Chad Kloepfer worked closely with us to bring this book to its final visual form. Our gratitude also extends to Editorial Caja Negra, who collaborated with us to produce the Spanish version of this volume. We appreciate everyone's collegiality and extraordinary work ethic.

We deeply acknowledge the work of former Cisneros Institute fellows Elise Chagas, Iberia Pérez González, Julián Sánchez González, and Horacio Ramos for their support in the research, public programs, and internal sessions, and their generous efforts to help us produce this publication. Our thanks also go to Daniel Pérez, Associate Chief Financial Officer and Director of Capital Projects, for his continuing assistance. We cherish the camaraderie and good spirit of our colleagues in the Department of Drawings and Prints, our home in the Museum, and would like to deeply thank Christophe Cherix, The Robert Lehman Foundation Chief Curator of Drawings and Prints, for his generous guidance.

Last but not least, we want to thank the artists whose works are discussed in these pages for the inspiration they provided. There are too many to mention here, but this book is not only indebted to them but dedicated to their work.

—Inés Katzenstein, María del Carmen Carrión,
and Madeline Murphy Turner, editors

INÉS KATZENSTEIN

MOMENTUM

Art and Ecology in Contemporary Latin America

This book brings together a wide range of ideas about how the visual arts have been metabolizing, confronting, and imagining alternatives in relation to the environmental crisis. Focusing on Latin America over the last five decades, it maps a long-developing and today wide and vibrant array of practices and ideas that have become urgent in both the arts and the broader culture and politics as we work to reckon with climate catastrophe.

The Cisneros Institute's foundational focus is Latin America, but beyond our dedication to the region, our geographical emphasis in *Momentum* arises from the continent's history of extractive colonialism, and its recent trailblazing contributions to global dialogues about art and ecology.[1] These contributions include, principally, the notion of *extractivismo* and the debates around the concept of nature as a commodity; the emergence of the rights of nature in the field of law (formally inaugurated by the plurinational constitutions of Ecuador and Bolivia in 2008 and 2009, respectively, as rights of the Madre Tierra or Pachamama); and last but not least, the increasing acknowledgment of Indigenous cosmovisions as models for an ethical relationship to the natural world. This book gathers together some of the most essential figures in the fields of art practice and theory, from anthropologists and Indigenous activists to artists, curators, and writers.

This publication—the Institute's first, edited with the Institute's project manager, María del Carmen Carrión, and former Cisneros Institute research fellow Madeline Murphy Turner—brings together sixteen newly commissioned essays on art along with eleven previously published interdisciplinary texts that serve to contextualize the issues at stake from diverse perspectives. It presents a range of political and aesthetic positions within the growing field of art dedicated to matters of ecology. But the discussions that unfold in this book also serve to document other crucial cultural shifts taking place in Latin American art, all inextricably linked to the environmental crisis. I refer to the shifts of artistic vocabularies toward the recuperation of vernacular techniques and histories; the expansion of the field to encompass the voices of Indigenous artists until recently excluded from consideration in the contemporary art world; and lastly, the transformation of modern, linear conceptions of time through radical critiques of development and progress as the dominant terms for understanding history.[2] Thus, in addition to offering straightforward positions on strictly ecological issues, this book indirectly addresses these interwoven historical transformations.

Analyzing artistic projects in different parts of Latin America, from Puerto Rico to Chile, from Amazonia to the Andes, most of the contributors here offer not only diagnoses of our current, fast-changing ecological situation but conceptual models for surviving and even perhaps easing the crisis. The terms are manifold, including the positioning of "imagination in a struggle for wholeness" (Carla Macchiavello Cornejo), the adoption of specific practices of care (Joanna Page and Mara Polgovsky Ezcurra), the construction of postdevelopmental futures (Julieta González), the idea of postextractive imagination (Lisa Blackmore), refutations of concepts of "future" and "hope" (Miguel A. López and Helena Chávez Mac Gregor), the practice of a postanthropocentric law (Paulo Tavares), and the relinquishment of the idea of the autonomy of art in pursuit of experimental alliances of survival among artists, activists, scientists, citizens, and nonhumans (Jens Andermann). Despite this multiplicity of perspectives, and the broad scope of the artistic practices and territorial challenges discussed here, all of the contributors, while imagining better futures, share an understanding of our need to recompose a lost integrity by engaging in alternative forms of relation to, care of, and cooperation with nonhumans. In order to reachieve this integrity we must replace old ideas of "production" with "practices of engendering," to use Bruno Latour's terms, in which origin and gender are considered from the start.[3] Last but not least, most contributors seem to agree that art, insofar as it has historically been a privileged vehicle for the fetishization of nature, offers the possibility of imagining such forms of engagement anew, envisioning alternatives to the organization of life as we conceive it today.

The newly commissioned essays are organized in three sections. The first, "Interspecies Thinking," explores different perspectives on the question of interspecies relationships, disputing the Cartesian divide that separates nature from culture. It highlights early, speculative experiments with technology and cybernetics in the late 1960s and '70s, poetry as reparation for a lost sense of communion with the natural world, and the work of emerging Indigenous artists who think beyond the binaries of human/nonhuman. The second, "Land Disputes and Colonial Legacies," addresses extractivism, presenting artists who critique the colonial and neocolonial systems that have so seriously damaged the land, the waters, and living species in general, and that threaten worse damage still. A commitment to investigating hidden histories of land misuse, colonial appropriation, and social dispossession is central to these texts. The final section, "Proposals for the Future," gathers texts about artists and collectives whose practices focus on ecology-centered ways of living, absorbing proposals for radical subjective or social change—from ecofeminist projects to artistic neo-shamanisms.[4]

These commissioned essays are followed by a selection of existing interdisciplinary texts that speak to the most crucial theoretical and political debates in Latin American environmental thinking from the end of the 1990s to the present. Featuring contributions by both Western anthropologists and Indigenous activists, these texts discuss the centrality of Indigenous thinking in the region and suggest ways of imagining a world beyond ecological debacle, including the perspectives introduced by newly coined terms such as the "pluriverse," "nonhuman rights," and "environmental justice." The global infrastructure of extractivism, and the actions of brave and powerful ecoterritorial feminisms in the region, are also highlighted.

STEPPING IN FROM AFAR

This book is the result of a multifaceted process of collective research and exchange undertaken from 2019 to 2023 across various platforms. Led by the team from The Museum of Modern Art's Cisneros Institute, established in 2016, this process made the book possible, even if it is not mirrored directly in all of the essays that follow.

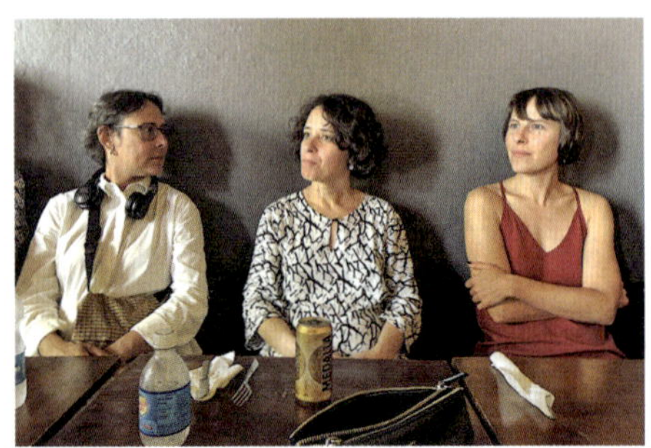

1 Left to right: curators Michy Marxuach, the late Tamara Díaz Bringas, and Camila Marambio at a think tank in San Juan, Puerto Rico, organized by the Cisneros Institute, The Museum of Modern Art, New York, in May 2019, in preparation for the research project leading to this book

The idea of exploring the intersection of art and ecology came out of an independent think tank that gathered ad hoc in San Juan, Puerto Rico, in April 2019. Over a week of discussions with a group of curators from Latin America, our ways of relating to planet Earth emerged as one of the most pressing and potentially enriching topics that would anchor the Institute's initial engagement with the region.[5] The conversations in Puerto Rico were followed by the formation of a study and reading group that went on to meet weekly for two years.[6] The readings and intense discussions integral to this effort established the foundation for the Institute's programs and inspired the commission of the essays included in this book. The readings offered us new conceptual tools for a more ethical reconnection with lifeworlds, and also made clear to us that but for a few key texts, there was little written scholarship explicitly dedicated to the question of art and the ecological crisis in Latin America.[7]

The dystopian global halt created by the COVID-19 pandemic, with its unexpected conditions of lockdown, social distancing, and, importantly, our inability to travel to the places where the art in question had originated, defined the most significant phase of the book's preparation. But since the pandemic was itself the result of environmental distortions, the sense of purpose that had imbued the project from its beginnings became even more urgent. To generate the parallel programs that eventually led to the book, we engaged in dialogue with a score of scholars, curators, and artists. We began with two conferences. The first, "*Cumbre Aconcagua*" (Aconcagua summit), dealt with water, a crucial topic in the ecological debate and one that immediately after became a central trope in activism, books, and global art exhibitions related to the environment.[8] Focusing on issues related to water and the rights of nature in Chile, a territory defined by the privatization of its natural resources, this conference was originally planned to take place in the Chilean Andes. Due to the pandemic, however, it became our first virtual program, held online in August and September 2020. Through dialogues with the Colombian artist Carolina Caycedo, the Chilean poet and artist Cecilia Vicuña, and the Brazilian artist Maria Thereza Alves, it examined how water has been owned, used, and debated in the Americas. Conceived by Chilean guest curator Camila Marambio, the conference included these three conversations and an experimental written forum, "The Water Is the Law."[9]

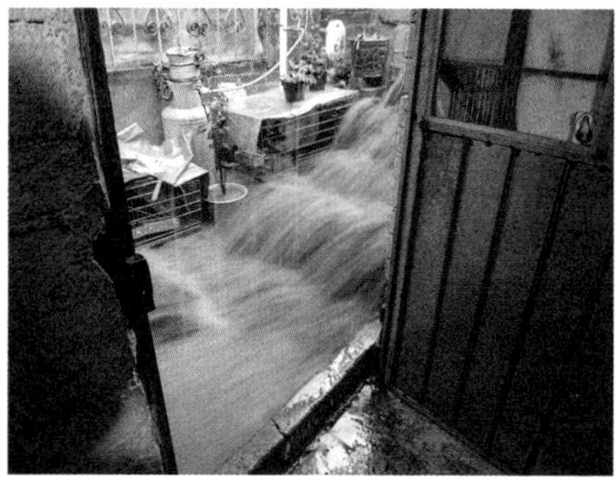

2 Maria Thereza Alves. *Il regreso de un lago/The Return of a Lake*. Cologne: Walter König Verlag, 2012. Image from artist's book published to accompany Alves's installation at DOCUMENTA 13, Kassel, Germany, 2012. A typical day during the rainy season in Chalco. Collection the artist

Our second conference, "Reweaving Ourselves: Contemporary Ecology through the Ideas of Juan Downey," took a historical perspective, positioning the ideas and work of the late Chilean artist as a referent and thread in understanding how artists have addressed the environment and technology from the 1960s to the present. The Venezuelan curator Julieta González, who guest-curated the conference, wrote, "Looking back at Downey's conception of ecology from the perspectives of cybernetics, anthropology, and the ways in which his work and thought posed challenges to the onto-epistemic formations of capitalist modernity, the conference aim[ed] to assess the shifts that have taken place and are currently shaping the relations between technology and ecology, the political space constituted by the natural environment of Indigenous communities, and the epistemologies of the South as alternatives to development and modernity."[10]

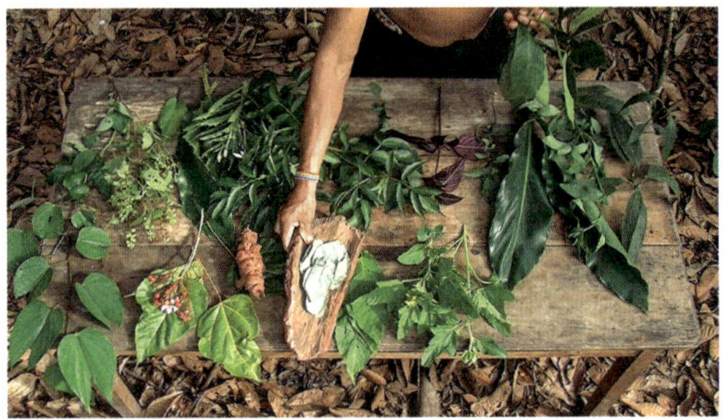

3 Ursula Biemann and Paulo Tavares. *Forest Law*. 2014. Image from synchronized two-channel video installation, 38 min. Centre national des arts plastiques, Paris

While designing these conferences in collaboration with Marambio and González, we were also conducting a weekly series of "virtual studio visits" with an expansive group of artists in the region, all of whom think through their artistic practices in diverse relation to topics of ecology, extractivism, and the interspecies. These more than thirty conversations clarified the breadth and nuance of contemporary artistic practice in the area, the sense of which this book tries to encapsulate.

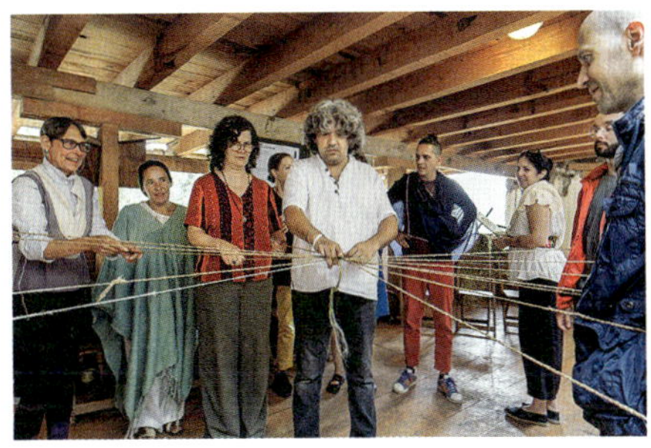

4 Jorge González conducting the workshop *Hilando Juntxs. Espacio de estar* (Spinning together. Space of being), which focused on *soles de Naranjito*, a traditional weaving technique of Puerto Rico. Part of "*El Canto de la yerba bruja*" (The song of the witch grass), a workshop organized by the Cisneros Institute, The Museum of Modern Art, New York, in Valle de Bravo, Mexico, June 2022

In 2021, we expanded the scope of the disciplines under investigation on the occasion of a collaborative seminar organized with scholars from Harvard University. Architecture, urban planning, and design were discussed alongside art in terms of their engagement with environmental aesthetics and politics. In that seminar, "Recreating Territories: Art and Urban Imaginations," we heard about fog catchers installed in a village in the dry Peruvian Andes to collect clean water, and about an artist's efforts to heal, through sound baths, the traumas of displacement felt in the bodies of Central American migrants arriving in New York City. We learned about the growing urbanization and interconnection all over the planet caused by the never-ending flow of data and commodities, connecting, invisibly and inextricably, a rural village in northern Argentina, small and remote, to pig farms in China. And finally, in conversations about resource extraction in the Ecuadorian Amazon, we discussed the vital role of Indigenous sovereignty.[11]

YEARNING PRACTICES

Central to the Institute's initial ideas about how to conduct research from its geopolitical position within an institution such as The Museum of Modern Art was the creation of spaces for different forms of dialogical exchange both with and within Latin America. From the start, it was important for us to create strategies for relativizing the parameters and constraints implicit in producing content in English from our position in New York, parameters and constraints connected to the standards of American academia and the expectations of North American institutions in general. We wanted to open the platforms of the Institute to a wide range of voices, including not only diasporic scholars based in the North but also the thinking produced under both institutional and parainstitutional conditions in the South. The pandemic put a temporary halt to this idea, at least in terms of its realization within Latin America. After the lockdown was lifted, in 2022, we were finally able to resume travel and to put situated methodologies into practice.

Some of the most insistent terms that we had seen emerging in artists' discursive and performative practices during this period were "healing" and "repair" in the context of ecological threat, on the one hand, and a reconnection with spirituality and local traditions, on the other. Interested in exploring the implications of these concepts, we invited curators Michy Marxuach and Mauricio Marcín to work with us in planning an in-person convening. Those questions came to fruition in "*El canto de la yerba bruja*" (The song of the witch grass), a research workshop held in Mexico City and Valle de Bravo in June 2022. Through case studies of the work of four artists—Jorge González (Puerto Rico), Bernardo Zabalaga (Bolivia/Argentina), Alia Farid (Puerto Rico/Kuwait), and Ariel Guzik (Mexico)—the workshop explored what we participating in it collectively defined as "yearning practices" (*prácticas anhelantes*), including healing, artisanal, and sound works.

As climate change intensifies, the ecological situation of Latin America has evolved. At least in progressive media, there is finally a consensus about the severity of climate change and the critical situation in certain areas of the region—most visibly the Amazon basin, which is threatened by deforestation, monocropping, and illegal mining. Public discussion, however, has been dominated less by this acceptance of the facts than by the challenges of energy transition. Along these lines, in the North, states and multilateral blocks have set public goals and timelines for themselves and have activated a new "climate diplomacy" to guarantee resources and partnerships; corporations have strategically updated their discourses and begun to expand rapidly into green energies; and the goals of decarbonizing power and transportation networks and of designing systems for carbon capture and storage are gaining traction. Yet it nonetheless seems to us that little discussion has been dedicated to the other end of the system, to reducing our consumption of energy as opposed to finding alternative ways to produce it. In the discourses of prosperous countries, the narratives of transition rarely consider the urgency of a radical change in our existing modes of intense, hyperconnected patterns of consumption. In the meantime, in the South, where, for many, access to basic systems of energy is still an unrealized promise, the extractivist model perpetuates itself as the only way the establishment seems to envision an escape from extreme deficit and debt.

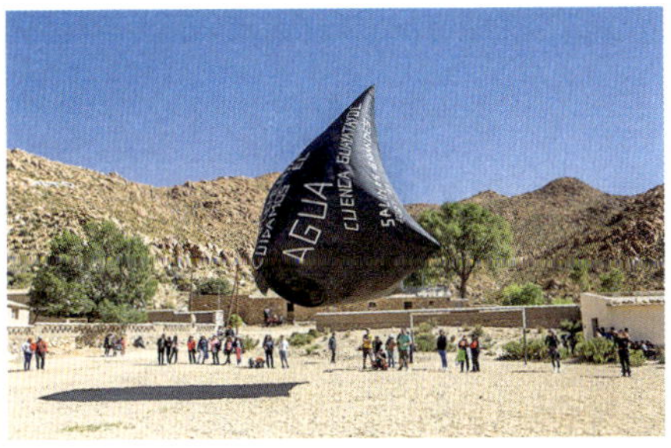

5 *Encuentro Salinas Grandes* (Salt flats meeting), Aerocene-movement event at El Alfarcito, Argentina, January 2023. The balloon bears a message written collectively with the communities of Salinas Grandes and Laguna de Guayatayoc: "*En complementariedad cuidamos el agua*" (In complementarity we take care of the water)

In this context, another problem has emerged: the actual ecological and social tolls of energy transition. As Thea Riofrancos has observed, "One of the most recent innovations in *extractivismo* discourse targets the forms of extraction and dispossession that accompany efforts to confront climate change. These are the extractive frontiers of green technology supply chains and the land-use patterns of large-scale solar and wind farms."[12] According to Riofrancos and many other analysts (including some artists), energy transition, as it is seen today, paradoxically entails a new phase of extractivism and injustice. The Pacto Ecosocial e Intercultural del Sur, one of the most active transnational and transdisciplinary ecological collectives, is denouncing the new, green pressures on the Global South, pointing to the consequences of the extraction of cobalt and lithium (key for the production of green batteries), to the problems of land use for solar panels, and to the extraction of balsa wood to supply a new demand for it in the construction

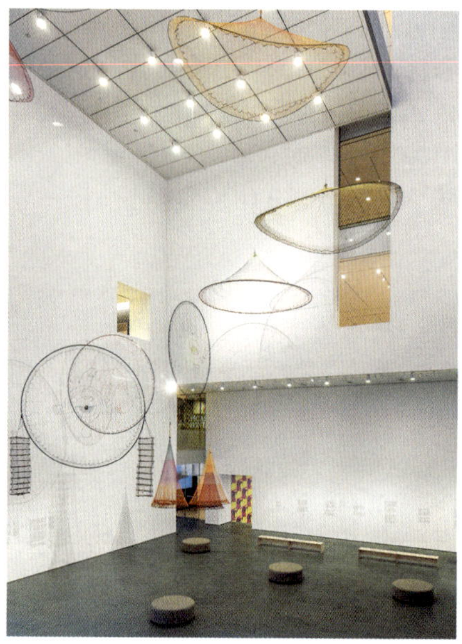

6 Carolina Caycedo. *Spiral for Shared Dreams* from the series *Cosmoatarrayas* (Cosmonets). 2022. Nylon, polypropylene, hemp, dye, and metal, dimensions variable. Gift of the Fund for the Twenty-First Century, Latin American and Caribbean Fund, and Adriana Cisneros de Griffin. Installation view of the exhibition *Carolina Caycedo: Spiral for Shared Dreams*, November 11, 2023–May 19, 2024. The Museum of Modern Art, New York

of wind-turbine propellers. According to Maristella Svampa and Breno Bringuel, both members of the Pacto Ecosocial, what it calls "new energy colonialism" "condemns the peripheral countries to be zones of sacrifice, without changing the metabolic profile of society nor the predatory relationship with nature."[13]

In February 2022, I participated in an intercultural exchange with Indigenous communities engaged in ecological activism on the occasion of the *Encuentro Salinas Grandes* (Salt flats meeting), organized by artist Tomás Saraceno in Jujuy, northern Argentina. There we saw hundreds of Indigenous peoples gathering in the tiny town of El Alfarcito with a group of lawyers, environmental activists, filmmakers, and writers to strategize on how to confront the new rush for lithium used in the manufacture of batteries. These tensions between the emergence of green-transition projects in the North and the persistence of "zones of sacrifice" in the South was one of the problems we addressed in our last conference, "Energies and Imaginations: Hypotheses for a Present in Transition," held at Harvard University in 2023 and featuring guest presenters from different contexts and disciplines.[14]

At that conference, confronted with the question of how to think about energy transition, the Brazilian artist Jota Mombaça proposed, tongue in cheek, to embrace blackout, darkness, myth—challenging the very notion that we need or even want to transition toward another, superior, greener phase of modernity. That moment of the session saw an opening to an entirely different dimension of possibility, very much expressing what the Argentine scholar Graciela Speranza thinks art does: daring beyond the impoverished realism of economics or politics.[15] *Momentum* narrates different ways in which the move away from realism has been unfolding in Latin America at this crucial moment of public reckoning with climate change. By proposing a blackout as a performative halt to the violence of modernity, imagining different forms of rebirth coming up from ruins, and producing a reenchantment of our relationship to nonhuman species, art is expanding our ecological imaginaries toward experiences of empathetic, communal, and nondestructive uncertainty.

In respect to the life of our ecosystems, the political landscape of Latin America is staging an unequal and often violent confrontation between opposing views of the future. In this volatile context of ideological polarization, we can only hope that the ideas expressed in this book, and embedded in the artworks discussed, serve as tools for a much-needed political reeducation of our imagination.

1 "Latin America" is the name generally used in the Western world to refer to the entire continent of South America in addition to Mexico, Central America, and the Caribbean islands. In recent years, in an attempt to decolonize the naming of the land, a growing number of people have begun to use the name "Abya Yala" to refer to the region. The practice dates back to at least 2004, when the II Cumbre Continental de los Pueblos y Nacionalidades Indígenas de Abya Yala (Second continental summit of the Indigenous peoples and nationalities of Abya Yala) was held in Quito, Ecuador. In the Indigenous Kuna language the term means "land in its full maturity." For the sake of consistency, I use "Latin America" here, while remaining aware of the complexity of the designation.

2 For more on these shifts see Inés Katzenstein, "Chosen Memories: Contemporary Latin American Art from the Patricia Phelps de Cisneros Gift and Beyond," in Katzenstein, *Chosen Memories: Contemporary Latin American Art from the Patricia Phelps de Cisneros Gift and Beyond*, exh. cat. (New York: The Museum of Modern Art, 2023), 10–97.

3 See Bruno Latour, "Production or Engendering?," in Nick Axel, Daniel A. Barber, Nikolaus Hirsch, and Anton Vidokle, eds., *Accumulation: The Art, Architecture, and Media of Climate Change* (Minneapolis: University of Minnesota Press, e-flux Architecture, 2022), 241–46.

4 The brief texts introducing each of the sections were written by Madeline Murphy Turner.

5 The idea that our research focus should be both attuned to the priorities of The Museum of Modern Art but reliant on the experience of curators and scholars from Latin America was key from the beginning of the process. Participants in the San Juan think tank were María del Carmen Carrión, the late Tamara Díaz Bringas, Pablo Helguera, Sofía Hernández Chong Cuy, Camila Marambio, Michy Marxuach, Rodrigo Moura, Iberia Pérez González, and myself.

6 Our study group ran from June 2019 to June 2022 and included, on and off, Carrión, Turner, Ruth Halvey, Sarah Lookofsky, Iberia Pérez, Yudi Rafael, and myself. In November 2019, the topic was presented to and approved by the Cisneros Institute's Advisory Board, which is composed of Adriana Cisneros Phelps de Griffin, Stuart Comer, Thomas B. F. Cummins, Mimi Haas, Jay Levenson, Harper Montgomery, and Gabriel Pérez-Barreiro.

7 For one such text, e.g., see Macarena Gómez-Barris, *The Extractive Zone: Social Ecologies and Decolonial Perspectives* (Durham, NC, and London: Duke University Press, 2017).

8 In the arts, most prominently, we can mention the thirteenth Shanghai Biennale, titled *Bodies of Water*, curated by Andrés Jaque, April 10–July 18, 2021, and the 23rd Biennale of Sydney, titled *rīvus*, curated by José Roca, March 12–June 13, 2022.

9 "*Cumbre Aconcagua*" was organized in three sessions: "Part One. *Digna Rabia* [Worthy rage] and Moral Hazard," a dialogue between scholar Ignacio Valero and artist Carolina Caycedo (August 11, 2020); "Part Two. *El Robo* (Theft)," a conversation between scholar Denise Ferreira da Silva and artist Maria Thereza Alves (August 25, 2020); and "Part Three. *La memoria del agua* (The Memory of Water)," an exchange between anthropologist Marisol de la Cadena and artist and poet Cecilia Vicuña (September 9, 2020), all moderated by Marambio. The forum "The Water Is the Law" included contributions

by Bárbara Saavedra, Francisca Fernández, Catalina Valdés, Nancy Yáñez, and Maristella Svampa. See "The Aconcagua Summit: The Water Is the Law," available online at www.moma.org/momaorg/shared/pdfs/docs/research/cisneros/agua-brochure-english-v2.pdf (accessed May 13, 2024).

10 "Reweaving Ourselves: Contemporary Ecology through the Ideas of Juan Downey" included four sessions: "Part One. Juan Downey: Cybernetics, Ecology, and an Ontology of Becoming," a conversation between curator Julieta González and scholar Javier Rivero Ramos (November 10, 2020); "Part Two. Technology for a Disoriented Humanity," with presentations by architectural historian Felicity Scott and artist Eduardo Kac (November 18, 2020); "Part Three. The Cosmopolitics of the Forest," with presentations by architect Paulo Tavares, Sarayaku community member Franco Viteri Gualinga, and curator María Belén Sáez de Ibarra (November 24, 2020); and "Part Four. Beyond Technology and Development," with presentations by anthropologist Arturo Escobar and artists Minerva Cuevas and Beatriz Santiago Muñoz (December 1, 2020), all moderated by González. See www.moma.org/research/cisneros/extended-research-art-and-ecology (accessed May 13, 2024).

11 This initiative took place in 2021 as a collaboration between the Cisneros Institute at The Museum of Modern Art; the Dumbarton Oaks Research Library and Collection; the Art, Film, & Culture Program, David Rockefeller Center for Latin American Studies, Harvard University; the Harvard Mellon Urban Initiative; and the Harvard University Department of History of Art + Architecture. It was organized by Carrión, Cummins, Bruno de Carvalho, Patricio del Real, Thaïsa Way, and myself. The presenters were organized in pairs: Felipe Correa and Esperanza Martínez; C. J. Alvarez and the partnership of Caycedo and David de Rozas; Ana María Durán Calisto and Jorgge Menna Barreto; Ximena Garrido-Lecca and Gastón Gordillo; Guilherme Wisnik and Héctor Zamora; and Leann Andrews and Guadalupe Maravilla.

12 Thea Riofrancos, "Extractivism and Extractivismo," *Global South Studies: A Collective Publication with "The Global South"*, November 11, 2020. Available online at https://globalsouthstudies.as.virginia.edu/key-concepts/extractivism-and-extractivismo (accessed April 13, 2024).

13 Breno Bringel and Svampa, "Del 'Consenso de los Commodities' al 'Consenso de la Descarbonización,'" *Nueva Sociedad* 306 (July-August 2023): 51–60. Available online at https://nuso.org/articulo/306-del-consenso-de-los-commodities-al-consenso-de-la-descarbonizacion/ (accessed April 13, 2024).

14 Speakers included Pedro Ignacio Alonso, Cara Daggett, Jota Mombaça, Marina Otero Verzier, Naufus Ramírez-Figueroa, and Elizabeth Wagemann.

15 "It is clear that in the discourses of politics, economics, and sometimes even the social sciences, a crass realism reigns, incapable of imagining the future. But this impoverished notion of realism is less at home in art, even in art that has no political vocation but becomes political when it reveals the limits of the imagination and generates realistic fantasies that at first sight are impracticable." Graciela Speranza, *Lo que no vemos, lo que el arte ve* (Barcelona: Anagrama, 2022), 24.

PART I

Interspecies Thinking

The opening section of *Momentum* addresses contemporary artists in Latin America who are using their practice to rethink our relationship with nature and the living things, animate and inanimate, that constitute it. Artists at the forefront of these ideas are exploring interspecies thinking as a way to reimagine the intersection of the human and the nonhuman, a relationship characterized by the Western, colonial, and capitalist world as one of domination and control. In this sense, imagining horizontal engagements between humans and nonhumans provides an opportunity to better understand the entanglement of humans and what is often dubbed the "natural environment."

Interspecies thinking has gained momentum in the new millennium, yet it also has a long history. In a groundbreaking move in 2008, Ecuador became the first country in the world to codify in its constitution the rights of nature, which are based on the intrinsic value of nature rather than on its usefulness to humans. Such decrees, even when contested, reflect a broad legal, cultural, and ontological shift that is unfolding throughout Latin America. At the same time, for centuries continuing into the present, nonanthropocentric understandings of the world have been fundamental to the worldviews of the over 500 distinct Indigenous groups of Latin America. Challenging the vertical power structures of colonial knowledge, these peoples generally see humans as guardians rather than owners of the land.

While propositions for combatting the ecological crisis will be addressed at length in the third section of the book, this first part invites related conversations, meditating on how the relegation of all things nonhuman to the category of less-than-human, a consignment both historical and ongoing, has shaped an oppressive and self-destructive relationship. At the same time, it also shows that many artists and thinkers have been conceiving alternative paths to theorize and visualize how organisms—be they animal, plant, human, or technological—exist in relation to one another.

JULIETA GONZÁLEZ

FROM CYBERNETICS TO THE TECHNOLOGIES OF REENCHANTMENT

Some Notes on Contemporary Art and the Environment in Latin America

We were once the masters of the earth, but since the gringos arrived we have become veritable pariahs. . . . We hope that the day will come when they realize that we are their roots and that we must grow together like a giant tree with its branches and flowers.
—Francisco Servin, Pãi-Taviterã, at the Congress of Indians, Paraguay, 1974, quoted in James Clifford, "Introduction: The Pure Products Go Crazy," *The Predicament of Culture*, 1988

Our true wealth lies in the forest: its waters, its fish, its game, its trees and fruits. It has nothing to do with merchandise.
—Davi Kopenawa, *A Queda do Céu*, 2010

"The Amazon is gone," reads an advert for the Trans-Amazonian Highway collaged into a recent painting, *Itarawatakhetti* (2022), by artist Denilson Baniwa, along with other ads featuring catchphrases such as "The trail that leads you to a gold mine," "Enough of legends, let's make money," "To unite Brazilians, we tore through the green inferno," and "This road will pass by an agency of the Bank of London." Baniwa's recent series of paintings lays bare the ethos of mercenary exploitation and extractivism that propelled the agenda of *desarrollismo* (developmentalism) in the Amazon rainforest in 1970s Brazil. This advertising campaign embodies what many scholars have described as the darker side of modernity, one sustained by the colonial logic of occupation, exploitation, extraction, and the systematic erasure of the Indigenous cultures that inhabited the territory before the arrival of the Europeans. Baniwa's works resonate with the words of Francisco Servin, a leader of the Pãi-Taviterã people quoted by James Clifford in his introduction to *The Predicament of Culture* (1988).[1] Like the Pãi-Taviterã, who live in the Mato Grosso and Pantanal regions of Brazil and Paraguay, the Baniwa people have been subjected to the expropriation of their land, displacement, and acculturation for centuries, all in the name of progress.

As his last name indicates, the artist belongs to the Baniwa people, who inhabit a territory in the Rio Negro basin, at the intersection of the Brazilian, Venezuelan, and Colombian borders. He was born in the state of Amazonas, Brazil, and is part of a generation of contemporary Indigenous artists who are redefining the terms of the discussion around nature and culture—voicing from their own perspectives their relation to the land, the environment, and its centrality to their artistic production. For these artists the wholesale destruction inflicted on the Amazon forest, historically and today, is an everyday reality. Moreover, many of the Indigenous artists now working throughout the South American continent are spearheading an epistemic shift toward a post-Cartesian and post-development approach that transcends the nature/culture divide. To address the issue of ecology in contemporary art without their voices is today unthinkable.

In his essay "The Pure Products Go Crazy," Clifford proposes a departure from the notion that Indigenous peoples are "tied to traditional pasts, inherited structures that either resist or yield to the new but cannot produce it."[2] For Clifford we must look in the opposite direction, that is, toward Indigenous peoples' own strategies for circumventing the forces of "progress" and "assimilation" and building their own futures, as well as ours. What Clifford envisioned in 1988 has indeed become a reality, as Indigenous voices have increasingly taken an active role in their self-representation and have also provided their perspective on the human relationship with nature, a crucial one if we are to reverse the course of self-destruction that modernity has set out for us as a species.

In this conversation, the West and the Westernized peripheries have everything to learn from Indigenous peoples, who may help us to disassemble the Cartesian epistemic framework that is embedded in our ontological and epistemological formations.

This essay thus takes Baniwa's work as a point of departure to go beyond artists' resistance to the rhetoric of *desarrollismo* in the 1970s, attempting to chart a major shift in the discussion of art and the environment provided by the more recent entrance of Indigenous voices into the conversation. It is as if we were to trace an imaginary line between Juan Downey's mid-'70s image of a Yanomami holding a video camera and the voices of Indigenous artists active today, who have tamed technology and are using it as a tool and a discursive weapon in the struggle against the forces of environmental destruction. Perhaps Downey's encounter with the Yanomami can be taken as an inflection point where both cybernetic theories and Indigenous cosmovisions can reunite, an idea posited by Claude Lévi-Strauss as early as 1962, in his book *The Savage Mind*: "we have had to wait until the middle of this century for the crossing of long separated paths."[3] Lévi-Strauss also wrote about how Indigenous societies conceptualize their world, "treating the sensible properties of the animal and plant kingdoms as if they were elements of a message"—an approach that lacked the precision afforded by technological tools, he allowed, but that nonetheless gave them insights at which Western societies only arrived after the invention of "telecommunications, computers and electron microscopes."[4] The intersections between cybernetics, ecology, contemporary art, and the Indigenous mindset may in fact be considered fundamental to an assessment of the shifts currently shaping the relations between technology and ecology, the political space constituted today by the natural environment of Indigenous communities, and the role of the Indigenous peoples in the formulation of alternatives to development and capitalist modernity. I summon Mary Louise Pratt's concept of the contact zone, undertaking this analysis from a perspective that "emphasizes how subjects get constituted in and by their relations to each other," treating "the relations among colonizers and colonized, or travelers and 'travelees,' not in terms of separateness, but in terms of copresence, interaction, interlocking understandings and practices, often within radically asymmetrical relations of power."[5]

CYBERNETIC AFFINITIES: TECHNOLOGY AND INFORMATION IN THE CONTACT ZONE

> *Ironically, the man-nature chasm can only be closed by technology. The process of reweaving ourselves into natural energy patterns is Invisible Architecture, an attitude of total communication within which ultra-developed minds will be telepathically cellular to an electromagnetic whole. . . . Human beings would share with all other species the benefits of natural cycles: communicant balance.*
> —Juan Downey, "Technology and Beyond," 1973

> *For a theory of information to be able to be evolved it was undoubtedly essential to have discovered that the universe of information is part of an aspect of the natural world. But the validity of the passage from the laws of nature to those of information once demonstrated, implies the validity of the reverse passage—that which for millennia has allowed men to approach the laws of nature by way of information.*
> —Claude Lévi-Strauss, *The Savage Mind*, 1962

Technology and information play a significant role in this account. The contact zone is where the Indigenous Other is introduced to technologies outside their realms and cosmogonies, and where they must negotiate these asymmetrical relations of power by incorporating and utilizing these technologies. In a recent essay on the artist Joseca Yanomami, Denilson Baniwa introduces an argument that transforms the West's prior readings of Indigenous art in a postcontact context (fig. 1).[6] Here Baniwa explains the Indigenous incorporation and use of technologies, ranging from tools and weapons to writing, and to art itself, as a domestication or taming of these external entities that enable them to upset the balance of power with the "people of the merchandise," as Davi Kopenawa has referred to White people.[7]

Pratt comes up with the concept of the contact zone in her analysis of the Inca Felipe Guamán Poma de Ayala's lengthy manuscript *El primer nueva corónica y buen gobierno* (The first new chronicle and good government), written in the early seventeenth century.

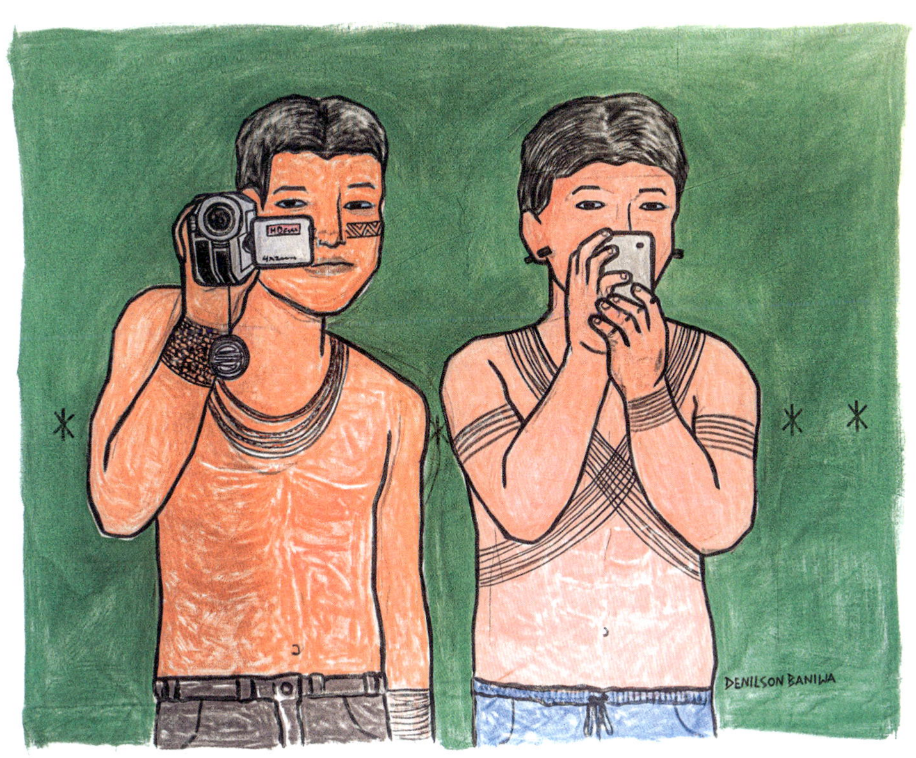

1 Denilson Baniwa. *Awá uyuká kisé, tá uyuká kurí aé kisé irü* (He who lives by the sword, dies by the sword). 2018. Acrylic on canvas, 63 × 6 ft. 6 ¾ in. (160 × 200 cm). Collection the artist

Viewed from the perspective of Baniwa's analysis, the epistle itself can be seen as the result of the domestication of tools introduced by the Spaniards—in the case of the *Nueva corónica*, writing, history, graphic representation, cartography, the manuscript—by an Inca, whose Quechua name, Guamán Poma, means "Hawk Puma" in English. His *Nueva corónica*, the first written history of Andean cultures by an Inca, chronicles the history of the region before the arrival of the Spanish—the Inca rule, its society, economy, laws, religious beliefs—and offers a detailed critique of the mistreatment of the Incas by the Spanish colonizers. Guamán Poma proposes a "good government," one that would make good use of the land, resources, seasons, and the labor it took to produce sustenance for the people—a regime through which both the Inca and the Spaniards, through the use of their respective technologies, could build a mutually beneficial future.[8] Addressed to King Philip III of Spain, though it apparently never reached him, the *Nueva corónica* attests to Guamán Poma's profound understanding of the substantial value of text and image, and of their narrative and discursive organization as tools of self-representation; both of which were instrumental to negotiating the livelihoods and survival of his people with the colonizers in the contact zone. In fact, Guamán Poma's domestication of technologies introduced by the Spanish—his use of writing to translate and codify the history of a people for Western reference, shifting the terms of the discussion—seems very much akin to the notion of information as action posited by second-order cybernetics in the mid-twentieth century, and deployed in the 1960s and '70s by artists working with the reflexive nature of video feedback to potentialize agency and autonomy.

In this sense, technology and information also provide another starting point for this narrative, for it was at the dawn of the information age in the 1940s that cybernetics—the interdisciplinary study of systems and mechanisms of control in humans, animals, and machines, to put it briefly—rocked the Cartesian foundation upon which Western modernity had been built, calling into question the "subject" of liberal humanism. In the systems perspective of cybernetics, humans are but another element in an assemblage of human and nonhuman beings and machines, in constant feedback relation between themselves and the environment to ensure the equilibrium, the homeostasis, of a given system. Like all living organisms, humans are basically compounds of cells, organs, and neurological impulses, organized in specific patterns, and as Norbert Wiener wrote, "It is the pattern maintained by this homeostasis which is the touchstone of our personal identity. Our tissues change as we live ... we are but whirlpools in a river of ever-flowing water. We are not stuff that abides but patterns that perpetuate themselves."[9] If we are but organized patterns, then the Cartesian dualities mind/body, nature/culture, and human/animal are instantly dissolved in this "river of ever-flowing water." Moreover, the systems perspective of cybernetics puts the ecological question at the forefront: these patterns of organization, these ever-changing biological and neurological assemblages—human, animal, and vegetal—are in constant interaction and exchange with the environment in such a way that the environment is essentially indistinguishable from them.

The notion of reflexivity that accounted for the observer's awareness of the act of observation, and for the observer's influence over the system being observed, was developed mainly by Gregory Bateson, Humberto Maturana, Warren McCulloch, Margaret Mead, and Heinz von Foerster. In second-order cybernetics' shift from homeostasis to reflexivity, the observer is also influenced by the observed system, in a recursive feedback loop. This in turn changed the perception of information as a thing to an understanding of information as action; in Bateson's phrase, information is "the difference that makes a difference."[10] This differential view of information had an enormous influence on the social and ecological experiments of the techno-utopian counterculture of the 1960s, as well as on artists working with communications, technology, language, and information.

Among the artists active in the late 1960s and '70s who deployed the potential of information as action and positioned their practice within an ecological framework, Juan Downey stands out, in terms of the centrality he gave not only to the ontoepistemic formations of Indigenous societies but also to the idea of producing systems sustained by this dynamic of the observer, the observed, and the act of observing. His work challenged not only the ontoepistemic formations of the West but also ethnographic authority. Many of his Latin American peers working within the matrix of communications, information, and cybernetics, especially those affiliated with the Centro de Arte y Comunicación (CAyC) in Buenos Aires, looked toward the cultures of the Indigenous peoples of the Andean region and critiqued the extraction of natural resources there, from the silver mines of Potosí, Bolivia, in colonial times to the copper mines owned by multinationals in Chile.[11] Downey, however, went a step further. In several bodies of work of the 1970s that crossed the threshold of alterity to deploy video as a political and ideological tool, he aimed to counteract 500 years of colonial oppression and erasure of Indigenous knowledges, in effect connecting cybernetics, anticolonialism, ecology, anthropology, and the Indigenous mindset.

Downey's feedback undertaking was unique in his generation, for the works he produced were the result of his direct experience among the Indigenous peoples he visited. Contact was essential to these works. The intersection between modern technology and the ancestral cultures of the Americas indeed lies at the center of Downey's epic *Video Trans Americas* (1976), perhaps his most ambitious project involving video feedback. The sociocultural empowering potential of diverse signals, multidirectional networks, and feedback were central to his project, which he carried out in three expeditions from New York to Central and South America between 1973 and 1976.[12] A highly utopian undertaking that sought to integrate the Indigenous peoples of the Americas through video feedback, *Video Trans Americas* was also Downey's way of taking the role of the active observer prescribed by second-order cybernetics' vision of information as a form of agency; he saw the act of playing back one culture in the context of another, and a culture itself in its own context, as an instrument of political, social, and cultural empowerment. His agency and role as an observer, however, were called into question in a way that would change his subsequent encounters with Indigenous peoples and his own perception of the underpinnings of a Western culture:

> The Video Trans Americas black and white expeditions have been completed. Like a chemical catalyst I expected to remain identical after my video exchange had enlightened

> many American peoples by the cross-references of their cultures. I proved to be no real catalyst, for I was devoured by the effervescence of myths, nature and language structures. Pretentious asshole leveled off!! Only then did I grow creative and in manifold directions. Me, the agent of change, manipulating video to decode my own roots. I was forever deciphered and became a true offspring of my soil, less intellectual and more poetic. An unexpected level had been reached among the strange roads of the heart![13]

With his three-year expedition across the Americas and subsequent stay among the Yanomami, in the Amazon, in 1976–77, Downey aimed to rewire the structures of communication set in place by centuries of colonial rule, which he held accountable for the erosion and alienation of Indigenous cosmologies, languages, and knowledges, as well as for the isolation of one culture from another. While operating within and making visible the contact zone, with its topological infolding of time and space, its bridging of past and present, *Video Trans Americas* can be regarded as what Pratt calls a virtual "safe house," that is, a space "where groups can constitute themselves as horizontal, homogeneous, sovereign communities... [and] construct shared understandings, knowledges, claims on the world that they can then bring into the contact zone."[14]

Moreover, *Video Trans Americas*, and the drawings, writings, and videos that Downey produced as a result of his stay with the Yanomami, can be inscribed in the framework of the critiques of anthropology and of its place in the colonial enterprise that critical anthropologists developed in the late 1970s and the '80s. In fact they foreshadow, from an artistic and poetic perspective, many theoretical developments in the field: notably, Clifford's call for the centrality of Indigenous voices in the construction of a future after contact has been made; Pratt's concept of the contact zone; Johannes Fabian's analysis of traditional anthropology's denial of coevalness; and Eduardo Viveiros de Castro's theory of Amerindian perspectivism, as well as the revisions of the notion of animism in present-day anthropology.

Far from documenting or portraying the Yanomami in his videos *Guahibos* (1976), *The Singing Mute* (1978), *The Laughing*

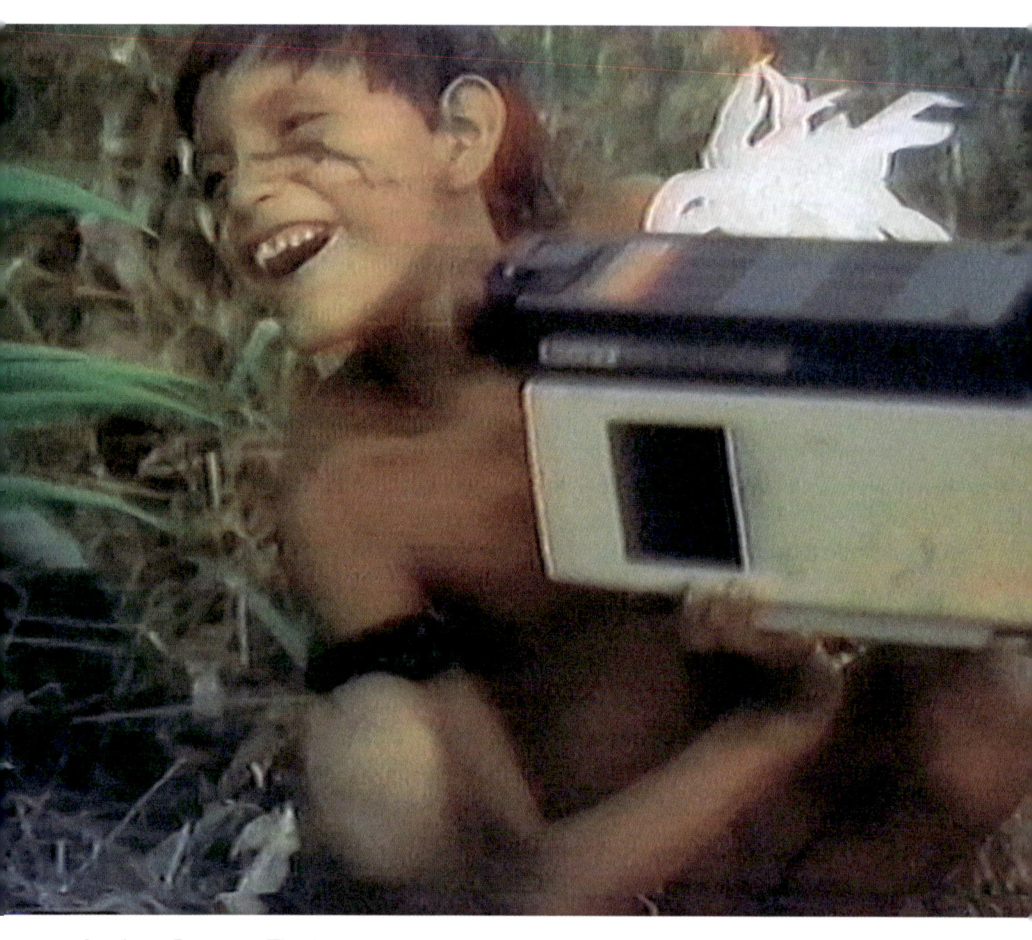

2 Juan Downey. *The Laughing Alligator*. 1979. Standard-definition video, black and white and color, sound, 27 min. The Museum of Modern Art, New York. Gift of the Gloria Kirby Conahy Fund

Alligator (1979; fig. 2), and *The Abandoned Shabono* (1978), Downey produced footage in collaboration with them, which he then edited in a narrative and style that unsettled the conventions of ethnographic representation, from a humorous take on the ethnographic documentary to an autobiographical, confessional approach. These strategies in turn allowed him to articulate a critique of the West's evaluation of Amerindian societies as archaic, as temporally, and thus historically, removed from industrialized societies. The *shabono*, the communal dwelling of the Yanomami, revealed itself to Downey—architect, ecologist, and techno-utopian artist—as the most perfect expression of a cyclical and ecological architecture. This circular lean-to construction, built with the leaves and branches of trees felled to make a clearing in the forest, naturally disintegrates every two or three years. Aside from its function as shelter, the *shabono* regulates the social structure of the Yanomami: there are no hierarchies, and families are distributed around hearths placed along the building's circular frame (the "circle of fires," in Downey's phrase), while collective and ritual activities take place at the center of the structure, open to the sky and the elements.

Cybernetics, too, afforded Downey a perspective on his encounter with the Indigenous communities that he visited on the *Video Trans Americas* expeditions. While the root of his enterprise lay in feedback, his understanding of the radical departures from the Cartesian mindset that were proposed by cybernetics was clarified by his experiences in the Amazon, especially among the Yanomami, and enabled him to pose challenges to the ontoepistemic formations of capitalist modernity. Downey grasped the essence of cybernetics, in its systems-based view and its dissolution of subject and object positions, as an ontology of becoming. It is perhaps this aspect of his work and thought that best connects to the way in which Indigenous epistemologies are relevant to environmental and ecological thinking in Latin America today.

The Yanomami's expanded and undifferentiated view of the self gave Downey a full perspective on the cybernetic ontology of becoming. Transcending the duality of self and other, it saw animals and the forest as beings endowed with human attributes, while humans could also possess animal qualities. Downey's experience of traveling and living among Indigenous cultures confirmed his deep-seated conviction that other ways of being and knowing were not only possible but superseded Western notions of state and capital. He wrote that only technology would enable a "disoriented humanity" to close the "man/nature" chasm.[15] After his experience among the Yanomami, however, his concern with technology was displaced by an interest in semiotics. Perhaps this reflected a disenchantment with the technologies of progress, since in the Amazon forest he had now witnessed a perfect cybernetic society, one that had attained a seamless balance with nature without the aid of technologies. Almost fifty years later, though, Indigenous artists are perhaps proving him right, and are in fact using technologies to reweave the "man/nature" chasm.

FROM COSMOPOLITICAL AESTHETICS TO PUSSANGA: CLOSING THE NATURE/CULTURE CHASM WITH THE TECHNOLOGIES OF REENCHANTMENT

> *The principles on which technology is based are universal; their application is not. The unthinking adoption of North American technology in Mexico has produced no end of misfortunes and a progressive degradation of our way of life and our culture. This is not nostalgic obscurantism; the only real obscurantists are those who cultivate the superstition of progress at any price. I know that we cannot escape; we are condemned to "development," but let us make the penalty less inhuman.*
> —Octavio Paz, *Children of the Mire*, 1974

> *If modern Western multiculturalism is relativism as public policy, then Amerindian multinaturalism is perspectivism as cosmic politics.*
> —Eduardo Viveiros de Castro, *Cosmological Perspectivism in Amazonia and Elsewhere*, 1998

> *As an agent of enchantment (technology of taming), the artist employs the means that ensure survival in Western art:*

> *image and text. Having caught the White man's attention, he casts yet another spell, that of the* pussanga *(a spell using matter, in this case drawing and writing on paper); an incantation of linguistic difference.*
> —Denilson Baniwa, "Art to Tame the *Napë* Heart: Cosmologies of Contact and Enchantment of the People of the Merchandise," 2022

Today, Latin America is witness to the failed promises of development and the shortcomings of an imposed modernity, which have embroiled it in a vicious cycle of dependency, corruption, and foreign debt. In this context the intersecting concerns that guided Downey's work in the 1970s have gained renewed currency among artists and thinkers of the past fifteen years or so, many of whom have sought to form alliances with and give visibility to Indigenous communities and activist groups working in this direction. The global ecological crisis that Downey and his contemporaries foresaw in the 1960s and '70s has significantly aggravated. Meanwhile, the neural network of communications that he and his peers imagined is today a reality, shaping our world in manifold ways, but instead of proving instrumental in the struggle against ecological disaster, it has intensified the West's alienation, its surrender to social media–induced consumer trends and cryptocurrencies—yet another extractive activity, a quest for wealth in the virtual realm that has a real ecological impact.

Since the mid-2000s, a generation of artists from different countries in the Americas, born between the 1960s and the 1980s, have looked toward the cultural and social resources, the forms of knowledge and relationality, that Western thought has historically suppressed. Within the regimes of visuality and discourse in which they operate, these artists are drafting a blueprint for a post-Cartesian, postdevelopment future, whether by questioning Cartesian dualities and the hegemony of science or by working in collaboration with Indigenous and rural societies to build a common front against the systematic destruction of natural habitats. Nature and information are at the center of their undertakings.

One such artist is Carolina Caycedo, who works with rural and Indigenous communities in her native Colombia, especially in the

Upper Magdalena River region, and has focused on rivers and on hydroelectric dams, which affect vast ecosystems and their inhabitants. Her performative and choreographic actions center on the body as a tool for generating political consciousness and environmental awareness. Another is the Peruvian artist Ximena Garrido-Lecca, whose work can be seen as operating in the space of the contact zone, with its asymmetrical power relations between colonizer and colonized and its infliction of violence on Indigenous populations, erasing forms of knowledge derived from the natural world. Garrido-Lecca's multipart installation *Insurgencias botánicas: Phaseolus Lunatus* (Botanical insurgencies: Phaseolus lunatus, 2017; fig. 3) features a variety of Lima bean—*pallares Mochica*—that bears distinct black-and-white patterns. In the Moche culture of pre-Incaic Peru, these patterns were interpreted as signs of some kind; the Moche people also developed sophisticated irrigation systems, which Garrido-Lecca reprises in her installation by growing the beans hydroponically. This work comes close to Lévi-Strauss's observations about how Indigenous societies "treat the sensible properties of the animal and plant kingdoms as if they were the elements of a message."[16]

At the center of Paulo Tavares's and Ursula Biemann's collaboration with the Sarayaku people of Ecuador lies the "cosmopolitical forest." Using information technologies, such as satellite images and maps, and the language of film, *Forest Law* (2014) is a journey into the struggle of the Indigenous inhabitants of this region to defend their ancestral land, one of the most biodiverse territories in the country but also one of the richest in minerals, putting the Indigenous communities at the center of a conflict with the state and multinationals. *Forest Law* also centers on the mobilization of Indigenous groups and environmental activists to file a lawsuit against the Ecuadorian state on behalf of the rights of nature for allowing mining and oil extraction in the Cordillera and Amazonian regions of the country. While the Western capitalist mindset has always regarded nature as a trove of resources to be extracted, the Sarayaku and other forest-dwelling peoples have taught us that the forest and everything within it, including the animals and humans who inhabit it, compose a sentient living system and as such have constitutional rights as a juridical entity.

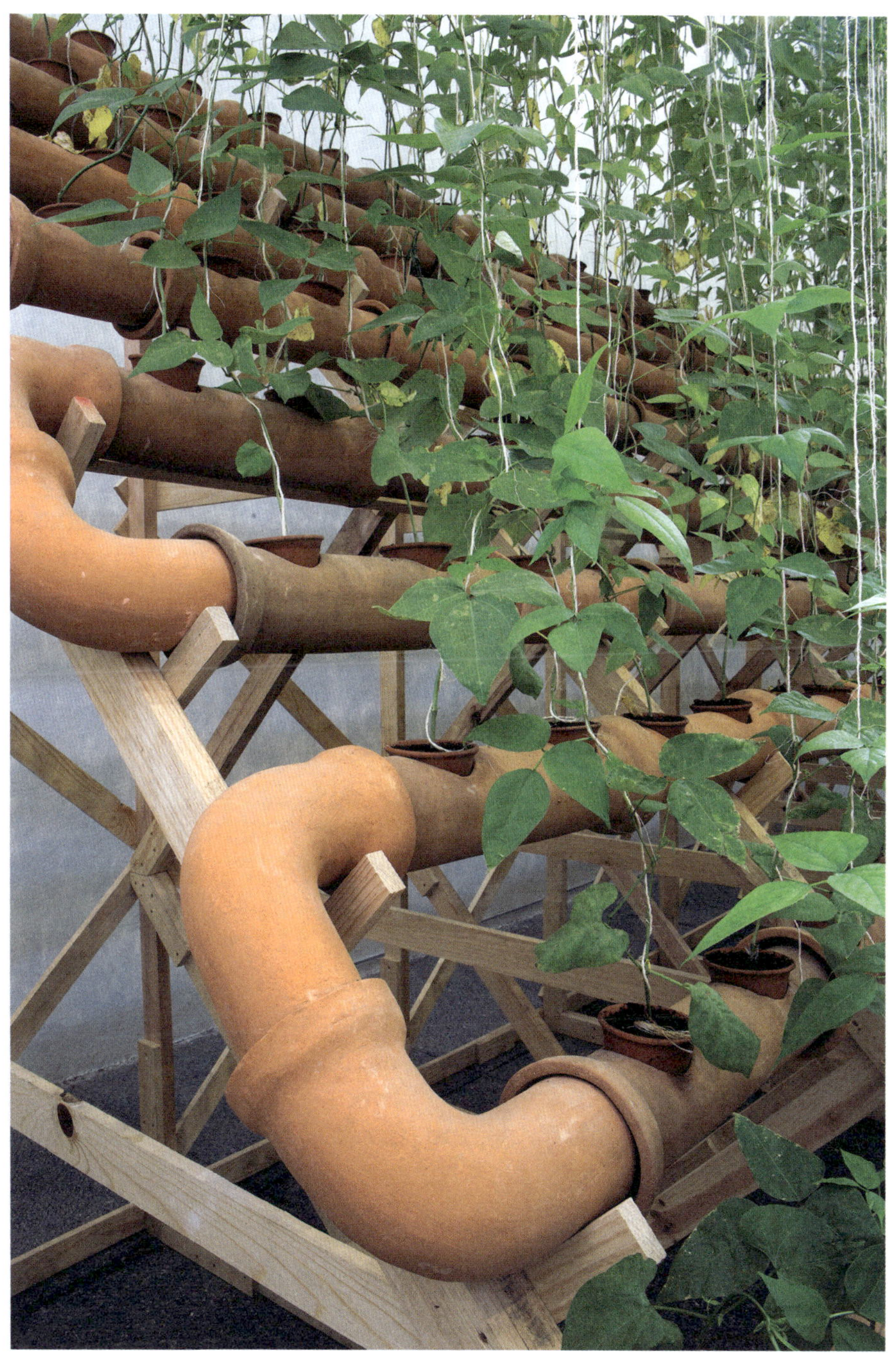

3 Ximena Garrido-Lecca. *Insurgencias botánicas: Phaseolus Lunatus* (Botanical insurgencies: Phaseolus lunatus). 2017. Installation, dimensions variable. Collection the artist

Tavares and Biemann invoke Isabelle Stengers's concept of "cosmopolitics" as an ontological and epistemic framework that can reconcile multiple entities and forms of knowledge.[17] As the word implies, a politics of the cosmos cannot be reduced to humans and their technologies but must account for all living systems and their realities. A progressive awakening from the spell of progress has been taking place among Western philosophers and anthropologists. At the dawn of the twentieth century, Max Weber spoke of a disenchanted world, one that had abandoned myth in favor of objective scientific reality.[18] In the 1940s, Theodor Adorno and Max Horkheimer stated that "the disenchantment of the world means the extirpation of animism."[19] Since the 1960s, a number of philosophers, reacting to the objectivity of science, have called for transcending the Cartesian divide between nature and culture and have instead proposed various holistic conceptions that would integrate all human and nonhuman entities: Gregory Bateson's "ecology of mind"; Félix Guattari's "ecosophy"; Bruno Latour's dismantling of the modern cult of "factishes"; Viveiros de Castro's "Amerindian perspectivism"; and Stengers's refashioning of Kantian cosmopolitanism by way of Guattari, Latour, and Bateson.

It is perhaps in this as yet undefined scope of cosmopolitics that cybernetics and Indigenous cosmovisions can coexist. The systemic view of cybernetics dissolved the mind-body duality, suggesting instead an integration of mind and body in which the body thinks, the mind feels, and the body is connected to its natural environment. Transcending the mind-body duality makes it possible to dismantle the nature/culture divide. For the peoples of the Amazon forest and other Indigenous peoples around the world, humans and nonhumans, trees and rivers, are all part of the same entity: *urihi* for the Yanomami, the "immanent mind" for Bateson, while Viveiros de Castro calls it "multinaturalism," that is, "a spiritual unity and a corporeal diversity."[20]

Viveiros de Castro has situated the fundamental difference between a European and an Indigenous perspective—the basis for his theory of Amerindian perspectivism—in a passage from Lévi-Strauss's *Tristes Tropiques* (1955) in which he quotes Gonzalo Fernandez de Oviedo's sixteenth-century account of a tribe in the Caribbean, ostensibly the Taínos, who "used to kill off any captured

Europeans by drowning. Then they would mount guard for weeks around the dead men to see whether or not they were subject to putrefaction." Lévi-Strauss drew "two conclusions from the differences between the two methods of enquiry: the white man invoked the social, and the Indians the natural, sciences; and whereas the white men took the Indians for animals, the Indians were content to suspect the white men of being gods. One was as ignorant as the other, but the second of the two did more honor to the human race."[21] The Amerindians in Lévi-Strauss's account apprehend and come to know the world through their bodies and through an understanding of the body. An Indigenous metaphysics resides in this difference in perspective, as the "savage mind" apprehends the world through the sensorium.

How are we then to bridge this ever growing chasm between nature and culture? How are we to reenchant the world if our brains have been wired, so to speak, since our acquisition of language, to the binary dualistic configuration of Cartesianism? Even if, as artists, philosophers, anthropologists, and thinkers, we are able to envision this fallacy of modernity, and its "factish" cults of scientific objectivity, progress, and profit, it seems impossible to escape it without the voices of those whose minds have not been domesticated, have not been conditioned by a Western ethos designed in functions of production, yield, and return, and whose insights can open up unexpected possibilities of meaning. Stengers speaks of the modern world's aversion to the fetish and the *pharmakon*, but it is through the Indigenous *pussanga*, which is fetish and *pharmakon* at the same time, that Indigenous artists are expanding our horizon of possibilities.[22]

A jaguar/shaman (*onça/pajé*) roamed uninvited through the 2018 Bienal de São Paulo. The exhibition had essentially turned its back on the significant epistemic and ontological ruptures that have been taking place throughout the global South, instead presenting a series of shows organized by artists as curatorial auteurs and featuring their personal obsessions with myth, overt homages to the canon of geometric abstraction, and an insider vision of today's New York and other avant-gardes. The presence of the jaguar/shaman introduced an element of disruption into an otherwise tepid biennial, in which one artist/curator had fashioned an imaginary museum presenting blown-up images of the Selk'nam people. In an exhibition where Indigenous artists were otherwise absent, the jaguar/shaman's critique of the artist's appropriation of the images of Indigenous peoples threw a spanner into the works of the well-oiled machine of the art world, which has systematically excluded those outside the Western canon.[23]

The jaguar/shaman/hacker was Denilson Baniwa, who has been deploying the domesticated technologies of the West, from writing to art to video and information networks, to subvert the art system from within. Like his predecessors who used the information technologies of the Europeans, notably the seventeenth-century hawk/puma Guamán Poma de Ayala, Baniwa believes that it is "necessary to grasp these technologies to understand how they can be used as tools of resistance and not of alienation."[24] In his essay on Joseca Yanomami, Baniwa lays the groundwork for a new understanding of Indigenous art. This art has entered the exhibition space in recent years, but without the proper tools to understand the Indigenous mindset, Western or urban Latin American critics have appreciated it for its formal qualities and naiveté without making an effort to understand its Otherness.[25] Yanomami artists such as Sheroanawe Hakihiiwe and Joseca Yanomami, for example, work with what the Western-trained eye may perceive as abstract or figurative inflections but are in fact intricate codes. Hakihiiwe's abstract drawings and paintings are related to aspects of Yanomami cosmovisions, but rather than represent specific entities they invoke their power through images in the form of messages, incantations of sorts. Joseca Yanomami's deceptively simple drawings appear to represent scenes of Yanomami daily life and myth, but in reality they acknowledge postcontact realities and are a way of dealing with the technologies and objects introduced into their world by Westerners. Thus, contrary to what the conditioning of Western art history allows us to see, both artists are operating at complex levels of semantic organization.

Both Baniwa and the late Macuxi artist Jaider Esbell have invoked the figure of the *pajé* (shaman), with its inherent performativity, in their work. The *pajé* has the ability to transit between spiritual dimensions and physical bodies; it can be a jaguar, a hawk. As an operator of signs, the artist is also a *pajé* and moves between worlds, as Baniwa has said, taming the technologies of the West in order to cast spells, *pussanga*.[26] The word has various interpretations: it can mean "fetish" (*feitiço* in Portuguese, meaning "fetish" and "spell" at the same time) but also, like the *pharmakon*, "cure," "remedy," and "poison." Viveiros de Castro describes shamans as "multinatural beings by definition and office, ... always capable of transiting the various perspectives, calling and being called 'you' by the animal subjectivities and spirits without losing their condition as human subjects, and accordingly they alone are in a position to negotiate the difficult 'paths' ... that connect the human and the non-human Amazonian worlds."[27] In this sense, and as Baniwa himself has stated, this transit between planes is a means to ensure survival, to upset the asymmetrical balance of power in the contact zone. Thus contemporary artists/*pajés* have mastered technology as a weapon and a tool of resistance; contrary to the West, which has succumbed to the mind-numbing and alienating spell of social media and digital technologies, the artist/*pajé* uses technology to move between the worlds of the forest and that of the *napë* (the foreigner, the white person).

Zahy Guajajara, of the Tentehar-Guajajara people of Brazil, uses the language of video and the genre of science fiction in her video *Karaiw a'e wà* (The civilized, 2022) to question the notions of civilization and technological progress and to imagine an Indigenous futurism. Performing rituals as a *pajé yawareté* (shaman jaguar), Baniwa walks through spaces of conflict and contact, be they cities, exhibitions, or vast cornfields on former forest lands, casting his *pussanga* in order to reactivate the dormant natural spirits and entities that lie under the concrete and the monoculture. As part of his ritual he includes a technology made possible by the printing press: books, specifically art-history books of the abridged kind that cater to a general audience. Baniwa tears up these books while voicing the exclusion of Indigenous artists and the simultaneous appropriation of their images, consigned to a remote past with no possibility of building a future.

In 2019, 400 years after Guamán Poma (Hawk Puma) wrote his epistle from the New World to the king of Spain, Esbell wrote another letter to the Old World. In his words, the letter took the form of "contemporary Indigenous art," and as "pure anthropophagy" it articulated an "open denunciation of the centuries of devastating colonization in the Americas," focusing on the present-day "global drive for development" and its "impact on nature and the genocide of the last native peoples of the Pan-Amazonian region."[28] Like Baniwa, Esbell began the ritual of this performance with an art-history book, this one purchased in a used-book store in northeastern Brazil. The artist/*pajé* intervened in its 400 pages of text and plates of European paintings with "the force of resignification," sending "messages laden with the forest's energy," enabling an encounter between contemporary Indigenous art and Eurocentric art in the pages of the book. For Esbell, this invasion of the book, and of the historical account within its pages, was a way of "demonstrating the sentiment of native peoples when their forests (which is to say themselves and their ways of knowing) are violently trespassed. Does anyone ask for permission to invade and destroy?" The artist, "a Makuxi from Raposa, Serra do Sol, Indigenous land in Roraima," exhibited the book in an art space west of Paris.[29] The work was subsequently included in the Bienal de São Paulo in 2021.

At the beginning of this essay I suggested a temporal arc ranging from *napë* Juan Downey's mid-'70s image of a Yanomami holding a video camera to the contemporary work of Indigenous artists/*pajés* and their shamanic transits between entities and dimensions. I stand corrected: the temporal arc goes all the way back to the Inca Guamán Poma, the Hawk Puma who tamed text and image, technologies brought into his world by contact with Europeans in the sixteenth century. Guamán Poma created the matrix from which twentieth-century anthropology could reverse its point of view, if we are to give Clifford and Pratt credit for their words. In this story (or is it history?), technology, information, and contact are inextricably bound; the incantation cast by Hawk Puma was a code that waited four centuries to be deciphered when the *napë*, or at least some of them, awoke from the spell of progress. Esbell's letter to the past, injected with the energy of the forest, is *pussanga*, both a spell and a cure. It is addressed to the Old World because, as these contemporary artists/*pajés* have made clear, they are alive and the forest lives within them. They will not be consigned to a remote primeval past but will contribute alongside the *napë* to the configuration of a new cosmopolitics, reenchanting the world with their technologies and closing the chasm between nature and culture. The future is theirs.

1 James Clifford, *The Predicament of Culture: Twentieth-Century Ethnography, Literature, and Art* (Cambridge, MA, and London: Harvard University Press, 1988), 1.

2 Ibid., 5.

3 Claude Lévi-Strauss, *The Savage Mind*, 1962 (Eng. trans. London: Weidenfeld and Nicolson, 1966), 269.

4 Ibid., 268.

5 Mary Louise Pratt develops the concept of the contact zone from her study of one of the first autoethnographic texts produced in the New World, *El primer nueva corónica y buen gobierno*, written in 1631 by the Inca Felipe Guamán Poma de Ayala. See Pratt, *Imperial Eyes: Travel Writing and Transculturation* (London: Routledge, 1992), 7.

6 Denilson Baniwa, "Art to Tame the Napë Heart: Cosmologies of Contact and Enchantment of the People of the Merchandise," in Adriano Pedrosa and David Ribeiro, *Joseca Yanomami: Our Forest-Land*, exh. cat. (São Paulo: Museu de Arte de São Paulo, 2022).

7 Karl Marx's concept of commodity fetishism comes extraordinarily close to the Yanomami's definition of Western societies' obsessive fascination with merchandise. See Davi Kopenawa and Bruce Albert, *A queda do céu. Palavras de um xamã Yanomami* (São Paulo: Companhia das Letras, 2015).

8 Guamán Poma warns that the greed of the Spaniards, and their overexploitation of both natural and human resources, are conducive to the end of the world: "Si no defendemos lo han de acabar todo, y si no nos guardamos de las enfermedades y riesgos y peligros moriremos y nos acabaremos" (If we don't defend [against these things], they will finish off everything, and if we don't protect ourselves from diseases and risks and dangers, we will die and we will end). In Guamán Poma, *Nueva corónica y buen gobierno*, between 1600 and 1615 (Caracas: Biblioteca Ayacucho, 1980), 2:476.

9 Norbert Wiener, *The Human Use of Human Beings: Cybernetics and Society*, 1950 (repr. ed. London: Free Association Books, 1989), 96.

10 See N. Katherine Hayles, *How We Became Posthuman: Virtual Bodies in Cybernetics, Literature, and Informatics* (Chicago and London: The University of Chicago Press, 1999).

11 For reasons of the scope and length of this essay I will not focus on CAyC's particular intersection between technology, ecology, and Indigenous knowledges, laid out in writings such as *Myths and Magic: On Fire, Gold and Art* and "The End of the Second Middle Ages," which respectively accompanied the exhibition presented on the occasion of the 1st Bienal Latino-Americana de São Paulo in 1978, *Mitos e Magia* (an exhibition that built on CAyC's participation in the XIV Bienal de Sao Paulo the preceding year), and the exhibition *CayC Group at the Musée Cantonal des Beaux Arts*, Lausanne, in 1981, parallel to the colloquium "Dialogue between America (North and South) and Europe." See *Myths and Magic: On Fire, Gold and Art*, Art Criticism Briefs (Buenos Aires: AICA, 1979), and *CAYC Group at the Musée Cantonal des Beaux Arts* (Buenos Aires: Ediciones Centro de Arte y Comunicación, 1981).

12 Juan Downey wrote in a travelogue, "Many of America's cultures exist today in total isolation, unaware of their overall variety and of commonly shared myths. This automobile trip is designed to develop a holistic perspective among the various populations inhabiting the American continents, thus generating cultural interaction.

"A videotaped account from New York to the southern tip of Latin America. A form of infolding in space while evolving in time. Playing back a culture in the context of another, the culture itself in its own context, and, finally, editing all the interactions of time, space and context into one work of art.

"Cultural information (art, architecture, cooking, dance, landscape, language, etc.) will be mainly exchanged by means of video-tape shot along the way and played back in the different villages, for the people to see others and themselves.

"The role of the artist is here conceived as a cultural communicant, as an activating aesthetic anthropologist with visual means of expression: videotape." Downey, "Video Trans Americas," *Radical Software* 2, no. 5 (Winter 1973): 4. Available online at https://www.radicalsoftware.org/volume2nr5/pdf/VOLUME2NR5_0006.pdf (accessed July 21, 2023).

13 Downey, "Travelogues of Video Trans Americas, 1973–75," *Journal of the Centre for Advanced TV Studies* 4 (1976): 22–28.

14 Pratt, "Arts of the Contact Zone," 1990, in *Profession*, 1991, 33–40. Available online at www.jstor.org/stable/25595469 (accessed October 15, 2023).

15 Downey, "Technology and Beyond," *Radical Software* 2, no. 5 (Winter 1973): 2.

16 Lévi-Strauss, *The Savage Mind*, 268.

17 See Isabelle Stengers, "The Cosmopolitical Proposal," in Bruno Latour and Peter Weibel, eds., *Making Things Public* (Cambridge, MA: The MIT Press, 2005), 994–1,003.

18 See Max Weber, "Science as a Vocation," 1918–19, in Weber, *The Vocation Lectures*, ed. David Owen and Tracy B. Strong, trans. Rodney Livingstone (Indianapolis: Hackett, 2004), 13 ff.

19 Max Horkheimer and Theodor W. Adorno, *Dialectic of Enlightenment: Philosophical Fragments*, 1947, Eng. trans. Edmund Jephcott, ed. Gunzelin Schmid Noerr (Stanford, CA: Stanford University Press, 2002), 2.

20 Eduardo Viveiros de Castro, *Cosmological Perspectivism in Amazonia and Elsewhere: Four Lectures Given in the Department of Social Anthropology, Cambridge University, February–March 1998* (Manchester: Hau Masterclass Series, 2012), 46.

21 Lévi-Strauss, *Tristes Tropiques*, 1955, Eng. trans. John Russell (New York: Criterion Books, 1961), 80.

22 In philosophy the *pharmakon* simultaneously denotes remedy and poison. Jacques Derrida has analyzed writing as *pharmakon*. Stengers returns to the pharmaceutical connotation of the term to dissect our time's contradictions between truth and fabrication, science and magic. For Stengers, "The lack of a stable and well determined attribute is the problem posed by any pharmakon, by any drug whose effect can mutate into its opposite, depending on the dose, the circumstances, or the context, any drug whose action provides no guarantee." In Stengers, "Culturing the Pharmakon?," *Cosmopolitics I*, trans. Robert Bononno (Minneapolis: University of Minnesota Press, 2010), 29.

23 Baniwa's critique was aimed not only at the cultural appropriation of these images but also at the fact that Indigenous presence was until that moment very much limited to appearances in non-Indigenous artists' works. In 2020, coincidentally, the Maya Kaqchikel artist Edgar Calel walked through the Pavilhão Ciccillo Matarazzo, the

historical venue of the Bienal de São Paulo, wearing the skin of a jaguar, a sacred animal in Mayan cosmovisions.

24 Baniwa, quoted in Alanis Reis, “A arte antropófaga de Denilson Baniwa: sobre o uso da tecnologia como meio de descolonizar e resistir,” *Arte, Mídias e Tecnologias*, January 2021. Available online at https://artemidiastec.wordpress.com/2021/09/22/a-arte-antropofaga-de-denilson-baniwa-sobre-o-uso-da-tecnologia-como-meio-de-descolonizar-e-resistir/ (accessed October 24, 2023).

25 Baniwa, “Art to Tame the Napë Heart,” 42–43.

26 Ibid., 44–45.

27 Viveiros de Castro, *Cosmological Perspectivism in Amazonia and Elsewhere*, 150. For his use of the word “path” in this passage Viveiros de Castro cites Graham Townsley, “Song Paths: The Ways and Means of Yaminahua Shamanic Knowledge,” *L’Homme* 33, no. 126–28 (1993).

28 Jaider Esbell, *Carta ao Velho Mundo*, March 20, 2019. Available online at http://www.jaideresbell.com.br/site/2019/03/20/carta-ao-velho-mundo/ (accessed July 23, 2023).

29 Ibid. The exhibition in France was at the Espace Philippe Noiret, Les Clayes de Sous Bois, April 19–May 22, 2019.

JOANNA PAGE AND MARA POLGOVSKY EZCURRA

PAST ECOLOGICAL FUTURES

Revisiting Luis Fernando Benedit, Marta Minujín, and the Work of CAyC

In December 1977, the Brazilian artist Frans Krajcberg dismantled the works he had presented at the XIV São Paulo Bienal and threatened to burn them. His protest was motivated by the news that the group show *Signos en ecosistemas artificiales* (Signs in artificial ecosystems), by Argentina's Centro de Arte y Comunicación (Center for art and communication, known as CAyC), had been awarded the Gran Premio Itamaraty, the Bienal's highest honor. In his view, this highly conceptual series of installations—featuring wired potatoes, two-dimensional caged birds, and plastic lamb carcasses hanging above a dining table—detracted attention from more critical visions of deprivation in Brazil under the country's military dictatorship that were developed both in his work and in the work of Grupo Etsedron, from Bahia, Brazil.[1] Giving the prize to a country also under military rule was for Krajcberg not only an act of censorship but also a politically motivated decision.[2] CAyC's art did not make it to São Paulo in 1977 through any support afforded by Argentina's junta of 1976–83, however, but thanks to critic and entrepreneur Jorge Glusberg's efforts to promote innovative forms of both Argentine and Latin American art.[3] A decade prior, in 1968, Glusberg had founded CAyC (first named Centro de Estudios de Artes y Comunicación, CEAC) as a space for interdisciplinary dialogue between the sciences, the humanities, and the arts. CAyC's inaugural exhibition, *Arte y cibernética* (Art and cybernetics, 1968), defined the center from the outset as a haven of experimental art, new technologies, and new philosophies of art that would call into question the very relationship between art and society.

If CAyC's series of installations at the 1977 Bienal seemed to lack relevance to pressing socioeconomic concerns in Brazil, as several critics in addition to Krajcberg have argued, many of the pieces included in the show addressed issues that are now at the forefront of ecological and environmental art in the region. Jorge González Mir's *Factor interespecífico* (Interspecific factor) drew attention to the erasure of natural habitats, with artificial birds invoking the specter of extinction, while Vicente Marotta's *Más y mejores alimentos para el mundo* (More and better food for the world) denounced practices of intensive cattle-ranching and industrial agriculture. The attention paid by CAyC members to the intersection of cultural, social, ecological, and geopolitical systems closely matches the multidisciplinary approaches taken by Argentine scholars and institutions that made significant contributions to emerging global debates on ecology in the 1970s. Paradigmatic in this regard was the development of the *Modelo Mundial Latinoamericano* (Latin American world model, 1972–76) by Fundación Bariloche, which demonstrated that the limits to human growth were not physical, relating to the exhaustion of natural resources, but principally sociopolitical, linked to underdevelopment, poverty, and inequality.[4] CAyC's interest in the relationship between social and environmental justice also resonates with a number of today's "artivist" projects, projects that combine art and activism. In the same Bienal, Luis Pazos's *Arte como forma de vida* (Art as a way of life), which involved members of the public in the creation of a *huerta* (vegetable garden), anticipated contemporary artistic practices that have turned to the sowing and cultivation of crops, either as a form of protest against monocrop agriculture or to lend visibility to alternative farming methods and philosophies of nature.[5]

Signos en ecosistemas artificiales was part of the "*Arte Não Catalogada*" (Uncatalogued art) section of the Bienal, an apposite name given the diversity of the practices it included and the difficulties they presented for categorization. Each of the ten male artists who participated in the section—Glusberg, González Mir, Marotta, Pazos, Jacques Bedel, Luis Fernando Benedit, Víctor Grippo,

Leopoldo Maler, Alfredo Portillos, and Clorindo Testa—made artworks markedly different in appearance, medium, and materiality. Yet they partook in a shared vision of art that, as Johanna Gosse and Timothy Stott have recently argued, "prioritizes process over product, relations over autonomy, and . . . signals a 'turn outward' to the world."[6] In a seminal article published in *Artforum* in 1968, the American art historian Jack Burnham theorized this vision as "systems aesthetics," a notion that Glusberg, in his resolve to position Argentine art firmly within the international sphere, appropriated both to describe CAyC artists' multiple approaches to conceptualist art and to place them in dialogue with developments in the rest of Latin America, the United States, and Europe.[7]

Glusberg's influential writings on the works of CAyC members have often shaped the way in which they have been read, effectively confining them to an almost exclusively anthropocentric, Western (as opposed to Indigenous), and patriarchal vision. Although Glusberg—following Burnham—borrowed his "systems" concept from biology, emphasizing the openness of biological systems and their complex interactions with other systems around them, his focus was unfailingly directed toward those systems that he considered uniquely human, namely symbols and language. He did not explore the complex forms of semiotics and social organization developed by other animals, and in retrospect he may be seen to have underestimated the extent to which animals coevolve with their environments.[8] More recent findings by zoologists and neurobiologists have challenged this assumption of human exceptionality, and contemporary artists are accordingly more likely to emphasize the complex social behavior of many other species, as well as to offer evidence of their creativity and intelligence, characteristics that, in the post-Enlightenment West, have been largely associated with human animals. Glusberg's rather narrow understanding of the potential scope of "systems aesthetics" has not, however, limited the ways in which the practices of the 1970s continue to resonate in today's art.

Not only do systems aesthetics enable us to comprehend the complex interactions between cultural, political, and technological systems, they also provide a way to explore ecological relationships between living and nonliving organisms and human and more-than-human subjects.[9] Ecology was indeed at the heart of early systems art. It emerged in the 1960s as part of an effort to maintain "the biological livability of the Earth, producing more accurate models of social interaction, understanding the growing symbiosis in man-machine relations."[10] It responded to what Burnham describes as a "superscientific culture," in which power came to lie less in the ideological control of traditional symbols than in the technocratic control of information. Systems art aimed to reclaim and redirect these information flows, opening routes of communication and exchange between unusual subjects or objects and unveiling relationships between entities that remained out of sight. The "specific function" of systems art was therefore accomplished by its own capacity, in Burnham's words, "to show that art does not reside in material entities, but in relations between people and between people and the components of their environment."[11] These ideas resonate with the interest of many artists and theorists today in environmental aesthetics, as well as in non-anthropocentric understandings of agency and communication. Yet the study of systems art in Argentina and elsewhere has largely dismissed this ecological dimension, instead almost exclusively privileging artists' engagement with the technologies and scientific ideas of the 1960s and '70s. This has resulted in what Gosse and Stott describe as "technodeterministic" readings that leave aside the larger ecological and even decolonial implications of this artistic tendency, as well as its relationships with recent forms of relational, participatory, and ecological art.[12]

In this essay we wish to revise these readings through a discussion of two key artists closely associated with CAyC. The first, Benedit, was active in the institution from its foundation and remained a central figure in many of its exhibitions; the second, Marta Minujín, worked in close collaboration with CAyC after 1974. She has not been extensively discussed in relation to CAyC as a result of the primarily masculinist approaches often taken to the history of the center, approaches instigated by Glusberg himself. A close analysis of works from the 1970s by these artists will allow us to move beyond technodeterministic readings of CAyC. In doing so, we will discuss ways in which its members were exploring a future-oriented ecological vision capable of encompassing more-than-human forms of life, communication, and intelligence. These works may retrospectively be read as significant precursors to the more recent interest in nonhuman modes of perception and being among artists in and beyond Latin America. Our vision, which positions Minujín center stage in the history of CAyC, seeks to provide both an ecological and a feminist re-reading of this important group of Argentine artists.

BENEDIT: "LIVING IN A CROWDED WORLD"

Benedit's best-known works of the 1970s were artificial habitats designed to allow human viewers to observe the growth and behavior of plants and animals (fig. 1). Both at the time of their first exhibition and since, these works have largely been valued for their symbolic meanings. Glusberg, for example, described Benedit's labyrinths for animals and plants as "an almost literal metaphor for the behavior of man in our societies: the pitfalls, the labyrinthine paths to obtain a reward, success, or failure"; in this way, they allude to "the conditions of life defined in our contemporary societies."[13] While the works certainly open themselves to this reading, to interpret Benedit's art—and that of CAyC artists more generally—primarily as allegories of the human condition is to overlook the insights they lend us into the lifeworlds of animals and plants, and in particular their staging of the complex and often problematic ways in which more-than-human forms of life are inscribed within human culture.

Some of Benedit's habitats, such as *Biotrón* (Biotron, 1970) and *Fitotrón* (Phytotron, 1972), were large installations, while others—including *Pecera para peces tropicales* (Fish tank for tropical fish, 1970), *Hábitat para caracoles* (Habitat for snails, 1970), and *Evaporador de Sachs* (Sachs evaporator, 1972)—were miniatures designed to house a single plant or a group of small invertebrates. All of them re-created the minimal conditions each species needed to survive (nutrients, water, and light, as required) and used transparent plexiglass to permit observation from every angle. From today's perspective, Benedit's habitats might be criticized for reproducing the dubious ethics of the zoo. We will argue here, however, that they bear much greater affinities with the post-anthropocentric

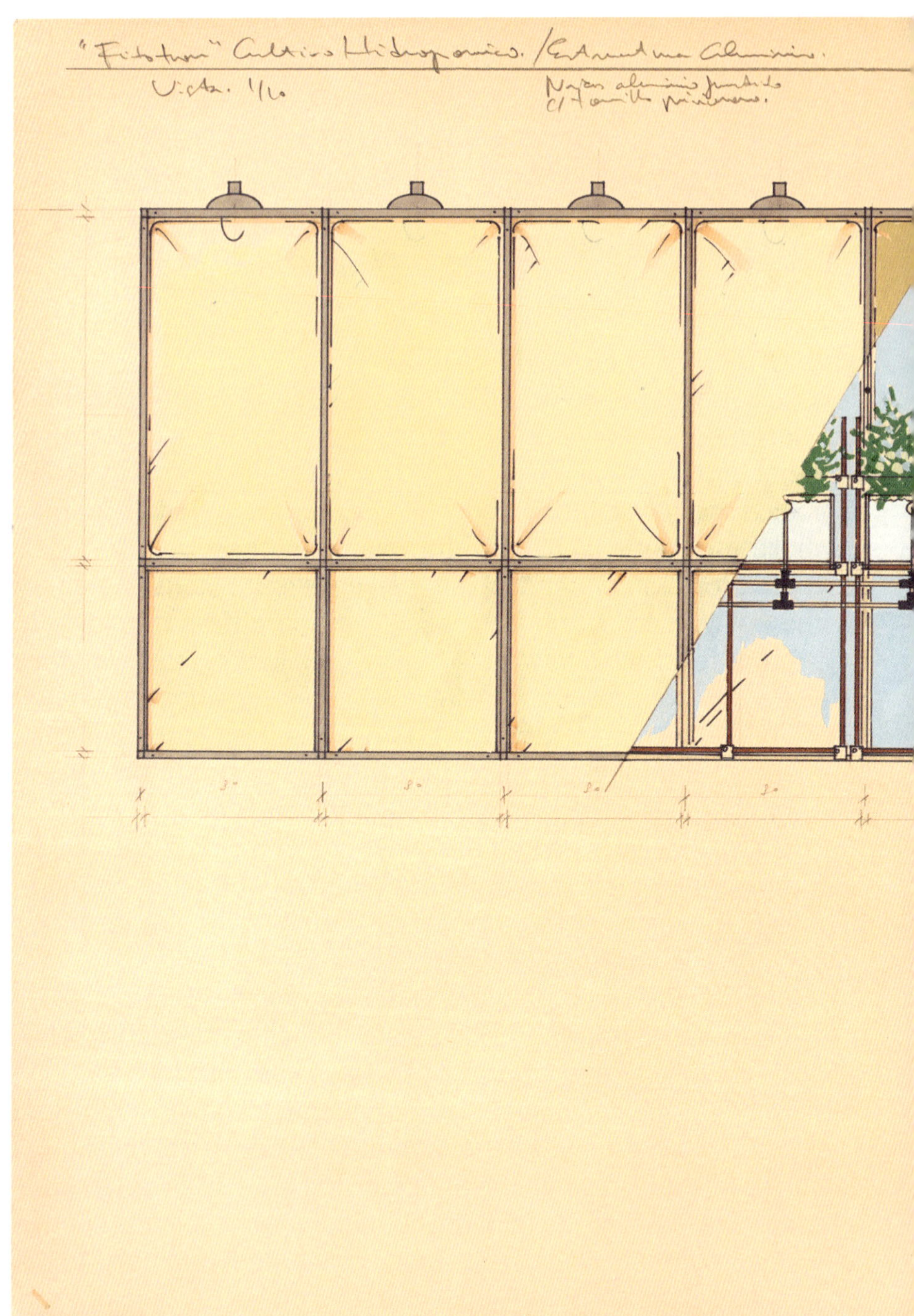

1 Luis Fernando Benedit. *Drawing for Phytotron: Hydroponic Environment for Plants*. 1972. Gouache, synthetic polymer paint, felt-tip pen, and pencil on paper, 22 ⅛ × 30 in. (56 × 76 cm). The Museum of Modern Art, New York. Gift of the artist

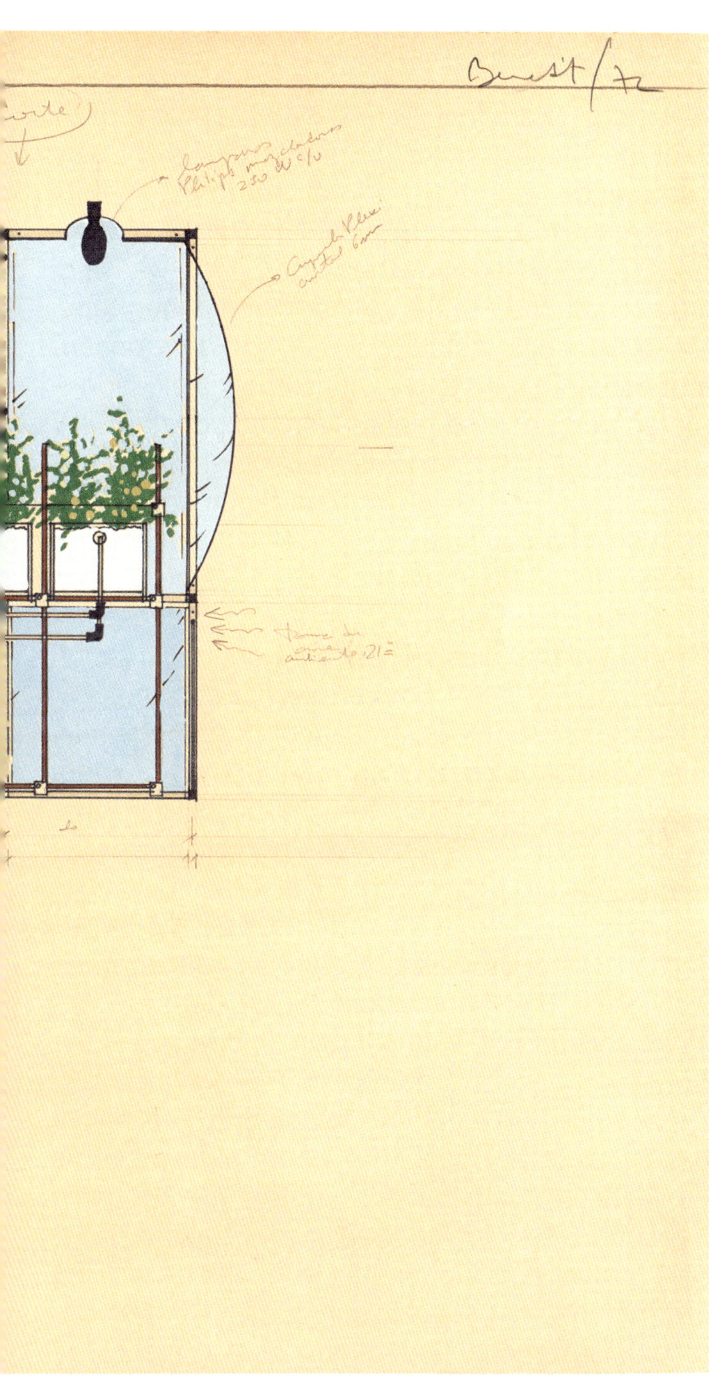

perspectives of twenty-first-century artists working with live animals and plants.

Minibiotrón (Hábitat para arañas) (Minibiotron [Habitat for spiders], 1970) is one of the smaller works in Benedit's series. It consists of a plexiglass cylinder with tiny perforations to allow air to flow in and out, a plug at the top, a central pillar (also made of plexiglass) with branches that would facilitate the spider's web-weaving, and a magnifying glass attached to the cylinder with a long tube. Marcela Castelo finds Benedit's habitat "*demasiado antropocéntrico*" (overly anthropocentric), pointing—with some justification—to the human-centered perspectives that govern the construction. Although she acknowledges that the inclusion of other organisms was a radical step in art (an exclusively human realm until then), she argues that Benedit's habitat was not conceived with consideration of the spider's needs and interests: it provides no shelter, no relief from light, and no food sources.[14] In his original design, however, Benedit imagined that flies would be introduced through the plug at the top of the cylinder.[15] Shelters would be unnecessary, due to the absence of predators; and although Benedit may not have known this, recent studies have shown that many spiders are drawn to light sources as good sites for web-building.[16] If the spider's exposure to visibility in its plexiglass habitat makes us uncomfortable, perhaps we are projecting on it a human desire for privacy and our own need for darkness in order to rest.[17]

The larger point here is that Benedit's *Minibiotrón* was designed to provoke precisely such questions about the spider's sense-world and how it intersects (or not) with our own. In creating it, the artist stated, "I did not want anything to remind them of their natural habitat."[18] To grasp how the work was intended to function—he acknowledged that "there are people to whom all this seems cruel"—we need to bear in mind that the context for Benedit's designs was not a world of ecological health and natural diversity, but one in which "nature is disappearing" and "cities are filling everything."[19] If his structures may seem to imagine a callously confined existence for spiders, we must understand them in the light of his belief that there was no future for organic life outside technology, and that it was necessary to embrace the artificial and build new spaces in which humans might dwell with other species. These are speculative works, then, produced in a register that we might even recognize as science fiction. In this evolution toward the artificial, Benedit believed, "It is feasible to achieve a balance between men and nature, but to achieve it you must have prior knowledge of nature."[20] Yet in his claim that his designs would produce a "*mejoramiento*," an amelioration, we might well detect the overweening ambitions that have fueled the humanist embrace of technology and the Western domination of nature.[21] Contemporary artists who have designed artificial habitats—including Gilberto Esparza, from Mexico, and Ana Laura Cantera, from Argentina—are more likely to emphasize the regenerative capacities and the resilience of other species than the redemptive role of human technology.[22] But they share with Benedit an interest in using such technology to draw attention to the lifeworlds of other organisms, as a prerequisite for finding ways to balance their needs with ours.

The integration of the magnifying glass in the design of *Minibiotrón (Hábitat para arañas)* does mark its central purpose as one of human observation. But if the plexiglass allows us to observe the spider, it also provides the protection from human activity that the creature requires to survive: let loose in the museum, the spider would be accidentally stepped on, brushed away, or even deliberately destroyed. It is only its framing in a container that allows us to see it as something other than a pest or intruder, to recognize it as a member of a species with a right to live. In important ways, then, the isolation of the spider within its artificial container encodes two crucial truths: first, that the survival of many animals' lifeworlds is dependent on our adoption of specific practices of care; and second, that we can only begin to understand the sense-worlds and social behavior of many species through the use of technologies of visualization and amplification.

Such technologies are at the core of recent art projects by the Argentine artist Tomás Saraceno, on spiders, and the Ecuadorian artist Kuai Shen, on ants. The extensive research conducted by Saraceno and Shen on the social lives of invertebrates, and on these animals' cooperations with other species, brings their work into a sustained dialogue with recent scholarship on multispecies interactions.[23] Their use of acoustic technologies allows us to grasp something of the particular sense-worlds of spiders and ants, who communicate to a great extent through vibration. Benedit's *Minibiotrones* may only be observed with a magnifying glass, but he shares an interest with Saraceno and Shen, not only in deploying technology to help us to understand how other species perceive the world but in forging techniques of DIY biology that equip human viewers to make their own explorations.

Benedit's production of multiple *Minibiotrones* for spiders and snails, intended for people to purchase and take into their homes, is echoed in the work of the Argentine artist Joaquín Fargas, who makes sealed *Biosfera* (Biosphere) globes containing water and plants. In 2007, Fargas began to create tiny versions of these globes and to distribute them to individuals to keep at home, highlighting the fragility of Earth's systems and the urgency with which we need to adopt practices of care. Arguably, Benedit's *Minibiotrones*—which would have involved a much more active commitment on the part of their human owners to keep the spiders fed with meals of flying insects—represent a fuller modeling of such practices.

Benedit extended his experiments with animals and plants in the series of labyrinths he created for ants, cockroaches, mice, plants, and, ultimately, humans. Zoologists often use mazes to measure animals' problem-solving capacities; indeed, many studies have found that ants, working collectively, are much quicker than humans at finding the shortest route from one end of a maze to the other. Critics past and present have consistently read Benedit's labyrinths as metaphors for the increasing control and surveillance of human life. Although Mari Carmen Ramírez does pay attention to these works as experiments in animal behavior, she too refers to their status as "metaphors of the conditions experienced by Argentine citizens in the highly volatile political climate of the period between 1966 and 1976."[24] It is difficult to find a basis for such readings in the works themselves, or in Benedit's account of them. Even *Laberinto invisible* (*Invisible Labyrinth*, 1971), a system of mirrors, lights, and electronic alarms designed for humans to try to navigate, seems to lend itself more to a scientific reading than to a political one. Exhibited alongside *Laberinto para ratones blancos* (Maze for white mice, 1971), it put its human participants in the position of animals performing an experiment, encouraging them to reflect on their own problem-solving capacities as if they were mice trapped in a maze.

Laberinto invisible can easily be read from a posthumanist perspective, as an experiment eroding the distinctions between human and nonhuman animal behavior, or perhaps even revealing our own disadvantage compared to other animals' ability to orient themselves in space. Benedit's *Laberinto vegetal* (Vegetable maze, 1972), which demonstrates the capacity of plants to sense and move toward light, also prefigures a number of more recent art/science projects that explore nonhuman forms of intelligence, including plant cognition and communication.[25] Many of Benedit's works appear to have been designed to erase divisions and hierarchies between humans and other animals; as he stated, "Since all of us are members of the whole of nature, ecological laws are valid for us as well, intertwining everyone with everyone and everything with everything."[26]

We would contend that Benedit's habitats and labyrinths are primarily models that cast light on how organisms—human and nonhuman—learn to navigate new environments. Rather than acting as metaphors of the political present, they situate themselves in the ecological future, anticipating a time when expanding human populations and growing urbanization require us to confront, as David Elliott suggests, the effects of overcrowding and confinement.[27] As Benedit stated, "We'll be living in a crowded world and we must learn to think in totalities."[28]

This emphasis on the pressing need to find more equitable forms of coexistence with other species brings Benedit's work into direct dialogue with art projects conceived a full fifty years later. His "assimilation of technology as an indispensable tool for the preservation of life on the planet" corresponds closely to the approaches of many twenty-first-century Latin American artists, such as Esparza and the Brazilian artist Ivan Henriques, who have designed technological interventions into ailing ecosystems.[29] These works, like those of Benedit, do not defend the hegemonic models of "green" technology or try to solve the problems of climate change through geoengineering projects on a planetary scale. Instead, they adopt amateur, do-it-yourself technologies that focus our attention, as Anna Lowenhaupt Tsing might put it, on possible forms of "collaborative survival."[30]

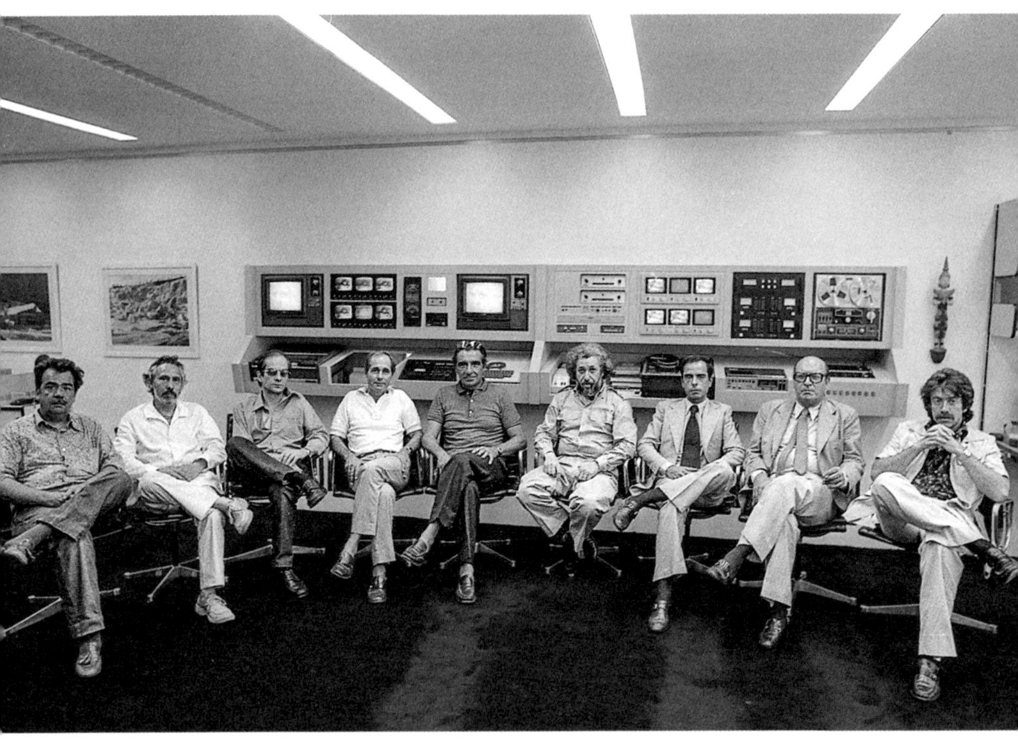

2 The CAyC Group around 1980, in Jorge Glusberg's office in Buenos Aires. Left to right: Luis Fernando Benedit, Alfredo Portillos, Víctor Grippo, Jorge González Mir, Vicente Marotta, Jorge Glusberg, Jacques Bedel, Clorindo Testa, and Leopoldo Maler. Archivo Roth

MINUJÍN: UNEARTHING A NEST FOR THE FUTURE

The sense of a multispecies futurity so poignantly felt in Benedit's more-than-human architectures also appears in the work of other CAyC artists, including that of the women who regularly participated in CAyC exhibitions, although often without due recognition and visibility. Such is the case with Marta Minujín, who in a recent interview described CAyC as "the only place I could work after [the Centro de Artes Visuales of the Instituto Torcuato] di Tella closed [in 1970]."[31] To discuss Minujín as part of CAyC is therefore to propose a revisionist reading of the overwhelmingly masculinist approaches that have been adopted in writing the organization's history. These are yet another consequence of Glusberg's insistent framing of CAyC on his own terms, and of his interest in primarily promoting the work of thirteen male artists known as the "Grupo de los Trece" (the group of thirteen)—a term that conveys a misguiding sense of permanence given that the members of CAyC changed over the years, and for a brief period female artist Mirtha Dermisache belonged to it.[32]

One of CAyC's most iconic images, taken by Pedro Roth around 1980, portrays, from left to right, Benedit, Portillos, Grippo, González Mir, Marotta, Glusberg, Bedel, Testa, and Maler sitting next to one another in a semicircle in front of numerous computers neatly mounted on a wall in CAyC's futuristic premises (fig. 2). The photograph powerfully conveys Glusberg's vision of CAyC as a cutting-edge, cosmopolitan, all-male institution enabling a dialogue between art, technology, and society. As we suggest here, however, this is a partial perspective on the actual history of CAyC, with respect both to the ecological vision beyond technology that is present in the art of a number of CAyC's artists and to the participation of women in the shaping of this vision.

Minujín was not in the Grupo de los Trece, but when she returned to Argentina in 1974, after living in New York for several years, she immediately began to collaborate with Glusberg, who had included her in international group shows such as the major exhibition *Kunstsystemen in Latijns-Amerika* (*Art Systems in Latin America*), in Antwerp in 1974. The pieces Minujín developed at CAyC include *La academia de fracaso* (The academy of failure, 1975), conceived with "philosophy professor and futurologist"

Agustín Merello, in which filmmakers, musicians, actors, curators, and artists were invited to discuss their failures;[33] *Comunicando con tierra* (Communicating with earth, 1976; fig. 3), a multistep action using soil to create a system of communication and exchange across Latin America; and *3,000 naranjas* (3,000 oranges, 1979), part of a series, *Arte agrícola en acción* (Agricultural art in action), in which various performers dressed in white and, with their heads covered with buckets, sang phrases like "*La teoría es la práctica*" (Theory is practice) and "*naranjasjas*" as they touched, moved, and played with thousands of oranges placed on the floor inside a demarcated square.[34] The performance of these actions at CAyC, together with Minujín's participation in group shows either at CAyC or under its auspices, such as *Arte en cambio II. ¿Hay vanguardia en Latinoamérica?* (Changing art II. Is there an avant-garde in Latin America?, 1976), *América Latina '76* (1977), and *Veinte artistas argentinos* (Twenty Argentine artists, 1977), are sufficient reason to consider her an artist who significantly shaped what we describe as CAyC's ecological aesthetics.

Dubbed by Minujín in one of her sketches as an "*obra de arte total*" (total artwork), *Comunicando con tierra* combines spirituality, mysticism, and the desire to interconnect different regions of Latin America through the materiality of its earth, producing a system of soil exchange that is at once performative and allegorical. The piece is in this sense paradigmatic of a systems art that works to surpass the techno-scientific impulse to encompass a larger ecological vision, and in this case also a sense of interrelation between the living and the spiritual worlds ("I interconnect mystical symbolisms and existential alchemies with conceptual intentionality," Minujín writes in one of CAyC's bulletins).[35] As Chus Martínez suggests, the conception of this piece draws on systems and, we could add, cybernetic thinking whereby soil becomes a sort of "black box" that on the one hand enables a system of communication and on the other "contains history, information, place, event, testimony, code, energy, identity."[36]

The work began in Machu Picchu, Peru, the Incan sanctuary that Minujín considered the metaphysical center of Latin America, and evolved to produce a network interconnecting all of the continent, as the artist showed in an embroidery intervention in a map

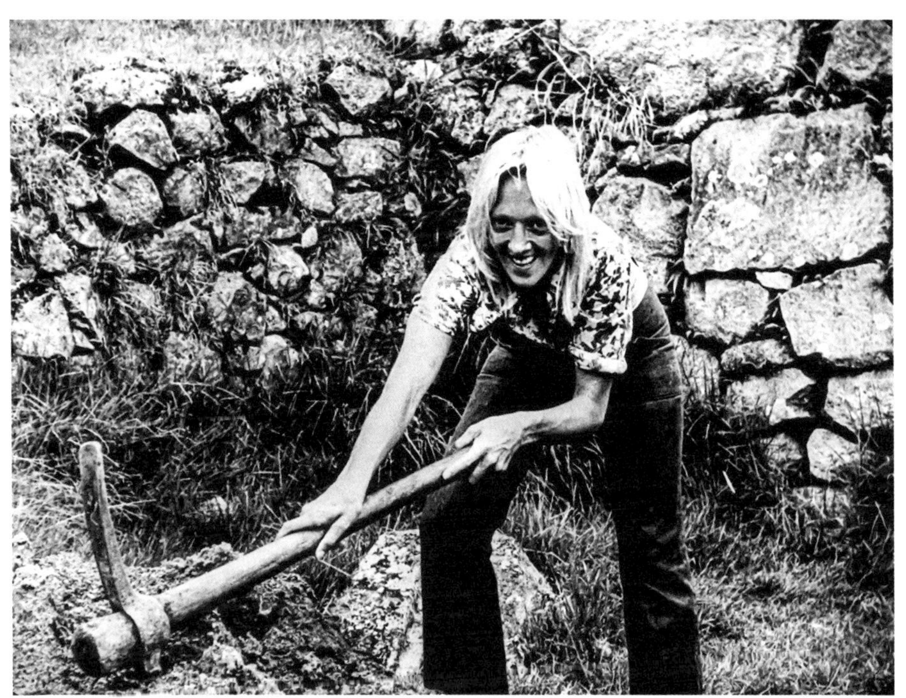

3 Marta Minujín. *Comunicando con tierra* (Communicating with earth). 1976. Black and white photograph, 3 15⁄16 × 9 7⁄16 in. (17.5 × 24 cm). Archivo Marta Minujín

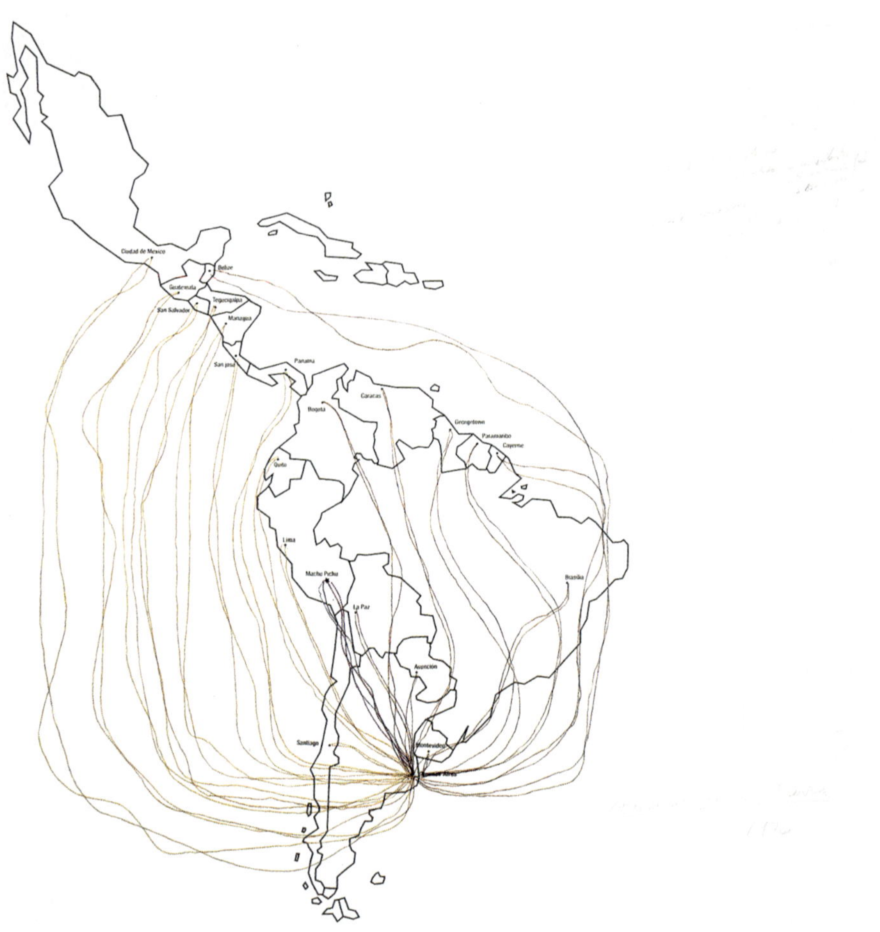

4 Marta Minujín. *Comunicando con tierra (mapa)* (Communicating with earth [map]). 1971/1976. Drawing and embroidery on paper, 37 ⅜ × 37 ⅜ in. (95 × 95 cm). Archivo Marta Minujín

of the region (fig. 4). The initial step in this network's production was the "extraction" of a sample of soil from Machu Picchu that was then taken to Buenos Aires. There, Minujín used some of the sample to build an enormous, human-scale reconstruction of the nest of the ovenbird, or rufous hornero, a species of warbler native to South America. She packed the rest of the soil into one-kilogram bags, which she exhibited alongside the nest in *Arte en cambio II*, together with documentation of the extraction of the soil that shows Minujín digging the land with a mattock. The soil packages were sent to artists from each Latin American country (then numbering twenty-three) with instructions to mix the content with soil from sites located as near as possible to their studios. Having produced this organic mixture, the artists were asked to return a kilo of the soil in a bag to Minujín's address in Buenos Aires so that she could bring this heterogenous Latin American soil back to the place in Machu Picchu where the original extraction had taken place. This process of reparation took close to a decade.[37]

In asking twenty-three artists to engage synchronically in the mixing and gathering of soil across different locations, *Comunicando con tierra* returns to some of the themes that run through Minujín's iconic *Simultaneidad en simultaneidad* (*Simultaneity in Simultaneity*, 1966), in which she used television, telephones, mail, radio, teletype machines, and satellite transmission simultaneously to "invade" the lives of people living in Buenos Aires and to send instructions to the artists Wolf Vostell in Berlin and Allan Kaprow in New York so that they could do the same (even if their access to technology appears to have been more limited and they failed to do so).[38] For Catherine Spencer, this event, exploring a kind of "transnationalism fostered through mass media technologies," moved beyond the prior ephemeral, spontaneous logic of Happenings to produce circuits of communication and feedback, conveying a sense that the challenges of "brute geography" could be overcome through increasingly dematerialized forms of connectivity.[39] Yet the technological matrix of the piece, with its deployment of the most advanced means of communication, is markedly different from the exclusively organic materiality of *Comunicando con tierra*, where soil becomes the universal communicant. The speed of *Simultaneidad en simultaneidad*—geared, according to

Niko Vicario, to eliminate the "lag" between Latin America and the rest of the world, in an analogy with the modernization theory of the 1950s and '60s—was also left behind in the later piece in favor of slow communicative circuits: the main technology became the postal system, just as in the mail-art movement that was rapidly growing during the 1970s.[40]

These material and temporal shifts from *Simultaneidad en simultaneidad* to *Comunicando con tierra* suggest a return to locality and the land, with its geographical, material, and sacred significations—a process coinciding with Minujín's own return from the United States to live in Argentina. Moreover, they reveal a markedly different approach to technology from that of the *arte de los medios* (media art) developed at the Instituto Torcuato di Tella in the 1960s. In this later moment of her practice, Minujín—attuned to the work of Benedit and Grippo—was developing a more ecological stance, no longer aspiring to generate some kind of technological sublime that would situate her alongside new-media artists in the United States and Europe. By contrast, she was interested in exploring forms of relationship, communication, and coexistence. Here, technology figures less as a series of complex apparatuses, or as a metonym for the power of the industrialized world, than as process, what Bernard Stiegler (following Gilbert Simondon) would describe as "a new economy of desire that could lead to the invention of new ways of life, new modes of individual and collective existence."[41]

Minujín's *Nido de hornero* (ovenbird nest) is striking precisely in relation to this interest in the relationship between technology and modes of existence, and, more broadly, to its connection with Benedit's creation of not-exclusively-human dwellings, habitats, and nests through which we might anticipate and imagine an ecological future. In a press release announcing the presence of the ovenbird nest at the annual Exposición de Ganadería, Agricultura e Industria (Livestock, agriculture and industry exhibition) at La Rural in Buenos Aires, the seemingly warm and snug organic structure is described both as a "*habitat del futuro*" (habitat of the future) and as the immemorial inspiration of the "*tradicional rancho de barro argentino*" (traditional Argentine adobe ranch). The object fulfills everyone's dream of entering an ovenbird's nest, the press release adds, while displaying a kind of art that "*existió siempre*" (always existed).[42] The rhetorical operation here is noteworthy in its dissolution of the division between *physis* and *techne*, positing art not as an emulation of nature but simply as "nature," and the putative author of the work as equally possibly an artist or a bird. Following the writings of Vinciane Despret, Mercedes Claus argues that the piece can be understood as an "anthropo-zoo-genic" practice, producing both humanity and animality: "the bird's refuge is anthropomorphized, allowing us to inhabit it, but in doing so we're also zoomorphized, as our bodies are affected by the sensorium of the animal world."[43]

Minujín exacerbates this logic by placing in the interior of the gigantic nest a TV monitor showing her video *Autogeografía* (Autogeography, 1976). Here the artist lies seminaked on a large bedsheet, appearing at times uncomfortable and at times more at ease as soil is thrown on her, in a scene evocative of a burial or some other kind of ritual involving direct contact with the smell, texture, and, for some, the spiritual symbolism of Pachamama, the "Mother Earth" of North Argentine and Andean cultures. The "transterritorial" and transspecies utopianism of *Comunicando con tierra* is thus confronted with an at times uneasy juxtaposition of skin and soil, body and landscape. The artist's desire to return to her national and Latin American origins—what Juan Cruz Pedroni has called "*las raíces*" (the roots) and "*el espacio uterino del nido*" (the uterine space of the nest)—acquires a tense sensorial dimension as Minujín welcomes the soil but prevents it from entering her mouth.[44] This tension is enhanced by her simultaneous embodiment of the cosmopolitan pop star: throughout the video she wears her signature sunglasses, a gesture suggesting an almost humorous rejection of the idea of finding any kind of Latin American "existential alchemy," to use her words, in the Incan sacred site from which the earth to build the nest came.[45]

A second version of the video, *Autogeografía (con máscaras)* (Autogeography [with masks], 1976), is even more explicit in its exploration of these contradictions. In this piece the camera movements are reduced to alternating between shooting Minujín's feet being covered in soil—from a frontal angle that foregrounds the scene's mortuary connotations and the passivity of the body—and moving upward to show her face, on which she wears not only her sunglasses but also a series of exaggerated silicone masks: "an antiaging mask, a Puerto Rican mask, a Miss Universe mask, a [Henry] Kissinger mask . . . a Fidel Castro mask."[46] The same soil that seems to conjure a Latin American spiritual origin in *Comunicando con tierra* appears now as an element that would bury the multiplicity of different cultures, themselves portrayed as unstable facades. The nature-culture divide that organizes the symbolism of this piece is ultimately dissolved, however, as the dark soil originally thrown on Minujín becomes a yellow powder (possibly corn flour). As the video progresses, the seemingly natural and seemingly artificial elements of the piece are rendered coconstitutive, and a sense of aliveness emerges from this entanglement of organic and inorganic agents.

Comunicando con tierra's integration of biological and technological habitats distinguishes it from Ana Mendieta's *Siluetas* (Silhouettes, 1973–78), in which a nostalgia for both cultural and natural origins is more prominent. The piece situates Minujín instead as a predecessor of contemporary installations and performances in which organic and technological elements are put into dialogue, both to generate a range of affects and to produce systems of communication and signification. *Rizosfera FM* (Rhizosphere FM, 2016), for example, by the Argentine artist duo Colectivo Electrobiota, also focuses on the communicative capacity of soil while proposing the appropriation of mass media by hybrid networks of humans and other species.[47]

Uterine, feminized, and at times sacralized approaches to the earth (*la tierra*) appeared in the work of other CAyC artists, including Portillos and Carlos Ginzburg. Some of these pieces have been revisited in recent years, as, on the one hand, the Anthropocene era has made visible the extent to which extractive capitalism has brought the ecological assemblages that sustain life to a point of no return, and on the other, Indigenous epistemologies have gained salience as an effective response to the current environmental crisis. In *Madre Tierra* (Mother Earth, 1971), Ginzburg proposed excavating a three-kilometer-wide uterus in the pampas of Buenos Aires to investigate the "geo-psychoanalytical dimension" of the earth.[48] He imagined himself living there, at the creative locus of

Pachamama; he also envisioned that plants would begin to sprout after the excavation and that animals would visit, so that the earth appears in the work as capable of preserving its generative capacities even as powerful technologies are used to reshape it.

While Ginzburg never realized this clearly animistic project, he returned to it in 2019 to create *Post Madre Tierra* on the grounds of the Facultad de Psicología of the Universidad Nacional de La Plata. This work also involved an act of excavation—this time twenty-five meters, over eighty feet, in width—but left aside any imagining of the earth as generative, or of himself as inhabiting a space potentially fully capable of nurturing human and more-than-human lives. The excavation was by contrast filled with dead flowers taken from a nearby cemetery and with waste materials generated as a result of local practices of fracking and oil extraction. The work was furthermore surrounded by metal sheets carrying banners that alternated the word "*futuro*" (future) with arrows pointing in a single direction, thus producing a loop. The future here seems stuck, failing to download.[49] The suggestion moreover is that while refuges from dystopian landscapes—such as Minujín's ovenbird nest—are more necessary than ever, the organic materials with which they could be made have been polluted to the point of irrecoverability. The earth, as another banner reads, has turned "*rara*" (rare).

CONCLUSIONS

In his book *Retórica del arte latinoamericano* (Rhetoric of Latin American art, 1978), Glusberg includes "conceptual art, zoological art, magical art, ecological art, social art, action art, catastrophic art" in the corpus of systems art.[50] Recalling a Borgesian taxonomy, this description is telling in its oddity. The conditions of production of these different kinds of art, Glusberg adds, are similar, even if they rely on different codes: "Its creators are interested in what's happening in the society where they live, not in surrealist and academic representations of still lifes or technological geometricalisms. They want to communicate with the audience, and they are sure to do so without the involvement of traditional 'middlemen.'"[51] The implicit suggestion here is that Latin American systems art was an art of *doing*, rather than of *saying* or *representing*; that it had a penchant toward the real, including the natural world; and that it was oriented, first and foremost, toward staging its own conditions and processes of production. Its Latin American character relied on the singularity of these conditions and on the questions they posed, including questions around the role of art in the face of what the Argentine artist Nicolás García Uriburu, a collaborator with CAyC, sketched in 1977 as the dark "ecological balance of the world."[52] While Glusberg often emphasized the primacy of *human* signs and systems in CAyC work, an understanding of systems art in the fullness of its transdisciplinary approach to the interlocking systems of the natural and social worlds sheds new light on the ecological dimension of the art of CAyC. In doing so, it brings to the fore CAyC's relevance as a predecessor of, and interlocutor with, contemporary relational, feminist, and ecological practices.

Previous scholarship on CAyC has primarily highlighted its role in enabling interdisciplinary dialogues between art and technology. The readings we propose in this text take a closer look at the ways in which CAyC artists often engaged with the natural world, rejecting the idea of an opposition between technology and nature. Instead they essayed approaches that on the one hand make visible what Stiegler describes as humans' "originary technicity"—whereby technology is seen as coconstitutive of the idea of humanity[53]—and on the other share points of intersection with the "social-ecological assemblages" theorized by Eduardo Gudynas.[54] These draw on Andean and Amazonian thought to explore how the social organization of humans and nonhumans has evolved within specific landscapes, shaping and being shaped by its particular contours and conditions.

Some of CAyC's work was speculative and even utopian. Although Glusberg framed its vision in fundamentally anthropocentric terms, our readings open interpretative possibilities in which a more-than-human sense of futurity is already present. Moreover, we believe that CAyC's call to "think in totalities" could not be more relevant to the kind of pluralist, multispecies, decolonial, and planetary thinking needed to address the unfolding ecological crisis.

Mara Polgovsky Ezcurra wishes to thank the Leverhulme Trust for the support provided to carry out this research.

1 See "Escándalo en la Bienal," *Punto de Vista* 1, no. 1 (March 1978): 13.
2 See Julia Detchon, "Signs, Systems, Contexts: The Centro de Arte y Comunicación at the São Paulo Bienal, 1977," *ICAA Documents Project Working Papers* no. 5 (December 2017): 10.
3 The accusation that Jorge Glusberg sought or received support from the military government is carefully refuted in Adrián Gorelik and Graciela Silvestri, "Invierno argentino del 81: ¿Qué Argentina?," in Gorelik and Francisco Díaz, eds., *Tafuri en Argentina* (Santiago: Ediciones ARQ, 2019), 139–40. The Argentine government did send an official entry to the Bienal, which was curated by Roberto del Villano. See Detchon, "Signs, Systems, Contexts," 10.
4 The *Modelo Mundial Latinoamericano*, led by Amílcar Herrera, was a direct response to *The Limits to Growth* (1972), the first report on resource consumption and human social systems commissioned by the Club of Rome.
5 Many of these are now more likely to take the form of collaborations between artists and Indigenous or rural collectives, as is the case with *Plantío Rafael Barrett* (Rafael Barrett plantation, 2015–), a series of seed-planting actions led by Mónica Millán and Adriana Bustos with Conamuri Paraguay and the Movimiento Nacional Campesino Indígena, Argentina.
6 Johanna Gosse and Timothy Stott, "After the Breakdown: Sixty Years of Systems Art," in Gosse and Stott, eds., *Nervous Systems: Art, Systems and Politics since the 1960s* (Durham, NC: Duke University Press, 2022), 8.
7 Jack Burnham, "Systems Esthetics," *Artforum* 7, no. 1 (September 1968): 30–35. On conceptualism in Latin America see Mari Carmen Ramírez, "Tactics for Thriving on Adversity: Conceptualism in Latin America, 1960–1980," in Luis Camnitzer, Jane Farver, and Rachel Weiss, *Points of Origin, 1950s–1980s* (New York: Queens Museum of Art, 1999), 53–71, and Camnitzer, *Conceptualism in Latin American Art: Didactics of Liberation* (Austin: University of Texas Press, 2007).
8 See, for example, Glusberg, *Retórica del arte latinoamericano* (Buenos Aires: Ediciones Nueva Visión, 1978), 70.
9 See Gosse and Stott, "After the Breakdown," 13.
10 Burnham, "Systems Esthetics," 31.
11 Ibid.
12 See Gosse and Stott, "After the Breakdown," 8.
13 "Una metáfora casi literal sobre la conducta del hombre en nuestras sociedades: los escollos, los trayectos laberínticos para obtener una recompensa, el acierto o el fracaso"; "las definidas condiciones de vida en nuestras sociedades contemporáneas." Glusberg, *Retórica del arte latinoamericano*, 103–4.
14 Marcela Castelo, "Minibiotrón (Hábitat para arañas)," *Simbiología. Prácticas artísticas en un planeta en emergencia*, 2022. Available online at https://simbiologia.cck.gob.ar/hashtags/inter-agencias/ (accessed August 7, 2022).
15 Luis Fernando Benedit's study for the piece, showing how its features were designed to meet the spider's basic needs, may be viewed on the website of The Museum of Modern Art, New York: https://www.moma.org/collection/works/35696 (accessed August 7, 2022).
16 See, for example, Nikolas J. Willmott et al., "Guiding Lights: Foraging

Responses of Juvenile Nocturnal Orb Web Spiders to the Presence of Artificial Light at Night," *Ethology* 125, no. 5 (May 2019): 289–97.

17 Research shows that spiders do not appear to have circadian rhythms of the kind that regulate human sleep. See, for example, Viviane Callier, "Spiders reset body clocks to avoid 5-hour jet-lag every day," *New Scientist*, November 15, 2017. Available online at https://www.newscientist.com/article/2153407-spiders-reset-body-clocks-to-avoid-5-hour-jet-lag-every-day (accessed November 27, 2022).

18 "No he querido que nada les recordara su habitat natural." Benedit, in Alicia Dujovne Ortiz, "Luis Benedit y la caja de cristal," 1970, in Roberto Amigo, Silvia Dolinko, and Cristina Rossi, eds., *Palabra de artista. Textos sobre arte argentino, 1961–1981* (Buenos Aires: Fondo Nacional de las Artes and Fundación Espigas, 2010), 234.

19 "Hay gente a la que todo esto le parece cruel"; "La naturaleza está desapareciendo"; "Las ciudades lo llenan todo." Ibid., 235, 234.

20 "Es factible conseguir un equilibrio entre hombres y naturaleza, pero para lograrlo hay que tener un conocimiento previo de lo natural." Ibid., 234.

21 Ibid., 235.

22 See Joanna Page, *Decolonizing Science in Latin American Art* (London: UCL Press, 2021), 181–86, 193–99.

23 See Eben Kirksey, *Emergent Ecologies* (Durham, NC, and London: Duke University Press, 2015), and Anna Lowenhaupt Tsing et al., eds., *Arts of Living on a Damaged Planet: Ghosts and Monsters of the Anthropocene* (Minneapolis: University of Minnesota Press, 2017).

24 "Metáforas de las condiciones vividas por la ciudadanía argentina bajo el clima político extremadamente volátil que dominó el período entre 1966 y 1976." Ramírez, "Noción paradójica de la imagen: Los 'sistemas vivos' de Luis Fernando Benedit," in María Torres, ed., *Benedit: Obras 1968–1978* (Buenos Aires: Fundación Espigas, 2020), 48.

25 These often draw on the startling experimental findings related in books such as Stefano Mancuso, *The Revolutionary Genius of Plants: A New Understanding of Plant Intelligence and Behavior*, trans. Vanessa Di Stefano (New York: Atria Books, 2018), and Monica Gagliano, *Thus Spoke the Plant: A Remarkable Journey of Groundbreaking Scientific Discoveries and Personal Encounters with Plants* (Berkeley: North Atlantic Books, 2018).

26 "Todos los hombres somos miembros del conjunto de la naturaleza. De ahí que también sean válidas para nosotros las leyes ecológicas, el entrelazamiento de todos con todos y todo con todo." Benedit, quoted in Enrique Molina and Glusberg, *Luis Fernando Benedit: Memorias australes desde el Río de la Plata hasta el Canal del Beagle* (Buenos Aires: Ediciones Philippe Daverio, 1990), 31.

27 David Elliott, "Dibujos en las arenas del tiempo. Poética y política en la obra de Luis F. Benedit," in Torres, ed., *Benedit: Obras 1968–1978*, 66.

28 "Vamos a vivir en un mundo apretado y debemos aprender a pensar en totalidades." Benedit, quoted in Molina and Glusberg, *Memorias australes*, 31.

29 Ramírez, "Noción paradójica de la imagen," 45.

30 Tsing, *The Mushroom at the End of the World: On the Possibility of Life in Capitalist Ruins* (Princeton: Princeton University Press, 2015), 4.

31 "El único lugar donde podía trabajar

después del cierre del Di Tella." Marta Minujín, in an interview with Mara Polgovsky, March 31, 2022.

32 Glusberg invited only one woman, Mirtha Dermisache, to become part of the Grupo de los Trece. Although Dermisache participated in some initial activities and discussions, she decided that she would rather be a guest artist at CAyC than belong to the core group. Like Minujín, she featured in many CAyC exhibitions. See Annalisa Rimmaudo and Giulia Lamoni, "Entrevista a Mirtha Dermisache," in Cecilia Cavanagh, *Mirtha Dermisache: publicaciones y dispositivos editoriales*, exh. cat. (Buenos Aires: Pontificia Universidad Católica Argentina Santa María, Pabellón de las Bellas Artes, 2011). Available online at http://hipermedula.org/2017/08/entrevista-a-mirtha-dermisache/ (accessed August 11, 2022). See also Detchon, "Work-Around: Lea Lublin, Marie Orensanz, Mirtha Dermisache, and Margarita Paksa, 1968–1983," PhD diss. (University of Texas, Austin, 2022).

33 Anonymous, "El triunfo de los fracasados," *7 Días*, Buenos Aires, October 17, 1975, quoted in Victoria Noorthoorn, *Marta Minujín: Obras 1959–1989* (Buenos Aires: MALBA, 2010), 152.

34 See ibid., 150.

35 "Interconecto con la intencionalidad conceptual Simbolismos Místicos y Alquimias Existenciales [*sic*]," *Comunicando con tierra*, CAyC bulletin, n.d. Marta Minujín Archive.

36 Chus Martínez, *Comunicando con tierra* (Buenos Aires: Henrique Faria, 2014).

37 Minujín does not recall the identities of all twenty-three artists involved. Interview with Polgovsky.

38 See Rodrigo Alonso, "On Technological Tactics," 2005. Available online at http://www.roalonso.net/en/videoarte/tacticas.php (accessed August 13, 2022). Originally published in *11e Biennale de l'Image en Mouvement* (Geneva: Centre pour l'Image Contemporaine, 2005). Wolf Vostell made a poster of *Three Country Happening* (1966), as the joint action between Minujín, Vostell, and Allan Kaprow is called, by drawing a triangle between the cities of Buenos Aires, New York, and Berlin on a map. This image resonates with the Latin American map sewn by Minujín years later.

39 Catherine Spencer, *Beyond the Happening: Performance Art and the Politics of Communication* (Manchester: Manchester University Press, 2020), 3.

40 Niko Vicario, "The Matter of Circulation: Teletype Conceptualism, 1966–1970," in Christian Berger, ed., *Conceptualism and Materiality: Matters of Art and Politics*, Studies in Art & Materiality 2 (Leiden and Boston: Brill, 2019), 219. Modernization theory, developed by Walt Whitman Rostow among others in the United States, posited that economic growth would allow "developing countries" to reach a level of industrialization equal to that of the so-called "developed world."

41 Bernard Stiegler, in Pieter Lemmens, "'This System Does Not Produce Pleasure Anymore': An Interview with Bernard Stiegler," *Krisis: Journal for Contemporary Philosophy* 1 (2011): 35.

42 "Nido de hornero gigante ¿hábitat del futuro?," press release signed by Minujín, n.d. (c. 1976). Marta Minujín Archive. For a description of the exhibition and peoples' responses to the "warm atmosphere of the nest" see the newspaper clipping titled "Con mucho frío y poco público continuaron hoy en Palermo las actividades de la 28 muestra rural," n.d., n.p., Marta Minujín Archive.

43 "El refugio del pájaro es

antropomorfizado, al permitirnos ingresar y habitarlo, pero al hacerlo somos también zoomorfizados, en el sentido en que nuestros cuerpos son afectados por el sentir del mundo del animal." Mercedes Claus, "Nido," *Simbiología. Prácticas artísticas en un planeta en emergencia*, 2022. Available online at https://simbiologia.cck.gob.ar/hashtags/nido/ (accessed August 14, 2022). See also Vinciane Despret, "The Body We Care For: Figures of Anthropo-Zoo-Genesis," *Body and Society* 10, no. 2–3 (June 1, 2004): 111–34.

44 The rufous hornero is one of Argentina's patriotic symbols. For a discussion of the nest as both a transterritorial utopia and a uterus see Laura Lina Alonso and Juan Cruz Pedroni, "A propósito de *Comunicando con tierra*, de Marta Minujín," *Simbiología. Prácticas artísticas en un planeta en emergencia*, April 23, 2020. Available online at https://simbiologia.cck.gob.ar/publicaciones/marta-minujin-x-rodrigo-alonso-laura-lina-y-juan-cruz-pedroni/ (accessed August 14, 2022).

45 It might be argued that Minujín adopts this position to reflect her awareness of the processes of cultural appropriation that underpin her act of directly engaging with an Incan sacred site as an Argentine, white artist.

46 Text taken from the description of the work in the leaflet "Marta Minujín: *Comunicando con tierra*" (Buenos Aires: Herlitzka + Faria, n.d.).

47 See the artists' website page for *Rizosfera FM*, online at https://colectivoelectrobiota.wordpress.com/proyectos/rizosfera-fm/ (accessed August 14, 2022).

48 Claus, "Madre Tierra, 1971," *Simbiología. Prácticas artísticas en un planeta en emergencia*, 2022. Available online at https://simbiologia.cck.gob.ar/hashtags/animismo/ (accessed August 14, 2022).

49 The project's leaflet is available online at www.centrodearte.unlp.edu.ar/wp-content/uploads/2020/04/Post-Madre-Tierra.pdf (accessed November 23, 2022).

50 "El arte conceptual, el arte zoológico, el arte mágico, el arte ecológico, el arte social, el arte de acción, el arte catastrófico." Glusberg, *Retórica del arte latinoamericano*, 133.

51 "Sus creadores se interesan por lo que ocurre en la sociedad donde viven, no por las representaciones surrealistas y académicas de naturalezas muertas o los geometricalismos tecnológicos. Desean comunicarse con la audiencia, y están seguros de lograrlo sin la participación de los tradicionales 'intermediarios.'" Ibid., 105.

52 These words are written in capital letters in a CAyC bulletin announcing the presence of Nicolás García Uriburu in the exhibition *Latin America 76* at the Fundación Joan Miró, Barcelona, in February 1977. CAyC collection of press releases and ephemera, 1969–1977, The Museum of Modern Art Library, New York.

53 Stiegler, *Technics and Time*, vol. 1, *The Fault of Epimetheus*, trans. Stephen Barker, Richard Beardsworth, and George Collins (Stanford, CA: Stanford University Press, 1998), 16.

54 See, for example, Eduardo Gudynas, "Deep Ecologies in the Highlands and Rainforests: Finding Naess in the Neotropics," *Worldviews: Global Religions, Culture, and Ecology* 21, no. 3 (January 1, 2017): 262–75. Doi https://doi.org/10.1163/15685357-02103005.

ARNAUD GERSPACHER

EDUARDO KAC'S ANIMAL FUTURES

Some critters; a stuffed pig confined to a crate; a nonexistent wild boar; labyrinths for _______; a territorial anaconda; ten chickens (burned alive); a multitude of cow's bones; innumerable ants; a punctuating parrot. This is not some new Borgesian taxonomy but a selective bestiary of nonhuman animals found in Latin American art since 1960, be they real, fictional, living, or in remains. In fact this list limits itself to Argentina, Chile, and Brazil.[1] With respect to the latter, one finds amphibians playing a catalytic role in a primal scene of the Brazilian avant-garde. In 1928, the painter Tarsila do Amaral and the poet Oswald de Andrade took some friends to a São Paulo restaurant specializing in frogs. As the dishes were being served, de Andrade gave an impromptu speech on the evolutionary kinship between humans and amphibians. This prompted Tarsila to note that, in a way, they were in the process of eating their own kind—of being anthropophagous. This epiphany, be it serious or cheeky, centered on the fact that humans and frogs live not in discrete purity from each other but along a continuum of limbed, sentient, and minded being. Days later, this same dinner party would reconvene to discuss these ideas and de Andrade would write his celebrated *Manifesto Antropófago* (Anthropophagic manifesto).[2] Although these cannibalized amphibians and the other nonhuman animals inhabiting my list point to different cultural, historical, and political contexts, they each offer important ways of interpreting these contexts; nonhuman animals always tell us more than we think.

One of the most notorious nonhuman animals in art was an albino rabbit, born in 2000, who it was said could glow green under the right lighting conditions. Her name was Alba. Fittingly enough, in the wake of all-out globalization post-1989, she was born in France at the behest of an expat Brazilian artist who called Chicago home:

Eduardo Kac. Alba is different from the aesthetized animals in my list above in at least one respect: she is the first whose medium was, in part, her genetic makeup (the work in question is called *GFP Bunny* [2000; fig. 1], so named after the bioluminescent jellyfish protein introduced into her body). Unlike the pig, cows, chickens, and ants in the artworks alluded to above, biopolitical control was imposed on Alba not only from the outside but also from the inside. She is therefore symptomatic of the transgenic revolution in biotechnology, with all its attendant promises and pitfalls.

There have been plenty of scholarly partisans and detractors with no shortage of an equally riven public reaction to Kac's *GFP Bunny*. Much of the reception, rightly, has been preoccupied with questions concerning the ethical ramifications of such Bio Art practices. Without obviating the need for ethical questioning, in what follows I take a less traveled interpretive route, which is twofold: 1) while Kac's Bio Art is often deracinated from specific art histories and geographies in favor of a more universalizing Western scientistic/aesthetic discourse bearing on "life," I attempt to ground his work within certain trajectories of Latin American art since the 1960s, especially precedents from Kac's native Brazil. This is not intended to claim his work for some essentializing notion of Brazilian art but rather to uncover affinities between his Bio Art practice and cultural anthropophagy, the Neo-Concrete movement, the *poema/processo* movement, and legacies of body art, participatory art, and media practices in Latin America—all of which already exceeded national barriers in their own time. 2) With the admittedly shallow distance of some twenty years, I attempt to periodize the 1990s and early 2000s and to question the ways in which Kac's works incorporating animals relate to social and cultural issues that have given way to different (though related), pressing issues today—most notably the decolonial imperative and ecological politics.

ARTE PORNÔ

Before turning to nonhuman animals as subjects for art in the 1990s, Kac focused on the animal who is fond of making lists: homo sapiens. The *Movimento de Arte Pornô* (Porn art movement), founded

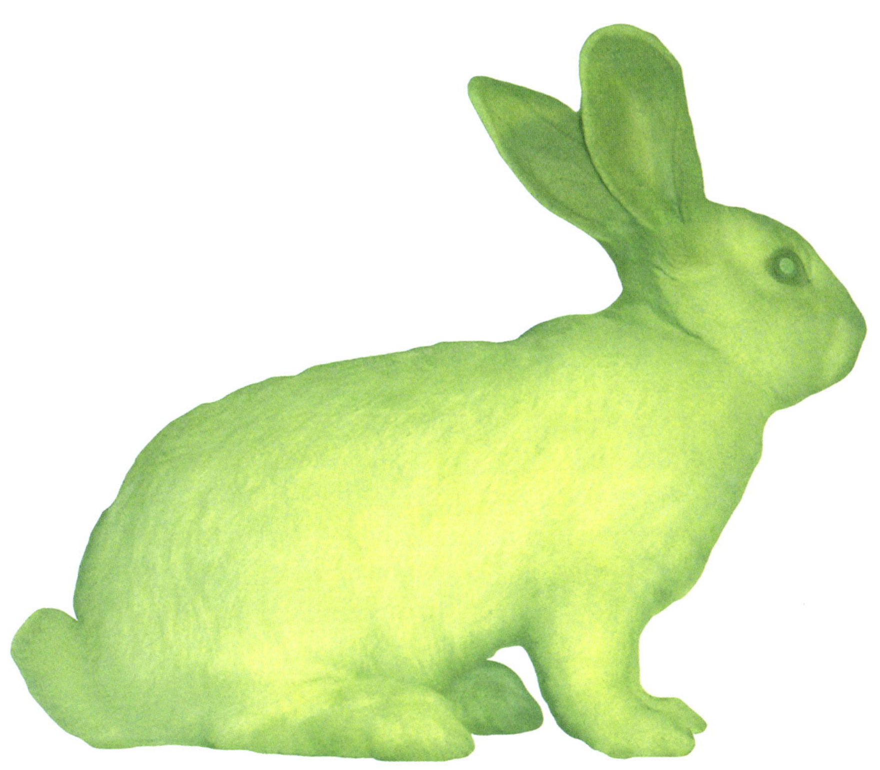

1 Eduardo Kac. *GFP Bunny*. 2000. Transgenic artwork featuring Alba, the fluorescent rabbit. Collection the artist

by Kac in February 1980 and lasting exactly two years, embraced the naked human body and critically interrogated the codes of behavior that control it. The movement's multimedia activities included public interventions, participatory performances, poetry, photography, graffiti, mail art, and zines and other publications.[3] Arte Pornô can be linked to earlier avant-garde movements in Brazil such as the Neo-Concrete movement (with its emphasis on participation and phenomenological experience) and the poetry movements that expanded what poems could be, notably the Noigandres group and the *poema/processo* movement. Arte Pornô conceived the human body in itself to be, through and through, a poem.[4] As Kac has written retrospectively about the movement, the human body was not treated merely as a poetic theme but was taken more fundamentally to be a "praxis, from its verbal drives to its textual pleasures, from its sounds to its movements, from its form to its function, from its common behaviors to its transgressions, from the individual to the social body, from its carnality to its carnivalization, from its surface to its organs."[5] In Lacanian terms, the ambition of this body-as-poetry encompassed the imaginary, the symbolic, and the real. In other words, Arte Pornô showed how the body is simultaneously visualized, informed by language, and manifest as raw, vital material that can never be exhausted by images and words alone.

The activities of Arte Pornô were an instigation against the military dictatorship still in place in Brazil in the early 1980s and against the hegemonically conservative Catholic sense of morality and propriety.[6] In their public displays of nudity, Kac and his collaborators, who called themselves "Gang," were not simply speaking truth to power but embodying truth to power (fig. 2). At this same moment, in a compelling historical convergence, the German philosopher Peter Sloterdijk was writing his important book *Critique of Cynical Reason* (1983), in which he diagnosed the onset of modern cynicism and resignation in contrast to ancient forms of cynicism (the classical Greek *kynicism*), whose bawdy forms of social critique involved a more active and mocking repudiation of cultural normativity.[7] The most famous classical cynic, Diogenes of Sinope, was notorious for thumbing his nose at power and societal norms by, among other things, masturbating in public, living a threadbare existence, identifying with dogs, throwing a plucked chicken during a lecture at the Academy so as to refute Plato's definition of "man" as a featherless biped, and receiving preferential favors from Athenian sex workers. Arte Pornô's cheeky irreverence enacted similar forms of corporeal social critique by placing the human body's frank desires at the center of its countercultural program. When reading the sexual innuendos and calls for bodily freedom in the movement's 1982 manifesto *Movimento de Arte Pornô*, hearing about the ribald "yellpoems" recited to passersby in the street, and learning about the group's embracing of social outcasts and its naked public interventions on Ipanema beach, one cannot help but understand Arte Pornô as a Brazilian renewal of kynicism. Like Diogenes turning social mores on their head in order to reveal the hypocrisy underneath, Kac and Gang showed that it was the military dictatorship and its lust for power and violence that were the true obscenities, not the naked human body or its natural drives for pleasure and freedom.[8]

For all its emphasis on the human body in its naturalness, at first blush Arte Pornô seems far removed from Kac's subsequent work employing digital communication and biotechnology. Indeed, contradictions exist, yet so do telling continuities. If Arte Pornô challenged the orthodox view of what is "proper" or "obscene" with respect to human sexuality and desire, Kac's subsequent work does the same in relation to scientific dogmas and digital protocols. From his earliest works onward, Kac refused to accept commonly received notions and chose to work from inside a field so as to scramble its operational codes—in the case of Arte Pornô, by envisioning a nonerotic mode of pornography going against the grain of sexual exploitation and normative gender roles and sexuality;[9] in the case of his subsequent telepresence works, by subverting anthropocentric perceptual norms in favor of nonhuman experiences of the world; and in his more recent Bio Art practice, by literally altering genetic codes via transgenic biotechnology. In his essay on Arte Pornô, Kac himself seems to make this case for continuity in using language that forecasts transgenic splicing and recombination of genetic sequencing in living bodies: "Often to sharp and sardonic effect, I turned formulaic and consumerist pornography on its head and subjected words to operations—*recombination*,

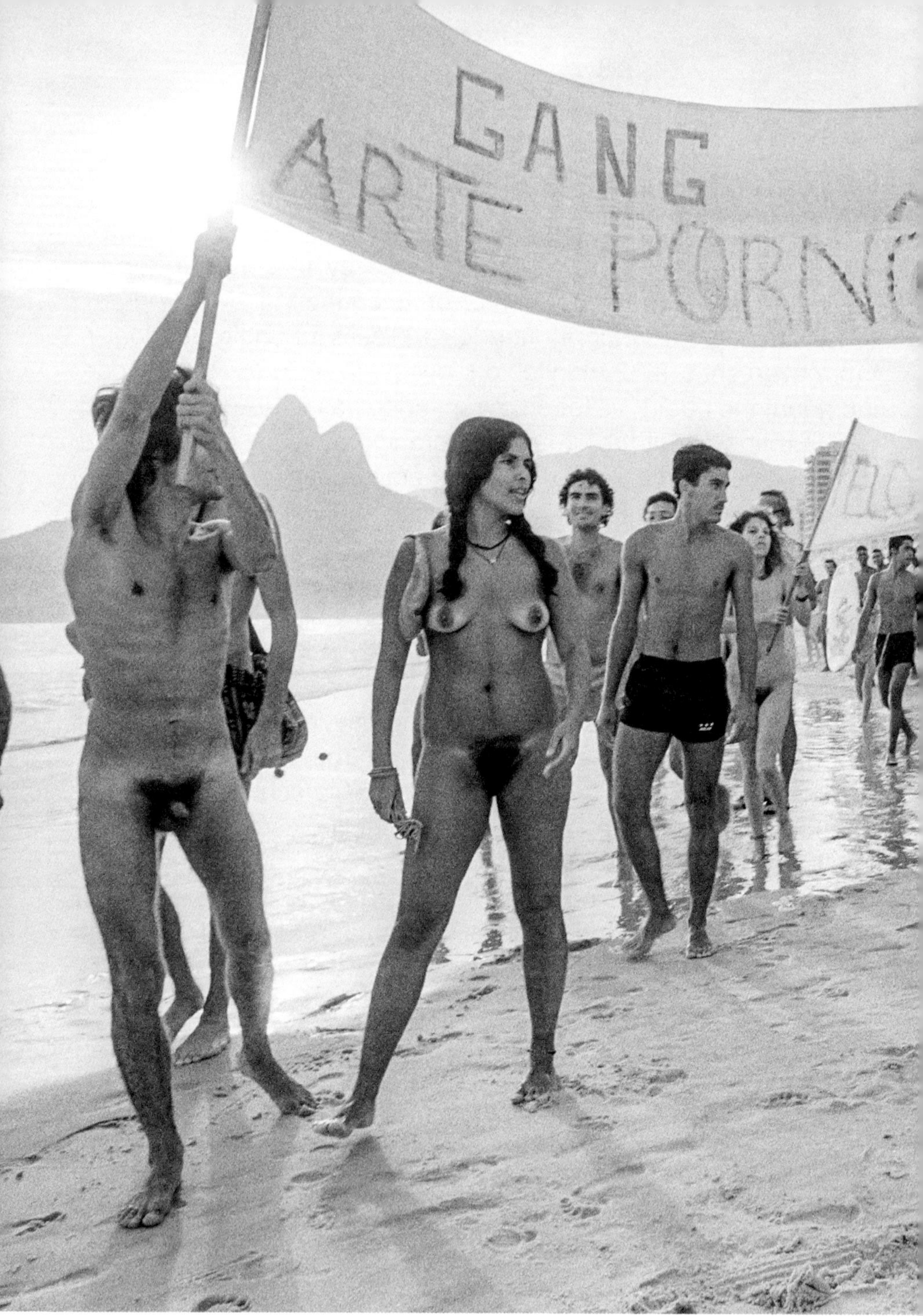

2 Eduardo Kac (center, holding banner) performing at Ipanema Beach, Rio de Janeiro, 1982. On the left, under the banner, Cynthia Dorneles; to the left, a few steps behind Kac, Teresa Jardim, also holding a banner. Gelatin silver print, $7\frac{7}{8} \times 9\frac{13}{16}$ in. (20 × 25 cm). Collection the artist

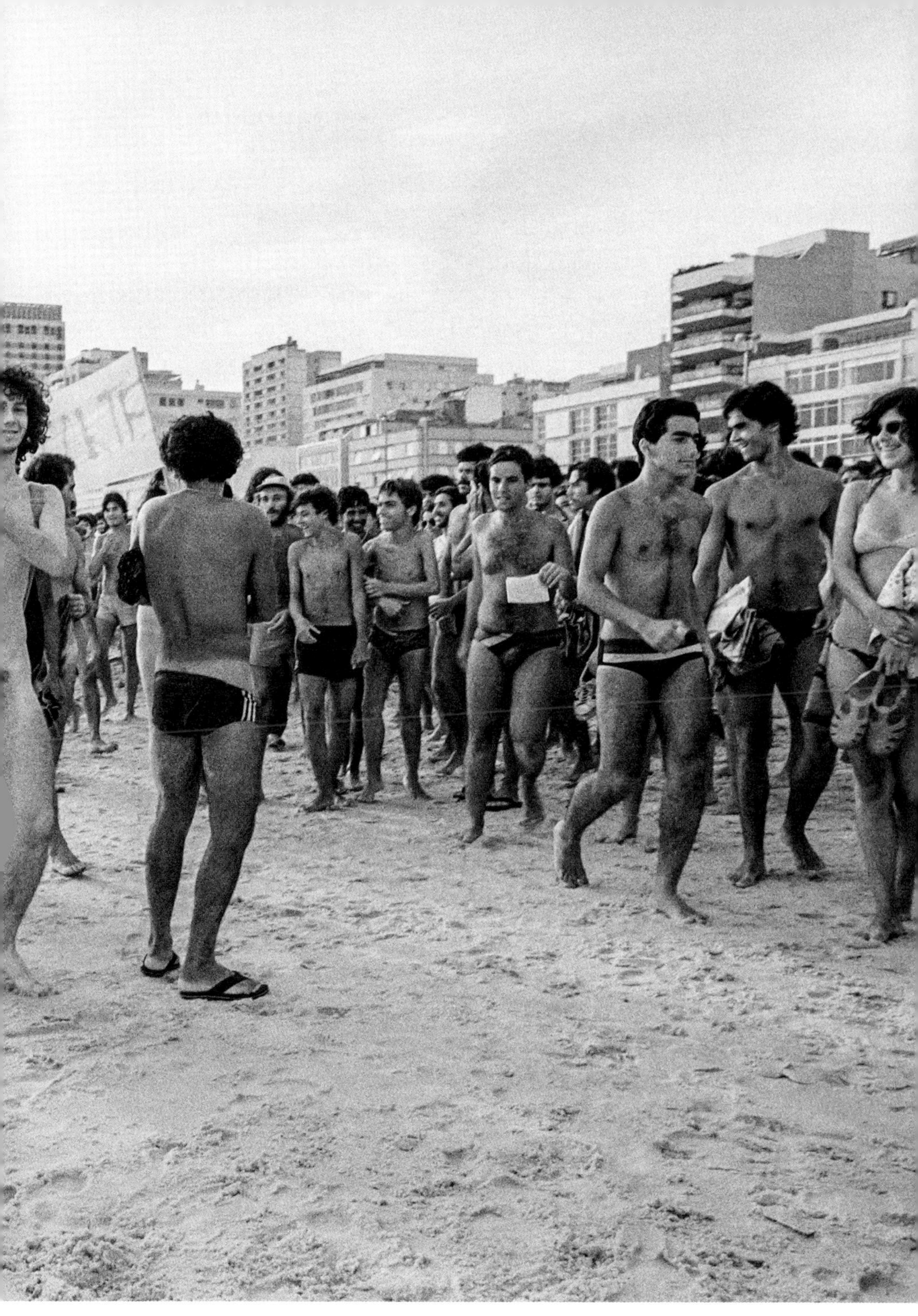

recontextualization, *excision*—that deconstructed their meaning and subverted their message."[10]

There are additional ways in which Arte Pornô augurs Kac's latter practice with animals, telepresence, and biotechnology. In each case there is an emphasis on participation, interaction, and dialogue, be it in the flesh or mediated by technological distance. In each case, one finds a noninstrumental view of the body, be it human or nonhuman (animal or even vegetal). And in each case there is a postnatural impulse to go beyond what is "naturally" given. With respect to Arte Pornô, this postnatural impulse entailed going beyond the purportedly static codes of sexuality and gender toward a conception of identity that is fluid and changeable. (In this way, Arte Pornô was highly paradoxical, emphasizing the body in all its naturalness while also questioning this very naturalness.) With respect to Kac's telepresence works (which I turn to next), this entailed exceeding the purportedly given limits of human perception and ontological fixity in time and space. With respect to his Bio Art works, it entailed going beyond a conception of "life" as "naturally" given by evolutionary processes through a reworking of the plasticity of genetic material. In all three instances—Arte Pornô, telepresence, and Bio Art—Kac pushes the limits of what a body (be it human or nonhuman) is and can do, and challenges purportedly set codes of behavior, be it through ludic and irreverent activities, as in his earliest works, or through the more sober (if also ludic in its own way) technological and scientific apparatus of his later work.

TELEPRESENCE

Kac's first employment of nonhuman animals arrived with his telepresence works of the 1990s.[11] A major impetus was his interest in telecommunications, computers, robotics, and systems theory, the latter of which, beginning in the 1940s with cybernetics (or the first-order systems theory of Norbert Wiener and others), ushered in a destabilization of the long-held Western divisions between human, animal, and machine. Fittingly, then, Kac's first critter was not a nonhuman animal at all but a robot named Ornitorrinco (the Portuguese word for "platypus," in keeping with the artist's fascination with hybridity and genre-bending—in this case, of a species betraying both avian and mammalian characteristics). Ornitorrinco, developed in 1989 with the hardware designer Ed Bennett, was the central interface in Kac's installations in a number of exhibitions.

The basic protocol of Kac's telepresence art was to allow for public participation through the telematic possibilities of the Internet—telematics entailing the dissemination of computerized information remotely over long distances. During these very early days of dial-up connection, this involved using telephone keypads that allowed the public, however rudimentarily, to modulate Ornitorrinco's behaviors from multiple locations. For Kac, telepresence was not simply about "being there" remotely; instead, he sought to investigate "how the fact that we are experiencing this remote site in a given way (i.e., through a particular telerobotic body, with a given interface, and over a specific network topology) modulates the very notion of reality we conjure up as we navigate the remote space."[12] Ultimately, Kac's ambition for telepresence was to make palpable the ways in which we bring forth the world, and to facilitate sympathetic imagination with nonhuman forms of bringing forth the world. As such, his conception of telepresence runs counter to disembodied notions of cyberspace by offering a neocorporeality, a synthetic phenomenology of being human and robot akin to Donna Haraway's contemporaneous techno-feminist cyborg manifesto.[13] Nonhuman animals appear in four telepresence works dating from 1994 to 1999. The main objective in these installations was twofold: making works of art tailored to nonhuman experience, or *Umwelten* (that is, "lifeworlds," a term Kac borrowed from the early-twentieth-century German biologist Jakob Johann von Uexküll), while creating conditions of engagement for the human viewer/participant to enter into empathic relation with nonhuman perspectives. In short, these works were intended to instantiate multispecies conviviality within a network ecology. The first telepresence work involving a nonhuman animal was *Essay concerning Human Understanding* (1994), a multilocation installation at the Center for Contemporary Art at the University of Kentucky, Lexington, made in collaboration with the artist Ikuo Nakamura.[14] Here Kac displayed a living yellow canary in a cage fitted with a microphone, circuit boards, and a speaker, all wired to dial-up Internet. Six hundred miles away at the Science Hall in New York, Kac simultaneously displayed a philodendron fitted with an electrode on its leaves that allowed for prerecorded sounds to be emitted in response to the electrical activity of the plant. In bidirectional feedback between the bird's vocalizations and the plant's electrical activity, each were fed the other's sounds, which, Kac claims, allowed them to dialogue for several hours each day. This feedback loop was conditioned by the live human visitors/participations, whose presence affected the interaction, be they in proximity of the canary or of the philodendron.

In 1996, *Rara Avis* (fig. 3) premiered at the Nexus Contemporary Art Center in Atlanta.[15] The work comprised an aviary of thirty zebra finches living alongside a large telerobotic macaw, a Tropicália-tinged artificial bird that perched on a branch, motionless aside from its head, which would swivel to left and right. Its "eyes" were fitted with cameras. Through a virtual reality headset linked to the bird's head movement, gallerygoers could enter into macaw-perspective among the zebra finches in the aviary, as well as see themselves seeing through nonhuman vision. Via an Internet interface, remote participants could also enter into macaw-perspective—and additionally, with their computer microphones, trigger its built-in vocal apparatus so as to emit sounds in the aviary, thereby communicating with the zebra finches. *Ornitorrinco, the Webot, travels around the world in eighty nanoseconds going from Turkey to Peru and back*, from the same year, featured Ornitorrinco interacting with two live turkeys on the ground floor of the Otso Gallery in Helsinki, with gallerygoers upstairs at a teledistance and with the remote participation of online users.[16] Finally, in 1999, Kac installed *Darker Than Night* in a bat cave at the Blijdorp zoological gardens in Rotterdam.[17] The setup was much like that of *Rara Avis*: a robotic bat—aka the "batbot"—was hung inside the artificial cave, allowing the public to interact telematically with its Egyptian fruit bat denizens. Since the cave is continually shrouded in darkness, the telepresent aspect of the work entailed a visual translation of echolocation: participants wore a virtual reality headset connected to the batbot inside the cave, giving them the ability to "see" through its eyes and swivel its head; the headset translated bat experience

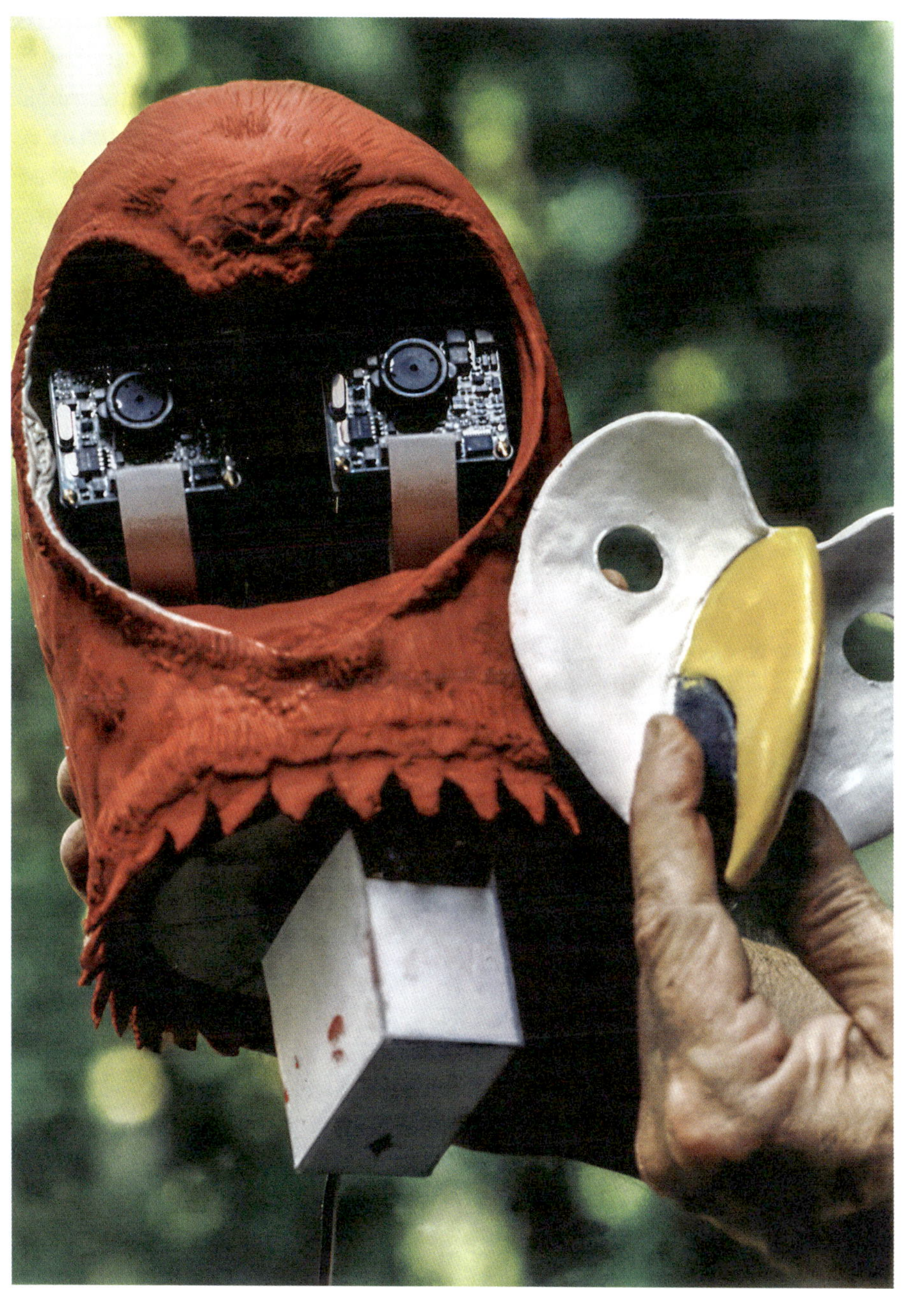

3 Eduardo Kac. *Rara Avis*. 1996. Online telepresence work, including telerobot, aviary, zebra finches, VR headset, and Internet (detail). Edition of 2 and AP. Collection the artist

by visualizing the echolocation calls coming from the bats, as well as ultrasonic sounds coming from the batbot itself—a 45 kHz frequency inaudible to human ears. The result was a highly mediated dialogue between two very different mammalian *Umwelten*.

Kac's telepresence works with nonhuman animals can be placed within a lineage of postwar art influenced by systems theory, such as the work of the Chilean artist Juan Downey and, most notably, the 1970s work of the Argentine artist Luis Fernando Benedit.[18] From bees to eels to mice and sundry other animals, Benedit constructed labyrinths as miniature ecological systems incorporating natural and artificial materials. His laboratory aesthetic often involved a more or less closed system in which nonhuman animals were controlled and tested, thereby evincing negative feedback in the system, with humans usually playing the role of "objective" observers. Kac's telepresence works by contrast were far more open-ended and implicated the observer-turned-participant, who no longer remained unimplicated on the outside. In Kac's work, systems, information, and indeed the world itself are no longer conceived as simply "out there" awaiting an observer, but are coemergent with and only made possible by observers—entities, whether human or nonhuman animal, plant, or machine, with the self-recursive ability to respond, react, or "make order from noise," bringing forth an environment in accordance with specific ways of perceptual engagement. In short, Kac reflects a shift from cybernetics to neocybernetics, which, simplifying to the utmost, entails a reconceptualization of information and meaning as self-generated, what is known as "autopoiesis," which will vary according to the variety of biological (and perhaps artificial) and cognitive entities with the ability to bring-forth a world.[19]

If a central ambition of Kac's telepresence is to make art for nonhumans, thereby scrambling the humanist codes that have ordered art history for centuries, there are moments when he ontologically flattens the different entities operative in his installations.[20] There are important ways in which human and nonhuman-animal cognitive engagement (evinced here by humans, birds, and bats) differ from a philodendron's biotic but noncognitive engagement (or alter-cognitive engagement, rather, in relation to creaturely cognition grounded in sentient nervous systems), and certainly from abiotic robotic and computer systems. Perhaps the most interesting question pertaining to Kac's telepresence works, however, has to do with the digital mediation of human perception.

The neocybernetic theorist Bruce Clarke provides an interesting periodization of telepresence contemporaneous with Kac's work. Clarke observes the evolution of telepresence in popular movie culture: whereas James Cameron's blockbuster film *Titanic* (1999) opens with a telematic submarine, which affords a remote view of the ill-fated ship's remains on the ocean floor, his even more blockbuster film *Avatar* (2009) went farther by imagining the remote inhabitation of an organic alien body, a member of the Na'vi people.[21] Clearly the former scenario is today literal/possible while the latter is only metaphorical/as of yet impossible (even virtual reality with multisensory feedback remains of the first variety). This being the case, can it really be claimed that digital mediation—in the form of a telematic robotic macaw or batbot—allows for human perception to mingle with nonhuman animal perception? Does this not presume an ontological or, at the very least, perceptual convergence between human technologies and nonhuman animals? Is it not more the case that nonhuman technologies are simply an extension of the human *Umwelt*, despite any experiential alterations elicited by these technologies that can be mistaken for, or overestimated to seem to converge with, nonhuman animal experiences of the world?

BIO ART

A recurring imperative in de Andrade's *Manifesto Antropófago* is to transform taboo into totem—a drive toward turning prohibition into a critical apparatus. In light of this sentiment, de Andrade alludes to the history of xenotransplantation through the figure of Serge Voronoff, a Russian-born French surgeon then well-known for his "rejuvenation" technique involving the grafting of chimpanzee testicles and ovaries inside human bodies.[22] This macabre and ultimately unsuccessful venture can nonetheless be claimed as part of the pre-history of Bio Art, a term coined by Kac to designate the use of transgenic techniques in order to "create unique living beings," the most famous of these being Alba.[23] Of course, a key difference between Voronoff and Kac is that the former violently instrumentalized nonhuman bodies for human gain while the latter saw his transgenic art practice as noninstrumental (even if, as I argue below, this can only be a quasi-noninstrumentality). Another key difference is that Kac's mixing of biological materials occurs at the genetic and embryonic level, not in already living, sentient, and minded bodies. Yet both cases manifest the anthropophagic impulse of contamination, of striving against purity, of consuming the other by reaching inside, and of passing beyond taboos. Kac's interest in the history of hybrids, teratology, and "atypical" forms of life—which, he argues compellingly, have often maintained the ideological fixities of what is deemed "normal" or "pathological"—parallels his practice of challenging these ideological fixities.[24] In many ways, Kac was already echoing de Andrade with Arte Pornô, which, as we have seen, sought to challenge the ideological fixities of sexuality and gender through the tabooed vernacular of pornography.

Kac's move from telepresence to Bio Art was a natural one, as in both instances the insides of nonhuman animals are at play. In the case of telepresence, getting inside other minded bodies is only figurative; one does not really experience the world through bat or bird mind, but only analogically, through technological mediation. By contrast, in the case of Bio Art, one really does get inside a nonhuman body. More precisely, one leaves a trace by setting the conditions for certain epigenetic expressions before birth—in Alba's case, by introducing the bioluminescent GFP gene of a jellyfish. Moreover, in both his telepresence works and his Bio Art, Kac's hope is to elicit an empathic and dialogic encounter—the first through telematics, the second through biotechnology. If this impulse toward empathy through telepresence remains in Bio Art, it can paradoxically be argued that the more we are able to get inside and even alter the biological components of a nonhuman animal, the less we are able to grasp its being and lifeworld. In other words, the more we attempt to "get at" life, sentience, and consciousness through invasive scientific techniques, the more we distance ourselves from the phenomenological reality of the subject under study.[25]

This impulse to get inside was already present in Arte Pornô through that movement's reimagining of poetry as a praxis reaching from the body's "surface to its organs." It can also be argued that Kac's Bio Art practice is a natural extension of his poetic influences, most notably Wlademir Dias-Pino and Neide Sá of the *poema/processo* movement, in which poems were often conceived of as informational code.[26] In this respect—much like Umberto Eco's "open work," an influential theory for many Latin American artists generating open-ended art objects that could be modulated by artist and audience alike—Alba can be described as a processual aesthetic subject. And if *GFP Bunny* can be likened to an open work, this is so not only at the level of Alba's modulated genetic material but, even more important, through the mutual modulation of her existence and the socio-ecological network she found herself in—the systems that created the conditions for her birth, including the French laboratory that ultimately refused to give her up and Kac's public campaign to claim her and take her back to Chicago, a campaign that, as we say today, ultimately led to her going viral. Kac is adamant that the social and participatory feature of this work is crucial and that Alba should essentially be understood as a conversation piece (in this respect, as a conversation piece, Alba was at least partly instrumentalized).[27] This emphasis on the socio-ecological envelope of *GFP Bunny* is a key part of Kac's aim for a nonreductive science, wherein genetic expression makes no sense outside of its lived context.

"The animals are to be loved and nurtured just like any other animal."[28] Kac makes this statement in his text on transgenic art and means to convince his reader that genetically hybrid creatures like Alba are not "freaks" but instead should be treated as subjects of care. There is a twofold radical optimism to this statement, which may border on delusion: first, it is far more likely that genetically altered animals will be disposed of and instrumentalized for various anthropocentric purposes; second, the vast majority of nonhuman animals under human "care" on this planet undergo traumatic forms of violence and shortened lives (the food industry alone kills more than 72 billion land animals a year). With this grim reality as backdrop, Kac's faith that transgenesis will lead to an ethical revaluation feels out of joint with reality. Although he has little to say about mundane "food" animals, he is not completely unaware of this and other grim realities. In his writings, he often points to the dangers and nefarious uses of new technologies—the use of telepresence in guided missiles, for example, auguring today's drone warfare, or genetically modifying pigs as a means for organ harvesting. And yet, on balance, Kac remains optimistic that his practice can provide a countervailing force. Throughout his career one gets a sense of Panglossian fiat: that art can recalibrate the codes of pornography; that telepresence can engender productive networks, allowing those plugged in to experience empathic connections to other humans and nonhuman animals; that Bio Art can lead to care and a social literacy about biotechnology. These past few decades have severely worn any such optimism: normative and exploitive pornography proliferates; the Internet has by and large congealed into a hegemonic apparatus of pseudo-telepresence, breeding alienation, misinformation, resentment, and hatred; and biotechnology has largely subsided as a subject of media or popular interest, as it had been in the late 1990s and early 2000s.[29]

Is Kac technophilic or techno-skeptic? Does his practice reflect a straightforward Promethean impulse or one buffered with irony? Is he posthumanist or transhumanist? Does he question human supremacy, in all its entrenched forms, or embrace our species' purportedly superior technological enhancements? I claim that the answer to each of these questions is an uneasy and contradictory "yes." The many claims of Promethean "creation" in Kac's writings toggle between the serious and the ironic, depending on the work or entity in question. Furthermore, one can find plenty of emphasis on hybridity and a critique of purity in his work, yet simultaneously an insistence on aesthetico-scientific "creation" that does not fit easily with hybridity and nonpurity. Furthermore, Kac often adopts both sides in his works and writings: he can be critical of the new while at the same time entering into it headlong. (This is especially true of his transgenic works, which might be described as science experiments without the scientism.)

This approach may be productively balanced and prescriptive of better relations between humans, nonhumans, and technology in the future. Yet it has to be emphasized that art will never be able to negotiate any such future by itself; only by entering into an ecosocial political program will aesthetic activity stand a chance of bringing about better animal futures. Perhaps this is the most productive failure of Kac's practice: it shows scant evidence of any concrete ecopolitical commitment, aside from vague and arguably naive entreatments for empathy and care. One of the less favorable readings of Kac's artistic trajectory is to find a socially engaged political commitment in Arte Pornô giving way to a depoliticized practice that fetishizes telecommunications and biotechnology to compensate for a dissatisfaction with given human and nonhuman limits; an overriding interest in network ecologies over actual ecologies; and an overestimation of the value of employing Western scientific techniques and paradigms to arrive at empathy and care, with little to no corrective of these techniques and paradigms through decolonial knowledge and practice. Living as we do in the wake of deleterious political necro-economies and their technologies—above all fossil-fuel industries and the intensive breeding and killing operations of the animal industrial complex, not to mention the heat wave of climate disaster and ecocide already upon us—it is unclear how telematics and Bio Art will save us: "us" in the multispecies sense of this term.

1 The artists and artworks alluded to in this list via their allusion to or incorporation of nonhuman animals in art include: Raúl Escardi, Roberto Jacoby, various of Luis Fernando Benedit's installations from the 1970s, Lygia Clark's *Bichos* (Critters) from the 1960s, Eduardo Costa's *Happening para un jabalí difunto* (Happening for a dead boar, 1966), Juan Downey's *Anaconda Map of Chile* (1973), Nelson Leirner's *O Porco* (The pig, 1967), Cildo Meireles's *Missão/Missões (como construir catedrais* (Mission/missions [How to build cathedrals], 1987), and Rivane Neuenschwander's *Contingent* (2008) and *Sunday* (2010).

2 For an account of this episode see Raul Bopp, "The Life and Death of Anthropophagy," 1965–66, in Pedro Neves Marques, ed., *The Forest and the School/Where to Sit at the Dinner Table* (Berlin: Archive Books, 2014), 135–50.

3 For an overview of the Arte Pornô movement see Eduardo Kac, "The Porn Art Movement: A Brazilian Avant-Garde, 1980–1982," *OEI Magazine* no. 66 (2014): 499–512, and Zanna Gilbert, "Transgressive Bodies," *Art in America* 103, no. 10 (November 2015): 118–25.

4 Antonio Manuel's intervention at the Museu de Arte Moderna, Rio de Janeiro, in 1970, when he publicly stripped naked to protest artistic censorship and the jury's refusal to accept the submission of his body as a work of art, is another important precedent for Arte Pornô. As Claudia Calirman argues, though, Manuel's action was a general call to artistic freedom, rather than a more specific subversion of gender and sexual politics, as Arte Pornô would be a decade later. See Calirman, *Brazilian Art under Dictatorship: Antonio Manuel, Artur Barrio, and Cildo Meireles* (Durham, NC: Duke University Press, 2012), 37–42.

5 Kac, "The Porn Art Movement," 499.

6 The military dictatorship ruled Brazil from 1964 to 1985.

7 Peter Sloterdijk, *Critique of Cynical Reason*, 1983, Eng. trans. Michael Eldred (Minneapolis: University of Minnesota Press, 1987).

8 In his retrospective essay on Arte Pornô, Kac is explicit about this subversive reversal of obscenity: "Living during a moment of political uncertainty, rampant police brutality, and spiraling inflation, I considered the authoritarian government and the social inequality it produced obscene and immoral—not one's body in its natural state or its effluvia and proclivities." Kac, "The Porn Art Movement," 500.

9 "Conventional pornography instrumentalizes sexuality to induce arousal; Porn art instrumentalizes pornography itself, undermining it from within and transforming it into an imaginative tool at the service of both experimental poetry and the invention of new realities." Ibid.

10 Ibid., 503. Emphasis mine.

11 Although Kac's earliest use of animals came in the mid-1990s, he had already begun to develop his telepresence works in 1986.

12 Kac, "Dialogic Telepresence Art and Network Ecology," 2000, in Kac, *Telepresence & Bio Art* (Ann Arbor: University of Michigan Press, 2005), 192.

13 Donna Haraway, "The Cyborg Manifesto," in *Manifestly Haraway* (Minneapolis and London: University of Minnesota Press, 1987), 3–90.

14 See Kac, "The Emergence of Biotelematics and Biorobotics: Integrating Biology, Information Processing, Networking, and Robotics," 1997, in Kac,

Telepresence & Bio Art, 219–21.

15 See Kac, "Telepresence Art on the Internet," in Kac, *Telepresence & Bio Art*, 162–66.

16 See Kac, "Dialogic Telepresence Art and Network Ecology," 194–98.

17 See ibid., 202–3.

18 For an overview of Benedit's work see Mara Polgovsky Ezcurra, "The Future of Control: Luis Fernando Benedit's Labyrinths Series," *post: notes on art in a global context*, available online at https://post.moma.org/the-future-of-control-luis-fernando-benedits-labyrinths-series/ (accessed November 20, 2022).

19 For a concise history of the move from cybernetics to neocybernetics see Bruce Clarke, *Neocybernetics and Narrative* (Minneapolis: University of Minnesota Press, 2014), ix–xvii.

20 In his introductory essay for the anthology *Signs of Life: Bio Art and Beyond*, for example, Kac writes, "Subjects are alive, free, and autonomous. From bacteria to bunnies, from frogs to flowers, living organisms grown or bred in unique ways, modified or invented by artists, are the elements of a true art of evolution." Kac, "Art That Looks You in the Eye: Hybrids, Clones, Mutants, Synthetics, and Transgenics," in Kac, ed., *Signs of Life: Bio Art and Beyond* (Cambridge, MA: The MIT Press, 2007), 14.

21 Clarke, *Neocybernetics and Narrative*, 163–64.

22 Oswald de Andrade, "Anthropophagic Manifesto," 1928, in Neves Marques, ed., *The Forest and the School/Where to Sit at the Dinner Table*, 104.

23 See Kac, "Transgenic Art," in *Kac, Telepresence & Bio Art*, 236.

24 Kac, "Art That Looks You in the Eye," 9.

25 This aligns with the philosopher Cary Wolfe's reading of Kac's work as a visual feint: "The more you look, the less you see." See Wolfe, *What Is Posthumanism?* (Minneapolis: University of Minnesota Press, 2014), 158.

26 Kac has written about Neide Sá's work. See Kac, "Neide Sá's process/poems," *OEI Magazine* no. 66 (2014): 292–96.

27 For an overview of the discourse generated by Kac's campaign to retrieve Alba see Kac, "Life Transformation—Art Mutation," in Kac, ed., *Signs of Life*, 165–71.

28 Kac, "Telepresence & Bio Art," in Kac, *Telepresence & Bio Art*, 243.

29 Around 2005, the visual studies theorist W. J. T. Mitchell could read Kac's work through what Mitchell deemed to be the most pressing issues of the day: cloning and terrorism. Today these two issues have largely receded from public attention. See Mitchell, "Vital Signs | Cloning Terror" and "The Work of Art in the Age of Biocybernetic Reproduction," in *What Do Pictures Want?* (Chicago and London: The University of Chicago Press, 2005), 5–27, 309–35.

ON CECILIA VICUÑA AND A NEW SPELLING FOR EVOLve:

Feel Love

The gesture is as simple as it is seemingly banal: AMOR. A sign written across a tree trunk to visually confirm a feeling. This time, though, the word hovers over the tree's surface and extends beyond it, suspended by leafless branches, its contours barely delineated with red thread. A bloodline whose wayward geometry seems caught by mistake in the forest, floating detritus of a bygone party, tenously articulating a desire, a call: my love.

Unlike other, more possessive marks made on trees, whether to record a fleeting human feeling, trace a forest map, or leave a rubbed scent on bark to attract others, as bears and deer do, this fragile written gesture acknowledges its own passing. This threaded word AMOR knows no knots, no fixed points of security, and its twisting forms will likely blow away. It just states its deep feeling for the forest and its inhabitants, their threatened disappearance, without claiming ownership of any. As a response to a land and beings that beckon with their own swaying arms, this linguistic sign in yarn encounters a plurality of languages and forms of communication and seems to ask of murmuring wind, fallen leaves, knotted rocks, spongy moss, passing ants, and half hidden birds: do we always hurt what we care for? How to respond to land's call? *Ay amor*.

Part of a series of ephemeral ritual gestures made by the poet and artist Cecilia Vicuña in the 1980s, and perhaps one of her less-known spatial weavings in yarn, the work is called *Amor. Tejiendo en todos los lugares equivocados* (*Amor. Weaving in All the Wrong Places*, 1981; fig. 1), a title evoking a sense of error, of going against the grain, as all love does. Weaving that happens where it shouldn't,

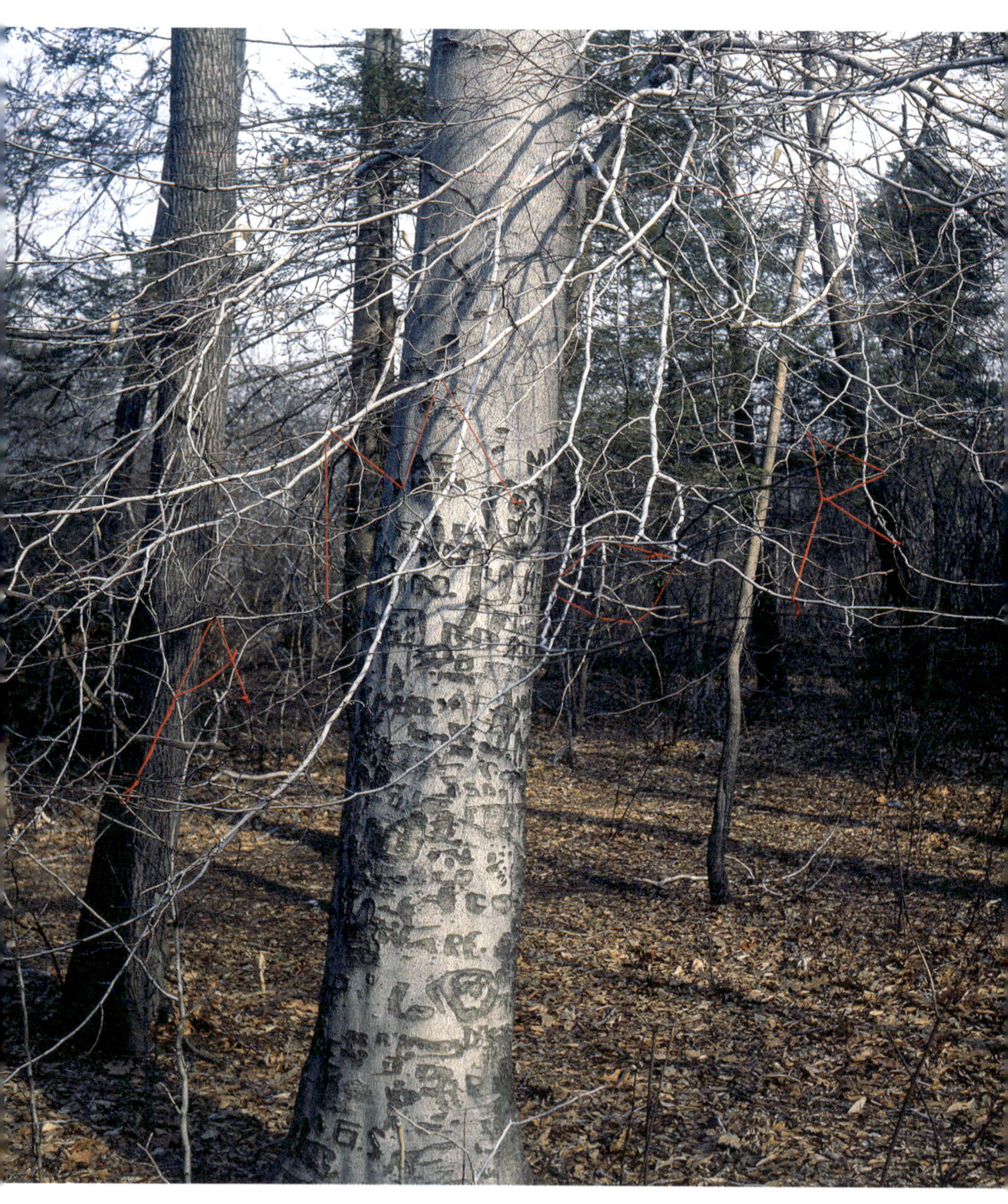

1 Cecilia Vicuña. *Amor. Tejiendo en todos los lugares equivocados* (*Amor. Weaving in All the Wrong Places*). 1981. Site-specific performative installation, Bronx, New York

that connects what has been severed and pulled apart. Perhaps an act of resistance and disobedience, too, even regarding what constitutes a loom.[1] An absurd minimal gesture of reconnection that can be disarming in its unexpectedness, even as the word AMOR, its plea, its reminder of being part of a larger web of life, hangs vertiginously by a thread. If, as Vicuña suggests in a poem, weaving may have originated in models provided by the worldlings of animals and plants, "the first thread coming out of fleece trapped in vegetation," or a cross made out of the "intertwining of branches and twigs," then this symbolic gesture of reconnection, enacted in a New York forest, can be taken as a reminder of more-than-human cultural memories.[2] The forest not only speaks, as Eduardo Kohn argues, it remembers, nudging us to recall how, in an inversion of modern legal understandings and proceedings, human actions can be and are measured by the environment (as the climate crisis painfully evinces), the earth acting in its own ways as final judge.[3] "The opinions of birds are very important. Would they accept the woven trees?," Vicuña asked regarding another spatial weaving made in New York State, *For the Trees and Birds* (1987). "In a few minutes they were singing like crazy."[4] Crazy in love?

Vicuña wove a similar gesture in the hills of Bogotá in 1981, barely leaving a trace. Having trodden many paths by the mountain's edge during the five years she lived there, she had noticed thin tracks made by slight changes in the grasses' depth and found out that these were ancient Indigenous trails woven "in the crests of the hills to join the villages of Cundinamarca and Boyacá."[5] Recalling that the Kogi people of the Sierra Nevada consider weaving a sacred practice, Vicuña picked up the remains left behind by other beings in the same hills—feathers, twigs—and walked these eroded lines creating small offerings on improvised altars, alongside crisscrossing pathways of red and yellow threads with which she wove the land, the air, the sun. "The sun spins the thread of life around the world the earth is a loom and the sun weaves the night and day," reads a continuous line of text in the book *QUIPOem*, of 1999, a thread of exhaled breath that stretches diagonally across the page, connecting two images of *Chibcha Trail* (1981) to the Kogi origin story to the mestizo woman, the loom moving swiftly from surface to surface.[6] The Colombian photographer Oscar Monsalve's photographs of Vicuña's ritual action emphasize a sense of ephemeral passage, with changes in scale that help unsettle the imagined dominance of the human perspective, transforming threads, sticks, and feathers into woven portals that open up other spaces, points of view, and temporalities. Knowing that she was in Bogotá to pack up her few possessions and migrate to New York, Vicuña invisibly tied herself up to the city, too. Or rather, by treading lightly on the contours of hills, following the trailing tales of Andean ancestors, and learning by attending to the land, she labored her love: ama/ara.[7] *Amarrando Bogotá* (Tying Bogotá, 1981).

Since then, more weavings have unfolded, linking places, peoples, and different forms of being. As Carina del Valle Schorske writes, in following desire lines Vicuña has "never stopped 'weaving in all the wrong places' (her words): mapping a childhood bedroom in blue, tracing words and shapes among the branches of graffiti-scarred trees, linking the bodies of strangers in live performance."[8] Rather than call Vicuña's multifarious production a total work of art, "*una experiencia poética totalizadora*," I think of these migrating ritual gestures as an earthbound poethics anchored in love,[9] a "spatial poetics" that is also a spiritual practice and that emerges from acts of listening to the sounding of the wor(l)ds and to the wisdom expressed by their caretakers,[10] a way of word(l)ing, *un palabrar* tied to an erotics of the earth, rooted in a search for connectivity that is deeply imbued with ritual gestures, words as sung prayers, singing lands and waters, dissolving materials and woven threads. Rituals emerging from specific territories, the massive earth being of the Andes and the shores of the Pacific coast, and responding to them and to the multiverses that they and their guardians, past and present, unfold. Becoming with them a devotee of the precarious—"and if I devoted my life to one of its feathers?"—that sees the divine in the small, and commits a life to the possibility of healing and transformative justice.[11] Starting at home.

BY THE SHADOWS OF MOUNTAINS: DIVINANZAS

Me ha dado la marcha de mis pies cansados
con ellos anduve ciudades y charcos
playas y desiertos, montañas y llanos
y la casa tuya, tu calle y tu patio

> [Life] has given me the march of my tired feet
> with them I walked cities and puddles
> beaches and deserts, mountains and plains
> and your house, your street and your garden
> —Violeta Parra, "*Gracias a la vida*" (Thanks to life), 1966

That love emerges over and again in Vicuña's works should not surprise us, especially if we consider that one of her first poetic experiments involved the word *enamorados*—AMOR spinning in a purple cloak. It was 1966 and Vicuña has retold this story many times, playing with her own biography in a practice that can be likened to the biomythology of Audre Lorde.[12] The story goes that an unruly schoolgirl was writing in her bed, in a house framed by mountains and water, "oriented towards El Plomo and Manquehue, the sacred mountains of the valley of Santiago, and two blocks from the Mapocho, the river that is born at El Plomo," when she suddenly saw a word unfold, dance, open up: "*de pronto* 'ví' *una palabra armarse y desarmarse, bailar y mostrarme sus partes, como si viniera de otra 'realidad,' la de su propia creación*" (suddenly I "*saw*" a word arming and disarming itself, dancing, showing me its parts, as if it came from a different "reality," that of its own creation).[13] The cracking word that sprouted metaphors leaping into unknown universes was followed by more that seemed to come from another time and place, "individual words opened up to reveal their inner associations, allowing ancient and newborn metaphors to come to light."[14] Metaphors as vehicles, spaceships into the unknown, "*un decir de frontera*" (a frontier saying).[15] Their playful antics also involved a somatic engagement with words as they were beheld and mouthed, as vibrations creating friction, rubbing each other in erotic play, bringing about laughter, a tenderly scolding parental face and a new/old form of poetry: *adivinanzas*/divinations.

> *¿Quiénes son los dulces enajenados de color morado?*
> los enamorados
>
> What errs sweetly?
> The Free Reign of Lovers[16]

Ad-divinus. The name recalls "*canto a lo divino*" (Singing to the divine), a popular form of sung poetry from the central zone of Chile that orients its improvised verses (its *décimas*) toward the spiritual. This rural, earth-bound, chanted tradition hybridizes the Indigenous with the colonial in a cosmic dialogue, one that Violeta Parra—a hero of Vicuña's, as evinced in an oil painting of 1973 dedicated to her—had threaded to her own songs in the 1950s and '60s. To divinate also speaks of the sacrality of language, of how it touches on the mysteries of life; *ad-divinus*, suggests Vicuña, positions the body/spirit toward the gods. It brings us back to the riddle that exerts its wondrous enchantment, an enigma dressed in literary veils, indicating an attitude toward life that is ensconced in questions rather than fixed answers. The full title of the collection of sixty-two *adivinanzas* that grew from that night is "*Adivinaciones de Relación Insólita*" (Strangely related riddles), pointing to a relational understanding of existence shaped by curious, unexpected crossings.[17] And to the magical power of language, its ability to hurt and heal, "touching in place of touch" performatively and thus acting "affectively on bodies" and in the world.[18]

It could be said that the birth of these animated divinations somewhat resembled Vicuña's experience of becoming aware of awareness, another story of origins involving earth beings that she has often recounted—the story of an adolescent thinking of studying architecture, lying on the beach of Concón, in central Chile. Vicuña has referred to this moment at the coast as one of experiencing sudden enlightenment born out of joyful play, somatic tuning, and feeling caressed by the person-wind, feeling seen by the person-ocean, maybe even longed for. An eco-erotics, a feeling-knowing that wind and ocean were alive "and had as much awareness as I do now."[19] Feeling/knowing that language has many forms, and that communication happens beyond words too. Perhaps this moment could also be said to be one of entering into a sensing-knowing relation in which "to feel the earth as your own skin,"[20] a relation that restores—even if only partly—and confirms a sense of threaded unity, of never having not been like other beings, where my breath and its carbon cycle, my catching and releasing and the moments in between, interlace and affect, touch and combine with yours, with the plants', with the ocean's.

A nonhierarchical relation, close to Indigenous experiences and ontologies in the Andes such as those described by Marisol de la Cadena in her work with the Turpo family in Peru, in which "there is no necessary difference between humans as subjects of awareness and places as objects of awareness, for many of the 'places' that Mariano and Nazario 'sensed'... were also 'sentient.'"[21] These sentient earth beings are co-constituted with human beings, they enable becoming collectively, like a woven textile in reciprocal, intertwined relations. "In Quechua," says de la Cadena, "the Indigenous language I am familiar with, there is a word for that: ayllu, a relation in which people and land *are* at the same time."[22]

One might think this was purely anthropocentric projection, yet Vicuña suggests that this awareness is less about human desire than about sensing the desires of water, air, and forest to be loved, cared for. "This is the way a communion with the sky and the sea began, the necessity to respond to their desires with a work that would be a prayer, a joy to the elements."[23] She responded to that windy oceanic beckoning by setting up an improvised altar, with materials found at the site—a minimal yet spontaneously familiar gesture: how many altars, castles, and cosmic cities are created and new/old beings summoned by children at the beach? "I felt that I needed to respond, to make a sign to indicate to the ocean that I understood. So I picked up a little stick that was just lying about. It was this beach that had a lot of debris. I stood it up, and once I stood it up, making it vertical, I knew that in that change—between horizontal and vertical—I had woven my place in the world."[24] That first ordering of "scattered bones, sticks and feathers," that act of offering by "listening to the elements" and following their will, is the origin of Vicuña's *basuritas*, little loved nothings that are also "metaphors in space," as Vicuña has described them, "scraps of stone, wood, feathers, shells, cloth and other human-made detritus [that] are gently juxtaposed" in fragile sculptural compositions—Vicuña's art of the precarious, her view from down below (fig. 2).[25]

To weave one's place in the world—not just to find it (or to be found, too) but to create it—can be taken as an act of love. bell hooks insisted on the importance of defining love as a verb, not merely a state, a feeling, a noun. Following Erich Fromm, she stressed the importance of the intention ("We choose to love") and the accompanying action: we love to "nurture growth."[26] Growth not in capitalist terms, but of a spiritual kind/kindred that makes space for others, stopping to listen to what is spoken and making an offering that might not be reciprocated. But precarity is not the same as scarcity, as, in a neoliberal system, we are often told to believe. In precarity there is divinity; in a fugitive sand altar, that deemed useless and disposable raises and rises in its beauty, a language of the gift to be taken back by the waves.[27]

> *Favelas, callampas, pueblos jovenes, villas miseria*,
> shantytowns by any name are *Pueblos de altares*.[28]
> —Cecilia Vicuña, *Unravelling Words & the Weaving of Water*, 1992

ELEVATED ALTARS: THE ENIGMA OF THE MUMMY

Weaving her art and life back to the mountain El Plomo, writing herself in the land, Vicuña, in her multiple performative biographies, both oral and written, has also recounted the importance of the location where the *adivinanzas* sprouted—perhaps because the unsettled stories thus (re)told manifest the cultural erasures inflicted by the modern nation-state of Chile on marginalized groups, particularly Indigenous peoples. Each performative biography is a new inscription of a story that diverges from official narratives, manifesting a will to remember otherwise while constructing its own precarious archive.

Vicuña grew up first in La Florida, an outskirt of Santiago, filled with *chacras* (orchards) and agricultural plots, where she was surrounded by animals and listened to the sounds and songs of toads and birds. When she was nine, the family moved inward and upward into the rapidly expanding upper-middle-class neighborhood of Canal San Carlos, a change that brought with it a more detached relation with nature. In an unpublished version of her text "*Dí vida*" (Now life speaks), she speculates that her father may have built the family's house as a veiled way to "'adore' and feel the proximity" of the mountains, but concedes that most Chileans at the time did not consider them to be altars.[29] Rather, and this remains true for many Santiaguinxs, the Andes and surrounding hills were "vistas,"

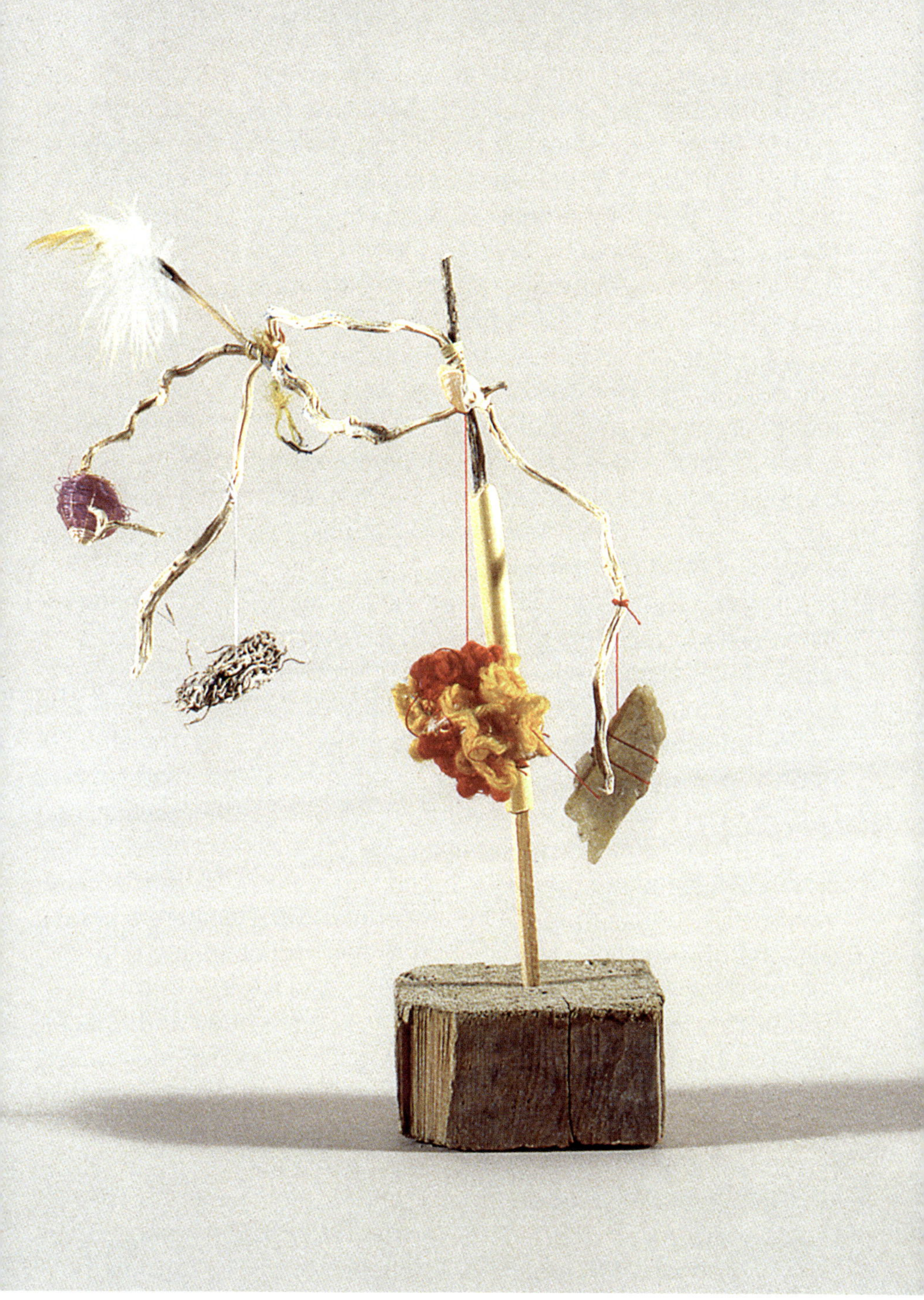

2 Cecilia Vicuña. *Árbol de vida* (Tree of life). 1984. Precarious object, mixed media, wood, twigs, wool, seashells, and horsehair, 8 × 4 ½ × 4 ½ in. (20.3 × 11.4 × 11.4 cm). Collection the artist

framing landscapes, majestic backdrops, even if everyday life circulated around and depended on them. The Andean legacies and the life forms and rituals currently performed in the mountains, including those of extraction—of materials from minerals to glacial water—were apparently invisible from that cultural distance; the view of the Mapocho River as the *río de la caca* (river of shit) is revealing of the ambivalent relation of citizens to the life source.[30] As Vicuña writes, "A tacit agreement made it impossible to name or feel a connection to the past. Santiago was not an Andean city and neither was our way of life. We were modern people and our modernity was based on erasing any past trace."[31] That modernity was built on the ideology of developmentalism, a politics of racial homogenizing, the dispossession and displacement of Indigenous peoples, and an urban expansion that also involved segregation along class and racial lines.[32]

It was only in the 1980s that Vicuña came upon the anthropologist and archaeologist Johan Reinhard's work on "*altares en altura*" (high-altitude shrines), and on his description of El Plomo as the southernmost highland sanctuary of the Inca world.[33] She had taken mountainous treks in Bogotá, had dived into the study of Andean ontologies once she was living in New York, had learned from Andean cosmologies on returning to Chile and Argentina, had produced an animated video based on Peru's ancient Paracas textiles, and had made works with thread and pigment on stones at the Inca site of Pumamarca; but her encounter with Reinhard's phrase was one of revelation. The veil that turned mountain into dramatic yet two-dimensional postcard was ripped, and another form of existence made itself manifest: "I saw it as a living body and I began to remember." Not animated but a living body, the mountain was a person that opened up for her the coexistence of worlds, radically different forms of inhabiting a territory and of being, and superimposed memories.[34] "I felt robbed," Vicuña has said.

Remember the robbery, connect to the theft: who robs whom? "To remember (*recordar*) in the sense of playing the strings (*cuerdas*) of emotion, the threads of the heart, *cor*."[35] Let's remember that when the first *adivinanza* opened up a portal to the mysteries of words, in 1966, it had already been twelve years since two impoverished *arrieros* (mule drivers) opened up a pirca, a stone structure, at the peak of El Plomo and took out, along with silver, textile, and ceramic accoutrements, the frozen body of a child, which they then sold to the Museo Nacional de Historia Natural in Santiago.[36] The body, cuddled into a small bundle, was that of an eight-year-old boy, his long hair neatly braided, his life sacrificed around 500 years before, according to a report published by the museum in 1957, on a "mountain that has been a sacred center to which the peoples of the last pre-Columbian era dedicated their religious activities."[37] Brought alive to the peak from his home in Coyasuyo (the southern part of the Inca empire), as could be told by the blue-toned skin of the child's three congealed fingers, his life was sacrificed to the Inca god Inti in a ceremony known in Quechua as *capacocha*.[38] Overfed and possibly drunk under the effects of chicha beer, the unconscious boy was left to freeze in a stone-covered pit at an altitude of almost 18,000 feet. The local press of 1950s Santiago at first misidentified the child as a girl and ran photographs of the little "princess" seated on a chair; to avoid deterioration of the fragile tissues as they warmed, the remains were later placed in a glass case to be studied and analyzed. Vicuña has repeatedly told the story of how she and other schoolchildren were taken to see the child, and how impressed she was, identifying with the boy-turned-girl: their smallness, their vulnerability, the exposure. And what she saw in them was not death but calm sleep, a placid image perhaps invoked in order to work through part of the trauma of seeing another child in a museum, violently sacrificed, forcefully removed from its resting place, then displayed under a colonial scientific and spectacularized gaze.[39]

It should be noted that the unearthing of the child on El Plomo affected other Chilean artists of Vicuña's generation, such as Eugenio Dittborn, who in the 1980s and '90s obsessively returned to drawn and printed images of the boy/girl and to photographic reproductions of the snowy peaks where the burial was located.[40] Producing conceptual works of mail art based on appropriated images, Dittborn associated the so-called "mummy" with absent bodies, particularly those disappeared during Augusto Pinochet's military dictatorship, and with journeys related to colonization and coloniality, the heterogenous temporalities of technological mediations, and the disjointures produced by modernization.[41]

Vicuña took an alternate path, seeking to restore and repair these colonial wounds through poetic performative gestures passing through the body, the "effaced memory of the rites and songs at the peak drawing dreaming."[42] To re-pair as in paying attention to the elements and to local histories, resignifying the idea of sacrifice (and even sacrificial zones), of holding together the pain of the fragmented with the possibility of wholeness in a space of uncertainty (*reparar* in Spanish is both "to fix, amend," and "to look carefully, to note"). To re/pair by weaving the complementaries of life and death into precarious offerings, whether small assemblages of debris that might fall apart and be taken by waves or larger installations of unspooled wool, tied and unfolded as imaginary khipus and ceque lines. To make works that evoke Andean textile-based practices and imaginaries—the knotted recording device, the invisible lines connecting huacas in the sacred land—and make a call to enter into con/versation (creating verses together) with many worlds (fig. 3).

Today other poets are asking for a similar gesture, on a plurinational scale. Speaking to the Chilean senate on December 21, 2021, the Mapuche poet Elicura Chihuailaf referred to love as central to Mapuche cosmogenies and epistemologies: "Our people say 'we must have love for everything that is around us,' because we are nature, but not the center. . . . There is a natural order, which is not what the so-called 'authority' intends to establish; the natural order is what the trees in the forest, the stones and—I repeat—the stars show us."[43] Chihuailaf also requested that the country's new constitution, drafted after the social uprising of 2019, center on conversation as an act of repair, taking "as its center the need for conversation, this conversation that has been lost, [it] should be on the table of the authorities . . . to resolve . . . conflicts, especially when it comes to the institutions of the State, which, let us remember, usurped a territory. And that requires repair and requires . . . a conversation, which we are waiting for."[44]

María José Barros has rightly asked whether Vicuña has romanticized or idealized the death of the child on El Plomo, ignoring "the intention of political dominance" that these religious rituals had.[45] Yet Vicuña has also recognized what she calls "*el pecado del Inca*" (the sin of the Inca), acknowledging the violence imposed by the Inca state through the *capacocha* ceremonies to exert control over its diverse populations and maintain authority.[46] Maybe the question is: how can violence, whether of ritual or of the state, whether past or present, be upended and transformed? How can the traumatic and catastrophic effects of ecological devastation, land usurpation, violent persecution, and even genocide be rallied into and beyond activism, worked upon, and molecularly transmuted into healing and restoring processes, personal and communal? And what does art have to do with it? Can art create actual forms of sanctuary, become a politico-spiritual force? Ayayayamor.

"If we want to continue existing, we need to create rituals that resist the force that has limited our ability to be in touch with the immaterial dimensions that permeate and support our existence," observes Camila Marambio in conversation with Vicuña.[47] Thinking of Gloria Anzaldúa's work as spiritual practice, and of her writing as embodied thinking that is also prayer and incantation—"words of power, delivered with intentionality, as acts"—Laura E. Pérez reminds us that spirituality can be political activism as long as the written (and perhaps we should add the chanted, drawn, woven, performed) words, signs, and metaphors call forth the imagination in a struggle for wholeness.[48] Words and signs, even gestures, teaching us ("*enseñas: you teach; en-señas, you put in signs*," says one of Vicuña's graphic *palabrarmas* [word weapons]), evoking sign language and the worlds of semiotics, moving us forward, sprouting like the futurity of dormant seeds, seeking connection and reintegration.[49]

FINDING THE WORDS

> *Finding the words is another step in learning to see.*
> —Robin Wall Kimmerer, *Gathering Moss. A Natural and Cultural History of Moss*, 2003

> *Tenemos una tarea, nos dicen, que es conocer lo que nos ha tocado, porque conocer es la única posibilidad de amar aquello y de . . . amarnos a nosotros mismos.*

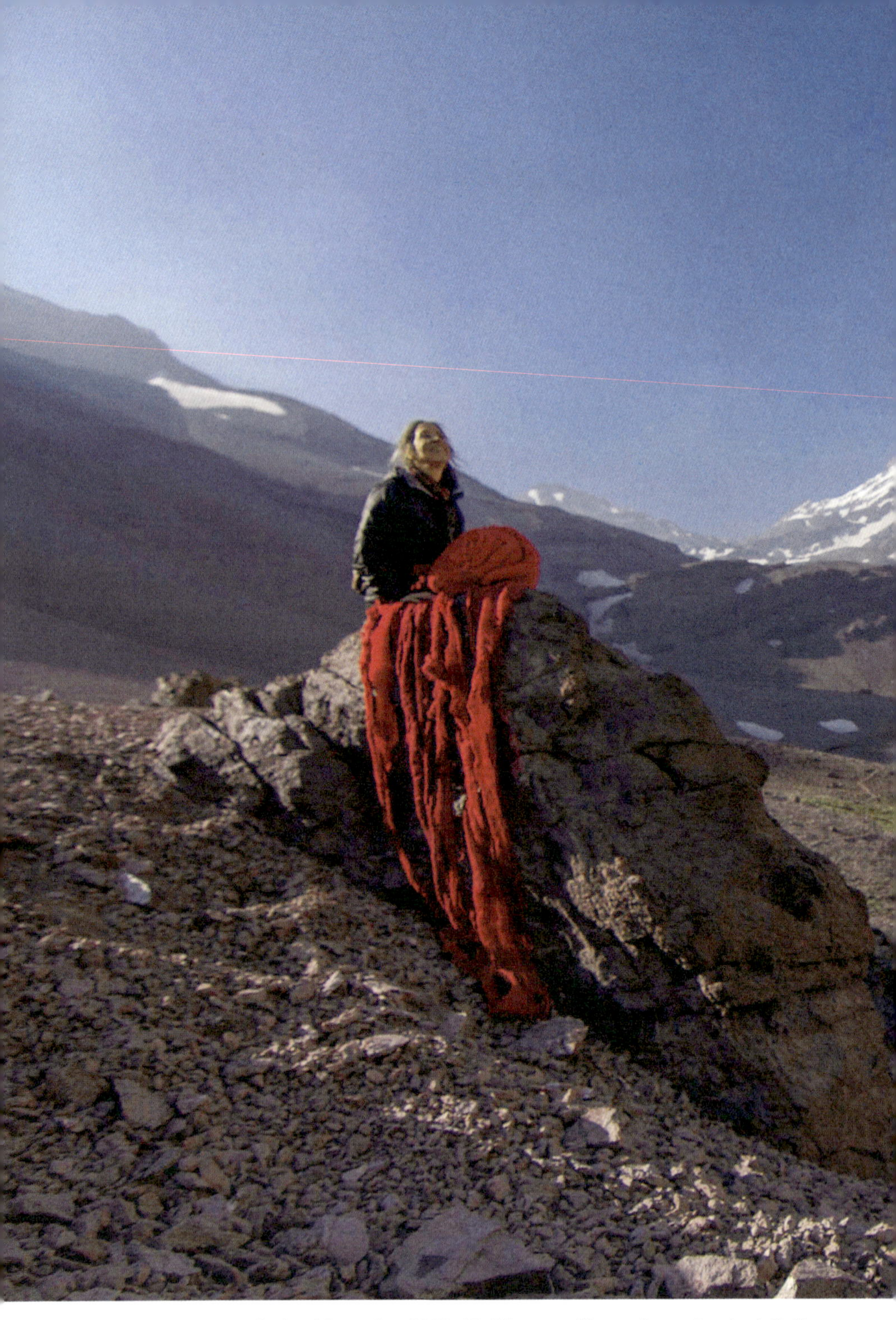

3 Cecilia Vicuña. *Quipu Mapocho*. 2016–17. Site-specific performative installation, Río Mapocho, Chile

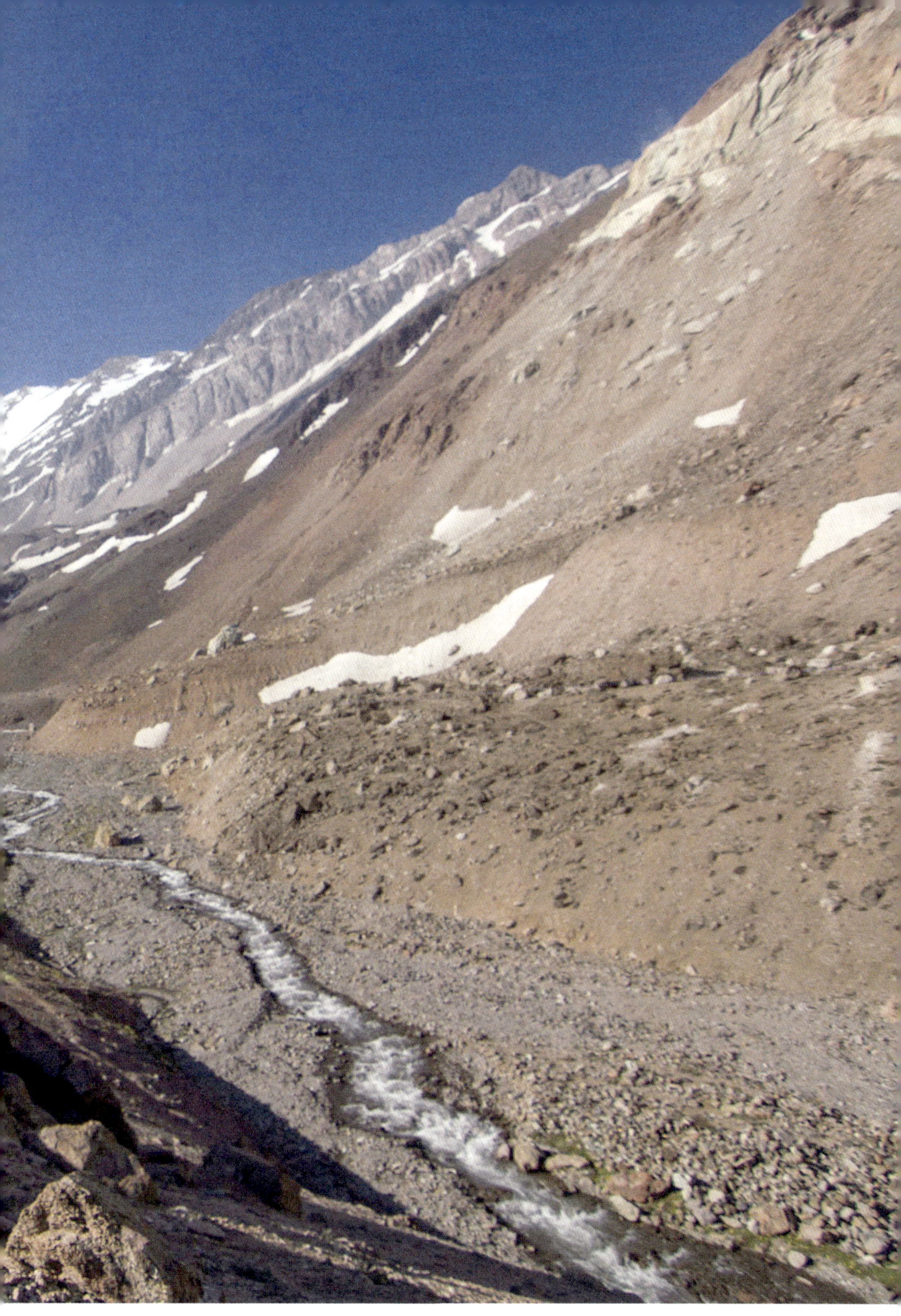

We have a task, they tell us, which is to know what we have been given, because knowing is the only way of loving that and of... loving ourselves.
—Elicura Chihuailaf, speech to the Chilean senate, December 2021

CON o SER
ser con

to know:
to be with
—Cecilia Vicuña, *Palabrarmas*, 1984

Vicuña's poethics were nurtured in the contradictory environment of Chile in the 1960s and '70s. In the same way that she encountered the works of Joyce Mansour, Latin American avant-garde poetry magazines such as *El Corno Emplumado*, and a Surrealist anthology in the home of her aunt, Chilean artist Rosa Vicuña, in 1964, and similar literature in Santiago's university bookstores, she also found a book of Guaraní poetry whose ecological consciousness would reverberate throughout her work in later years. "So you will watch over the fountain of mist where inspired words are born," say the Mbyá Guaraní, a people whose oracular wisdom Vicuña would weave into her poetry as warnings to learn from: "The Guaraní of the rain forest in Paraguay and Brazil say that when the mist and the forest are gone, we will all be gone."[50] This was also a time of countercultural critiques building from civil-rights movements, student and worker protests, and growing ecological consciousness worldwide, associated with a revaluation of Indigenous cultural traditions and the growth of new forms of spirituality and expanded consciousness, developments that in the 1970s would engage with Indigenous political organization and feminism. Even before Vicuña began to study art teaching at the Universidad de Chile, which she attended from 1967 to 1971, she was also writing poetry and devising corporal experiments and poetic actions along with her friends from Tribu No, a group of like-minded young friends who, according to Magda Sepúlveda, valued Indigenous groups in particular for their "form of living in community and of reconnecting with a magical and non-objectual vision of the world."[51]

Yet the force of cultural oblivion concerning Indigenous peoples, their anticolonial struggles, and their ancestral connection to and relational understanding of the environment was strong in modernizing Chile, despite reform movements beginning in the mid-1960s that were further advanced by Salvador Allende's Unidad Popular (Popular Unity) government of 1970–73. The agrarian-reform movement begun during the government of Eduardo Frei Montalva (1964–70) melded Indigenous people with farmers and peasants into a single cipher of identity, making the "popular" and the "people" all-encompassing and homogenizing notions. This trend deepened during the Unidad Popular period, even as demands for Indigenous sovereignty were in some cases gaining sharper focus. Meanwhile, the notion of a "vanishing race" was widespread,[52] especially concerning Indigenous peoples such as the Selk'nam in the archipelago of Tierra del Fuego, and the press constantly ran stories about the disappearance of the "last" member of an Indigenous community irrevocably producing the death of a culture (a performative case of a need to repeat in order to affirm an imagined occurrence, without any attendant reference to the state's involvement in this genocide). Partly influenced by these, Vicuña was also painfully aware of the sustained oppression faced by different Indigenous groups, particularly the Mapuche.[53] In her 1973 book *El Cuaderno Café* (The brown book), which she made as a migrant in London, shortly after finishing a scholarship to study painting at the Slade School of Fine Arts, she likens the Indigenous to a seed, quoting the apparently objective language of a popular encyclopedic publication in order to contradict its naturalized claims of progressive disappearance: "The Mapuche population has been calculated. . . . No, it has not been calculated. The Indians are only waiting to germinate, to conquer their own way of life. The Indian people will rise, enlightening thus the rest of the people."[54]

If the initial statement still bears traces of the "salvage paradigm" that aims to preserve an authentic culture from the ravages of modernization, Vicuña's reflection is also traversed by the parallel understanding that Indigenous peoples and knowledge are alive, have survived worldwide, and will continue to germinate.[55] It provides a sense of futurity embedded in the carnal image of the seed and the encoded knowledge it carries forth: "The seed is the word of the earth," says Vicuña.[56] It is thus not strange to find that in the oil painting *Sueño* (Dream, 1971), Vicuña brought together images of Indigenous couples culled from magazines and books, pointing their hunting technologies and weapons at the pope of the time, Paul VI.[57] *Sueño* represents an uprising of the Indigenous, the return of the repressed for Western empires through what looks like an organized global rebellion, armed with solidarity and seemingly precarious weapons: "maximum fragility," as exemplified by the stones a Selk'nam woman throws, "against maximum strength."[58] Reversing the violence of colonialism, and the Catholic church's evangelization of the wor(l)d in its attempt to vanquish the so-called pagan and idolatrous, the painting also echoes the guerrilla languages of armed struggle that had swept Latin America in the mid-1960s.

It should be no surprise that the *adivinanzas* returned to Vicuña's work after Augusto Pinochet's military coup, on September 11, 1973, cut short Allende's socialist path, establishing in its place a bloody dictatorship that spearheaded a massive neoliberal experiment in Chile. But they returned as *palabrarmas*, word-weapons responding to the violence, manipulation, and sickened words of the military and its civilian accomplices. Perhaps more important, they returned simultaneously as a practice, *palabrar*: to cultivate and work with language as with the earth. An agricultural image and an earth-bound metaphor. Precisely, one of Vicuña's *palabrarmas* of the late 1970s would evoke the Indigenous and popular forms of survivance: *existir es resistir*, resist by existing and also reexist.

Undoing those years of internalized forms of colonization became a life-long endeavor for Vicuña. It would start with territory, finding the words and what was in the names, an awareness of the knowledge inscribed, and obliterated or perhaps submerged, in the land. "Indigenous worldviews, our ontologies," Liisa-Rávná Finborg writes, "are shaped by our relation to land."[59] As Vicuña observes in *El Cuaderno Café*, Chile is a country built upon erasure of Indigenous history, a history that only needs to be paid attention to, attuned to, even in toponomy:

> The names of chile: in old times when I was colonised my mind was plagued with mysterious names coming from

> africa, asia, europe. none of the chilean names were in my head until I discovered chile on this trip. who could talk to me about puchuncaví? A virgin land in thought, film and poetry because nobody ever documented or touched it, except for the silent inhabitants, quiet visionaries who won't leave that sun. petorca & choncolco, pichí y chaulinec, everything is to be thought at llico de mataquito, nirivilo and toltén, all of them jewels resting upon violeta parra's bread, which are like her tapestries, songs and poems, food for the rest.[60]

From toponomy to topography to music (with Parra at the forefront again) and the sounds of the earth and its guardians, resonating with them, sources of spiritual nourishment. "Before being contaminated the river would like to be heard."[61] Yet learning to hear the sound of a place, the voices of an earth-being, a person-river, an ancestor, what positions must our bodies take? What rituals performed in response to which forms of touch, of sensing space, allowing oneself to listen to its vibrating chords and chants, to be lulled and embraced? What/who would remain inaudible, and why? Perhaps this is how one must understand Vicuña's early claim, made in 1973, "*yo me voy a volver india*" (I will become Indigenous), and the reference ten years later to her "shaman heart," as a re-Indigenization that involves reeducating oneself, both in loving ways and by listening to Indigenous wor(l)ds, paying attention by entering into con/versation with others, and not only as a reclaimed ethnic filiation.[62] This form of listening includes the breath-body, one's own sounds and truth, words as vibrations of multiple materialities engaging in what Leanne Betasamosake Simpson calls "the art of being together."[63] It is a listening that might not understand or hear, yet is willing to care and acknowledge that, holding the difference.

Perhaps reindigenization involves re-imagining language, too; thus the decolonial power of the *palabrarmas*, as they unlock the potentiality and futurity of Spanish by cracking it open and allowing for other languages and threads of planetary knowledge to come through. Learning, for instance, as Vicuña did, that riddling rituals existed among children and youth in the herding communities of the Andean highlands of Ayacucho, and that they occurred in contexts of amorous play. Functioning "as a discovery procedure as children expand their cognitive operative structures and semantic domains," bringing about new conceptualizations through unexpected relations, riddling was also part of juvenile sexual socialization, particularly in the ritual of *vida michiy*, "to pasture life."[64]

Word-sound-gestures as spiritual technologies, embodied metaphors in space.

A BRIEF INTERLUDE IN THE amaZONE

Imagine a small plane,
open windows, wooden crates in the back.
You take off, leave the verdant hills,
mist thick and softly palpable
until below the dense clouds
a presence manifests,
poetic beauty itself
your veins vibrate, pulsate
with the snaking water
its breath
magnificent scent
as you realize
you are the snake, the lung.

It was late December 1977 when Vicuña took that small plane and first saw the Amazon forest from the air, arriving at Leticia, Colombia. Returning to South America in November 1975, she had decided against going back to Chile in part due to her activism against the military dictatorship. While settling into a complex life of hardship and political activity in Bogotá, struggling to maintain a painting practice while writing poetry, developing more *palabrarmas* as a pedagogical project, Vicuña decided to visit a cousin in Brazil and to cross the Amazon by land, hoping to catch a glimpse of the recently started Trans-Amazonian Highway (she would travel only a section under construction and the rest by boat, foot, bus, and small planes). From that experience came more *palabrarmas*, a score of wordweapons and line drawings guided by the image of an Indigenous young woman, flying on shovels with wings. Vicuña unleashed a flow of imagery, of flowery-speech-drawn-words, *palabrarmas* nurtured by the sign languages of the Maya, South American petroglyphs and pictographs, the struggles of Huitotos, and of Black and Indigenous peoples.[65] Realizing that to "develop" was not a straight road, but a snake made of mist.

Desarrollo
desatar el arrollo

To develop
releases the stream[66]

In the *palabrarma* drawing, the woman squats and opens her legs, leaning to one side. With a bent shoulder she pushes a word away: *des* (negation), as if unlocking a dam. Let the river flow. Between her cupped hands, raised over her womb, the words begin to spill, flowing down and sowing the space of the page, with the lulling sounds of clotted blood: *arrollo. Deja que el arrollo te arrulle, mi amor*.[67]

RETURN TO EL PLOMO: LOVE YOUR GLACIER

So maybe we can let go of this spiral and take a flight by returning to the heights. To the clit, the nest of clouds. If ashes go to ashes, wool goes to its Andean source: water, and from there, cosmic gas. "In the Andes," Vicuña writes, "the energy of all potential, a form before form, is condensed in concepts such as *chanccani* (Quechua) 'to begin, to dwell in the unformed' and in ritual practices associated with unspun wool, *millma* (Quechua), a symbol of cosmic gas, the source of galaxies and stars."[68]

To remind Michelle Bachelet, the first woman president of Chile, of how accepting the Pascua-Lama open-pit copper- and gold-mining project in Diaguita territory would impact glaciers in the highlands above the Atacama Desert and the Huasco Valley, Vicuña enacted a private ritual at the mountain and a public performance in Santiago. A long thick woollen stream was placed at the entry of the presidential palace, accompanying an installation of hanging red wools in the cultural center built underneath, as if the building's interior and its supporting ideological infrastructure leaked through. There was no official response. More rituals and precarious prayers would be needed, more gestures evoked and repeated over the years, inviting other bodies to join, weave themselves in. If I argue that Vicuña's work is spiritual-healing work, that is unfinished work, as she has often said, as long as patriarchal colonial and heteronormative wounding persists.[69] Accompanied, always, with chants and rumbles and mumbles that will never, thankfully, translate well into the photograph or the page. "Twice behaved behavior," or "restored behavior," is how Richard Schechner defined performance; only this time it is of that which is not yet.[70]

> *A collective ritual is an act of learning. Together we go into a phase transition, an invisible metamorphosis, a different state of consciousness, in order to see.*
> —Cecilia Vicuña, "About to Happen," 2017

Marcel Mauss said that prayer is a form of communication with sacred beings and is intended to act on them.[71] Pascua-Lama ceased operations in 2020.[72]

Sol seco
y espacio de olor

la tierra
pidiendo amor

Dry sun
and scented space

the earth
asking for love.

—Cecilia Vicuña, "Antivero," 1981

1 Writing about another work from 1981, *Antivero*, in which threads joined the banks and stones of a river, Catherine de Zheger quotes Cecilia Vicuña: "the [Antivero] river is the warp, the crossing threads are the weft." De Zegher, "Cecilia Vicuña's Ouvrage: Knot a Not, Notes as Knots," in Griselda Pollock, ed., *Generations and Geographies in the Visual Arts: Feminist Readings* (London and New York: Routledge, 1996), 197.

2 Vicuña, "The Origin of Weaving," *Unravelling Words & the Weaving of Water*, trans. Suzanne Jill Levine and Eliot Weinberger (Saint Paul: Graywolf Press, 1992), 9.

3 "Life is constitutively semiotic." Eduardo Kohn, *How Forests Think: Toward an Anthropology beyond the Human* (Berkeley, Los Angeles, and London: University of California Press, 2013), 9.

4 Vicuña, *Precario/Precarious*, trans. Anne Twitty (New York: Tanam Press, 1983), 21.

5 Vicuña, "The Chibcha Trail," *Unravelling Words*, 16.

6 Vicuña, *QUIPOem/The Precarious*, ed. de Zheger, trans. Esther Allen (Middletown, CT: Wesleyan University Press, 1997), 66–67.

7 In her *palabrarmas*, Vicuña has often used the image of laboring with words as plowing (*labrar*, *arar*), inventing verbs like *palabrir* (to open words). Perhaps one can unloose the word to tie (*amarrar*) and arrive at acts of love (*ama*) that are also gestures of re/tracing the self into the earth (*arar*).

8 Carina del Valle Schorske, "Cecilia Vicuña's Desire Lines," *The New York Times Magazine*, August 25, 2022.

9 See Julieta Gamboa, "Cecilia Vicuña: Trama y urdimbre de la palabra: El tejido/texto," *Revista de literaturas populares* 2 (2012): 505–21.

10 Vicuña, quoted in Lucy Mercer, "The Erotic Socialism of Cecilia Vicuña," *ArtReview*, December 4, 2019. Available online at https://artreview.com/ar-december-2019-feature-cecilia-vicuna/ (accessed April 7, 2024).

11 Vicuña, *Precario/Precarious*, n.p.

12 See Audre Lorde, *Zami: A New Spelling of My Name* (Berkeley: Persephone Press, 1982).

13 Vicuña, *Palabrarmás* (Buenos Aires: RIL, 1984, second ed. Santiago: RIL, 2005), II. Trans. Vicuña, in oral communication with the author, May 7, 2024.

14 Vicuña, "Palabrarmás," in *Fire over Water* (New York: Tanam Press, 1986).

15 Fidel Sepúlveda Llanos, "Notas para una crítica alternativa," *Arte y Vida* (Santiago: Liberalia Ediciones, 2015), 123.

16 Vicuña, "Riddles/Divinations," in *New and Selected Poems of Cecilia Vicuña*, ed. Rosa Alcalá, trans. Alcalá, Esther Allen, Suzanne Jill Levine, et al. (Berkeley: Kelsey Street Press, 2018), 71.

17 Vicuña, oral communication with the author, May 7, 2024.

18 Anna Gibbs, "Language as Life Form," in Christopher Braddock, ed., *Animism in Art and Performance* (Auckland: Palgrave Macmillan, 2017), 95.

19 Vicuña, in Julia Bryan-Wilson, "Awareness of Awareness: An Interview with Cecilia Vicuña," in Vicuña, *Cecilia Vicuña: About to Happen* (New York: Siglio Press, for Contemporary Arts Center, New Orleans, 2017), 111.

20 Vicuña, "Entering," in *Precario/Precarious*, n.p.

21 Marisol de la Cadena, *Earth Beings: Ecologies of Practice across Andean Worlds* (Durham, NC, and London: Duke University Press, 2015), 101. Speaking of the inter-relatedness of the human and the

more-than-human in Vicuña's works, Lucy R. Lippard refers to Eduardo Viveiros de Castro's theories on Amazonian and Amerindian cosmologies and perspectivism in her essay "Floating Between Past and Future: The Indigenization of Environmental Politics," in *Cecilia Vicuña: About to Happen*, 130–37.

22 De la Cadena, in "Worlding Futurities. Marisol de la Cadena interviewed by Pedro Neves Marques," in Neves Marques, ed., *YWY. Searching for a Character between Future Worlds* (Madrid: CA2M, and London: Sternberg Press, 2021), 21.

23 Vicuña, "Entering."

24 Vicuña, in Bryan-Wilson, "Awareness of Awareness," 111.

25 See Lippard, "Spinning the Common Thread," in de Zegher, ed., *The Precarious: The Art and Poetry of Cecilia Vicuña* (Hanover and London: University Press of New England, 1997), 10.

26 bell hooks, *All About Love: New Visions* (New York: HarperCollins, 2001), 5.

27 Vicuña has recently said of this gesture: "And I realized that this was the beginning of a new kind of art, where I could build—because I was dreaming of architecture—a little city of *basuritas*, of debris, so close to the waves that when high tide would come, it would erase it. I continued to do that, and I suppose a moment came, in those early days, that I realized: this is *arte precario*." Vicuña, in "Cecilia Vicuña with Suzanne Herrera," *The Brooklyn Rail*, July–August 2022. Available online at https://brooklynrail.org/2022/07/art/Cecilia-Vicua-with-Suzanne-Herrera (accessed June 17, 2024).

28 "Pueblos de altares" are "towns of altars." Vicuña, *Unravelling Words*, 12.

29 Vicuña, "Dí vida," unpublished draft, courtesy the artist. This version differs from the text of the essay that appears in the second edition of *Palabrarmas* (Santiago: RIL, 2005).

30 On *río de la caca* see María José Barros, "Aguas y ríos: Activismo, descolonización y naturaleza en Cecilia Vicuña y Ana Tijoux," *Latin American Research Review* 55, no. 3 (September 2020): 548. See also Donald Kuspit, "The Triumph of Shit," *artnet*, n.d. Available online at www.artnet.com/magazineus/features/kuspit/kuspit9-11-08.asp (accessed April 8, 2024). I would say, though, that Vicuña's relation to "rubbish" is of a different kind from that of the artists Kuspit writes about, as it comes from an appreciation of everyday detritus from the gutter and from nature.

31 Vicuña, "Dí vida," unpublished draft.

32 See Claudio Alvarado Lincopi, *Mapurbekistán: Ciudad, cuerpo y racismo. Diáspora Mapuche en Santiago, siglo XX* (Santiago: Pehuén, 2021), 66. Recognizing this contradiction and her own privileged upbringing, Vicuña worked constantly in *poblaciones* in the 1970s, with displaced communities both in Santiago and Bogotá, and later with Indigenous and rural communities in pedagogical and memory projects and rituals. "In Chile, *poblador* refers to the poor who reside on the outskirts of large cities, or in *poblaciones*. Historically, *poblaciones* have been associated with land seizures that were conducted by left and center political parties since the 1950s, although nowadays the terms are used more widely to refer to poor or low-income neighborhoods." Nicolás Angelcos and Juan Pablo Rodríguez López, "Amplifying dignity in the neoliberal city: the Pobladores movement in Chile,"

Social Movement Studies, January 2023, n. 1.

33 See Johan Reinhard and María Constanza Ceruti, *Inca Rituals and Sacred Mountains: A Study of the World's Highest Archaeological Sites* (Los Angeles: Cotsen Institute of Archaeology Press, 2010).

34 See Arturo Escobar, *Pluriversal Politics: The Real and the Possible* (Durham, NC, and London: Duke University Press, 2020), 26.

35 Vicuña, "Entering," in *Unravelling*, 4.

36 The two men who excavated the pirca, Gerardo and Jaime Ríos, were led by Guillermo Chacón, who could not ascend to the summit due to his age and health. The Centro de Estudios Antropológicos de la Universidad de Chile contributed to the museum's acquisition of the mummy. See Cecilia Moraga Pinda, "Viejos actores, nuevas relaciones, nuevas prácticas. El niño El Plomo, Universidad de Chile," a master's thesis that includes a thorough bibliography and photographs from the time of the discovery. For a narrative from the perspective of one of the *arrieros* see Luis Ríos Barrueto, *El niño Inca. La verdadera historia del niño del cerro El Plomo* (Santiago: Pehuén, 2009).

37 "Esta montaña ha sido un centro sagrado al que dedicaron sus actividades religiosas, los pueblos de la última época precolombina." Luis Krahl T., "La momia del Cerro el Plomo," *Apartado del Boletín del Museo Nacional de Historia Natural* XXVII, no. 1 (1957): 95.

38 See Álvaro Sanhueza, Lizbet Pérez, Jorge Díaz, et al., "Paleoradiología: estudio imagenológico del niño del Cerro El Plomo," *Revista chilena de radiología* 11, no. 4 (2005): 184–90.

39 The body was displayed in the museum until the 1980s, when it was replaced by a replica to facilitate its preservation.

40 Besides reproducing a photograph of the boy and the site's location in the 1957 report, Dittborn also reproduced the drawing that Grete Mostny of the Museo Nacional de Historia Natural made when she saw the body for the first time.

41 See Valería de los Ríos, "Marks of Travel: Strategies in Eugenio Dittborn's Airmail Paintings," *Image and Narrative* no. 14 (2006). Examples include the silk-screen-on-cardboard *Mount Everest* (1979) and the airmail paintings *Trasandina, Airmail Painting no. 40* (1985), *The Corpse, the Treasure, Airmail Painting no. 90* (1991), and *Dust Clouds, Airmail Painting no. 99* (1992).

42 Vicuña, "Dí vida," unpublished draft.

43 "Nuestra gente dice 'debemos tener amor por todo lo que está en torno de nosotros,' porque nosotros somos la naturaleza, pero no el centro.... Hay un orden natural, que no es el que pretende instalar la denominada 'autoridad'; el orden natural es el que nos muestran los árboles en el bosque, las piedras y—reitero—las estrellas." Elicura Chihuailaf, speech to the Chilean senate, December 2021. Available online at https://cultura.fundacionneruda.org/wp-content/uploads/2022/01/cultura-fundacion-neruda-discurso-elicura-chihuailaf-homenaje-senado-2021.pdf (accessed April 9, 2024).

44 "Que... la Nueva Constitución, ponga como centro la necesidad de la conversación; esta conversación que se ha ido perdiendo, que tendría que estar sobre la mesa de las autoridades en primer lugar, para resolver, como ya se dijo, los conflictos, sobre todo cuando se trata de las instituciones del Estado, el cual, recordemos, usurpó un territorio. Y eso requiere una reparación y

requiere, sobre todo, una conversación, que estamos esperando." Ibid.

45 Barros, "Aguas y ríos," 544. My translation.

46 Vicuña, "Dí vida," unpublished draft. The choice of the word "sin," from the Christian ethical vocabulary, suggests that Vicuña is wondering how these ritual actions are to be valued today, using whose terms and definitions.

47 Camila Marambio, in "Cecilia Vicuña," interview, *Errata: Revista de Artes Visuales* no. 15, *Performance, acciones y activismo* (September 2017): 211. Available online at https://revistaerrata.gov.co/contenido/cecilia-vicuna (accessed April 9, 2024).

48 Laura E. Pérez, "The Performance of Spirituality and Visionary Politics in the Work of Gloria Anzaldúa," *Eros Ideologies: Writings on Art, Spirituality, and the Decolonial* (Durham, NC, and London: Duke University Press, 2019), 137.

49 Vicuña, *Enseñar: poner en señas*, 1978, from the series *AMAzone Palabrarmas*. Available online at https://www.moma.org/collection/works/429409 (accessed April 9, 2024).

50 Vicuña, *Unravelling*, 114.

51 Magda Sepúlveda, "Una Neovanguardia hippie: Cecilia Vicuña," in Meredith Gardner Clark, ed., *Vicuñiana. El arte y la poesía de Cecilia Vicuña, un diálogo sur/norte* (Santiago: Editorial Cuarto Propio, 2015), 66.

52 See Candice Hopkins, "On Other Pictures: Imperialism, Historical Amnesia and Mimesis," in Greg A. Hill, Hopkins, and Christine Lalonde, *Sakahàn: International Indigenous Art*, exh. cat. (Ottawa: National Gallery of Art, 2013), 30.

53 As late as 1995, Vicuña recalled the sound recordings of Lola Kiepja, supposedly "the last Selk'nam," made by the American anthropologist Anne Chapman in 1966, and lamented the culture's predicted disappearance. See the transcript of Vicuña's performance at the Poetry Project, St. Mark's Church, New York, on May 6, 1995, in *Spit Temple. The Selected Performances of Cecilia Vicuña*, ed. and trans. Rosa Alcalá (New York: Ugly Ducking Press, 2012), 129–43. Vicuña also sent me an unpublished text, "Selk'nam en mi vida," in which she mentions reading an article in *El Mercurio* in 1966 about the death of the "last" Selk'nam, and registering the disdain shown to the culture in the newspaper. Vicuña, email to the author, December 22, 2017.

54 "La población araucana está calculada, no, no está calculada. Los gérmenes indios van a brotar, los araucanos se van a levantar. Están esperando para empezar a conquistar su propia manera de existir." Vicuña, *Saborami*, 1973 (reprint ed. Oakland and Philadelphia: Chain Links, 2011), 44. *Saborami* includes *El Cuaderno Café* and other works.

55 See James Clifford, "The Others: Beyond the 'Salvage' Paradigm," *Third Text* 3, no. 6 (1989): 73–78.

56 "La semilla es la palabra de la tierra." Seeds were at the center of the exhibition *Semiya*, in Santiago in 2000, where Vicuña wrote, "¡Sólo un gesto colectivo de amor podría parar la destrucción, la tala y el incendio de los bosques!" (Only a collective gesture of love could stop the destruction, logging, and burning of the forests!) Vicuña, *Semiya*, exh. cat. (Santiago: Galería Gabriela Mistral, 2000), n.p. In the period of Salvador Allende's government, Vicuña created a seed bank at her parents' home and proposed to the presidency the idea of reforesting urban spaces. She also imagined storing autumn in a plastic bag as "an act

of love," and ended up filling a room in Santiago's Museo Nacional de Bellas Artes with dry leaves (*Otoño*, 1971), among other somatic experiments concerned with a coexistence of times and knowledge (what can a seed carry forth, and for how long? How long does plastic last?). At the time, she was also making paintings of her poems in which animals abound, especially sacred ones (the snow leopard, the jaguar).

57 The image comes from Herbert Wendt's popular history *It Began in Babel*, published in 1958 in German and in 1961 in Spanish. Vicuña, email communication with the author, 2017.

58 The phrase appears in *QUIPOem* and resurfaced as a poem scribbled on a wall in *Cecilia Vicuña: Spin Spin Triangulene*, an exhibition at the Solomon R. Guggenheim Museum, New York, in 2022.

59 Liisa-Rávná Finborg, "The Story of Terra Nullius: Variations on the Land(s) of Saepmie That Nobody Owned," in Finborg, Katya García-Antón, Beaska Niillas, et al., *Čatnosat: Indigenous Art, Knowledge and Sovereignty*, exh. cat. (Amsterdam: Valiz, for the Samí Pavilion, Venice Biennale, 2022), 4.

60 Vicuña, "Text of the Brown Book," in *Saborami* (Oakland and Philadelphia: Chainlinks, 2011), 44. The "trip" on which Vicuña says she discovered Chile was a "mental trip" to Tierra del Fuego as she worked on *El Cuaderno Café*. Communication with the author, 2022.

61 Vicuña, "Antivero," 1981, in *Precario/Precarious*, n.p.

62 As Barros notes, Vicuña performed a DNA test that "revealed her Diaguita origins." Barros, "Aguas y ríos," 545. I borrow the "not only" here from de la Cadena.

63 Leanne Betasamosake Simpson, *Dancing on Our Turtle's Back: Stories of Nishnaabeg Re-Creation, Resurgence and a New Emergence* (Winnipeg: ARP Books, 2011), 122.

64 Billie Jean Isbell and Fredy Amilcar Roncalla Fernandez, "The Ontogenesis of Metaphor: Riddle Games among Quechua Speakers Seen as Cognitive Discovery Procedures," *Journal of Latin American Lore* 3, no. 1 (1977): 19–49.

65 Edgar García refers to the Quechua-language concept of *quilca*, "a conceptual tool for understanding representational heterogeneity." Although he relates the term to khipus and to Vicuña's interpretation of them in particular, the term might apply to her practice as a whole. García, *Signs of the Americas. A Poetics of Pictography, Hieroglyphs, and Khipu* (Chicago and London: The University of Chicago Press, 2020), 193, 204.

66 Vicuña, *AMAzone Palabrarmas, 1977/1978* (New York: Circadian Press, 2018).

67 "Let the creek lull you to sleep, my love."

68 Vicuña, "About to Happen," in *About to Happen*, 102.

69 Vicuña has spoken repeatedly of this question of the unfinished work; see Bryan-Wilson, "Awareness of Awareness." The relevant installations and videos have been carefully analyzed by Macarena Gómez-Barris, Candice Amich, Regina A. Root, and others.

70 Richard Schechner, "Restoration of Behavior," *Between Theater & Anthropology* (Philadelphia: University of Pennsylvania Press, 1985), 36.

71 Marcel Mauss, *On Prayer*, 1909, Eng. trans. Susan Leslie, ed. W. S. F. Pickering (New York and Oxford: Durkheim Press/Berghahn Books, 2003).

72 In 2020, an environmental court ordered a "total" closure of the mine.

The ruling was confirmed by the supreme court in 2022. The Barrick Chile company has been engaged in a process of dismantling its facilities (a process that would take ten years) that it claims will return the site to its original state, while still considering alternatives for a new project at the same site. See Isabel Amos-Landgraf, "Chile's Pascua-Lama Mine Legally Shut Down, but Mining Exploration Continues," *State of the Planet: News from the Columbia Climate School*, January 15, 2021. Available online at https://news.climate.columbia.edu/2021/01/15/pascua-lama-mine-shut-down/ (accessed May 10, 2024).

GLIMMERS OF ANOTHER LIFE

"Compulsive users/perpetual customers/the planet exploded/the modern triumphed/Your mind will grow cloudy/with packaged dreams/Your strengths will atrophy/via designed services/... Your body will soften/with sweetened sodas./Your time will be extinguished/via motorized transit. There is nothing exterior to capital/condemned people," sings the artist Naomi Rincón Gallardo in *Insaciables* (Insatiable), one of the videos in her multimedia installation *Odisea Ocotepec* (*Ocotepec Odyssey*, 2014). Wearing a galactic suit that emulates an axolotl—a salamander, currently at risk of extinction, that is venerated by Mesoamerican Indigenous people—the artist sings in a landscape of clouds, accompanied by Latin American folk instruments: a guitar, a quena, a charango. Her voice is recorded to sound as if it has traveled from far away in space, like an echo of another world. Alongside her, three female figures with facial hair and sparkling suits form a carefully choreographed musical, as if they were embodying the choir of a futurist drag church.

The work is a hypnotic voyage. Through references to 1960s Mexican science fiction, Rincón Gallardo presents the journey of an intergalactic axolotl come back to Earth, inviting us to look at what we purposely choose to ignore: that the capitalist model of infinite growth is incompatible with a finite planet. The songs return constantly to the aridity of monoculturalism and to the way an extractive vision has taken hold of our way of creating connections in the world. At moments the work resembles a pilot for a children's program geared toward creating underworlds and subterranean forms of awareness and solidarity against the current neoliberal "common sense."

The election of the axolotl is not an accident. The name of this animal comes from Nahuatl, and is made up of the words *xólotl* (monster) and *atl* (water). In the sacred universe of the Nahua world, Xolotl is the god of twins, of twilight, of misfortunes, of sickness, and of deformities. The god's physical anatomy—an amphibian with the body of a fish, and with frog's legs that tend to remain at a larval stage—has generated fascination in literature and science; in recent decades it has also become an important icon for the queer community. The axolotl's capacity to change colors and camouflage itself, to regenerate tissue and develop lungs and gills, has made it an emblem of nonnormative forms of life and eccentric creatures that challenge the modern/colonial capitalist world system and heteronormative social structures.

Odisea Ocotepec has also taken the form of live concerts and a vinyl record. The title takes the name of the town of Ocotepec, a locality of Cuernavaca, Mexico, where the Austrian priest, philosopher, and educator Ivan Illich founded the Centro Intercultural de Comunicación (CIDOC) in 1966. In the late 1960s, Illich posited a withering critique of consumerist society, modern institutions—educational and medicinal—and the myth of a technology (with no limits) as liberator. He proposed that such tools should be at the service of the collective rather than of specialists. "Such a society, in which modern technologies serve politically interrelated individuals rather than managers, I will call 'convivial,'" Illich wrote in 1973, offering an alternative of a reality increasingly shaped by a technocratic paradigm.[1]

Through the CIDOC (1966–76), Illich promoted exchange among intellectuals and students with the goal of reimagining the relationship between teaching and learning. Rincón Gallardo's work establishes a performative dialogue with Illich's ideas and explores other genealogies of political and social experimentation that have taken place in Cuernavaca, like the development of psychoanalytic therapies using audiovisual stimulation and hallucinogenic drugs, or the ideas of the Brazilian philosopher and educator Paulo Freire about transforming the world through action.[2] In *Desescolarización* (Unschooling), another piece from *Odisea Ocotepec*, a children's chorus sings a song with a musical arrangement recalling protest songs by Latin American revolutionary folk musicians Víctor Jara and Mercedes Sosa: "White-collar criminals/in overeducated academies/obese from capital/define our curricula./We don't want education/no more obligatory school/or prefabricated instruction/for those who are not capable./School discourages us/from seeking important knowledges/classrooms with empty promises/useless certifications." The sweetness of the voices makes for an even more glaring denunciation in the face of education become merchandise. In the video *Verbo alienante* (Alienating verb), Rincón Gallardo revisits the premises of Freire's "pedagogy of liberation." Two bearded women appear singing and dancing on an unknown planet, vocalizing through a distorted voice the effects of what the Brazilian philosopher called the "banking system of education"—a capitalist educational paradigm in which students are seen as passive receivers of knowledge.

Odisea Ocotepec also confronts the perpetuation of colonial structures arising out of the violent process of conquest, extraction, and land dispossession in the territory we today call Latin America. That process, which began in the fifteenth century, resulted in the prevalence of the idea of race and in a new pattern of world power, the tangible consequences of which can be seen in the way racialized bodies have been considered undercitizens and still struggle today for the right to self-determination. They have also developed modes of protection and resistance. The Maya K'iche' thinker María Jacinta Xón Riquiac signals that the vitality of vernacular and Indigenous knowledges does not depend on verification from the so-called "true sciences" and their scientific method, but rather on what she calls an "endogenous epistemology"—that is, "the validity of knowledge of and for collectivities and future generations, underscoring its ontological nature within which there is no separation between the material and the natural, the social and the spiritual."[3]

In a similar way, Rincón Gallardo's work explores alternatives for the future of technology, as well as forms of cosmic speculation grounded in popular and Indigenous traditions that from the perspective of Western rationalism would be considered uncivilized and premodern. *Odisea Ocotepec* reminds us that the devastation of our planet is not a collateral effect of Western discourses around progress but rather the engine of its functioning. These videos and songs offer a psychedelic journey of surrealist delirium, rage, joy, and desire as a ritual for our collective reanimation, presenting environmental catastrophe not as a phenomenon restricted to

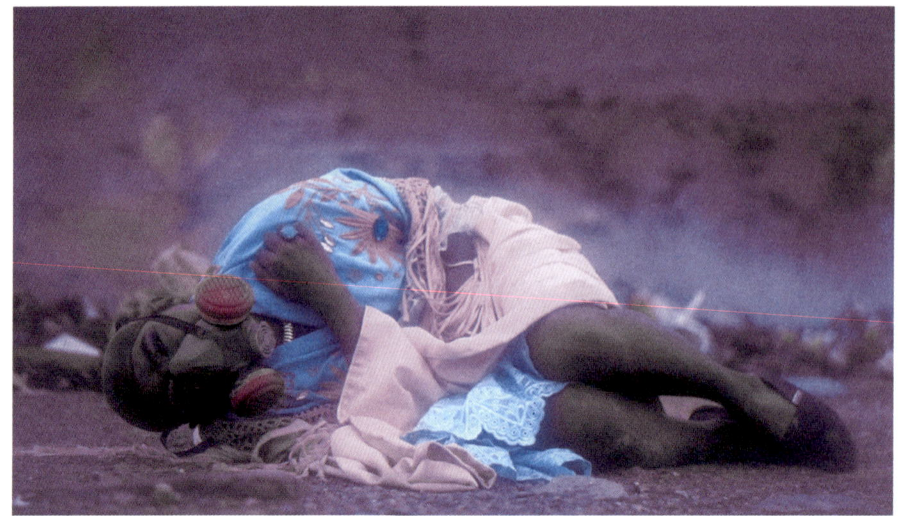

1 Bartolina Xixa. *Ramita seca: la colonialidad permanente* (Dry twig: the permanent coloniality). 2019. High-definition video, color, sound, 5:07 min. Collection the artist

the present but as a centuries-old process, first experienced by Indigenous peoples and other colonized communities.[4]

In 2017, the Argentine artist Maxi Mamani created the "diverse drag queen" Bartolina Xixa. The name pays homage to the Aymara leader Bartolina Sisa, who fought alongside Túpac Katari to liberate Peru from Spanish occupation in the second half of the eighteenth century. For Mamani, the creation of Xixa was a way to challenge Western LGBTQI+ discourse—to try to dismantle the fiction of total whiteness associated with Argentine identity, which erases histories of Andean migration from Peru or Bolivia (like that of Mamani's own Coya family) and hides the long-term effects of land dispossession resulting from settler colonialism.[5] "To bring [Bartolina Xixa] into the present from within my fag being [is an attempt] to anchor those struggles so we can say this isn't a fifty-year fight, but an ancestral one," Mamani has said.[6]

Xixa's choreography conveys the reality of a lost and devastated world. In collaboration with the audiovisual artist Elisa Portela, she created the music video *Ramita seca: la colonialidad permanente* (Dry twig: the permanent coloniality, 2019; fig. 1), a version of a song by the charanga player and composer Aldana Bello.[7] The original piece is a *vidala*, a musical form with Quechua roots that has existed in northern Argentina for more than two centuries. It is sung by Bello, the trans performer and writer Susy Shock, the singer and percussionist Mariana Baraj, and the hip-hop singer Sara Hebe. Bello's lyrics are compelling:

> The earth pocked with holes, imploded, tortured
> patriarchy, patriarchy, patrolled to privilege the few
> lots of money for just a few
> for everyone else nothing at all.
> Men and women born to ruin it.
> Dead land all through Argentina
> glyphosate even in our soup
> what I eat pollutes.
> This deformed humanity brings me shame and brings me
> laughter

take from it demand of it give it all they got as they can
When all this explodes another life will be heard.

In the video, Xixa wears the clothing of a Bolivian Andean chola—a pink skirt, an elegant pale blue poncho embroidered with flowers, a black porkpie hat, and two long braids decorated with tassels. The setting is a huge garbage dump. At the beginning of the song, Xixa appears lying on her back, while her body responds to the drumbeats as if they were her own heartbeats. The piles of steaming trash suggest an earth that is no longer habitable. She looks steadily at the camera before performing a series of movements as "ancestral remedies to colonial extractivist violence."[8] The gestures she makes with her hands lead to thoughts of drought, the devastation of the harvest, the land become wasteland. Xixa beats the earth to the rhythm of the percussion as if she were trying to reanimate an inert body, seeking to transfer to it the energy that barely allows seeds to germinate in the wake of capitalist ruin. In the final moments, she dances wearing a gas mask, then convulses on the ground.

Instead of offering a horizon of change and hope—all too often functional elements of green liberalism, with its false solutions in the face of climate catastrophe—Xixa embraces inequality and dispossession in order to underscore not just the consequences of human addiction to fossil fuels but also the relationship between the extraction and exploitation of feminized bodies and that of natural resources on our planet. What drag bodies do is "put the established world in doubt... even from within the most absolute lack of power," notes the transgender educator and activist Merlene Wayar.[9] Xixa uses dance as a ritual technology for communication with the earth, an empirical science that has functioned for millennia as an emotional and sensual language to express gratitude, offerings, and healing. Reclaiming dance as a technology, Mamami seeks not technical perfection but rather a recognition that our bodily movements are genuine forms of speaking and showing reciprocity with the land we inhabit.

A different form of dialogue with the earth appears in the work of Dan Lie. Insistently exploring the relationship between life and death, Lie questions the way the condition of dying is customarily separated from daily life, and how, from a hegemonic vision of the human, dying represents the end of our ability to sense the world. "But what if it's the opposite? What if, at the precise moment we die, the senses begin to expand in a crescendo? What if we manage to amplify our sensorial cells, if the floor becomes an extension of our skin, the wall our ears, the ceiling our vision, the neighboring city our taste, if the smell of a distant continent invades us, if we feel the touch of the deepest layer of the subsoil?" they wonder.[10]

Lie calls their works "living entities." To speak of "entities" moves us from the expectation of an inert object—the things that museums customarily collect and exhibit—toward an experience of art as cohabitation and codependency. Their work is a confluence of seeds, mud, flowers, fruits, natural fibers, water, terra-cotta pots, mineral remains, rocks, coal, and millions of biological entities. It is a dialogue with spiritual presences and energetic forms that are difficult to name from within the matrix of Western thought. Lie's installations unfold bodily sensations that move beyond the purely visual. The decomposition of the organic matter over days or weeks allows for a range of insects, spores, liquids, and odors to expand, generating unexpected dynamics for art spaces that are not accustomed to dealing with living organisms.

Lie created *Human Supremacy: The Failed Project* (2019; fig. 2) for Casa do Povo (House of the people), a São Paulo art space that works with cultural and social organizations to incubate collaborative projects. Invited to work in the garden, Lie opted to escape from that space, which they saw as domesticated and colonized, and instead to occupy the building's interior. They installed two works at the far extremes of the building, one in a disused basement theater, the other in a small room next to a terrace. Both works hung from the ceiling, activating erotic connotations. In the basement, a form made of hay tied with vegetable-fiber cords, and covered in mushrooms (*Lyophyllum shimeji*), floated in the center of the abandoned theater's stage, lit from the front, displayed before the empty seats. On the upper floor, two cloths dyed with turmeric hung around a large vertical bag of jute, covered in soil and flaxseeds some of which gravity had brought to the ground. With no ability to control or predict the results of the processes of nutrition,

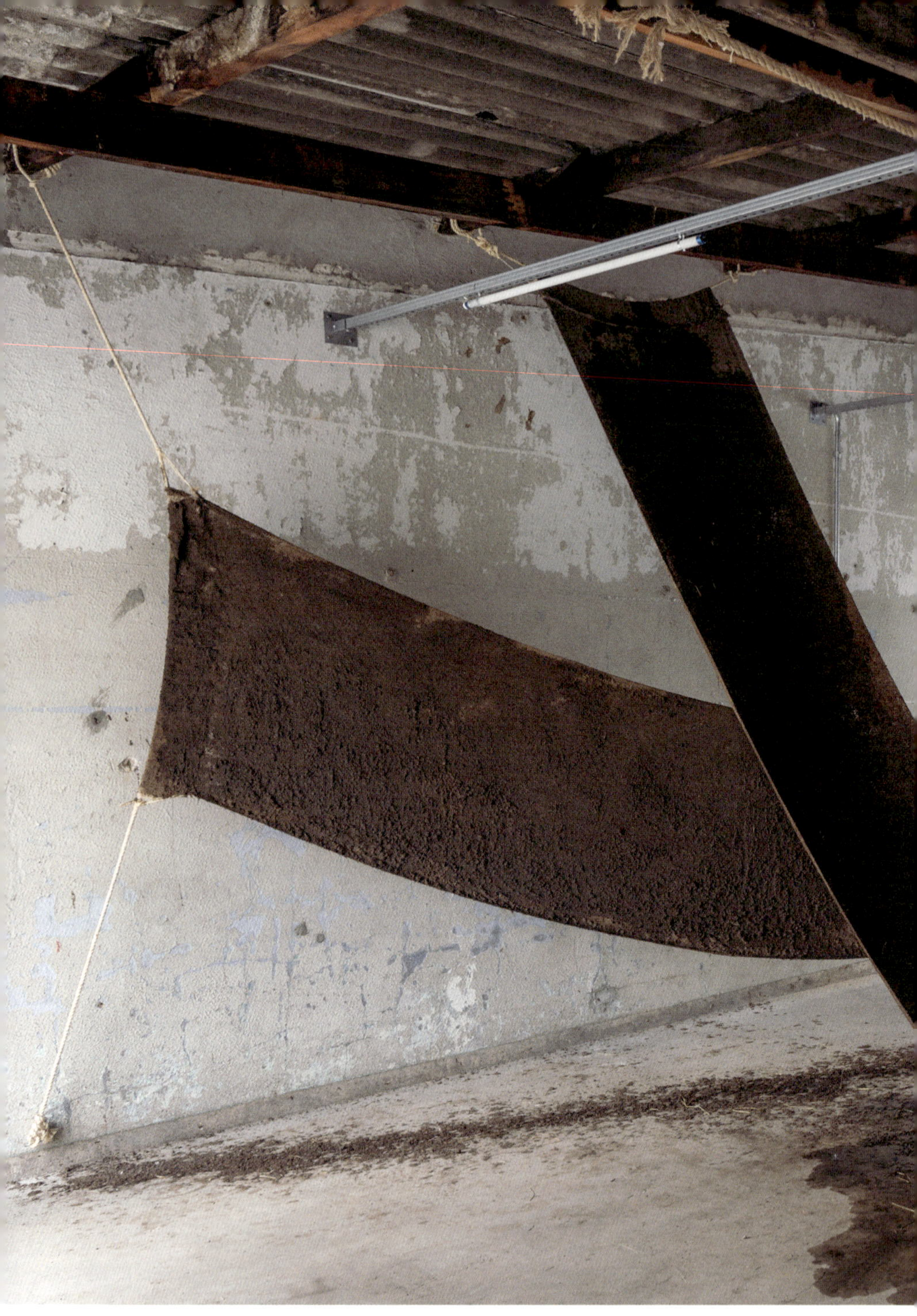

2 Dan Lie. *Human Supremacy: The Failed Project*. 2019. Cotton fabric naturally dyed with turmeric, jute fabric covered with mud and linseed, straw with fungi (*Lyophyllum shimeji*) and ropes. Site-responsive installation at Casa do Povo, São Paulo, Brasil, on the third floor by the terrace

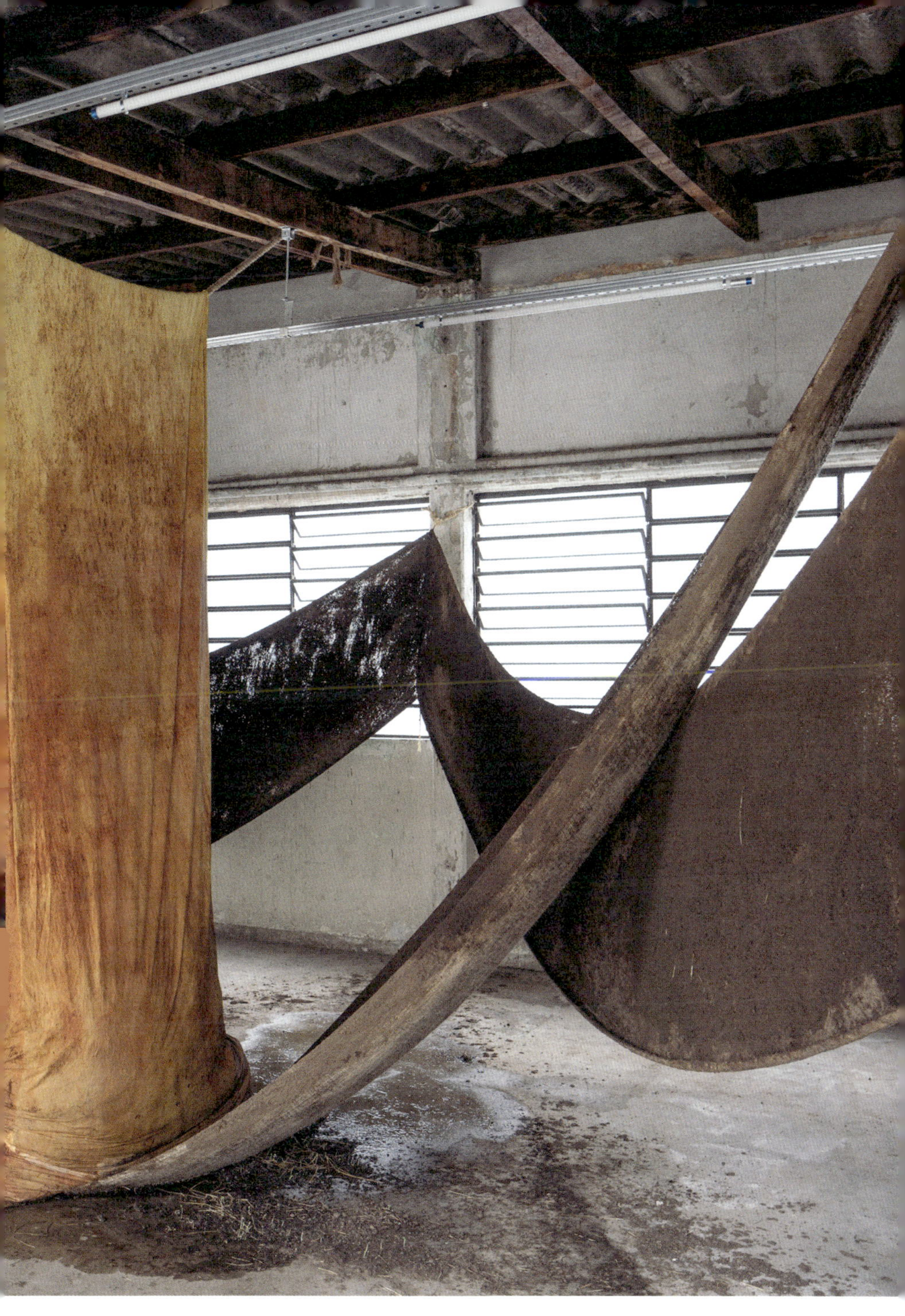

decomposition, and regeneration of organic matter, the artist created works that escaped their control and that of the institution.[11] While the fungal and microbial universes were imperceptible in terms of the day-to-day workings of the institution, both pieces drew attention to the various ecosystems and more-than-human forms of life that inhabit Casa do Povo. Subtly but effectively, Lie posed a question about which "people" it was that the Casa do Povo served: why not reimagine the institution's work horizon from a microbial perspective, from beyond the human viewpoint?

More recently, Lie has brought their living entities to the heart of metropolitan museums such as the New Museum in New York. Their installation *Unnamed Entities* (2022) challenged the authority of Western institutions through the simple poetic act of creating the conditions for organic activities of biodigestion, both metabolic and somatic. Could the proliferation of mold and bacteria be a way to exorcize the white colonial legacy of museums? Perhaps the dispersion of those microorganisms might open the possibility of making those spaces habitable not only for those who have historically been expelled from them—Indigenous, racialized, nonbinary bodies—but also for other forms of life. In this way putrefaction signals toward the personal, political, and sexual dimension of those human lives considered undesirable, sick, damaged, contaminated, useless, and disposable: all those bodies that do not efficiently comply with the productive demands of capitalism.

A number of Lie's works have included terra-cotta pots, as in *Death Center for the Living* (2017) and *Being Alone, Together* (2019). These handmade vessels, associated with medicinal, scientific, aesthetic, and ritual uses, are arranged in connection and in proximity to other organic elements. They usually contain a liquid mixture of fermenting rice, evoking memories of *masato*, an Amazonian beverage drunk in such countries as Peru, Colombia, and Brazil, and also the tradition of fermented foods such as Indonesian *tapai*. Developed by ancient cultures all over the world—from Asia to Africa, from the Middle East to South America—these clay vessels have been associated with agriculture, food preservation, traditions of eating, funerary rites, musical experimentation, and a range of different art forms. They function as an encounter with a "cluster of intelligences and not just human intelligence."[12]

In the face of the Western vision that understands technological development as a way of dominating the planet, Lie seems to envision an erotic vegetal-animal-bacterial-technohuman promise. They remind us in no uncertain terms that true emancipation will not come from the human hand but from the damp seed; liberation will not be shaped by the brain but by nonhuman technologies engendered by the spread of mold; real ecological reproduction and regeneration will grow not from ejaculation but from liquids let off by the oxidation, comminution, and mineralization of organic matter.

The artist Seba Calfuqueo, of Mapuche origin, has been exploring ways of identifying transgender and nonbinary bodies through the universe of water. In recent years their work has been devoted to researching the impact of neoliberalism on collective life. In the context of the Chilean state's refusal to recognize the colonial genocide of Indigenous communities—not only of the Mapuche but also of the Kaw'skar, the Selk'nam, and the Yánamas—Calfuqueo's work forms part of a broad process of resistance demanding respect and territorial self-determination. Diverse groups have risen up to defend the Pilmaiquén, Muco, Truful Truful, and other rivers in the face of the threat represented by dams and hydroelectric plants in the Araucanía region and elsewhere, even though these protests have been criminalized through an antiterrorist law inherited from the military dictatorship of Augusto Pinochet.

Calfuqueo's contemplation of water is also associated with recuperating the territorial memory of the Mapuche, who inhabited an area they called Wallmapu that extended from the Pacific to the Atlantic in what is today southern Chile and Argentina. Their works are paths toward affective recognition of those spaces. In the video *Tray Tray Ko* (2022; fig. 3), the artist returns to a waterfall carrying a large brilliant-blue cloth with small bells hanging at its farthest edge. Wearing a blue skirt made of the same material, Calfuqueo walks with their back to the waters, as if a magnetic energy were guiding their displacement. Their movement in reverse complicates the temporal and spatial meanings of the straight line of Western history and its expectation of ongoing progress. The video underscores the importance of the *trayenko*, a word in the language of

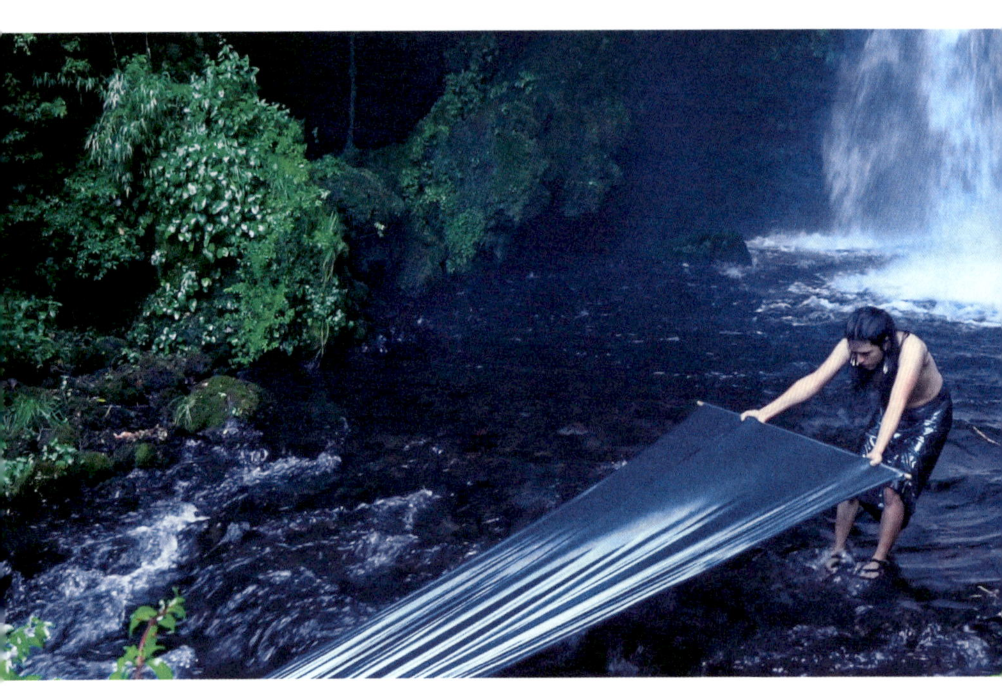

3 Seba Calfuqueo. *Tray Tray Ko*. 2022. Video performance, color, stereo sound, 6 min. Institute for Studies on Latin American Art (ISLAA), New York

the Mapuche people, Mapudungun, that means "cascade," "water running," or "stream"—sacred places for the Mapuche, functioning as stores of *lawen* or medicinal knowledge. The open composition and the sound of the river and the bells together construct a ritual atmosphere, framed by the rhythm of the artist's slow steps, which announces the reencounter between body and water. Like a great liquid veil, the cloth shimmers, while its elasticity allows it to take the shape of whatever it covers, be it grass, tree trunks, stones, or the river itself. Calfuqueo's reverse orientation proposes a kinetic deviation that impacts both perception and the construction of reality.

In *Kowkülen* (Liquid being, 2020), Calfuqueo represents themself once again adhered to the landscape, surrounded by trees, covered with the water of rivers and lakes. Unlike *Tray Tray Ko*, however, *Kowkülen* presents the artist's naked body as almost inactive, tied up and hung with blue cord. The video begins with a text that points to the enforcement of the *Código de aguas* (Water code), written in 1981, during the Pinochet dictatorship, which declares that water is an asset that can be privatized. In the first images we see their back illuminated by the sun's rays. Although at various moments the body is still enough to seem abandoned, at others this very immobility is charged with warm erotic energy, especially when Calfuqueo floats on the water, tied to the trunk of a tree, as if in some BDSM ceremony with nature. At the bottom of the screen, a text in a combination of Mapudungun, Spanish, and English speculates on the possibility of becoming liquid and escaping from the dominant Western heteronormative imaginary: "The history of the peoples/Settled historically/*Inaltu lafken mew*/Next to the water/I have been here/In a liquid state/Running through different basins. *Ngen ko, Arüm ko, Ngürü fily, Kay kay*,/Witnesses of this journey. . . . I want to be a fish/Without a sex to be reckoned with/As *shumpall*, interstice of man, of woman/Non-binary waters."[13] Their words also evoke the role of the Ngen-ko—spirit-possessors of water that accompany, fertilize, and defend wells, arroyos, rivers, lakes, and oceans, uplifting the healing and cleansing dimensions of water, which manifest despite human violence.

Calfuqueo's works suggest that to feel the water is a way of *hearing* the land. A communication technology appears in them that does not rely on verbal skill or Western scientific verification. To approach the language of the waters, which reverts time, might even allow us to imagine what the artist calls an "Indigenous futurism": "We've always been denied the present, and in this sense, the future isn't even a possibility."[14] Here the future is a body that returns to the earth to be with mushrooms, lichen, rocks, moss, and spores.

The recent work of the multidisciplinary artist Elena Tejada-Herrera has focused on the connections between coloniality, pollution, gender inequality, and economic violence. In the 1990s the artist worked in Lima, Peru, then in 2001 won a grant to study in Richmond, Virginia, and lived in the United States for almost twenty years before returning to Peru. Over these decades, her work responded to sexist and patriarchal visual culture through multiple and relentless forms of representation of her own sexual desire, advancing a feminist appropriation of technology as a counterinformation tool while exploring the notions of force and vulnerability as places to imagine a different ethical agreement.[15]

Exploring a genealogy of self-defense grounded in radical feminism, Tejada-Herrera has recently been paying attention to various forms of female physical resistance. The three-channel video installation *They Sing, They Dance, They Fight* (2020), commissioned for the eleventh Berlin Biennale, presents a dreamlike fiction that intertwines scenes of a self-defense workshop, a woman performing the scissor dance (a ritual dance from the Central South Andes that has its roots in the anticolonial Indigenous movement Taki Unquy, usually reserved for men), and a transgender mermaid singing to collective resistance. Tejada-Herrera presents self-defense not as an individual, corporeal reaction but as practices of self-organization that involve political fabulations and more-than-human entanglements.

The artist has been also addressing the devastating effects of extractive activities on bodies and land—a violence that weighs especially on BIPOC women—and corporate impunity from the consequences. The single-channel video *Las Bambas* (2019) is named after a copper mine in the Andean province of Apurímac, in central Peru.

The mine faced intense resistance from local campesino communities from the time its operations began, in November of 2015; these protests have only expanded—the company has more than once been forced to stop work. Tejada-Herrera recorded two scenes that are superimposed on each other in the video: on one side we see two mothers with their daughters, dressed in dazzling sequined outfits, caressing each other tenderly and playfully; on the other, images of a forest, an arroyo, a waterfall, and vegetation in flames, seen from a drone above them. The digital editing that mixes both registers does not obscure its artifice, showing the blurry edges of bodies that sometimes appear incomplete.

The women's gestures of affection and care are accompanied by an off-screen narration in which a girl's voice recounts the history of the mine. Her tale begins with the origin of the Chalhuahuacho district, where the mine is located; the district was created in 1994, during the dictatorship of Alberto Fujimori, with the goal of "favoring extractive ventures, such as mining."[16] Between 2007 and 2015, Chalhuahuacho attracted a large number of miners and other extraction workers and specialists, creating a context of labor disadvantage and pay inequity for local people, who additionally experienced limited access to health care and education. This process quickly devolved into a dynamic of exploitation, forced labor, and displacement that in turn led to the militarization of the area. The story told in *Las Bambas* underscores how men left their homes and moved to the area immediately around the mine, while women were responsible for the land yet lacked the right to own it, a situation that impacted their ability to support themselves through independent agricultural production. The mine also polluted the land, and the wealth it generated was not distributed and did not lead to positive changes in the lives of the workers. At the same time, the mutual caresses between mother and daughter bring attention to the social organization of care work, whose invisibility continues to be instrumental to the patriarchal structures and accumulative dynamics of capitalism. "Domestic work emerged as the key engine for the reproduction of the industrial workforce . . . according to the requirements of factory production," the Marxist feminist Silvia Federici reminds us, addressing how reproductive work has been subsumed to capitalist accumulation since the first phase of capitalist development.[17]

The contrast between the oral narrative about extractive violence and the shots from above of the women playing at getting to know one another with their hands is disturbing. The rich colors of their clothes—green, black, blue, and gold—stand in tension with the images of vegetation and water. Little by little we begin to understand that the mothers and daughters are cleaning each other's bodies: Tejada-Herrera progressively saturates the clothes' colors until they suggest radioactivity, the dazzling dresses coming to represent petroleum and precious metals against the women's skin. The shine of heavy metals—lead, mercury, cadmium, arsenic, chrome—appears in the video as a membrane that twinkles and seduces. The work seems to suggest that the aesthetics of glamour associated with the false neoliberal discourse of progress and entrepreneurial spirit perversely camouflage the ongoing process of exploitation and poisoning of bodies and lands, a foundational element of the modern colonial capitalist project.

Each in its own way, the work of the aforementioned artists delves into how territories, bodies, and epistemologies generate technologies and tools capable of opening and intervening in the world, whether through popular science fiction as a form of cosmic speculation or through other forms of empirical science associated with dialogue with the spirits of the waters, of nature, and of more-than-human forms of existence. In Rincón Gallardo's *Odisea Ocotepec*, Bartolina Xixa's *Ramita seca (la colonialidad permanente)*, and Tejada-Herrera's *Las Bambas*, body movement and choreography make us aware of the brutality of neoliberal capitalism and its recent extractivist offensive. Lie's living entities and Calfuqueo's reencounters with the waters demand an understanding of ecological justice free of Western binary logics, of oppositions such as masculine/feminine, life/death, organic/inorganic, and human/nonhuman. In Rincón Gallardo's words, "We claim a theory enmeshed with the body, a knowledge intertwined with nature and matter: a theory that can be danced, smelled, touched, away from the arrogant affirmation of superiority of certain forms of abstract knowledge that create epistemic and political erasure."[18]

The urgencies and desires that mobilize these pieces are important at a moment when the global hegemony around discussions of technology seems to center on a new space race—one that is white and patriarchal, a direct legacy of the colonial-era aspirations based on conquest, extraction, and slave labor. This development is clearly visible in the proliferation of magnates who dream of colonies on Mars, like Jeff Bezos and Elon Musk and their infamous phallic rockets. Plagued with false promises of capitalist technological liberation—as Illich noticed in 1973—this scenario comes hand-in-hand with imageries of deceptive harmony that conceal the colonial past and present. The Martinican philosopher Frantz Fanon described this type of violence precisely as a situation in which the levels of threat and risk that we are all exposed to are not the same for everybody and there is an established global order for those varying conditions of vulnerability. There are lives that are protected and whose defense is taken for granted and is unquestionable and there are others whose existence is permanently under attack. For Fanon, that dehumanization meant the constitution of two distinct species living on two different sides of the colonial reality: those in the *zone of being*, whose racial, class, sexual, and gender privileges entitle them to navigate the rule of law and civility; and those in the *zone of nonbeing*, who have been expropriated of their human condition and are subjected to acts of violence that would otherwise be unacceptable. Fanon saw, though, that the subaltern side of the colonial difference grants the possibility of revolutionary agency: "There is a zone of nonbeing, an extraordinarily sterile and arid region, an utterly naked declivity where an authentic upheaval can be born."[19]

The reality Fanon describes can be linked to what the Peruvian sociologist Anibal Quijano has called the "coloniality of power": the pattern of domination and forced exploitation forged since the invasion and colonization of the Americas, associated with European modernism and its processes of primitive accumulation.[20] As Quijano and other decolonial thinkers have signaled, modernity and coloniality are two sides of the same coin.[21] Against this backdrop, the works of Rincón Gallardo, Xixa, Lie, Calfuqueo, and Tejada-Herrera undermine the understanding of technology as a Western scientific universal, or as a tool designed to generate extraction and capitalist surplus value. These works reclaim a notion of technology as interwoven with community and in continuity with knowledges that emerge from the body and the land we inhabit. Nor do these artists offer abstract ontological reflections or totalizing interpretations of planetary collapse; rather, they stress that any political reflection on interdependence and relationality ends up irrelevant if it does not accurately account for the past and present impacts of Western colonialism.[22] Their works offer sensuous fabulations at the edges of human existence that invite us to discover in the fractures of colonial power the glimmers of another life.

1 Ivan Illich, *Tools for Conviviality* (New York: Harper & Row, 1973), xxiv.

2 Paulo Freire visited the CIDOC in 1968 and from there published his first book outside Brazil. See Freire, *Educación y concienciación: extensionismo rural* (Cuernavaca: CIDOC/ Cuaderno 25), 1968.

3 María Jacinta Xón Riquiac, “La revitalización de las epistemologías endógenas como proceso de reivindicación política de los pueblos indígenas,” *Ciencias* 111–12 (October 2013–March 2014): 132–41.

4 See Elizabeth A. Povinelli, *Between Gaia and the Ground: Four Axioms of Existence and the Ancestral Catastrophe of Late Liberalism* (Durham, NC: Duke University Press, 2021).

5 “Coya,” “kolla,” and “colla” are group names for the Indigenous Andean peoples of northeastern Argentina. Recent years have seen a reestablishment of differentiated Indigenous identities such as the Diaguitas, the Omaguacas, and the Chichas.

6 Casandra Sandoval, “Bartolina Xixa, la cuerpa de la Pacha no binaria: La danza ancestral que grita resistencia y lucha en movimiento,” *Sudaka TLGBI*, February 29, 2020. Formerly available at https://sudakatlgbi.com.ar/bartolina-xixa-la-cuerpa-de-la-pacha-no-binaria-la-danza-ancestral-que-grita-resistencia-y-lucha-en-movimiento.

7 The song originally appeared as “Ramita seca (vidalita riojana)” on Aldana Bello’s album *Puñuray* (2016).

8 Beatriz Lemos, “Bartolina Xixa,” in *The Crack Begins Within: 11th Berlin Biennale of Contemporary Art*, exh. cat. (Berlin: Kunstwerke and Berlin Biennale, 2020), 139. Available online at https://11.berlinbiennale.de/participants/bartolina-xixa (accessed November 14, 2023).

9 Marlene Wayar, quoted in Claudia Lorenzón, “Marlene Wayar: ‘Las travestis ponemos en duda el mundo establecido, desde el más absoluto despoder,” *télam Digital*, January 2, 2022. Available online at https://www.telam.com.ar/notas/202201/579690-marlene-wayar-diccionario-travesti.html (accessed November 14, 2023).

10 Daniel Lie, “Death as Expansion,” in Daniela Leykam and Christoph Tannert, *Daniel Lie: Scales of Decay*, exh. cat. (Berlin: Künstlerhaus Bethanien and Verlag Kettler, 2021), 9.

11 For more on these ideas see Miguel A. López, “Letter to Daniel Lie, the Living Entities that Co-Exist with They, and the Microbes, Fungi, and Spirit-Energies in Transformation,” New Museum website, May 2022. Available online at https://new-museum.s3.amazonaws.com/pdf/2022_Letter_to_Daniel_Lie_Miguel_A_Lo%CC%81pez.pdf (accessed November 14, 2023).

12 Dan Lie, in a Zoom conversation via the author, April 6, 2022.

13 For these quotes and a detailed analysis of *Kowkülen* see Florencia San Martín, “Sebastián Calfuqueo’s Kowkülen: Inhabiting Nonbinary Waters,” *Post: Notes on Art in a Global Context* (The Museum of Modern Art, New York), July 14, 2021. Available online at https://post.moma.org/sebastian-calfuqueos-kowkulen-inhabiting-nonbinary-waters/ (accessed November 14, 2023).

14 Seba Calfuqueo, “Una pedagogía al margen de la rigidez,” *La Escuela*, March 5, 2022. Available online at https://laescuela.art/es/campus/library/practices/una-pedagogia-al-margen-de-la-rigidez (accessed November 14, 2023).

15 See López, “When I Masturbate I am Hermaphrodite. The Re-Education of the Body According to Elena Tejada-Herrera,” in Florencia Portocarrero, *Elena Tejada-Herrera. Videos de*

esta mujer, 1997–2010, exh. cat. (Lima: proyectoamil, 2018), 159–73.

16 Elena Tejada-Herrera, *Las Bambas*, video, 2019.

17 Silvia Federici, "The Reproduction of Labor Power in the Global Economy and the Unfinished Feminist Revolution," in *Revolution at Point Zero: Housework, Reproduction, and Feminist Struggle* (New York: PM Press, 2012), 94.

18 Naomi Rincón Gallardo, “A Letter to Stained Subjects Who Persist in the Labor of Worldmaking Regardless,” in López, *And if I Devoted My Life to One of Its Feathers? Aesthetic Responses to Extraction, Accumulation, and Dispossession*, exh. cat. (Vienna: Kunsthalle Wien, and London: Sternberg Press, 2022), 50–57.

19 Frantz Fanon, *Black Skin, White Masks*, 1952, Eng. trans. Charles Lam Markmann (New York: Grove Press, 1967), 8.

20 See Aníbal Quijano, “Coloniality and Modernity/Rationality,” *Cultural Studies* 21, no. 2–3 (2007): 155–67. First published in Spanish as “Colonialidad y modernidad/racionalidad,” *Perú indígena* (Lima) 13, no. 29 (1992): 11–20.

21 See Arturo Escobar, “Beyond the Third World: Imperial Globality, Global Coloniality, and Antiglobalization Social Movements,” *Third World Quarterly* 25, no. 1 (2004): 207–30, and Walter Mignolo, “La colonialidad: la cara oculta de la modernidad,” in Sabine Breitwieser, et al., *Modernologías: artistas contemporáneos investigan la modernidad y el modernismo*, exh. cat. (Barcelona: Museu d'Art Contemporani de Barcelona, 2009), 39–49.

22 See Povinelli, *Between Gaia and the Ground*.

MUSEUM OF CONTEMPORARY NATURE

The Mexican writer Juan Villoro recounts that in August of 2016, hundreds of musicians, writers, and artists from Chiapas and beyond arrived at the Universidad de la Tierra (University of the Earth), on the outskirts of San Cristóbal de las Casas, to participate in the festival *CompArte por la humanidad* (CompArte for humanity), hosted by Zapatista communities. They were responding to an open invitation, grounded in the conviction that "in the face of the severe crisis that threatens the whole world... the sciences and the arts now represent the only serious opportunity to build a more just and rational world."[1] "Unlike politics," Subcomandante Galeano had written to Villoro a short while earlier, "art does not attempt to adjust or fix the machine. Instead, it does something more subversive and unsettling: it demonstrates the possibility of another world."[2] And he went farther: "And I'm not talking about 'reconstructing' in the sense of picking up what has fallen and putting it back together again, in the image and likeness of its previous version of disgrace. I'm talking about 'remaking'—that is, 'making anew.'"[3] That abstract ambition toward reconstruction and metamorphosis, Villoro further recounts with sharper focus, became three-dimensional inside a small shack at the Universidad de la Tierra, where artists and local residents turned a load of seeds, legumes, and leftovers from the market into a colorful insectarium. Beans and peanut shells were transformed into beetles, pine cones and gourds became dragonflies, other remains of exotic fruits and vegetables became bees, caterpillars, and even butterflies. "At the end of a few hours," Villoro wrote, no one knew who had set this antenna or that carapace in place. What mattered most was that the bugs took on convincing features of political animals.... The waste from the market had made space for a resistant species, an emblem of the transfiguration necessary for survival."[4]

In that shack, art became the heuristic ground for creative coexistence among the most varied species, a surreal synecdoche of a renewed dialogue with the environment, and with other forms of life, that has flourished through the art of the continent during the last few decades.

Sensitive to climate disaster and to the open debate about the Anthropocene Epoch, attentive to a new object-oriented ontology and to a non-anthropocentric understanding of the world present within Amerindian cultures, Latin American art has explored new forms of multispecies connection and coevolution as a prosperous model of effective relationships of hybridization, cohabitation, and interdependence with others: human and nonhuman, organic and inorganic. The philosopher Emanuele Coccia calls the space of art—open to all species capable of imagining nature beyond its limits, and of embodying a futuristic projection of the present, a metamorphosis—the "museum of contemporary nature." "Through art," he writes, "every society builds something that does not yet exist within it: it is no longer a harmonic reflection of its own nature, but an attempt to reproduce itself in a different way from what it is."[5] A gross realism, incapable of imagining the future, dominates the discourses of politics, economics, and indeed, at times, even the social sciences, but it is precisely within art that this impoverished notion of realism is least comfortable, even in art that has no political aims but becomes political when it reveals the limits of imagination and presents initially impracticable-seeming fantasies as realistic. And while contemporary art, energized by a desire to reconfigure the place of humans on the planet, has naturalized the relationships among different knowledges and crossed epistemological barriers, it has often managed to preserve its autonomy in works that do not merely illustrate the debates of their time, or reduce them to an ecological pronouncement, but rather renovate their methods with unprecedented forms, devices, and processes—possible works from the *museum of contemporary nature*.

Innovative forms of reconstruction and metamorphosis animate the work of Argentine artist Adrián Villar Rojas, who in one of his earliest installations—ephemeral, over the top, and, like almost all his installations to come, defying description—was already deliberately confusing nature and culture and was recomposing the world on a single plane. Separating itself from the post-Conceptual formalism of the 1990s, *Pedazos de las personas que amamos* (*Pieces of the People We Love*, 2007) shaped and assembled the remains of domestic objects and materials on a base measuring four meters by six meters (c. thirteen by nineteen feet), in a rhizomatic work that sprawled through a family home in Rosario until it occupied the space almost completely. With fantastical touches from Japanese anime, science fiction films, and the messiness of grunge, the young Villar Rojas's imagination laid out a futurist-romantic tale that bifurcated into hyperlinks and multiverses. Interconnected "narratively" by wires and a soundtrack of birds and songs by Radiohead and Nirvana, dozens of organic and inorganic objects proliferated in this scene, created with the help of family and friends. Very swiftly, the fantastical tale became post-apocalyptic in *Lo que el fuego me trajo* (*What Fire Brought to Me*, 2008), an installation in the basement of a Buenos Aires gallery. Here hundreds of pieces made from unfinished raw clay, cheap objets trouvés, a water tank, and a brick chimney splayed out atop a rubble-covered floor boldly contradicted the aesthetic mandates that propelled the work of his contemporaries. "Surrealism and romanticism struggle against conceptualism and Pop," wrote Inés Katzenstein in an early review. "*Lo que el fuego me trajo* recuperates figuration, fiction, and feeling."[6] But the work also prefigured an aesthetic illumination that over more than a decade would function in harmony with human discomfort in the face of the uncertain future of the Anthropocene. The real processes of matter, which constitute Villar Rojas's great formal discovery, *made* this dystopic fiction: raw clay—organic and inorganic sediment of Earth, predating humans on the planet—combined with cement, an essential material of the anthropic transformation of the landscape, shattered the pieces merely by taking form and created topological figures, "instant ruins" that crazed the arrow of time and offered a realistic theater of a somber future that now is already past. *Mi familia muerta* (*My Dead Family*, 2009; fig. 1), however, created a year later, proposed an even more ambitious plan via a giant

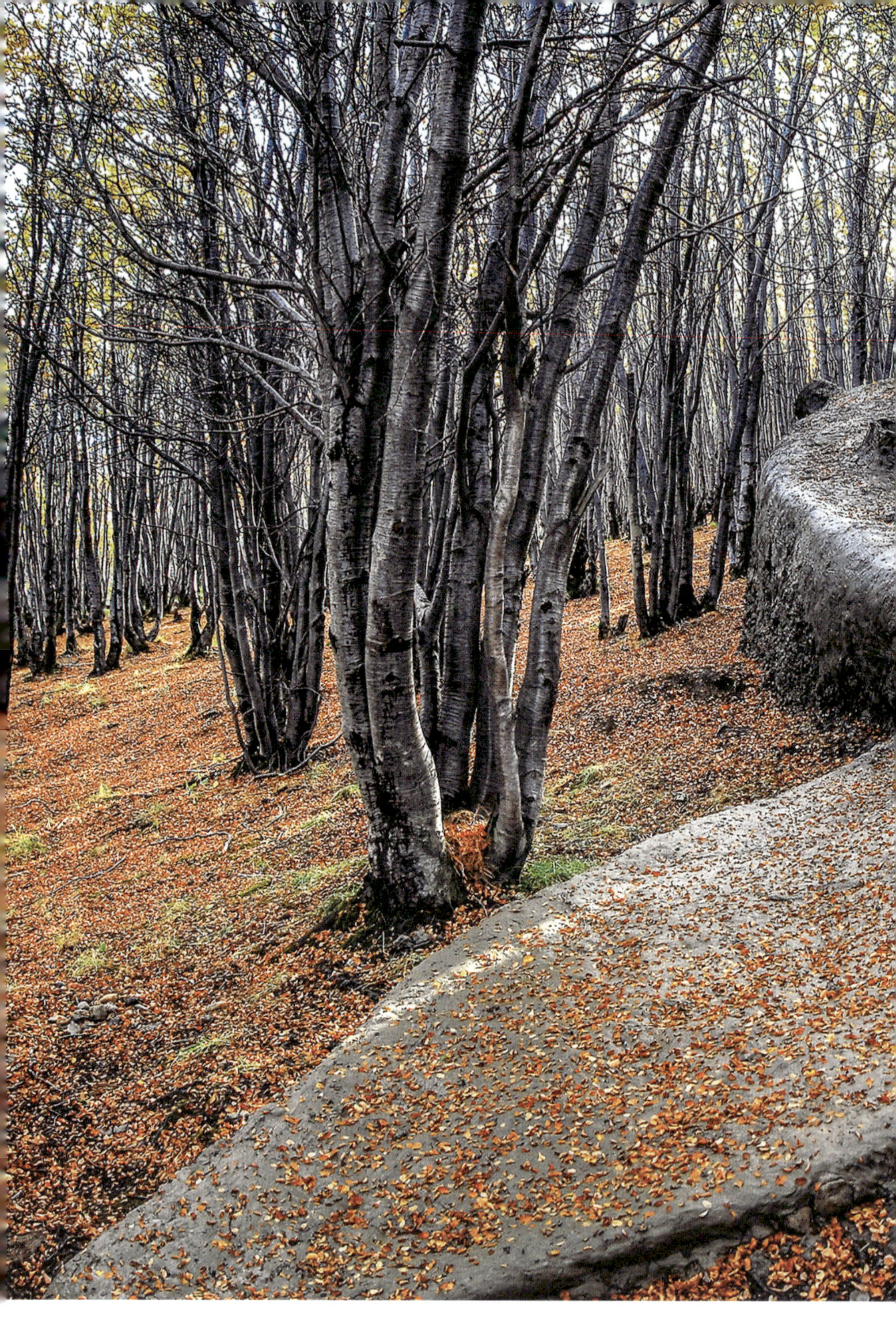

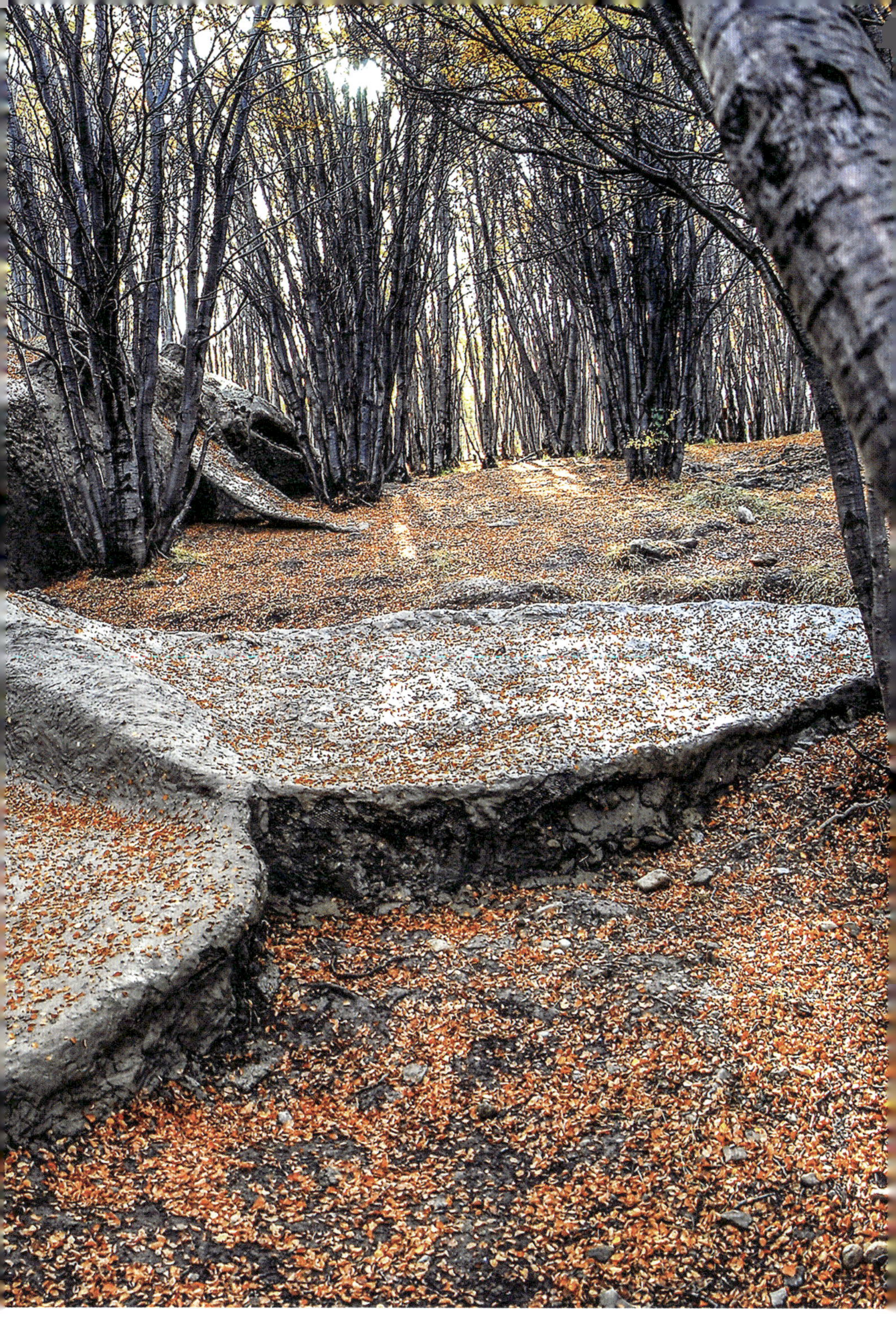

1 Adrián Villar Rojas. *Mi familia muerta* (*My Dead Family*). 2009. Unfired local clay and rocks, c. 9 × 88 × 13 ft. (3 × 27 × 4 m). Installation view, Bosque Yatana, 2nd End of the World Biennial, Ushuaia, Argentina

whale made of raw clay molded onto a wood structure in a forest in Ushuaia, Tierra del Fuego. The small universe mounted on a plane in Rosario began to unfold against the real surface of the globe, the futurist/romantic tale advancing toward a posthuman time (the world "seen with the naive eyes of an alien") and the improvised group in the family home becoming a nomad team of collaborators roaming the world like an itinerant theater company, building and dismantling ephemeral works that were ever more monumental.[7] "More than site-specific projects," Villar Rojas says, "I'm devoted to site-specific thought. This involves deep and sustained dialogical exchange with a variety of agents, human and nonhuman, organic and inorganic, human made or machine made."[8]

The planet and the work itself restarted, however, in *Today We Reboot the Planet*, an installation that Villar Rojas opened at the Serpentine North Gallery, London, in 2013. A surreal fence presided over by a female elephant, charging at the fence with her back to the gallery entrance, encircled the heart of the building and transformed one of its rooms into a Noah's ark of twenty-first-century culture in ruins. And if the female elephant were charging a British military bastion as a posthumous response from nature to imperial ambitions of conquest, in the here and now *Today We Reboot the Planet* invited viewers to experience the fossilized future of the Anthropocene, with the cycle of natural life beginning once again by seeding itself in adobe, in the aftermath of the fatal errors of our time. Life—stubborn, tenacious, resilient—was reborn on a rebooted planet and would be reborn from there onward among the organic remains strewn about the entire surface of a gallery in Mexico City, or in the fissures in vibrant material in eleven blocks of the High Line in New York. It was not by chance, then, that shortly afterward, in Turin, Villar Rojas launched a new series, *Rinascimento* (Renaissance, 2015), a celebration of live nature and its metamorphoses. At this point the brownish works made from raw material were left behind, having been delegated to a synanthropic species—the ovenbird, the Argentine national bird—whose abandoned nests, made using techniques kindred to his own, he would collect and then install, repaired or intervened in, in distant contexts and cities across the planet. In the large gallery of Turin's Fondazione Sandretto Re Rebaudengo, time would now expand into a hundred fossilized rocks and trunks brought from Anatolia and laid out theatrically in the half light. On these, Villar Rojas and his collaborators placed an unclassifiable range of elements, from dead crows, coral, clusters of grapes, artichokes, and melons to fishing nets, oriental amphoras, and all types of artisanal tools. Spared exposure to the elements in the cold, stripped-down space, these "still lifes," composed on each rock like Caravaggios brought to life, recuperated in three dimensions thousands of years of *natureculture* (the apt neologism comes from Donna Haraway)[9] and gave themselves over to the "sculpting tools of entropy" that transform them.[10]

That polyphonic assembly of kingdoms and species capable of dissolving the conventional categories and definitions of artistic media recalls the work of the Mexican artist Gabriel Orozco, who in the early 1990s was already making work that relied on living dialogue with his surroundings. A traveler by vocation and a voracious collector, photographer, sculptor, installation artist, and above all sensitive observer of the postindustrial landscape, Orozco knocked down the walls of the studio, disrupted natural taxonomies, and busted through geographic and cultural borders. Many of his works—inescapable reference points for a number of artists on the South American continent—recuperate materials from the organic world, bring them together with industrial residue, and arrange them into new structures that highlight the singularity of their forms, human neglect for the conservation of our planet, and the sculptural abilities of time itself. One open series, a formalist precursor, if you will, of Villar Rojas's *Rinascimento*, summarizes the creative encounter of nature and culture and the formal subtlety of Orozco's gaze. *Mesas de trabajo* (*Working Tables*)—hundreds of objects, unfinished sculptures, organic waste, and urban remains that the artist began to collect in 1991, then to compose and arrange on tables—bring together aesthetic discoveries and experiments, like a material record of the artistic process: its failures, its successes, and its open-ended aims. Terra-cotta bowls, tree bark adapted with geometric designs, shells, dried mushrooms, sheets of rubber foam packaging studded with bottle caps, all open a permeable network of relationships with the surroundings and redefine themselves constantly as they mutate.

It's no surprise to discover that Orozco is one of the anonymous creators of the insectarium at the Universidad de la Tierra in Chiapas.

But nature can also be reconciled with humans in two dimensions, on the still resistant plane of painting. This occurs in the work of the Dominican artist Hulda Guzmán, who has renewed pictorial landscape traditions with a deeply personal "reconstruction" of the Caribbean forest in a delightfully festive symbiosis—real, surreal, oneiric—with other species. Though Guzmán started out by alternating portraits with landscapes, the forest began to dominate in terms of scale, chromatic breadth, and variety of forms until it became the great protagonist of a series increasingly consisting of panoramic paintings in which human beings, decentered and shrunken within the totality, celebrate a joyful coexistence with plants and animals. In fact, her work seems committed to represent not the colonial vision of the idyllic landscape but rather the "metaphysics of mixture" to which Coccia aspires, integrating the natural world and human habitat increasingly intensely.[11] Plants "are not the landscape, they are the landscapers," Coccia affirms, "the major cosmogonic force on our planet."[12] To prove how the plants of the forest themselves become the real "landscape artist," it's enough to observe how in Guzmán's visual and narrative imagination the exterior permeates the tropical domestic interior until the classic opposition of interior and exterior is vanquished. With an attentive eye to the lines and materials of modern architecture, Guzmán frames interiors that increasingly open out toward the exterior, from the shadows of palms that dynamize the functional geometry of an empty house (*Behind the Wall*, 2016) through the minuscule circular window that allows a glimpse of a marine landscape (*Through That Little Window*, 2016) to the large picture windows of more recent works in which Guzmán plays with architectural openings and pictorial frames, often in a way recalling René Magritte. In *Pintando la almendra* (Painting the almond, 2020), for instance, the self-portrait of the exultant artist painting the multicolored flora that appear through the windows is multiplied to infinity via the Droste effect. The fusion becomes even more explicit in *Quarantine Visitor* (2020), one of the works Guzmán painted in Santo Domingo during the obligatory quarantine of the COVID-19 pandemic. Added to the small version of the painting seen on a table is the "visitor" of the title: an undulating palm that has burst into the house and seems to dance beside the artist, who is enthralled with this "visit."

We can experience the greatest celebration of nature reconciled with humans, however, in a series that opens at an early stage in *Fiesta en el batey* (Party in the batey, 2012) and then blossoms in the sequence that Guzmán brought together in 2020 under the title *My Flora, My Fauna*. In one of the countries most vulnerable to environmental crisis, nature perseveres in Guzmán's work, a sovereign force. The harmonic coexistence with the natural world of the Indigenous Taíno population, long thought extinct, revives in her festive batey (sugarcane-worker village), which re-creates the pleasure of Indigenous communities and their integration with other species. While forest plants fill two thirds of the painting with their exuberant form and color—innumerable greens, golds, and oranges—the batey, the typical people's housing space, now exhibits the attributes of contemporary Caribbean architecture and is populated with an animated frieze of small, barely covered or naked bodies in renewed dialogue with their Taíno ancestors. They lounge in hammocks, eat, drink, dance, or bathe in a stream. In *"come dance"—asked nature kindly* (2019–20; fig. 2), architecture disappears completely and forest plants rule the scene: the personification embedded in the title makes it clear that the forest is the undisputed hostess of the party. "I wish to convey to the viewer a sense of the natural environment," says Guzmán, "to transmit a perspective of synergy, to portray trees and plants in a glorified way and to inspire worship, respect, and reverence of nature."[13] And if the naturalist landscapes of eighteenth- and nineteenth-century travelers removed the bodies, Guzmán reinstates them in the bottom third of the canvas, where dozens of men, women, and children respond today to the "kind" call of the forest, dancing among domestic and wild animals. Guzmán herself, in a tiny rendering to one side, embraces the ancient trunk of a tree. Time seems to expand and dilate as well, in the "metaphysics of mixture" of the many pictorial traditions being recuperated. Hieronymus Bosch, Brueghel the Elder, the Douanier Rousseau, and Surrealism inject fantasy, detail-oriented furor, chromatic splendor, and dreamlike freedom

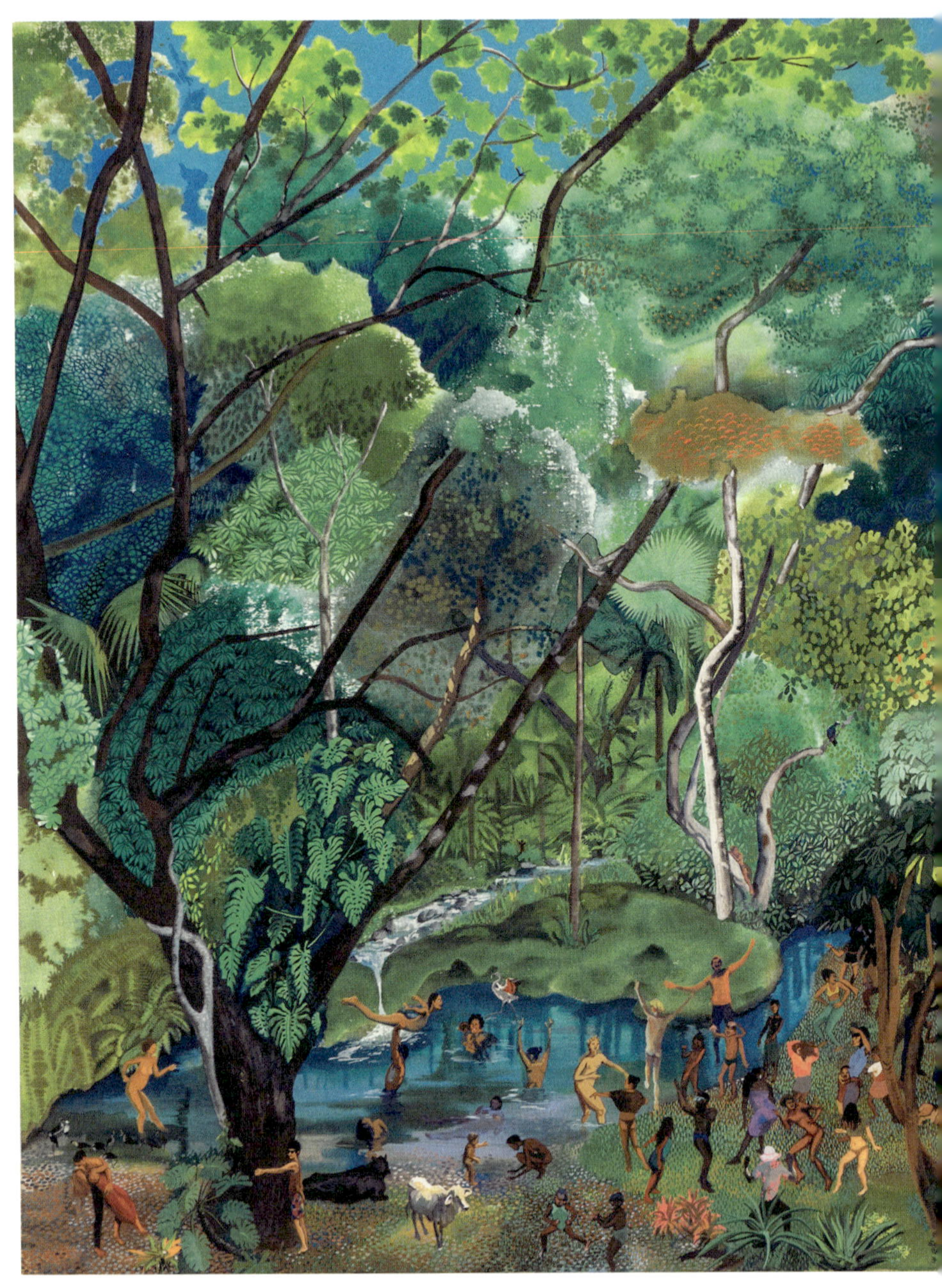

2 Hulda Guzmán. *"come dance"—asked nature kindly*. 2019–20. Acrylic gouache on linen, 48 × 78 in. (122 × 198 cm.) Museu de Arte de São Paulo Assis Chateaubriand. Gift of Rose Setubal and Alfredo Setubal, in the context of the exhibition *Histórias da dança*, 2020

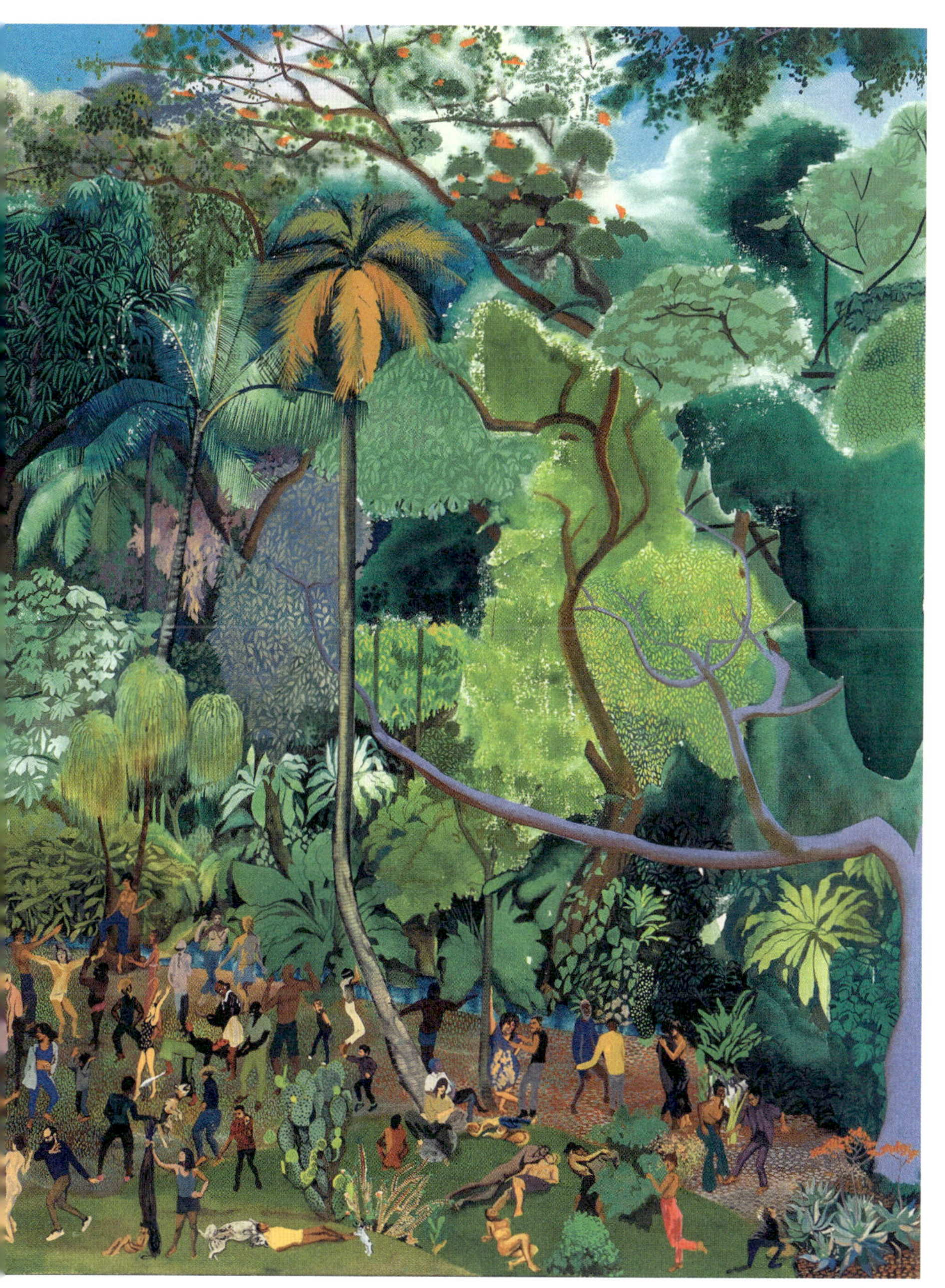

into this painting "after landscape."[14] And while today "*the world from which we live*" seems suddenly to invade "*the world where we live*" (the precise formulation is Bruno Latour's), that tension vanishes in Guzmán's fecund and auspicious recuperated forest.[15]

But art wants more. It wants to open up a real dialogue with the natural world, and to interpret its signs. A few statistics are sufficient to provide a view of the urgency of the call. Forests alone occupy a third of the surface of our planet, with more than 60,000 species, and it's likely that in the last 10,000 years we have reduced the vegetable kingdom by half. But what would they say if we offered them our ears? This question took on form and volume in *Sound Mirror*, an ingenious device made by the Argentine artist Eduardo Navarro at the 2016 Bienal de São Paulo with the purpose of acoustic interaction with nature. Navarro installed a large bronze horn, like that of an old-fashioned gramophone, against a lush palm in the city's Ibirapuera park, and connected it by a tube to the interior of the Bienal pavilion, where visitors were invited to sit on a stool to "listen" to it. Rhetorically sharp and with Navarro's habitual pataphysical humor, the work manifested a potent metaphor: human beings finally offered their ears to the trees, with the instrument's exaggerated form serving to remind them of the scale of the challenge. This occasional luthier managed to connect the park to the art space and to achieve an additional metaphorical benefit. With the crown of the palm in the park connected to the people in the art space (they faced each other, at almost the same height, through the space's glass exterior wall), the continuity of *natureculture* and *flat ontology* to which speculative realist philosophers aspire was figured graphically. "I'm collaborating with a living being," Navarro claimed in front of his *Sound Mirror*.[16] A beautiful instrument created to highlight the palm tree as star attraction, the work replaced the man and artist as the central figures of both sound transmission and performance, and at the same time, with echoes of John Cage, invited the palm tree to improvise at random, according to the intensity of the wind, the tap-tapping of raindrops, or the crackling of fronds in the sun.

A little earlier, in *Surround Audience*, the third Triennial at the New Museum, New York, Navarro had set himself the task not only of *listening* to nature but of *being* a living species in an experiment that was both life-oriented and aesthetic. In *Timeless Alex* he tried to move at the speed of a tortoise, assisted by the uncomfortable weight of a suit he had fabricated over a period of months, a tortoise-shell-like structure that forced his body to adopt a tortoise's position, perspective, and pace. He had a surefire model for the shell in Lonesome George, the last specimen of a species of giant tortoise from the Galapagos Islands that on its death was embalmed and exhibited at New York's Museum of Natural History. That model also introduced an echo of biodiversity under threat, yet more than anything else, the work was undertaken as an open investigation into other ways of experiencing time, from the Hindu meditation technique of Vipassana to the empathy with animals shown by some autistic people and studied by the ethologist Temple Grandin. With his body covered in leather and his shell made of glued-together pieces of cardboard, Navarro finally *was* a tortoise for two hours on a sunny terrace next door to the New Museum, visible from the museum's highest floor. He asserts that after forty minutes of moving extraordinarily slowly, with his movement limited by the shell and the suit, he nonetheless began to feel lighter, overtaken by a feeling of happiness and surrender, in a time completely foreign to contemporary acceleration.

Oído vegetal (*Vegetal Ear*; fig. 3), a sui generis flower installed in an eco-park in a Guatemalan forest in 2021, combines and expands on both experiences. If flowers are cosmic attractors that absorb the world and transform it into fruit, why not attempt to imitate them? Navarro built a 21½-foot-tall flower with a brilliant-red bell-shaped corolla, three filaments crowned with yellow anthers in the form of small bowls for birds to drink from, and a stem that uncurls toward a sort of large bulb that can be accessed by a small circular door. The dark interior offers visitors a space into which to retreat, sink into silence, set aside the sense of sight, sharpen their hearing, and attend to the faraway murmurs of the forest—the gusts of rain, or the wing-flaps of birds alighting on the anthers (and pollinating them even without intending to)—as a flower might perceive them. In the half-light inside the bulb, Navarro imagined

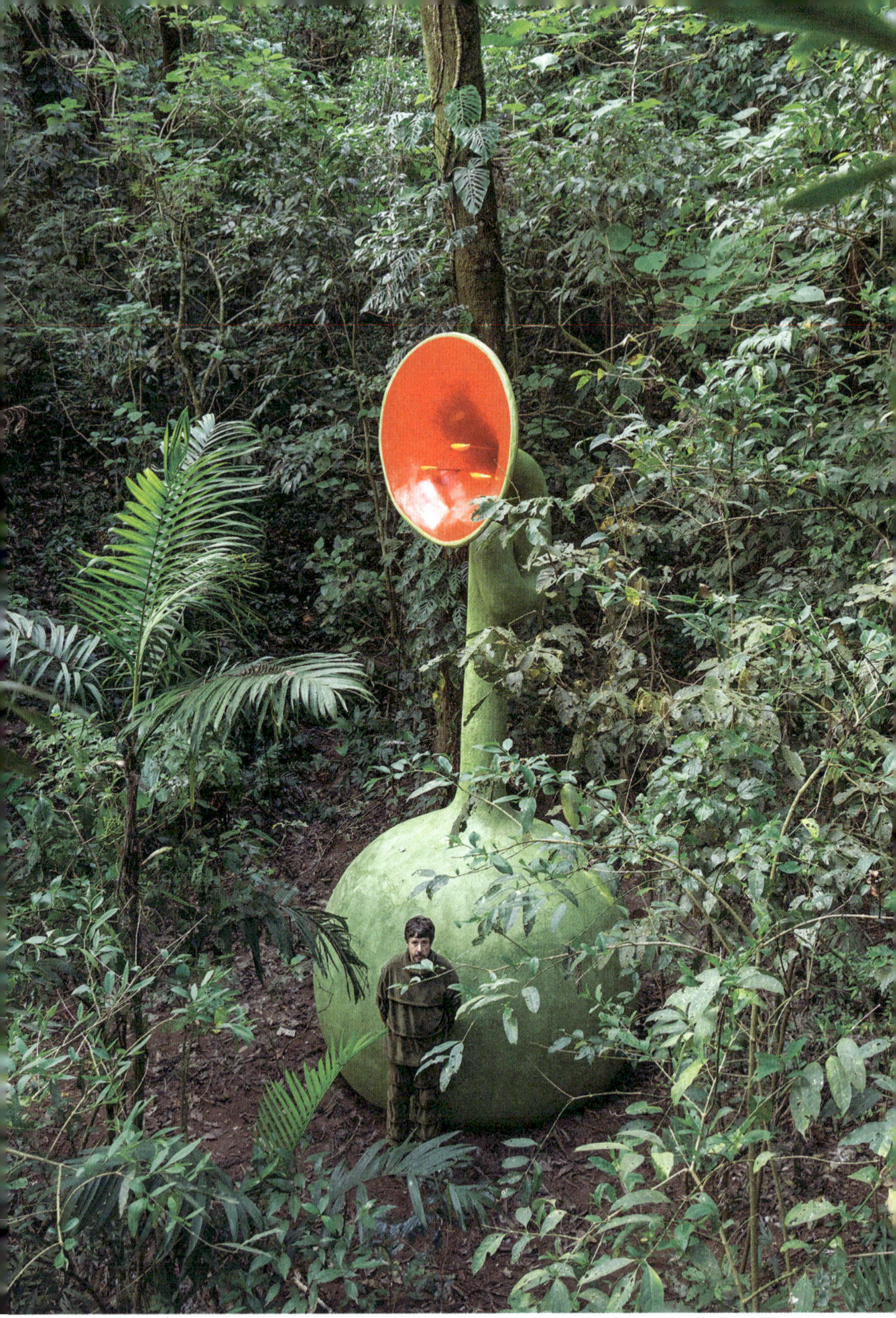

3 Eduardo Navarro. *Oído vegetal* (*Vegetal Ear*). 2021. Steel, ferrocement, pad, and paint, 22 ft. 11 ⅝ in. (7 m) high × 16 ft. 4 ⅞ in. (5 m) circumference. Installed at GreenRush ecological park, Villa Canales, Guatemala. The artist stands in front of the work

other ways of taking in our surroundings and other beings: the infinite variety of natural forms and appearances, like an Eden of genus diversity, with sexuality as a cosmic encounter.

Indeed the forest, and more precisely the Brazilian Mata Atlântica, as Daniel Steegmann Mangrané believes, is not just a privileged space in which to explore the interdependence of species but also a model for rethinking our conception of art and, even before that, the modern cultural paradigm that separates humans from nature, beginning with the Cartesian division between the res cogitans (thinking thing) and the res extensa (extended [material] thing).[17] Mythologized and then plundered by the colonizers, the Amazon rainforest remains one of the ecosystems with the greatest biodiversity on the planet, even under grave threat by the ravages of extractivist deforestation. It is no coincidence, then, that the Spanish artist Steegmann's early attraction to biology, his juvenile fantasy of losing himself in the virgin forest, and his fervent response to the interpellation of body and senses in the works by Hélio Oiticica and Lygia Clark that the Fundació Antoni Tàpies, Barcelona, exhibited in retrospectives in the 1990s would lead him to Río de Janeiro, where he has lived and worked since 2004. The engagement with Brazilian ecocide and also with the "Amerindian perspectivism" proposed by anthropologist Eduardo Viveiros de Castro inspired a radical transformation in his practice and an expansion of his media and languages.

If, as Viveiros de Castro posits, Amerindian cosmology conceives of humanity, not animality, as the original common condition of humans and nonhumans, multinaturalism would allow for the erasure of the classical division between subject and object, nature and culture, and the promotion of dynamic relationships of mutual transformation between works and viewers.[18] The anthropologist Eduardo Kohn, as well, after four years of field work in Peru's Alto Amazonas Province, invites us to "think with" the rainforest, attending to the more-than-human semiotic networks that the Indigenous Runa people have been interpreting for centuries.[19] From that starting point, Steegmann has set himself the task of researching the relationships of multiperspectivism and entanglement that the tenacity of the Mata Atlântica still offers to the attentive observer. But how to translate these into aesthetic experiences? All his work is dedicated to that attempt, but the forest is the indisputable protagonist of a series in which the image in movement offers a multisensory experience within the resilient density of the forest plants.

In *16 mm* (2007–11), an almost-five-hour film that recuperates the experimental impulses of structural cinema, the immersion is literal. Shot in a single take with a modified camera that moves at the same speed as the film along a cable hung at a height, it performs the fantasy of penetrating the impenetrable forest, cutting in a straight line across the multiform density of the thicket and hearing the myriad sounds harbored there. The ghostly promenade, distant and at the same time close, clears the forest of all idealist mythologizing and turns it wild again, pure forest. But the hyperrealist phantasmagoria, if there's space for such a paradox, takes on other dimensions in two further immersive works that use very different technologies. In *Spiral Forest (Kingdom of all the animals and all the beasts is my name)* (2013–15), a camera mounted on a universal joint allowing 360-degree rotation provides the material foundation to support a multiperspectivist eagerness for documentation. For ten hours and fifty-six minutes, the film traverses the undergrowth, the high crowns of the treetops against the sky, and even the joint that supports the camera, in a spiraling pan that disrupts the viewer's spatial coordinates and makes them strange within its disembodied gaze. That experience is taken to an extreme in *Phantom (Kingdom of all the animals and all the beasts is my name)* (2014–15), a virtual reality installation in which the viewer, equipped with an Oculus Rift visor, can traverse a black-and-white 3D re-creation of an actual section of the forest, at once vivid and spectral, hyperrealist and abstract. At the same New Museum Triennial where Navarro *was* a tortoise through the use of an artisanal device capable of altering his perception of time, Steegmann took recourse to a technological device to alter our perception of space. In this way he not only conceived of a form of "digital preservation" of the threatened rainforest but invited the viewer to dissolve into three-dimensional space, and *to be* in the Mata Atlântica on the Lower East Side of New York.

But it is perhaps -’— –‘- , the minimalist and poetic work with which Steegmann returned to the Fundació Antoni Tàpies in 2018, that best encapsulates his ambition to unite nature and culture in the art space and to transform that space in the attempt. For five weeks, with just the right conditions of light, humidity, and temperature, Steegmann “installed” a *Morpho helenor* butterfly in the museum—or, more precisely, he provided lodging for a chrysalis that after two weeks metamorphosed into a butterfly, which flew freely during the following weeks through the 50,500-square-foot museum, alighted on mature fruit laid out for that purpose, flew over the Tàpies works in the collection, and flaunted its diminutive grace before the formidable *Hope Hippo*, a life-sized hippopotamus, the work of Jennifer Allora and Guillermo Calzadilla. With its dazzling blue color, the ocellus spots on the brownish undersides of its wings, its undulating flight and slicing wing-beats, the *Morpho helenor*, an abundant species in tropical areas of Central and South America, could not possibly fail to astound visitors. This minimal intervention—it’s tempting to call it a *simbiopoiesis*—celebrated the discrete miracle of metamorphosis, a model of radical transformation of body and identity. At the same time, by leading the viewer’s attention toward a living being, it performed a reconsideration of the powers of art. Steegmann’s conclusion is simple, generous, and light-hearted, like the gesture that inspired his return to the Fundació Antoni Tàpies: “No work of art could ever compete with a butterfly.”[20]

No human sculpture, the visionary artist, architect, and environmentalist Tomás Saraceno might say in turn, *could ever compete with a spiderweb*. Among Saraceno’s many multidisciplinary endeavors dedicated to making the planet into a sustainable common space, he’s spent years attempting to convert the disoriented ship of the Anthropocene into a Noah’s ark, beginning with a tiny deaf and blind species that has managed to survive for 140 million years on our planet. “We are in a knot of species,” Haraway writes in *When Species Meet*, “coshaping one another in layers of reciprocating complexity.”[21] And if the question is one of learning to coexist with other species, why not begin with the spiders that make their nests in our attics, ceilings, and gardens, and be attentive to the social and practical abilities through which they have survived a number of extinctions? After converting part of his Berlin studio into a living archive of over 300 spiders and mapping and digitizing hundreds of spiderwebs, Saraceno re-created them three-dimensionally in *Galaxias formándose entre filamentos, como gotas de agua en la tela de una araña* (*Galaxies Forming along Filaments, like Droplets along the Strands of a Spider's Web*), an installation in a white-cube space at the Venice Biennale of 2009. In the process he not only demonstrated that it is possible to conjure the reticular structure of the universe at human scale by replicating those delicate filigrees, he also tried to offer spiders, his muses, the starring role in human dwellings that centuries of hygiene and anthropocentrism had denied them.

In Saraceno's first show in Argentina, *Cómo atrapar el universo en una telaraña* (How to trap the universe in a web, 2017), at the Museo de Arte Moderno de Buenos Aires, and since then in museums and galleries across the world, he converted the white cube into a black cube, a mix of artificial habitat and laboratory, in which thousands of spiders embroidered the air with silvery threads. With the help of arachnologists, biologists, and ethologists, he brought in 7,000 *Parawixia bistriata*—a social subspecies that forms cooperative colonies—from their natural habitat so that they could spend months weaving their webs in the large space, leaving behind them, once they had been returned to their place of origin, a visible demonstration of their practical wisdom. Together these spiders can ride air currents to deploy structural strands of silk more than thirty feet long. They can capture prey that's very large in size and guide themselves by the vibrations in their sisters' webs. With true community spirit, they can form colonies of up to 500 spiders, sharing a silk shelter during the day and leaving it at night to weave spherical web clusters that function as traps. The dimly lit exhibition space also made it possible to appreciate the aesthetic splendor of a work never before seen in an art museum. Thousands of silken strands phosphoresced in lacework that was simultaneously fragile and durable, with perfectly symmetrical forms and complex weave patterns. The now absent spiders were the indisputable coauthors of the work, making the result an unprecedented version of the "assisted ready-made." At the same time, it was ephemeral and unrepeatable, impossible to map in its entirety, only captured through direct experience. The presence of the artist, displaced from a position at the center of the creative act, was insinuated through a few nearly invisible supports, and art was redefined in a liminal space adjacent to scientific research. Science or art? An act of hospitality, Jacques Derrida might have said, can't help but be poetic.

Not content merely to give us a glimpse of the lacework these spiders have created, Saraceno, in an unprecedented bioacoustic experience, has set himself the task of attempting communication and making it audible. Beyond their astonishing structural engineering, spiderwebs are a vibratile tool that allows the species to communicate via seismic signals, strums, percussion, and even by tuning their silk filaments. The *Arachnid Orchestra* came into being, then, as a refutation of our overly human conception of music. In May 2018 in Berlin, the philosopher and musician David Rothenberg, invited by Saraceno to one of the many jam sessions held by the orchestra, played the clarinet with two spiders—a *Cytophora citricola* and a *Nephila inaurata*—and confirmed that their vibrations have rhythm and tonal complexity and that the trio could produce a new "music." "What does it sound like?" he wrote later in "Spider Music." "It sounds like the beginning of a beautiful friendship, with music reaching between two species of spider, one species of humans, but many individual humans sharing some untold possible magic linking science and art."[22]

Art, that magic that's re-created in a shack in San Cristóbal de las Casas, in Ushuaia, New York, or Barcelona, in the middle of the Dominican, Guatemalan, or Brazilian forest, and also in a museum gallery or artist's studio, can at this point be an auspicious laboratory where we can immerse ourselves in a world no longer centered on humans but rather open to coevolution with other species. "Nature," Coccia writes, "is not only the immemorial prehistory of culture, but its unrealized future; its surrealistic anticipation."[23]

Meanwhile in Chiapas, Villoro writes, the number of insects continues to grow.

1 Subcomandante Insurgente Moisés, "Convocatoria zapatista a actividades 2016," *Enlace Zapatista*, February 26, 2016. Available online at https://enlacezapatista.ezln.org.mx/2016/02/29/convocatoria-zapatista-a-actividades-2016 (accessed January 26, 2023).
2 Subcomandante Insurgente Galeano (formerly Subcomandante Insurgente Marcos), "Las artes, las ciencias, los pueblos originarios y los sótanos del mundo," *Enlace Zapatista*, February 28, 2016. Available online at http://enlacezapatista.ezln.org.mx/2016/02/28/las-artes-las-ciencias-los-pueblos-originarios-y-los-sotanos-del-mundo/ (accessed December 12, 2022).
3 Ibíd.
4 Juan Villoro, "El tianguis de los signos," *Reforma*, February 5, 2017.
5 Emanuele Coccia, *Metamorphoses* (Cambridge, UK: Polity, 2021), 168–69.
6 Inés Katzenstein, "Arte y sentimiento," *Otra parte* 17 (Fall 2009): 4.
7 Adrián Villar Rojas, in Robert Preece, "Kneading the World from Scratch: A Conversation with Adrián Villar Rojas," *International Sculpture* 35, no. 5 (June 2016). Available online at https://www.mariangoodman.com/usr/documents/press/download_url/133/sculpture-june-2016-.pdf (accessed December 12, 2022).
8 Villar Rojas, in Hugo Boss Prize 2020 Nominee, video, Guggenheim Museum. Available online at https://www.youtube.com/watch?v=RcatpdgMnkg&list=PLWt9nvDxzGOrCssrKDPqpz3850keCiTIW&index=4 (accessed December 12, 2022).
9 Donna Haraway, *The Companion Species Manifesto: Dogs, People, and Significant Otherness* (Chicago: Prickly Paradigm Press, 2003).
10 Villar Rojas, quoted in Carolyn Christov-Bakargiev, Hans Ulrich Obrist, and Eungie Joo, *Adrián Villar Rojas* (London: Phaidon Press, 2020), 119.
11 Coccia, *The Life of Plants: A Metaphysics of Mixture* (Cambridge, UK: Polity, 2018).
12 Coccia, "The Cosmic Garden," in Jens Andermann, Lisa Blackmore, and Dayron Carrillo Morell, eds., *Natura: Environmental Aesthetics after Landscape* (Zurich: Diaphanes, 2018), 20.
13 Hulda Guzmán, in Haley Mellin, "In Hulda Guzmán's Work, Nature Takes Precedence," *Garage*, October 30, 2020. Available online at https://garage.vice.com/en_us/article/akdz4b/hulda-guzman (accessed December 13, 2022).
14 See Andermann, *Tierras en trance. Arte y naturaleza después del paisaje* (Santiago de Chile: Ediciones Metales Pesados, 2018).
15 Bruno Latour and Peter Weibel, eds., *Critical Zones: The Science and Politics of Landing on Earth* (Cambridge, MA: The MIT Press, 2020), 13.
16 Eduardo Navarro, in Ariel Authier, "Eduardo Navarro," *L'Officiel*, 2017, 153. Available online at https://www.navarroeduardo.art/wp-content/uploads/2021/07/P26_22Eduardo-Navarro-introduce-lo-sensorial-y-la-percepcion-ensu-obra-misma.-Con-socios-inesperados-crea-obras-que-introducen-al-cielo-al-mar-a-la-tierra22_Ariel-Authier_2017.pdf (accessed December 14, 2022).
17 See *Daniel Steegmann Mangrané: A Leaf-Shaped Animal Draws the Hand* (Milan: Skira and Pirelli HangarBicocca, and Villeurbanne: IAC—Institut d'art contemporain, 2020), 116.
18 See Eduardo Viveiros de Castro, "Exchanging Perspectives: The Transformation of Objects

into Subjects in Amerindian Perspectives," *Common Knowledge* 10, no. 3 (Fall 2004): 463–84.

19 See Eduardo Kohn, *How Forests Think: Toward an Anthropology Beyond the Human.* (Berkeley: University of California Press, 2013).

20 Javier Díaz-Guardiola, "Daniel Steegmann Mangrané: 'Ninguna obra de arte puede competir con una mariposa,'" *ABC Cultural*, May 2, 2018.

21 Haraway, *When Species Meet* (Minneapolis and London: University of Minnesota Press, 2008), 42.

22 David Rothenberg, "Spider Music," *PAJ: A Journal of Performance and Art* 40, no. 1 (2018): 36.

23 Coccia, *Metamorphoses*, 169.

PART II

Land Disputes and Colonial Legacies

The second section of *Momentum* contains texts that show how artists are articulating the legacy of colonial occupation as it carries over into contemporary issues of environment and land use. Colonialism's processes of dispossession have led to the view of certain regions as valuable only for their natural resources, converting this "extractive zone," as Macarena Gómez-Barris calls it, into a source of capital gain at the expense of its life.

The extraction of natural resources in Latin America is not a recent phenomenon: it took hold over five centuries ago, with European conquest and colonization. As addressed in the previous section, the split of the human from the nonhuman gave rise to a view of nature as something to be studied, classified, and catalogued. Divorced from the human, nature became a field of inquiry, while the aesthetic genre of landscape contained and organized nature, turning it into an object of use, visual pleasure, and, finally, capital and consumption. In response to the long-term impacts of this process, many resistance movements have formed to fight for land rights. In the following essays, we see how artists shed light on these topics and more through histories of colonialism and neoextractivism.

FRANS KRAJCBERG

Articulating the Natural World

AN ARTIST IN MOTION

In their biography of the Brazilian artist Frans Krajcberg, Claude Mollard and Pascale Lismonde write that "ever since he remembers, Frans sees himself in the process of walking. Everywhere, on countryside pathways as well as in sidewalks of cities: it was a need of his temperament."[1] Krajcberg's early love of walking portended a life spent in constant motion. Born in 1921, forced to flee his native Poland when the German army entered in 1939, he passed through Vilnius, Minsk, Vitebsk, Saint Petersburg (then Leningrad), and Tashkent before settling in Kazakhstan, where he lived for over a year. In 1943, he joined a Polish contingent of the Red Army, participated in the liberation of Warsaw, and was one of the first soldiers to enter Berlin in 1945.[2] Realizing that his entire family had been murdered by the Nazis and disgusted by the anti-Semitism still prevalent in Poland, Krajcberg left for Germany and then France to pursue his dream of becoming an artist.[3] In Stuttgart he studied with Willy Baumeister, in Paris he met Fernand Léger and Marc Chagall, but, scarred by the war and seeing no future for himself in Europe, he boarded a ship to Brazil and arrived in Rio de Janeiro in 1947. Moving to São Paulo, he got in touch with the Brazilian artistic scene of the postwar period. He spent a year in the forests of the state of Paraná and lived in Rio de Janeiro between 1954 and 1957, becoming a Brazilian citizen in 1958.

In the late 1950s, Krajcberg moved back to Paris, where he met renowned artists and intellectuals of the time—Georges Braque, Sonia Delaunay, Yves Klein, Pierre Restany (who remained a lifelong friend), and many others—and got in touch with the main artistic trends of the period: tachism, *nouveau réalisme*, *nouvelle figuration*, and so on.[4] In 1959, tired of urban existence, he went to Ibiza, where, camping in a tent, he started to experiment with the use of rock and soil in his art. Back in Brazil, Krajcberg spent some time in Minas Gerais, fascinated by the landscape and by the intense colors of the natural pigments to be found in the area, but ended up settling in Nova Viçosa, in the south of the state of Bahia, where he built his Brazilian home and studio from 1971 onward.

Throughout the rest of his life (he died in 2017), Krajcberg divided his time between Brazil, where he traveled extensively—notably in Maranhão, Mato Grosso, and the Amazon, where he documented and protested the rampant destruction of the country's remaining rainforests—and Paris, where he established a studio in Montparnasse. His work was displayed in major venues in Brazil, in Paris, and throughout the world: in 1975 in Paris's Centre National d'Art Contemporaine (CNAC), which later became the Centre Pompidou; in 1992 in the Museu de Arte Moderna do Rio de Janeiro, as part of the cultural program accompanying the UN Conference on Environment and Development, also known as the Earth Summit; and in 2005 in the gardens of Paris's Parc de Bagatelle, to name only some of the most emblematic exhibitions. The year 2004 saw the opening of the Espace Krajcberg, Centre d'art contemporain Art et Nature, near the artist's studio in Montparnasse. The Espace Krajcberg currently houses a permanent collection of the artist's work, as well as an archive, and also mounts temporary exhibitions.

These brief notes on Krajcberg's eventful life suffice to show that displacement was a central part of his existence. His constant geographical dislocations can be traced back to a sense of nonbelonging to a specific regional, national, cultural, or religious community: "I am neither Polish, nor French, nor Brazilian, nor Jewish. I am a citizen of the world."[5] The "world" Krajcberg claimed to be a citizen of was sometimes that of the polis, of urban environments, especially Paris, where he found a community of artists with whom he shared exchanges that shaped his own path. But while he sought interactions with his peers, he always distanced himself from art movements or schools. As Mollard and Lismonde point out, "He flees from all kinds of dictatorship, in art as elsewhere. And he decides to trace a personal artistic itinerary."[6] Not only did Krajcberg refuse to settle within the parameters of a given artistic trend, his oeuvre constantly evolving, he also did not stick to one art form: "Initially a painter, he became a ceramist, a sculptor, a photographer, and even an architect" throughout his life.[7] This versatility testifies to his rejection of the boundaries and rules of specific genres, and to his commitment to continuous artistic experimentation.

Krajcberg's cosmopolitan mindset went beyond the narrow confines of any city, even such an ecumenical one as Paris. He regarded allegiance to the entire planet as the only viable form of political appurtenance, and the "world" that contributed more decisively to his art was the natural environment, from the forests of Paraná, in the south of Brazil, through Ibiza, Minas Gerais, Nova Viçosa, and, later in his life, the Amazon. If one were to identify the single major influence in his art, it would not be a particular movement, the work of a fellow artist or thinker, or even a historical event, decisive as World War II certainly was for his career. Rather, it was his ongoing engagement with more-than-humans, their modes of existence, movements, and shapes, which became Krajcberg's main source of artistic inspiration.

The artist told how the encounter with the natural world in his early years in Brazil gave him a new lease on life, both personal and artistic. "Since I left Stuttgart, I was a lost man. . . . I hated humans. I ran away from them. . . . But what is the point of living in isolation?," he said.[8] It was at this moment of profound despondency resulting from the traumas of war and the loss of his family that Krajcberg went to live in the forests of Paraná: "Nature gave me strength and the pleasure to feel, to think and to work. To survive. I walked in the forest and I found an unknown world. I found life. Pure life: to be, to change, to continue, to receive light, heat and humidity. True life: when I am in nature I think, speak and wonder truthfully. When I look at nature, I sense how it moves."[9] It was the motions of nature that the artist attempted to convey by a variety of means throughout his long career. As his friend Restany aptly phrased it, "Nature is his atelier. . . . it is his study and his medium."[10] In his constant geographical displacements, evolutions of style, and shifts in artistic means and materials—from painting through ceramics, imprints, tableaux composed of wood, soil, and stone, all the way to sculptures made from components including vines, wood, and charred wood salvaged after forest fires—Krajcberg kept searching for the best way to come closer to nature and to express the modes of more-than-human being in his work.[11]

Roberta Walters writes that in the last decades of his career, Krajcberg "sought to reduce his intervention in the appropriated forms he sculpted and run the risk that they would not be considered art."[12] As I have argued elsewhere,[13] his disregard for the labels attached to his creations stemmed in part from his activism, or, as Walters puts it, from the fact that "the aspect of the work most important to him was the political commentary on forest destruction. . . . seeing the destruction of nature, he came to the conclusion that art must be put to social action."[14] He increasingly regarded his artistic praxis as a means to denounce environmental crimes, especially the rampant destruction of rainforests throughout Brazil: "I became conscious [during the debates at the CNAC in 1975] that art for art's sake was finished and I wanted my sculptures to become witnesses of the disaster. My sculpture became engaged. I want to express my revolt."[15]

Krajcberg's indifference to the classification of his work as art was also tied to the process of creation that he fine-tuned throughout his life. Sylvie Depondt and Claude Mollard write that his goal was to "enter into a direct dialogue with the forest and the Brazilian landscapes," and his paintings, prints, and sculptures result from these ongoing exchanges.[16] Restany too highlighted the centrality of Krajcberg's ties to nature: while traveling with him and with the artist Sepp Baendereck along the Rio Negro, in the Amazon Basin, in 1978, Restany wrote in his diary that "nature was [Krajcberg's] culture."[17] To be sure, the artist felt a debt of gratitude to the natural world as the only cultural milieu he felt was worth engaging with, and consequently as the main inspiration for his art. But another of Restany's reflections on Krajcberg's work, formulated some years earlier, during the critic's trip to Brazil in 1974, better encapsulates the artist's connection to nature: Restany considered that Krajcberg's work expressed his "conscience of nature as culture."[18] This felicitous expression not only underlines Krajcberg's understanding of the natural environment as central to his own cultural development but also, and perhaps more significantly, suggests that for him, more-than-human forms of existence were themselves cultured. He believed that there was culture in nature. The devastation of the natural world therefore meant to him not only the annihilation of countless biological entities but also the destruction of more-than-human forms of cultural articulation. The artist saw his work as a vehicle for the expression of more-than-human cultural life and as a way to protest its loss at the hands of humanity.

When discussing his later sculptural work using tree trunks salvaged in the aftermath of forest fires, Krajcberg wrote, "This burnt bark, it's me. I feel myself in the wood and the stones. Animist? Yes. Visionary? No."[19] The artist's identification with charred tree-trunks evokes the link between acts of aggression inflicted by humans on other humans and human violence toward nature, a topic to which we will return. But his bond not only with wood but also with stones and the other natural elements he used in his work more broadly indicates that he saw more-than-human entities as beings in possession of many features, including culture, usually attributed only to humans. Hence his self-description as an animist, a term he was careful to distinguish from "visionary," understood as someone who makes often unreliable predictions about the future.[20]

Throughout Krajcberg's artistic development, I would argue, he strove to perfect his ability to express more-than-human culture through art. Depondt and Mollard hint at this desideratum when they write that "with Frans Krajcberg, there is no intermediary, no title, nothing superfluous... the elements express themselves on their own terms."[21] His aim was to create with the natural world and to let more-than-human entities speak through his art.[22] The artist's work thus involved a process of progressive self-effacement whereby his artistic intervention was perceived as a means to facilitate the articulation of natural elements. The direction of the movement in his artistic life in motion was toward foregrounding more-than-human beings, which increasingly take center stage as performers in his art. In the rest of this essay, I will examine some of the main phases in Krajcberg's work with the natural environment, from his early paintings to the majestic sculptures he made in the final decades of his career.

FROM PAINTING TO SCULPTING WITH THE ENVIRONMENT

In 1955 in Rio de Janeiro, Krajcberg made a series of paintings, *Samambaias* (Ferns), for which he was awarded the prize for best Brazilian painter at the Bienal de São Paulo in 1957. Inspired by his sojourn in the forests of Paraná, these were stylized depictions of the ferns and other plants of the south Brazilian forests, placed centrally in the canvases without any further context. In some of the *Samambaias* the leaves appear over a light background, detached from the soil, as if each painting were a study of the multifarious shapes of vegetation. Drawing attention to the ferns and the infinity of their possible forms, Krajcberg places plant life, literally and metaphorically, at the center of his work, a phytophily that would accompany him throughout his career.[23]

While continuing to focus on the motifs taken from the natural world that we find in the *Samambaias* series, Krajcberg was soon to turn his back on painting, owing to both conceptual concerns and a series of fortuitous events. On the one hand he suffered from turpentine intoxication, which precluded his ongoing handling of paint. But he also felt the need to move away from the representation of nature on canvas toward its direct presentation and incorporation in art.[24] He wanted to work with, not on, the natural environment, and to do that he had to invent new modes of expression: "I went to Ibiza. I had, for the first time, the need to feel matter, not painting."[25] He explained how his desire to be in touch with matter impacted his art: "I made prints of soil and stones. Then I took the soil directly and glued it. . . . There is no pictorial gesture. There are prints and reliefs. Pieces of nature."[26] Beginning with his first trip to Ibiza, in 1958, he used materials found in the natural environment in his compositions. This was when he created his first impressions of rocks and soils, for which he received the City of Venice Award in the 1964 Biennale. Back in Brazil, he continued to use found materials and turned to the multifaceted minerals from the region of Minas Gerais, where he also started to collect the natural dyes and pigments that he would continue to apply on his sculptures in the decades to come.

In his quest to convey the meaning expressed by pieces of the environment, Krajcberg reinvented the traditional setting of the rectangular canvas, filling it with natural elements. As Mollard and

Lismonde put it, "How can one be closer to nature than by working on transferring and combining its forms, and its colors, onto a surface meant to become a painting?"[27] In some of his works, such as an undated composition of Minas Gerais quartz assembled on wood, he carefully selected and arranged stones and applied natural pigments to parts of their surfaces. The crystals in this work seem to acquire a life of their own as they spring forth out of an ocher-and-gray surface reminiscent of the color palette that Krajcberg would later favor in his sculptures denouncing forest fires.

In other creations, Krajcberg used the classical shape of the canvas to record traces of the natural environment. In his home in Nova Viçosa, he made "print-reliefs" of the undulations left in the sand by the retreating tide, first creating a negative of the sand reliefs in plaster, then imprinting them on paper. The result evokes, in the words of Restany, "the wrinkles in the skin of the world."[28] This poetic formulation points to Krajcberg's view of his art as testimony to the fleeting beauty of every part of the natural world, parts chosen almost at random "among the unlimited virtuality of other possibilities, on that day among so many others."[29] As the artist put it, "My oeuvre is a long amorous struggle with nature; I could show a fragment of that beauty. And that's what I did."[30] A particular piece of sand in his print-reliefs only became an artwork because it was framed as such: "Taken out of its horizontality to be exhibited vertically, the object of nature becomes an artistic artifact that immortalizes an existence bound to disappear."[31] Art is, in this sense, a way to "mourn the face of the world" about to disappear, a facet of Krajcberg's oeuvre that was to become even more poignant when he started to work with elements from devastated natural landscapes.[32] But as Restany also wrote, the print-reliefs were a "matrix inscription, a drawing of nature"—that is to say, nature was the artist and Krajcberg functioned merely as an interpreter, translating the art of the natural world into a material form and an aesthetic language that could be understood by humans.[33]

Krajcberg's wish to place art at the service of nature's expression would take him ever farther from painting and from the straitjacket of square flat surfaces. As the artist put it, "I wanted to explode the square, get out of the frame. I had more than one reason to do so. Nature ignores the square, movement rotates. . . .

Life is not square and there are no fixed forms."[34] His *Cut-Out Shadows* series, which he began in the 1960s and worked on, at intervals, for over twenty years, can be seen, Malcolm McNee writes, as "intermediate works [that] indicate Krajcberg's shift away from the pictorial toward the sculptural as the preferred medium for working and exhibiting the selected natural elements and organic forms that had become his exclusive inspiration."[35] He made these works by capturing the shadows of branches, vines, mangrove and tree roots, and so on, cutting these shadows out in a wooden support, and then assembling them together with the element that originated the shadow. The ensemble was usually painted with the same mineral tint, creating a sense of continuity between the plant original and the shadow it projected.

While many of these compositions still relied on verticality and hung on walls, the "shadow" part adhering to a flat surface, the vegetal originals were veritable sculptures that rejected the flatness of painting to incorporate the fullness of form and movement to be found in plants.[36] These works can be seen as meditations on the artificiality of art. Krajcberg's cut-out shadows are a copy of a copy: the reproduction of a silhouette that originated in the projection of light passing through the contours of a plant. At the same time, the works underline an interchangeability within the process of artistic creation: the shadow and its "original" verge on being indistinguishable, and it becomes unclear whether the flat shadow determines the voluptuous shapes of the plant or vice versa. As Krajcberg wrote, "I wanted to unite the object and its shadow. I was looking for the object in its shadow."[37] Similarly, the position of the artist and of his object are also reversible. Did Krajcberg choose those particular shadows, among so many other arrangements of light, and those specific vegetal forms, or was he chosen by them?[38] Beyond a statement on the artistic potential of transience in the natural world, here frozen in one arrangement among infinite possibilities, the *Cut-Out Shadows* reveal the agency of more-than-human entities—wood, vines, light and shadow—that determine the artist's praxis.

Krajcberg's sculptures are the corollary of his artistic evolution toward embracing forms found in the natural world as the building blocks of his creations. His early sculptures of polished wood transform vegetal matter into art in a gesture to a certain extent echoing the readymades of Marcel Duchamp. But as Maria Justino points out, the acts of appropriation made by the two artists are not exactly symmetrical: "both [Duchamp and Krajcberg] were not interested in art as a product, but, rather, in its genesis, both questioning so-called art for art's sake... demanding that art participates in life, ... but while Duchamp, through scorn and irony, reaches antiart, Krajcberg... is, still... a man seduced by form, never fully breaking away from sculpture."[39] Krajcberg worked with materials found in the natural environment, but lightly altered them—polishing wood, adding colors, and so on. He made art *together with* his found materials, including plants and stones, *not on or about* them.[40]

In his use of found materials, Krajcberg came close to the *nouveau réalisme* theorized by his friend Restany. The artists of this movement took "a fresh look on the artificial and manufactured objects (created to meet the needs of urban life) and thus reveal[ed] their poetic reality."[41] Krajcberg acknowledged similarities between his work and theirs: they "were close to me because I was moved by the same desire to capture the real, making it the very matter of art: the industrial real for the New Realists, and the natural real for me."[42] But his focus on the natural world, and more specifically his work with vegetal materials—the basis of his sculptures in the final phase of his career—distanced his creations from those of the *nouveaux réalistes*. Restany suggested the term "naturalist realism" to describe the artist's work,[43] and in a manifesto he wrote with Krajcberg and Baendereck during their trip along the Rio Negro in 1978, they developed the notion of "integral naturalism" as a theoretical framework for understanding Krajcberg's art.[44]

The "Rio Negro Manifesto of Integral Naturalism," inspired by the natural environment of the Amazon, calls for a "practice [of] accessibility towards the natural datum" that would "admit the modesty of human perception and its limits."[45] In the face of the "excess of industrial and urban civilization"[46] and of the dematerialization and conceptualization of art as a reaction against the "tyranny of the object" in (post-)modern life,[47] the authors of the manifesto demand a return to nature: "original nature must be exalted as a hygiene of perception and as mental oxygen:

a whole and complete naturalism, a gigantic catalyst and accelerator of our faculties of feeling, thinking and acting."[48] For the authors of the manifesto, a deep engagement with the natural environment would trigger a renovation of artistic forms of expression, blunted by decades of consumerism in mass-culture societies. While the manifesto was not always favorably received—especially among nationalist Brazilian intellectuals, who accused the authors of neocolonialism in the Amazon—it was to prove key to Krajcberg's artistic development.[49] Mollard and Lismonde write that the manifesto "definitely convinced Krajcberg that his mission as an artist takes him beyond political and aesthetic frontiers... to denounce everything that endangers the health of the planet."[50]

Starting in the late 1970s, in the aftermath of the debates surrounding Krajcberg's solo exhibition at the CNAC and his presentation of the "Rio Negro Manifesto," his work became increasingly invested in denouncing environmental crimes: "My sculpture became engaged. I want to express my revolt. There is but one solution for the modern artist. Either his art participates in our third industrial revolution... or he fights against that pollution that is as fearsome as atomic bombs."[51] He went for the second option: "I chose to fight, to express myself no longer only with the beauty of forms of nature, but also with that nature that one is killing. My sculptures today are like a memorial of the disaster that I see and in the middle of which I am living."[52] Krajcberg continued to use materials found in the natural world, but instead of collecting these materials from relatively unspoiled landscapes, he sought them in areas devastated by human intervention.[53]

The most emblematic of Krajcberg's creations using materials found in ecological disaster-sites are his sculptures of charred wood salvaged in the aftermath of the human-induced forest fires that were and are decimating many of Brazil's old-growth forests, including the Amazon rainforest. Rather than framing his art as activism through, say, the use of photography, he encoded environmental degradation directly in the artworks.[54] The *Révolte* series encapsulates many of the central features of these pieces. In *La Révolte III* (The revolt III, 1994; fig. 1), for example, charred wood is dyed with natural red and gray pigments in a clear allusion to the fire that destroyed what was once a living tree. The tops of

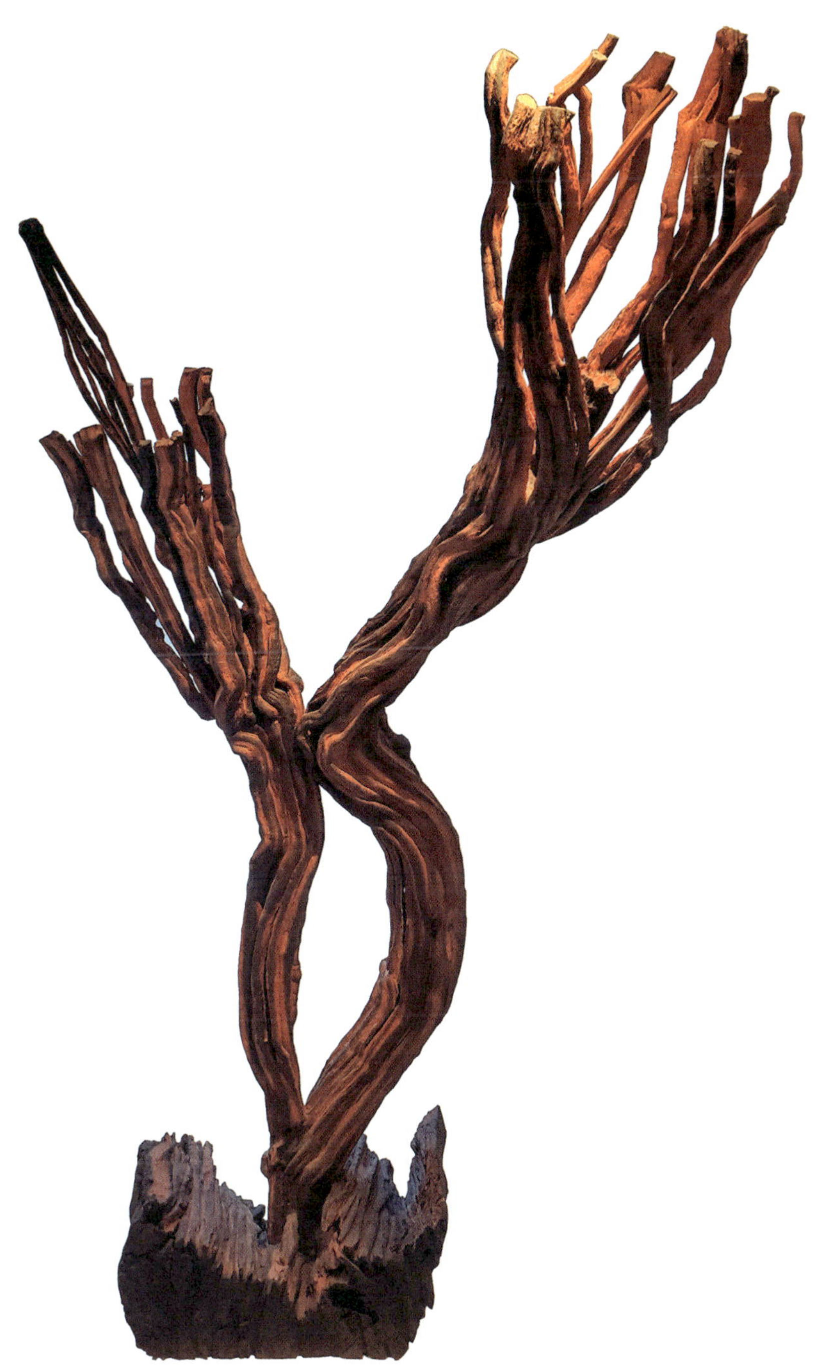

1 Frans Krajcberg. *La Révolte III* (The revolt III). 1994. Burnt wood and natural pigments, 7 ft. 10 ½ in. × 52 ¾ × 52 in. (240 × 134 × 132 cm). Gift of Frans Krajcberg to the City of Paris in 2003

the two wooden pieces recall flames, but at the same time bring to mind flowering buds, again drawing viewers' attention to the fact that these pieces of matter were once part of a vibrant vegetal being prematurely destroyed by human hand. The two upward-shooting wooden pieces stand for a series of dualities coexisting in an unresolved tension. They depart from the same root at the bottom, diverge in the middle, briefly touch again, and then move increasingly apart at the top of the composition. Beyond the contrast between a thriving plant and charred matter, life and death, the two wooden pieces, in coming together and falling apart, evoke the fraught relation between humans and the natural world. The entire composition also harks back to iconographic representations of lovers bound together in a coexistence that is not always easy, in the same way that the other dichotomous terms conjured up by the sculpture exist side by side, developing strained ties without the possibility of sublation.

The anthropomorphism of this and many other of Krajcberg's works is not fortuitous.[55] The artist often compared the destruction of the natural environment to the horrors he had witnessed as a young man during World War II: "every time I see the stacking of trees from the Amazon, burnt by humans, I can't help thinking about the ashes of the crematories."[56] The incinerated trees he encountered reminded him of Nazi atrocities; he described the destruction of the Amazon as a "Shoah of nature,"[57] and saw it as his responsibility to condemn this hideous offense: "My message is tragic. I show the crime. I bring documents on this crime, I collect them and I add to them; I want to give the most dramatic and violent face to my revolt. I want my works to be a reflection of the burns."[58] Krajcberg believed that while Prometheus's fire had once been a symbol of civilization, in late modernity humans had turned it against themselves and against the natural environment.[59] He saw the "world of ashes gaining ground every day," which is why he used them more and more in his sculpture.[60] His works in charred wood are his way to fight the necropolitics of ashes, to testify against environmental crimes, and to express his anger at the proliferation of this senseless violence.[61]

Krajcberg recognized that "it's not easy to create a work of art to shout in [my] place.... How can you make a piece of charcoal scream?"[62] Plants were his allies in denouncing environmental devastation through art. His use of incinerated tree trunks goes back to the notion of transience, in that wood is less durable than a material such as stone. In working with charred wood, he also had to respect the shape of the former plant, highlighting the collaborative effort between a human creator and vegetal life. Krajcberg explained the centrality of wood in his oeuvre: "You know that when I touch a marble sculpture it gives me a sensation of cold. Wood does not; one wants to touch it. And, you know: touching wood brings one luck. . . . It transmits heat. It seems as if it is transmitting the heat that existed when it was alive. That is very important to me. I like to feel that. . . . the heat of matter."[63] In burnt wood the artist recognized the plant that once was, imbued with the life-world of the rainforest, and he worked with its remains to create his activist art. Like Krajcberg himself, the charred tree trunks in his sculptures are survivors of human-made disasters, the continuous repetition of which he and the plants wish to warn against through their activist art.

1 Claude Mollard and Pascale Lismonde, *Frans Krajcberg: La traversée du feu* (Paris: Isthme Éditions, 2005), 16. This and all other quotes from an original in a language other than English are rendered in my translation.
2 Ibid., 65.
3 Frans Krajcberg's mother, Bina, a member of the Communist Party, was in prison for her political activities when the German army entered Poland and was murdered by the Nazis, together with all other Party members, when they reached the city of Radom, where she was being held. Krajcberg never knew for sure how the rest of his family died, but they most likely perished in concentration camps. The artist believed, "I am the last survivor of my family." Quoted in ibid., 157.
4 See Mollard and Lismonde, *La traversée du feu*, 94–95.
5 Krajcberg, quoted in ibid., 157.
6 Mollard and Lismonde, in ibid., 84.
7 Ibid., 85. Krajcberg designed the seven buildings of the museum that was to house his collection in Nova Viçosa.
8 Krajcberg, quoted in "Son histoire," n.d., available on the website of the Espace Krajcberg, https://en.espacekrajcberg.fr/histoire-de-frans-krajcberg (accessed November 3, 2023).
9 Ibid.
10 Pierre Restany, quoted in ibid.
11 As Mollard and Lismonde point out, "Is there a better way to be close to nature than to work on transferring its forms and colors onto a surface destined to become a painting?" *La traversée du feu*, 98.
12 Roberta Lanese Walters, "Frans Krajcberg: Art in Defense of the Forest," PhD thesis, Texas Tech University, 1999, 114.
13 See Patrícia Vieira, "Plant Art from the Amazon: Tree Performance in the Work of Frans Krajcberg," *Performance Philosophy* 6, no. 2 (2021): 83.
14 Walters, "Art in Defense of the Forest," 114.
15 Krajcberg, quoted in "Son histoire."
16 Sylvie Depondt and Mollard, eds., *Frans Krajcberg: un artiste en résistance* (Plaissan: Museo Éditions, 2017), 10.
17 Restany, in Mollard and Lismonde, *La traversée du feu*, 216.
18 Ibid., 177.
19 Krajcberg, quoted in "Son histoire."
20 On the history of the term "animism" and more recent debates on the subject, see Vieira, "Animist Phytofilm: Plants in Amazonian Indigenous Filmmaking," *Philosophies* 7, no. 6, 138 (2022): 4ff.
21 Depondt and Mollard, *Un artiste en résistance*, 10.
22 See Vieira, "Plant Art."
23 Krajcberg said, "For me, the history of humans is always disappointing. That of plants is much richer and happier." Quoted in Mollard and Lismonde, *La traversée du feu*, 115.
24 As Malcolm K. McNee puts it, "There is a turn away from representation of nature toward its revelation at the point where it literally overflows the traditional fields and boundaries of perception and representation. . . . That is, [Krajcberg's] works do not represent the land, or any other figuration of the real, but rather they integrate some piece of it into themselves." McNee, *The Environmental Imaginary in Brazilian Poetry and Art* (New York: Palgrave MacMillan, 2014), 106.
25 Krajcberg, quoted in "Son histoire."
26 Ibid.
27 Mollard and Lismonde, *La traversée du feu*, 98.
28 Restany, in ibid., 170.
29 Ibid., 179.
30 Krajcberg, quoted in Cristiana Bernardi Isaac, "Krajcberg e Oiticica:

Precursores da Arte na Paisagem no Brasil," *Paisagem e Ambiente: Ensaios* no. 31 (2013): 132.

31 Krajcberg, quoted in "Son histoire."

32 "Son histoire."

33 Restany, in Mollard and Lismonde, *La traversée du feu*, 179.

34 Krajcberg, quoted in "Son histoire." Framing his rejection of the square within the context of the history of twentieth-century art movements, he continues, "The abstraction of the square accompanied the revolutions of the beginning of the century, just like Expressionism accompanied its misery. I have always had an expressionist sensibility and I never saw myself as a member of Concretism. I did not want art for art's sake. I wanted to find new forms. And nature offered me thousands of those."

35 McNee, *The Environmental Imaginary*, 106.

36 See ibid.

37 Krajcberg, quoted in "Son histoire."

38 Krajcberg said that a large part of his research for the series consisted of "trying lightings to choose a shadow. There is an infinity of those." Ibid.

39 Maria José Justino, *Frans Krajcberg: A Tragicidade da Natureza pelo Olhar da Arte* (Curitiba: Travessa dos Editores, 2005), 14. Justino also distinguishes between Krajcberg and Yves Klein: Krajcberg "is also different from Yves Klein since, to find form, Krajcberg did not have to abandon all reference to matter. To the contrary, it is by diving into raw matter that he discovers it as a form of transcendence." Ibid.

40 Krajcberg's work also dialogued with other art trends of the period, namely arte povera and Land art: "But while Arte Povera worked in a spirit of provocation to reject Op-art, Krajcberg keeps from it the idea of movement. He believes that the materials inspiring him are not 'poor.' On the contrary! Their beauty speaks for itself when it comes to denouncing the destruction of nature by humans. And contrary to Land Art, he does not intervene in the landscape. He chooses natural elements that he transforms to magnify them, using the artistic codes he appropriates." "Oeuvre. Sa démarche: le naturalisme intégral," n.d., available on the website of the Espace Krajcberg, https://en.espacekrajcberg.fr/lenaturalismeintegral (accessed November 3, 2023).

41 Ibid.

42 Krajcberg, quoted in Mollard and Lismonde, *La traversée du feu*, 95.

43 Restany, in ibid., 99.

44 Restany, Baendereck, and Krajcberg, "Manifeste du Rio Negro du Naturalisme Intégrale," 1978. Repr. in Depondt and Mollard, eds., *Un artiste en résistance*, 74–76. Also available online on the website of the Espace Krajcberg, https://www.espacekrajcberg.fr/le-manifesto-do-rio-negro (accessed November 3, 2023).

45 Restany, Baendereck, and Krajcberg, "Manifeste du Rio Negro du Naturalisme Intégrale," 74. Krajcberg explained the centrality of the Amazon for the writing of the manifesto: "Amazonian nature makes me question my sensibility as a modern man. It also questions the scale of traditionally recognized aesthetic values. The current artistic chaos is a consequence of urban evolution. Here [in the Amazon] we are faced with a world of forms and vibrations, with the mystery of continuous change. We should know how to make the most of it. Integral nature can give new meaning to individual values of sensibility and creativity." Krajcberg, quoted in "Son histoire."

46 Restany, Baendereck, and Krajcberg, "Manifeste du Rio Negro du Naturalisme Intégrale," 75.

47 Ibid.

48 Ibid., 76.

49 Walters notes that when the manifesto was presented in Rio de Janeiro, São Paulo, and Brasília, some Brazilian members of the audience accused Krajcberg, Baendereck, and Restany of being foreigners who were trying to dominate the Amazon, even though Restany was the only one of them who was not a Brazilian citizen. Walters, "Art in Defense of the Forest," 106. The manifesto was more favorably received outside Brazil in North America and Europe. For an in-depth description of its reception see McNee, *The Environmental Imaginary*, 110, and Mollard and Lismonde, *La traversée du feu*, 137–39.

50 Mollard and Lismonde, *La traversée du feu*, 139. With Mollard, Krajcberg went on to write a "New Manifesto of Integral Naturalism," published in 2013, which "reaffirmed and radicalized" the call for the environmental engagement of the arts of the first manifesto. Available online at https://en.espacekrajcberg.fr/nouveau-manifeste-du-naturalisme-in (accessed November 4, 2023).

51 Krajcberg, quoted in "Son histoire."

52 Ibid.

53 McNee wonders to what extent this change points toward a deeper transformation of the role of nature in Krajcberg's oeuvre: "Is there a recognizable shift in his work . . . from a notion of nature as a force for the renovation of art to art as an instrument in the struggle to protect and preserve natural environments or to otherwise alter the way humans perceive of and engage with the nonhuman world?" McNee, *The Environmental Imaginary*, 100–101. I would argue that there is a continuum in Krajcberg's production, his latter work simply highlighting an activist streak that was implicit in his art from early on. This environmental activism was a response to the onslaught on the natural world that he witnessed firsthand in Brazil.

54 See ibid., 12.

55 Walters notes that "in [Krajcberg's] photographs of trees, as in his sculpture, the semblance of a figure is often apparent." Walters, "Art in Defense of the Forest," 30.

56 Krajcberg, quoted in Mollard and Lismonde, *La traversée du feu*, 64.

57 Ibid., 152.

58 Krajcberg, quoted in "Son histoire."

59 As Krajcberg himself put it, "I am a burnt man. Fire is death, the abyss. Fire has always been with me." Quoted in ibid.

60 Krajcberg, quoted in Mollard and Lismonde, *La traversée du feu*, 153. He also said, "The absolute gesture would be to dump, like that, in an exhibition, a truckload of charred wood, collected in the field. My work is a manifesto. I do not write: I am not a politician. I must find an image. If I could put ashes everywhere, I would be the closest to what I feel." Quoted in "Son histoire."

61 Krajcberg said, "My thing is to denounce. . . . I feel a duty to defend nature, and I'm still doing it today." Quoted in Alex Collontonio and Christian Cravo, "Screams from the Forest: Frans Krajcberg Uses His Art to Protest Man's Destruction of the Planet," *Americas* 57, no. 4 (2005): 44.

62 Ibid.

63 Krajcberg, in Walter Salles, *Krajcberg, o poeta dos vestígios* (Rio de Janeiro: Videofilmes/Rede Manchete, 1986), documentary film, 48 minutes.

64 Restany mentions that "these tree-sculptures, as I call them by instinct, contain the vital energy of the forest, that is to say, they articulate the rainforest itself." Restany, in Mollard and Lismonde, *La traversée du feu*, 169.

VÍCTOR MANUEL RODRÍGUEZ-SARMIENTO

ACTS OF VISION

Colonial Visuality and the Recurrence of the Telluric in Contemporary Colombian Art

What happens when people look, and what emerges from that act? The verb "happens" entails the visual event as an object, and "emerges" the visual image, but as a fleeting, fugitive, subjective image accrued to the subject.
—Mieke Bal, "Visual Essentialism and the Object of Visual Culture," 2003

Beginning in 1783 and intermittently for about thirty-four years following, the Viceroyalty of the New Kingdom of Granada—the Spanish Empire in northern South America—mounted an expedition, the *Real Expedición Botánica*, run by the botanist and mathematician José Celestino Mutis and covering around 8,000 square kilometers (over 3,000 square miles) in the Magdalena River basin of present-day Colombia. Its goal was both to record American flora and to map the kingdom's natural resources; the result was the multivolume compendium *Flora de la Real Expedición Botánica del Nuevo Reyno de Granada*.[1] In 1816, by order of King Fernando VII, the expedition's archive was sent to Spain in 109 containers. Although important documents went missing during the crossing, a good part of the archive is preserved in Madrid's Real Jardín Botánico;[2] it comprises 4,000 documents, 6,000 engravings, a 20,000-folio herbarium, a collection of seeds and woods, and other elements.[3]

Given the voluminous production of images by painters and draftsmen, both foreign and local, in relation to the *Expedición*

Botánica, the Colombian art historian Marta Fajardo de Rueda proposes considering the project as not only a scientific undertaking but an artistic one, entailing "an admirable integration of science and art hitherto unknown on the American continent."[4] For the Colombian painter Beatriz González, the expedition "changed art in many ways. One of these was learning how to draw from nature." González goes on to trace the project's influence on contemporary Colombian artists such as Nicolás Paris and Alberto Baraya, whose work reflects an interest in "approaching the image from the scientific point of view."[5]

This essay will review the work of contemporary Colombian artists who address cultural strategies such as the *Expedición Botánica*, examining not so much the aesthetic achievement of these artists' works as their articulation with the discursive strategies that *imagined* America. The *Expedición* along with other related visual and cultural dispositives were parts of a discourse that made America visible and articulable in the colonial context: they associated America with nature. Meanwhile the global powers claimed a *cultural* authority, justifying European supremacy and its colonial will to power. On the basis of this nature/culture dichotomy, Europeans constructed a system of truth involving both need and repudiation: America was needed to give Europe a sense of completeness, but it was also repudiated as an other that was "almost cultural, but not quite," to paraphrase Homi K. Bhabha's famous proposition "almost the same, but not quite," regarding mimicry as the sign that underlines the ambivalence of colonial discourse.[6] On the one hand, the discourse represents the colonized as similar to the colonial self in order to justify the implementation of economic, pedagogical, and cultural devices to discipline the other under the promise of becoming civilized. On the other, it represents it as inappropriate and as a menace in order to justify the implementation of military intervention and surveillance to control it.

Joaquín Barriendos has argued that "the permanent permutation of the racializing visual regimes produced after the 'invention' of the 'New World'... is constitutive of the heterarchic matrix of power out of which the coloniality of seeing and epistemological racism presently operates."[7] I would similarly like to deal not so much with seeing as with visuality, detaching seeing from images and the apparatus of vision, as well as from a knowing, centered, observing subject. And I intend to approach visuality as a component running transversal to social life, that is, as a regime or apparatus of power that structures observer and observed and proscribes other ways of seeing, making, and signifying. That is, following W. J. T. Mitchell, I aim to pass from the social construction of the image to the visual construction of the social.[8]

In this regard, Mieke Bal posits visuality as a social field, a field constituted not by the materiality of the object seen but by performative acts of vision.[9] These acts break the Enlightenment dichotomy between object and subject, showing how they mutually transform each other through the complex network of historical and social meanings that shapes them. Citing Michel Foucault, Bal argues that "knowledge ... is constituted, or rather, *performed*, in the same acts of looking that it describes, analyses and critiques."[10] At least three qualities of performativity underlie her formulation, which I wish to emphasize here:

1. Performative acts are social acts repeated in time.
2. As James L. Austin argues, some linguistic formulations are performative; that is, they create reality to the extent that they name it.[11] Such acts do not *re-present* the outside world but are *representations* in the sense proposed by Foucault—active constructions of power relations.
3. I follow Jacques Derrida in seeing performativity as highlighting a condition of language and of the production of meaning. The meaning is provisionally put together in relation to difference—in other words, the meaning is defined by what it is not. "Difference" emerges in a dual sense: the meaning "differs," that is, it opens itself incessantly, and it "defers," that is, it comes in stages, never fully arriving. As a social construction, the meaning, far from illuminating the ontology of the object, stages conflictive and heterogeneous forms of signifying the object that emerge from political and social agendas in conflict.

I propose to think through colonial visuality within the framework of a broader discourse structured around the nature/culture dichotomy, which configured colonial desire, as well as representations including images of the racialized "other" and of a "natural" America ready for colonization. Three cultural instruments are examples of the ways in which culture shaped the colonial visuality that I address here, and constructed that America as natural: the founding of the Museo Nacional de Colombia, Bogotá; the accounts of the travels of Alexander von Humboldt; and the *Real Expedición Botánica*. I will establish a critical dialogue between these events and some contemporary Colombian artistic projects that convert these colonial acts of vision into objects of analysis and performativity; that is, they stage acts of vision that challenge coloniality's acts of vision. The survey I am proposing is not based on either a chronological sequence or colonial projects or instruments. It is structured by a revision of the analytics of power by gathering archaeological and genealogical strategies, along with the places and struggles of subjectivity and resistance to the colonial visual regime.

Colonial visuality gave Europe the momentary tranquility of counting on America as a complement that could provide it with the illusion of plenitude. However, the complement becomes a threatening supplement, to the extent that Europe was made aware of its incompleteness and revealed its historical and contingent location. The artistic projects I will explore propose a kind of visual activism that does not demand positive images in the face of the negative images of America. Rather, repeating colonial acts of vision, they utilize the law of the supplement; that is, they expose the image's performative condition by turning it into the dangerous supplement that deconstructs coloniality's visual equation.

PRELUDE: NO RETURN TO THE MALOCA

Miguel Ángel Rojas makes intensive use of the colonial and colonized landscape, using it as a privileged mechanism for assembling a visual archaeology of coloniality. His interest in landscape has shaped several of his recent projects and installations. *El camino corto* (The short path, 2010), for example, is a kind of giant *mappa mundi*—it measures almost fifteen by thirty-nine feet—showing 700 names, cocaine-addicted celebrities in the upper part, drug traffickers, identified by their nicknames, in the lower. The names are written in coca leaves and dollar bills, coca leaves on the upper part—the northern part, in conventional map orientation—and dollars on the lower, southern part; this inversion of the landscape invites the viewer to connect the points on the meridians and parallels that create the shortest path in a geopolitics of power and desire. That path indexes not only the lives that vanish in the war against drug trafficking, and through addiction, but also those who demand the extermination of ancestral communities living in the Amazon forest and the possibility of life on the planet.

Nuevo El Dorado (Economía Salvaje) (New El Dorado [Savage economy], 2018; fig. 1) is made from satellite photographs of the landscape of the Amazon. *Nuevo El Dorado* measures over thirteen feet by fifty-nine feet, *Economía Salvaje* over thirteen by fifty-two feet. This two-part work is a print, but instead of using the inks and pigments of conventional screen printing, it constructs the Amazon landscape out of *mambe* (dried and ground coca leaves), gold leaf, local clay, and silver leaf. The rivers and their tributaries, as well as the wetlands, are traced in gold leaf. In *Economía Salvaje*, made in the style of *Nuevo El Dorado* and showing the same landscape, satellite records of tropical storms are superimposed in gold while chainsaw blades, the saws used in the deforestation and appropriation of Amazonian lands, appear in silver.

These mural-scale prints reveal the intricate map of economic and political interests that weave through Amazonia. As their titles indicate, Rojas insists on showing how that river basin, the only surviving originary territory in America, is becoming the target of a new conquest and colonization of its fauna and flora, along with the extermination of the peoples who inhabit it. In this regard Rojas has pointed out, “Industries are increasingly invading the borders of territories that were once full of life and conducive to human life. . . . I find no difference from the plundering and expropriation that began five centuries ago with the European conquest in America. . . . The whole situation makes me think there’s no return . . . no return to the *maloca* [the Amazonian communal house].”[12]

The prints repeat the landscape as the code that gave form to the European representation of America. By supplementing the code, they historically situate a vision in which nature emerges as

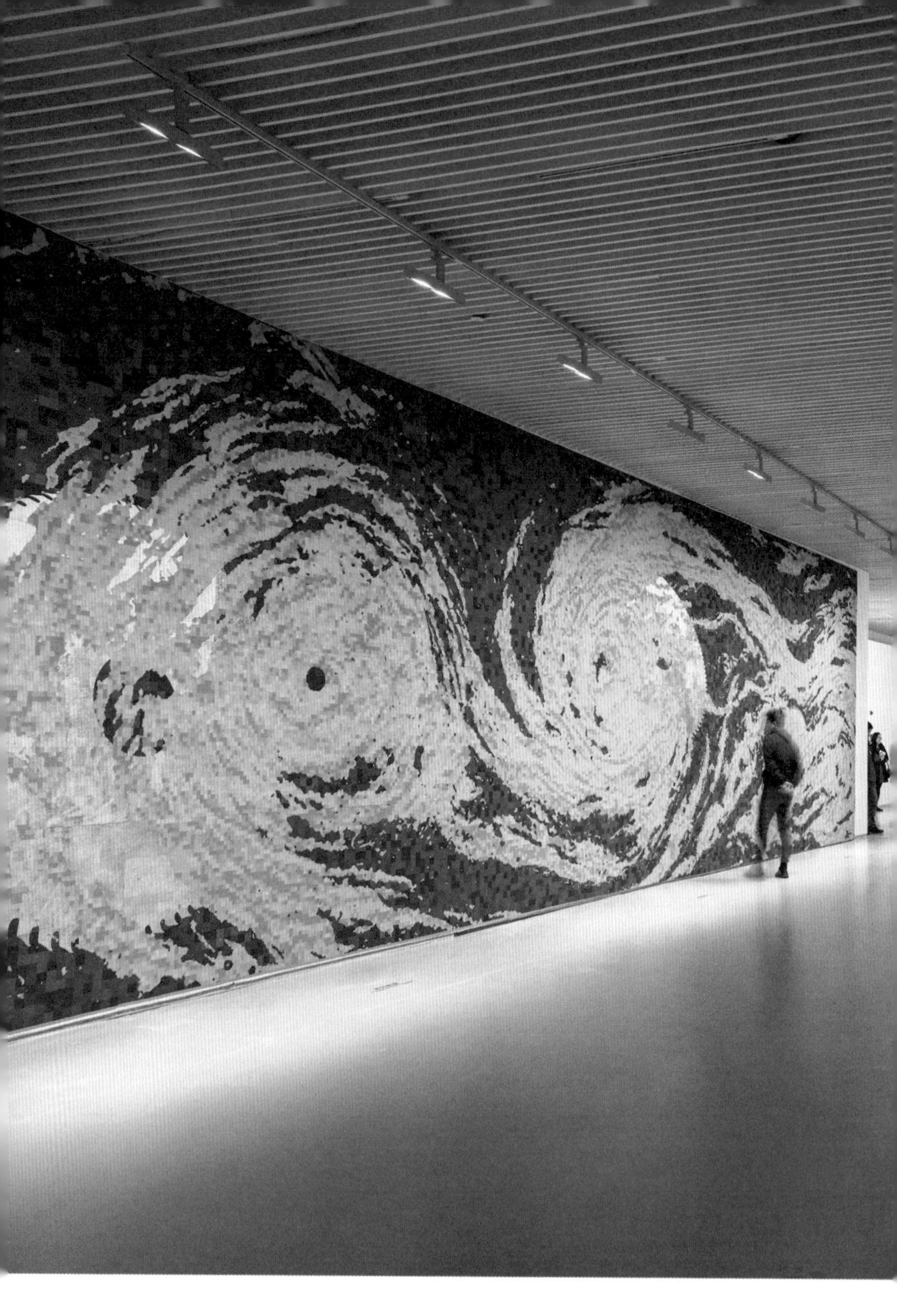

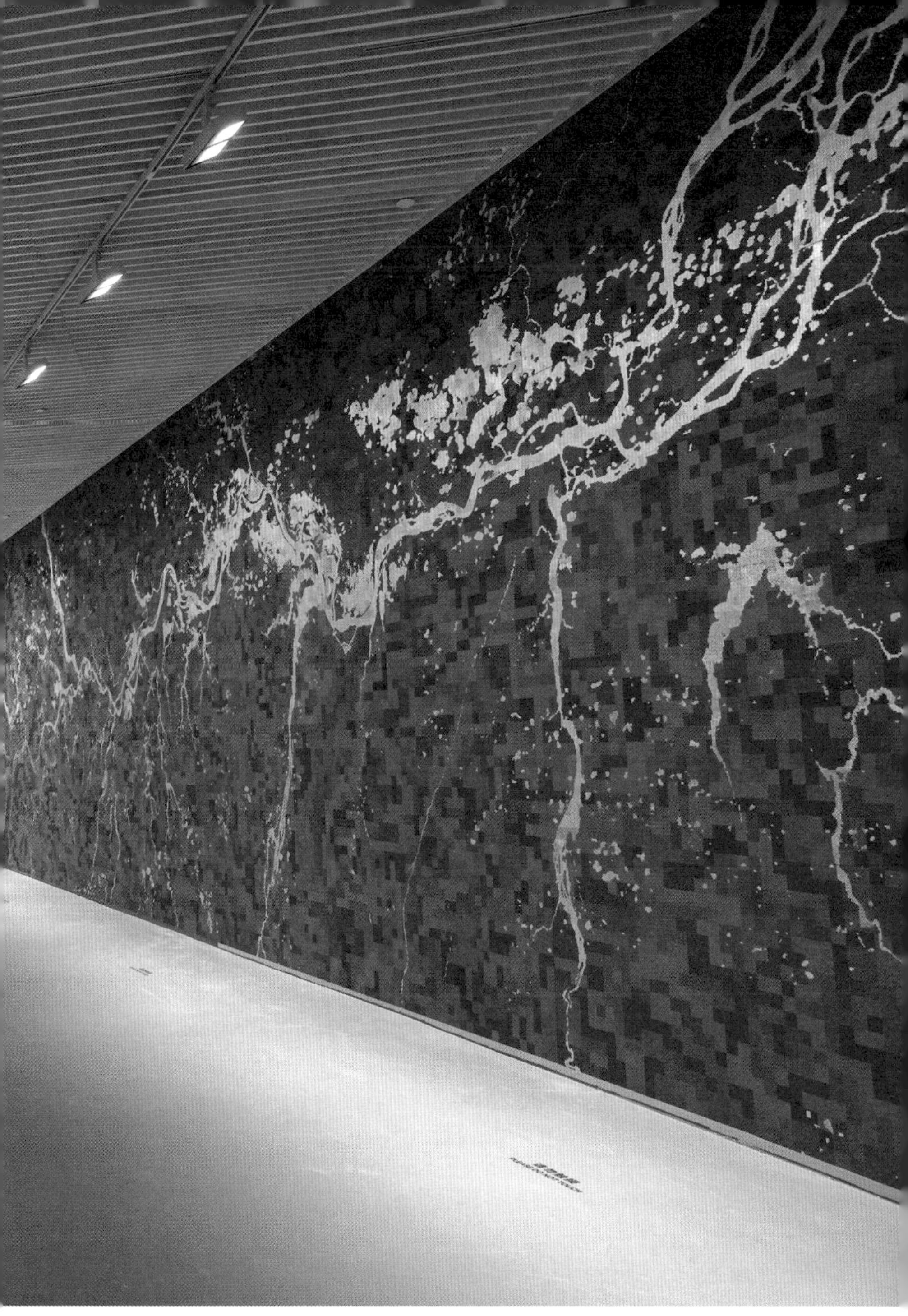

1 Miguel Angel Rojas. *Nuevo El Dorado (Economía Salvaje)* (New El Dorado [Savage economy]). 2018. Coca-leaf powder, neutral silk-screen base, clay, mineral pigments, and gold and silver leaf on fiber paper on foam board, 13 × 59 ft. and 13 × 52 ft. Collection the artist

a cultural and political construct suitable for colonization. Various greens and ochers evoke destruction and death, but they also recall military camouflage, setting up the Amazon as a crossroads: the goal of a new colonial undertaking that is protected as a goal of exploitation but unprotected in its biodiversity and in terms of the survival of the planet. The rivers and waters in gold leaf call to mind Indigenous communities' sacred narrative of water, challenging the Western gaze that considers water a source of wealth. If the map in *El camino corto* inverts north and south, *Nuevo El Dorado (Economía Salvaje)* disrupts the place of the gaze. Their object is the threatening condition of colonial visuality: we are no longer observing a "natural" landscape; rather, the landscape as a place of plunder and extermination is looking at us.

THE NATIONAL BUST

The Museo Nacional de Colombia was inaugurated in July 1823.[13] One year later, in its issue for July 18, 1824, the government newspaper *Gaceta de Colombia* described the collection representing the newborn republic: "In its infancy, the museum already possesses some rarities: . . . a collection of minerals. . . . some pieces of meteoric iron. . . . Many bones of unknown animals. . . . A mummy found near Tunja. . . . Some insects. . . . several mammals, reptiles, fishes, and several instruments."[14]

For the political scientist Benedict Anderson, the nation is an "imagined community" that uses cultural instruments such as the census, the map, and the museum to give homogeneity to the actual heterogeneity of "national" times and places.[15] While the collection of the Museo Nacional de Colombia inscribes America in the world order as a source of natural wealth, it ignores its people. This oblivion, as Anderson maintains, conveyed the fear and contention of the local population on the part of the creoles, those born in South America but of European descent. He has insisted that it was not resistance to Spain that paved the way for independence but the need of the creole population to contain the internal difference represented by enslaved and Indigenous communities. As Bhabha points out, the national grammar is written in a syntax based on the oxymoron forgetting/remembering, since in order to remember the nation, it is necessary to forget the histories of domination and extermination at its origin.[16]

In 1970, the artist Antonio Caro carried out an action in the XXI Salón Nacional de Artistas, an annual exhibition which that year was held in the Museo Nacional de Colombia. *Homenaje tardío de sus amigas y amigos de Zipaquirá, Manaure y Galerazamba* (Belated tribute from his friends from Zipaquirá, Manaure, and Galerazamba, 1970), also known as *Cabeza de sal* (Salt head) or *Cabeza de Lleras* (Head of Lleras), was a head made out of salt and housed in a glass box. The action consisted of the artist and visitors sprinkling water on the head, which, wearing glasses, evoked Carlos Lleras Restrepo, the president of Colombia until earlier that year. The glass box had leaks and the salt water flooded onto the floor.

Acknowledged as Caro's first project, *Cabeza de sal* inaugurated his interest in Indigenous communities and in the use of native materials such as achiote, maize, salt, papaya, and others. By using these natural media, Caro restated the stereotype of America as "natural" and deconstructed the Colombian nation's drive toward homogeneity. According to Caro, he was in part inspired by Gabriela Samper's twelve-minute documentary *El hombre de la sal* (The man of salt, 1969), which narrates a man's experience of the artisanal production of salt and explores the place of salt in both pre-Columbian and present-day peasant culture, its construction of an intimate relationship among culture, community, and territory. The divide between artisanal practices and the industrial methods used by the state reflects political tensions that date back to colonial times, but today those tensions have spread, less because artisanal techniques have been defended than over the ownership of Colombia's most important salt deposits, which lie in Zipaquirá, Manaure, and Galerazamba.

In 1970, at the time of the XXI Salón Nacional de Artistas, the collection of the Museo Nacional de Colombia was located on three floors of the building that had housed it since 1946, a panopticon-style structure originally designed as a prison. On the first floor were found archaeological materials and several mummies from ancestral communities; on the second, a large collection of portraits of ex-presidents and dignitaries of the Republic; on the third, the spirit of the nation was extolled through art. Indigenous and Afro-Colombian communities were represented as vestigial and their participation in republican and contemporary life was passed over. The national narrative was told as if Indigenous people and the descendants of enslaved Africans had disappeared after the conquest and during the colonial period to make way for the modern state, embodied in portraits of the rulers of Colombia and the illustrations of modern culture embodied by art.

Caro said that the production of *Cabeza de sal* led him to visit the salt deposits of Zipaquirá and to explore the work of the salt artisans there.[17] The head was made by an artisan who deposited salt and water on a clay mold of a head, then dried it with fire, imitating the use of evaporation in the mode of salt production specific to Indigenous communities. In supplying the head with the type of spectacles worn by an ex-president of the modern Colombian state (as documented in photographs of *Cabeza de sal*), Caro echoed the pedagogical and visual resources used by the museum to create the nation as a continuous, uniform space-time.[18]

Caro deconstructed the national narrative by inverting the process of salt production, dissolving the salt in water rather than crystallizing it out of water. As an "artwork" exhibited in the Museo Nacional, *Cabeza de sal* evoked the third floor of the museum's collection, its representation of the nation through art. As the bust of an ex-president, it recalled the second floor's series of portraits, making the republic visible; Lleras became a metonym for all the Colombian presidents who, in the museum's narrative, had given birth to the nation. And finally, in dissolving as water was applied to it, salt emerged in its "natural" form, connecting it to the peoples whom the national narrative wished to forget and becoming a metonym for the Indigenous population, which the museum represented as vestiges and a collection of natural objects on its first floor. In this way the museum itself was inverted and, as a metonym of the nation, was dislocated and decentered. In pouring water on the head, Caro and the participating visitors make the "national bust" disappear, revealing the nation's incompleteness and forgotten histories. The project's full title paid homage to ancestral cultures and underscored the importance of salt to their ways of life.

"Being obliged to forget," writes Bhabha, "becomes the basis for remembering the nation."[19] Caro disrupted what has to be forgotten in order to remember the nation and emphasized what has to be remembered in order to forget the modern nation and colonialism. The national bust disappeared, leaving the spectacles alone remaining as the vestige of a situated, contingent act of vision. For the Venezuelan painter and poet Juan Calzadilla, a juror of the XXI Salón, *Cabeza* "contains an original idea, cleverly resolved in an anti-artistic form that corresponds to the political art of our times, namely, a kind of *arte povera*... with the sole aim of challenging and annoying, far removed from any aesthetic purpose." And he went on, "This head of salt sporting dark glasses is, more than that, an ephemeral artwork intended to be destroyed and whose meaning resides in its own disappearance."[20] The anti-artistic and political purpose of this action resides in its appearing to translate the piece's ephemeral character into the very idea of the nation destined "to be destroyed and whose meaning resides in its own disappearance."

THE GAME OF RUSSIAN DOLLS

Between 1799 and 1804, Alexander von Humboldt traversed a large expanse of Latin America, creating a voluminous legacy of scientific observations, personal letters, and sketches. In his *Vues des cordillères, et monumens des peuples indigènes de l'Amérique* (*Views of the Cordilleras and Monuments of the Indigenous Peoples of the Americas*, 1810), Humboldt recounts his travels through what is now known as Mexico, Peru, Ecuador, and Colombia. In the first chapter he describes his experience crossing the Quindío mountain (part of what is now known as the Cordillera Central), which, according to him, is recognized as one of the toughest and most painful passes in the Colombian Andes. Besides spanning landscapes rich in native flora and fauna, the Quindío pass offered Humboldt archaeological sites and *lieux de mémoire* of the cultures that had inhabited that territory. In September 1801, en route to Popayán, he described the landscape as follows:

> Upon entering the Quindío cordillera... a highly picturesque vista was presented to the sight. The truncated code of [the volcano] Tolima... looks out above a massif of

> granitic rocks. In the background could be discerned . . . the great basin of the Magdalena [river] and the eastern chain of the Andes. . . . Here we found the *Ceroxylon andicola* palm. . . . It reaches the incredible height of 200 feet, and the traveler is surprised to find a plant of this species in a near-frigid zone, at more than 9,000 feet above sea level.[21]

This sight appears to have been the departure point for Humboldt's engraving *Paso del Quindío en los Andes* (The Quindío pass in the Andes, 1810), from which José Alejandro Restrepo in turn developed the video installation *Paso del Quindío I* (1992; fig. 2), after traveling through the pass using the diaries as a guide. Restrepo has said of the engraving,

> It's the result of a game of Russian dolls: Humboldt personally made a sketch in 1801; almost ten years later, he gave it to Joseph Anton Koch to make a definitive drawing in Rome, and Koch in turn gave it to Christian Friedrich Traugott Duttenhofer to make an engraving in Stuttgart. Humboldt himself participated by activating his memory and his compositional criteria, giving precise instructions: put a fallen trunk here with orchids here and there, backlit plants in the foreground, etc. . . . It's a palimpsest, a bricolage, a practice of montage.[22]

Restrepo's *Paso del Quindío I* consists of seventeen monitors arranged to simulate a mountain. Sound is integrated into the installation through a melody of just four notes, played on cello by Eduardo Valenzuela, as Restrepo states. He has said that he took a particular interest in the figure of the bearer, the man who had to carry "illustrious" people on his back for twenty days. In the video, the palm that seduced Humboldt appears insistently, sporadically, and discontinuously. The black and white images of the vegetation and the bearers seem titillating, incomplete, and diffuse, simulating the texture of the "original" engraving.

I think of *Paso del Quindío I* as a reflection on the untranslatability of cultural difference. Beyond the discussion around the precision of the European imagination of America, the project makes visible the limits of representation and its social and environmental effects. In this context, Humboldt's vision of the Paso del Quindío evidences the failed dissemination of the European imagination, interspersed with forms of representation that combine fantasy and fetish: America is what is desired, but it is also the impossible. Dissemination is an ambivalent maneuver of language in which a signified oscillates between the desire to remain faithful to itself and its possible death in cultural translation. The universalist drive of coloniality becomes its own ruin: local customs make it evident that, although all cultures are constantly subject to translation, cultural difference emerges as the untranslatable. Of the enunciated signifier, only shreds and patches remain.

Paso del Quindío I thus describes the sinuous, unpredictable trajectory of representation: it is the "*paso*" (in the sense of "step," "passage") from the frustration of not being able to dominate an "original" meaning to the fantasy of the fetish. The terms of the colonial nature/culture equation are sinisterly displaced between flora—the palm tree—and the bearers. These titillating and diffuse images are almost indecipherable signs underscoring the impossibility of desire. In the colonial operation of imagining America, the representation of the "untranslatable" landscape becomes a powerful mechanism of domination. From Humboldt's desire and, through him, from the European colonial drive, there remains devastation but also, and above all, the resistance that occupies the untranslatable interstices in the interchange of cultures.

VISUAL EPIPHANIES OF COLONIALITY

Here I would like to return to *Flora de la Real Expedición Botánica del Nuevo Reyno de Granada*, which I will discuss as a counterpoint to María Elvira Escallón's *Nuevas Floras* (2003–). Although Escallón does not directly relate her work to the *Expedición Botánica*, there is a way in which the two are in dialogue about the visual construction of coloniality. To the extent that the *Expedición Botánica* was linked to the production of knowledge about the South American land, and so to amplifying the wealth of the European empires, its pictorial practice functioned as a strategic visual instrument of colonial expansion. As Daniela Bleichmar writes, "In the eighteenth-century

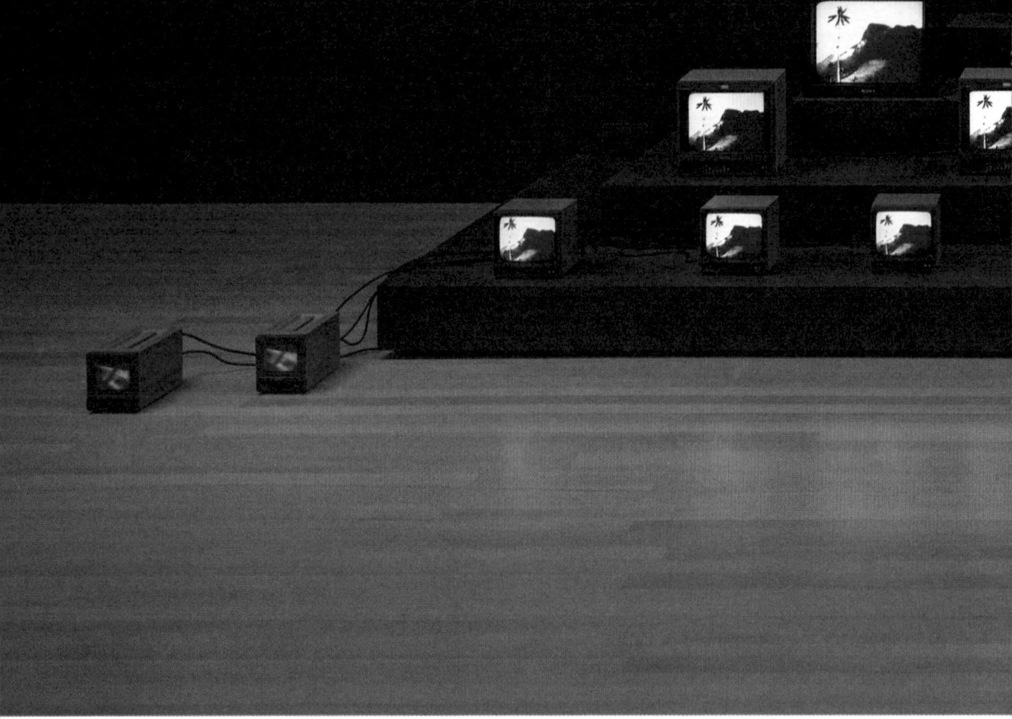

2 José Alejandro Restrepo. *Paso del Quindío I* (Quindío way I). 1992. Three-channel video installation, black and white, sound, dimensions variable. The Museum of Modern Art, New York. Gift of Patricia Phelps de Cisneros through the Latin American and Caribbean Fund

study of nature, seeing was intimately connected to both knowledge and owning. Images of plants and animals were more than pleasant, secondary by-products of exploration: they were instruments of possession."[23]

As in Humboldt's case, albeit with different modes of operation, the iconographic production of the *Expedición* was articulated through European ways of seeing. Its unfaithfulness to the original does not make it false; its condition of truth lay not in the ontology of its "materiality" but in its political and social effects. In this way, an imagined American flora was created, available for the colonial enterprise and making botany into a source of knowledge and power.

As various writers have argued, the images resulting from the *Expedición* stemmed from specimens removed from their context, from travelers' accounts, and even from observation of the plants in the studio greenhouse.[24] More than a space of scientific research, the laboratory was an atelier, a space of artistic creation, and a kind of "Mutis style" began to emerge in the representation of American flora. On the basis of a complex process of harvesting, handling, and dissecting plants, a standardized procedure was created for the production of the sketches and final plates. As Bleichmar recounts, the representations were altered according to at least three criteria. The first was time: the drawing of the model was altered in order to demonstrate different phases of growth. The second was size: the branches and leaves were adjusted to best occupy the space of the plate. The third was color: the brightness and tones of the flowers, leaves, and stalks were modified according to European taste. This visual domestication of the plants calls to mind the processes of colonial domination based on the nature/culture dichotomy, which encompassed not only peoples and their cultures but nature itself.[25]

Escallón's project *Nuevas Floras* (New flowers; fig. 3) was created in the Bogotá savanna and won the city of Bogotá's Premio Luis Caballero in 2003. It has also been carried out elsewhere in Latin America and Europe. The initial project consisted of interventions on native trees of the Colombian Andes, which Escallón inscribed in "the great processor that is nature, and in time, which is the great sculptor.... In turn, these pieces are inscribed in two spaces: the domestic landscape and the colonial landscape."[26] *Nuevas Floras*, she says, emerged from a visual epiphany. Looking at the trees of country houses in the municipalities of Sopó and La Calera, adjacent to Bogotá, she imagined working on the tree at a scale recalling both colonial furniture and European Baroque architecture. She initially considered working with digital alterations of photographs, but abandoned the idea because that approach would fail to show how nature would change as a result of the intervention. Instead, with the assistance of local woodworking artisans, she collected fallen branches that she then inserted into the "scars of the tree" like prostheses. When intervening in live trees, she took care not to cause irreparable damage.

If the modus operandi of the *Expedición* moved the production of American flora toward a more "natural" register, Escallón's, by intervening in native trees, moves it from nature toward culture and politics. Repeating the act of "domestication" visually used by the *Expedición*, her project creates the space for an act of vision or epiphany revealing the visual, cultural, and political construction of nature. For the initial staging of the project, photographic records of her interventions on native trees were exhibited in the Galería Santa Fe, Bogotá. This brings us back to the *Expedición*: what we perceive in both *Flora de la Real Expedición Botánica* and *Nuevas Floras* is a visual register of an object we cannot access. Its materiality is presented to us performatively. But whereas *Flora de la Real Expedición Botánica* celebrates the European imagination of seeing and possessing, *Nuevas Floras*, by calling attention to the sinister operation of "domesticating" native trees, reveals to us the traumatic experience of the colonial enterprise as a collective historical wound.

After being "domesticated," the tree transforms itself through its own resilience and resistance to domestication. Weeks after Escallón's intervention, the wounds begin to scar up and the trees produce new branches and leaves. Already in the time of the *Expedición*, the use of mestizo, Indigenous, and enslaved hands for the harvesting and maintenance of the herbarium, as well as for the production of the images, threatened Mutis's will to style. Bleichmar notes, "Captured by the power of the image, the viewer easily misses a most telling detail." Like an epiphany, an act of vision emerges:

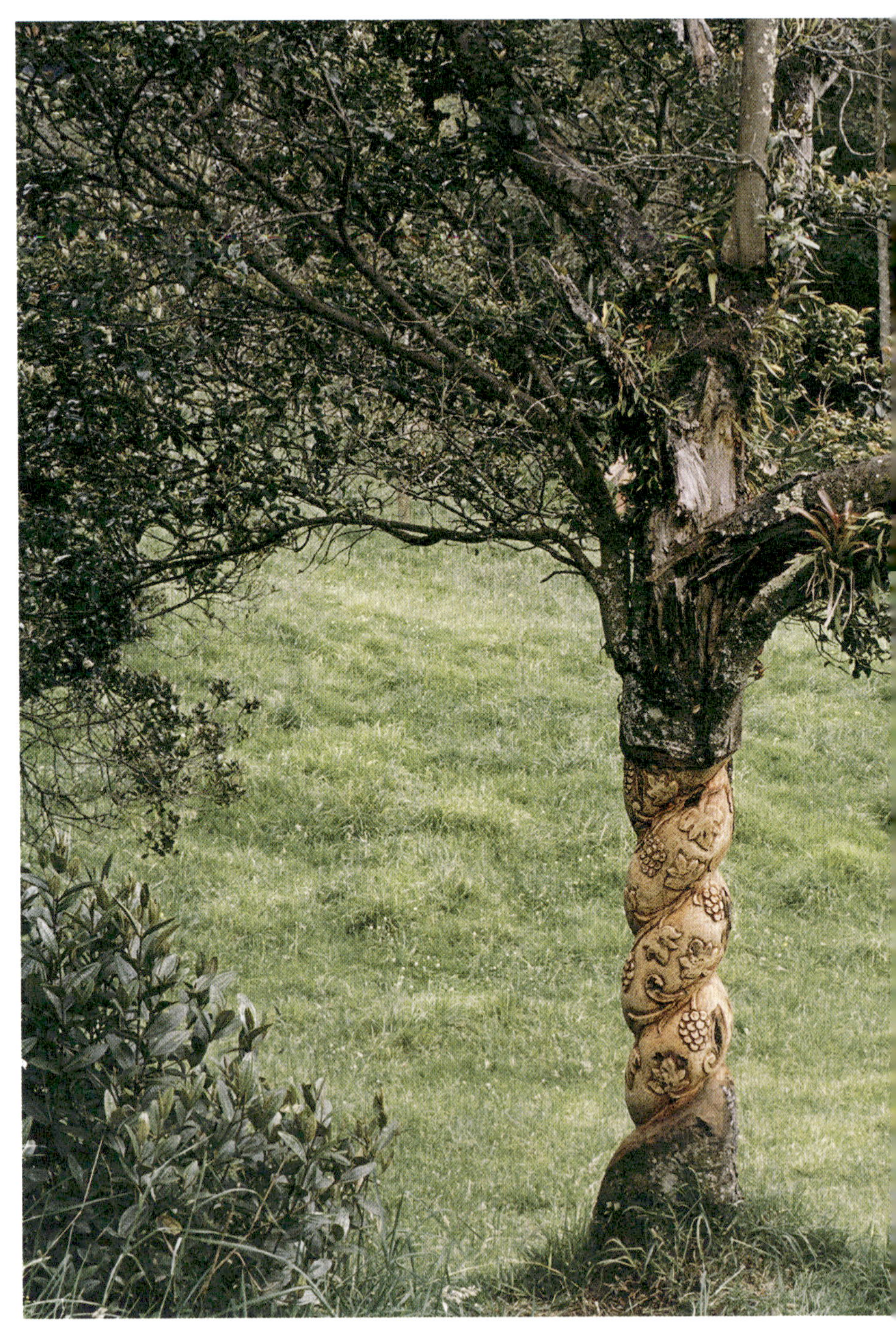

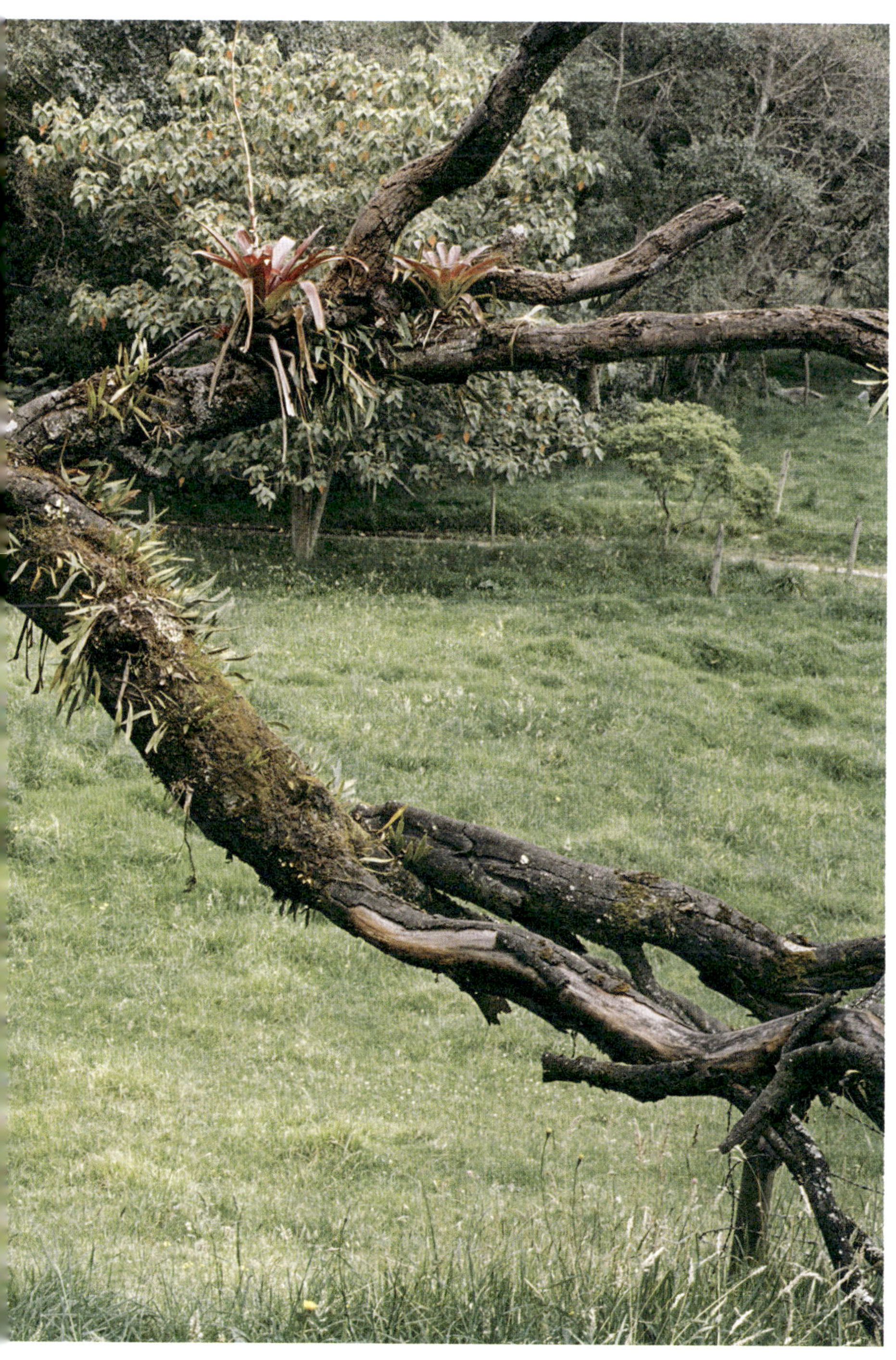

3 María Elvira Escallón. From the series *Nuevas Floras* (New flowers). 2003. Chromogenic print, 31 ½ × 47 ¼ in. (80 × 120 cm). Collection the artist

"In the bottom left corner, the painter has not only signed his name, Cortés, but also added something extremely unusual: a denomination of origin. The tag '*Americ. Pinx.*,' an abbreviation for *Americus pinxit*, declares 'Cortés the American painted this.' By not only signing his name but also pointing out his origin, and by extension the image's and the plant's, the painter shows how European botany and European illustration were appropriated and reinterpreted by the American artists who participated in Mutis's expedition."[27] The subterfuges of local imagination undermined the "reality" of the style, and the illusion of "drawing from nature." Through unofficial processes of representation, local draftsmen deliberately altered the canon, and in the interstices, if you will, they left traces of their own artistic affirmations, ignoring details that frustrated the desire of the European vision. *Nuevas Floras* not only evokes the colonial condition of visuality but underscores how vision's strategies of power are always accompanied by strategies of resistance that testify to supplementary ways of living in adversity.

The visuality of coloniality is constructed on the basis of the nature/culture dichotomy and is mobilized on the basis of the European visual drive to represent its other as the necessary yet repudiated element of the colonial equation. The examination of acts of vision as situated events enables us to understand its images within the world of social relationships, and as a disputed space in which power is at stake. By calling attention to the performative conditions of such projects as *Flora de la Real Expedición Botánica*, and inscribing them in the social world, we see how their images, beyond any aesthetic achievement they may attain, place themselves at the service of cultural and political struggles and activisms. Here we return to Mieke Bal, and the idea that acts of vision are linked to the historicity of visuality, its social anchoring, and its openness to analysis of its qualities, which are not only visual but acoustic, tactile, olfactory, and more. Through supplementary strategies, the projects of the contemporary artists discussed here suspend the discursive operations of visuality and place in question their condition of truth, opening new paths toward the decolonization of visuality that always emerges at the marginal borders of the colonial equation.

1 In 1952, the governments of Colombia and Spain began to publish a modern edition of the results of the expedition and its iconography under the title *La colección Flora de la Real Expedición Botánica de José Celestino Mutis*. Thirty-six volumes have been published, out of the ultimate fifty-one.

2 The collection of the Real Jardín Botánico, Madrid, may be accessed online at www.gbif.es/instituciones/cslc-real-jardın-botanico/ (accessed October 3, 2023).

3 See G. Fonnegra, *Mutis y la Expedición Botánica* (Bogotá: El Áncora Editores, 2008).

4 See Marta Fajardo de Rueda, "La obra artística de la Real Expedición Botánica del Nuevo Reino de Granada en el Siglo XVII, 1783–1816," *Ensayos: Historia y Teoría del Arte* 1 (1995): 104–30. Available online at https://revistas.unal.edu.co/index.php/ensayo/article/view/46328/47921 (accessed October 3, 2023).

5 Beatriz González, quoted in Pablo Correa, "Mutis revolucionó las artes," *El Espectador* (Bogotá), October 19, 2013. Available online at https://www.elespectador.com/educacion/mutis-revoluciono-las-artes-article-453290/

6 Homi K. Bhabha, "Of Mimicry and Man: The Ambivalence of Colonial Discourse," in *The Location of Culture* (London and New York: Routledge, 1994), 86.

7 Joaquín Barriendos, "La colonialidad del ver. Hacia un nuevo diálogo visual interepistémico," *Nómadas* (Col) no. 35 (2011): 15. Available online at www.redalyc.org/articulo.oa?id=105122653002 (accessed October 4, 2023).

8 See W. J. T. Mitchell, "Showing Seeing: A Critique of Visual Culture," *Journal of Visual Culture* 1–2 (2002): 165–81.

9 See Mieke Bal, "Visual Essentialism and the Object of Visual Culture," *Journal of Visual Culture* 2, no. 1 (2003): 11. Available online at http://vcu.sagepub.com/cgi/content/abstract/2/1/5 (accessed October 4, 2023)

10 Ibid., 11.

11 See James L. Austin, *How to Do Things with Words* (Cambridge, MA: Harvard University Press, 2003).

12 Miguel Ángel Rojas, quoted in "Miguel Ángel Rojas: Regreso a la Maloca," *Artishock*, May 20, 2021. Available online at https://artishockrevista.com/2021/05/20/miguel-angel-rojas-regreso-a-la-maloca/ (accessed October 5, 2023). According to Oscar Freire, "the maloca is a divine archetype, the uterus of Mother Earth, the house of the sun and moon or the receptacle of the heavenly ray." Freire, "La *maloca* de los sabedores," Part I. Available online at https://dokumen.tips/documents/la-maloca-de-los-sabedores.html (accessed October 5, 2023).

13 Some of the ideas in this section appeared earlier in Victor Manuel Rodríguez-Sarmiento, "La Fundación del Museo Nacional de Colombia: Gabinetes de curiosidades, órdenes discursivos y retóricas nacionales," in Santiago Castro-Gomez, ed., *Pensar el siglo XIX. Cultura, biopolítica y modernidad en Colombia* (Pittsburgh: Instituto Internacional de Literatura Iberoamericana, 2004), 165–84, and "El *un-art* de Antonio Caro y las políticas del conceptualismo desde América Latina," in *Acción Plástica: Homenaje a Antonio Caro*, exh. cat. (Bogotá: Museo Nacional de Colombia, 2022), 10–33.

14 *Gaceta de Colombia* (Bogotá) no. 144 (July 18, 1824). Available online at https://babel.banrepcultural.org/digital/collection/p17054coll26

/id/4403 (accessed October 5, 2023).

15 Benedict Anderson, *Imagined Communities: Reflections on the Origin and Spread of Nationalism* (London and New York: Verso, 2006).

16 Bhabha, "DissemiNation: Time, Narrative and the Margins of the Modern Nation," in *The Location of Culture*, 139–70.

17 See Juliana Flórez Luna, *Caro es caro*, documentary, 57 min. (Bogotá: Escuela de cine y televisión Universidad Nacional de Colombia, Interferencia Films, 2005); Rodríguez, "Entrevista a Antonio Caro," *Valdez* 2 (2007): 338–51; Rodríguez, "Entrevista pública," in Rodríguez, ed., *Prácticas artísticas/enfoques contemporáneos* (Bogotá: UNAL e IDCT, 2003), 49–70; and Rodríguez, "Antonio Caro," *Bomb* no. 110 (Winter 2010):16–24.

18 Caro made no more versions of *Cabeza de sal*. In the exhibition *Acción Plástica: Homenaje a Antonio Caro*, which I curated for the Museo Nacional de Colombia in 2022, I presented a version made by a salt artisan from Zipaquirá.

19 Bhabha, "DissemiNation," 161.

20 Juan Calzadilla, "Soy espectador de un funeral," *El Espectador*, October 21, 1970.

21 Alexander von Humboldt, *Vues des cordillères, et monumens des peuples indigènes de l'Amérique* (Paris: F. Schoell, 1810), 14–15.

22 José Alejandro Restrepo, in Restrepo and Madeline Murphy Turner, "Histories of Landscape during the Colonization of the Americas," *Magazine* (The Museum of Modern Art, New York), October 6, 2020. Available online at www.moma.org/magazine/articles/431 (accessed November 4, 2023).

23 Daniela Bleichmar, "Painting as Exploration: Visualizing Nature in Eighteenth-Century Colonial Science," in Norma Jay and Sumathi Ramaswamy, eds. *Empires of Vision* (Durham, NC: Duke University Press, 2014), 65.

24 This discussion of the Expedición combines information and ideas from many sources. In addition to Bleichmar, "Painting as Exploration," see especially José Antonio Amaya, *Bibliografía de la Real Expedición Botánica del Nuevo Reyno de Granada* (Bogotá: Instituto Colombiano de Cultura Hispánica, 1983); Marcelo Frías Núñez, *Tras El Dorado vegetal: José Celestino Mutis y la Real Expedición Botánica del Nuevo Reino de Granada (1783–1808)* (Seville: Diputación Provincial de Sevilla, 1994); and other authors cited in these notes.

25 See Bleichmar, "Painting as Exploration," 74.

26 María Elvira Escallón, in an interview with A. M. Ochoa. Available online at https://galeriasantafe.gov.co/gsfradio-maria-escallon/ (accessed October 13, 2023).

27 Bleichmar, "Painting as Exploration," 83.

JOSÉ FALCONI

THREE ENCHANTED POSTCARDS

Mimicking Nature, Landscape Simulacra, Territorial Sampling (Or the Work of Vivian Suter)

> *The authentic artist cannot turn his back on the contradictions that inhabit our landscapes.*
> —Robert Smithson, "Frederick Law Olmsted and the Dialectical Landscape," 1973

MIMICKING NATURE

Nesting its waters between two huge volcanoes, Lake Atitlán, Guatemala, overwhelms its visitors by conjuring the four states of matter in perpetual suspension: it makes you feel like you're in front of an infinite pool of water, framed by an unending cliff, covered by clouds, and crowned by everlasting fires. Soundtrack such a view with some Earth, Wind & Fire (to double up on the basic states of matter) and you will have the closest thing you can get to a DIY low-tech Gesamtkunstwerk.

Aldous Huxley, who warned against futuristic societies, then traveled extensively around the Americas doing psychedelic drugs, could not help but compare Atitlán with Lake Como, in northern Italy. He concluded bucolically, "Lake Como, it seems to me, touches on the limit of permissibly picturesque, but Atitlán is Como with additional embellishments of several immense volcanoes. It really is too much of a good thing."[1]

Huxley's Shakespearean disqualification of Atitlán is one of the politest formulations I can recall of one of the most enduring tropes of the American continent: its exuberance overflows the European mold and thus disqualifies the region, setting it outside of the norm.

"Too much of a good thing" is another way of saying that it is not good at all, because its unrestrained dimensions render it cloying.

It was probably such excess that drove Agustín Iriarte, member of the artists' group TRIAMA, to attempt to capture the misty shores of Lake Amatitlán—a smaller lake, only a little over sixty miles away, with very similar characteristics to Atitlán—and balk at how properly to depict it.[2] Between 1900 and 1916, Iriarte presented the lake in a number of ways, ranging from realist modes to modernist ones (literary modernism, that is to say, as proposed by Rubén Darío), with nymphs in its waters, kick-starting a national landscape tradition with a gambit: the lakes and torrid vegetation will be there alright, but the styles in which they will be represented, all of them, will be foreign.[3] It should not surprise us that this formulation followed the path of most things South American up until that point (and arguably up until the present day): the raw materials were ours, but the grammar and the technique belonged to others.

It should not be surprising, then, to encounter and feel the same sort of impasse at the center of the discussion surrounding the arresting paintings of Vivian Suter (fig. 1). Coincidentally, these paintings were produced on the very same shores on which Huxley felt the excess of Atitlán, but manage to register such excess differently from the way most pieces in the Guatemalan landscape tradition that Iriarte balkingly founded in the early twentieth century have channeled and expressed it.[4] What follows is, to a large extent, an exploration of the conditions through which Suter's paintings have become the latest iteration of the tension that seems to define any sort of pictorial production emerging from these shores. This, of course, implies attempting to elucidate how and why her pieces have become so arresting, and so ubiquitous, in the last decade. Why have works produced, almost indiscriminately, over the last twenty years (Suter has been producing similar types of work since at least the early 2000s) suddenly become iconic, representative of an era? What changed if the work didn't? In other words: what shift in the conditions of reception made it possible for the same pieces, in a matter of years, to become so palpable, so appallingly appealing, to a metropolitan audience that they have come to define a sense of place (marginal, barely touched by civilization) in our times?

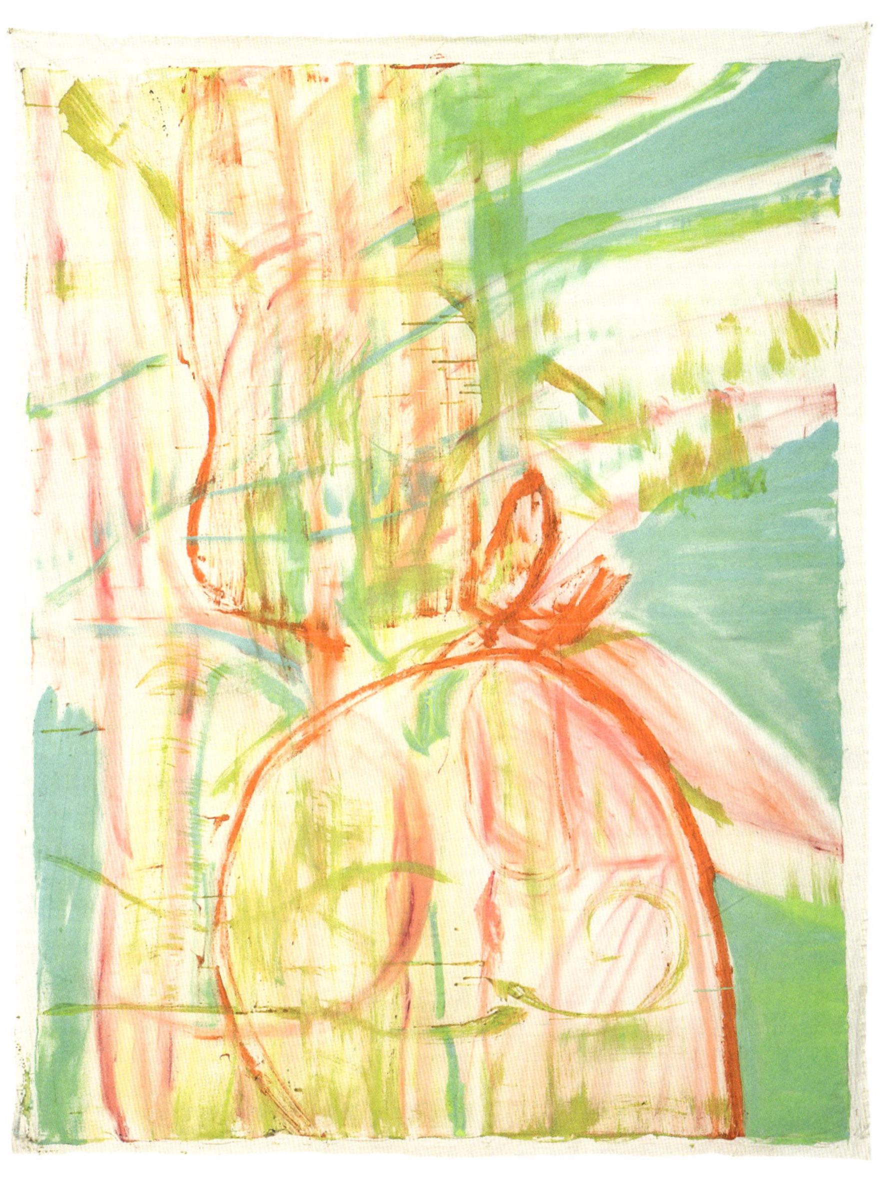

1 Vivian Suter. *Untitled*. n.d. Acrylic and natural elements on canvas, 7 ft. 4 in. × 69 in. (228 × 180 cm). Collection the artist

But do they? Or more precisely: why (and how, in the event that they have actually become representative of the region) have they actually become paintings "representing a place"? Because, as I think anyone even mildly acquainted with Suter's painting would clearly recognize, they are not, at least not straightforwardly, landscapes. And here is where the trouble (and this essay) begins.

In fact, let's start by upping the ante: these works might have been produced on the shores of Atitlán, and might well apprehend and register the excess of the territory where they were produced, but strictly speaking they are neither Guatemalan nor landscapes. Perhaps it is precisely this condition that makes them particularly poignant these days, as they seem to contain a particular tension between what they supposedly represent and what they actually present; between what a local terrain might refuse to relinquish to a national tradition, and what the international art scene might be able to value at this point in time—the many ways in which the local might escape and transcend the national by revealing new articulations of terrain and geographies.

So, if not landscapes, what kind of painting has Suter been developing over all these years? What else can these paintings be—highly colorful but raw, "earthy," usually large format and mostly abstract, although birds and, yes, terrain can occasionally be distinguished in some of them? The first clue might be contained in a preposition that she slipped in while responding in an interview to the American artist R. H. Quaytman, who asked if her works could be considered landscapes. Here is the key passage (the stress is in the original):

> RQ: I was thinking that if I had to pick the type of genre my paintings are in, then it would be landscape painting, and I think the same could be said of your work.
> VS: Yes, I often try to make works with the landscape.
> RQ: And I feel like they themselves are *actual* landscapes.
> VS: Yes, exactly, they are geographical things.[5]

Notice the slip between Suter's response that she wants to "make works with the landscape" and Quaytman's statement that her works are "actual landscapes." Making something *with* an entity is not the same as the work *being* that something—in fact the opposite is true: one does things with (works with) that which is precisely what one is not. Insofar as Suter does not think her paintings are landscapes, she enters into a collaborative relation with them; she decides to "make works with the landscape"—whatever that might mean.[6]

A possible interpretation of the exchange might be: Suter states her interest in "making works with the landscape," which might mean anything from painting it to using the elements of the territory, incorporating remnants of the environment and her surroundings. Instead of apprehending Panajachel and Lake Atitlán through mere representation, she does it through indexicals. This is, I believe, the sense in which Quaytman meant that Suter's paintings were "actual landscapes."

Obviously, it is the use of the term "landscape" here that might be a little fuzzy, and might be responsible for the confusion. As we all know, landscapes are not "apprehended" but composed; what is apprehended is "nature" (the visible natural environs), which is then recomposed, framed, and organized on the canvas by the artist. Landscapes don't exist in nature; they are representations of nature. In that sense we can complete the senses of the word "landscape" in Suter and Quaytman's exchange. When Suter uses the expression "I often try to make works with the landscape," she is referring to producing works of art in close relation to the natural environs, whether representing them pictorially or presenting them indexically. And when Quaytman uses the expression "I feel like they themselves are actual landscapes," she is referring precisely to the indexical quality that some of Suter's works exhibit: residues and debris from the rainforest, the rain, and other environmental conditions, captured by the canvases in their exposure to the forces of nature. (Suter, as we will discuss later, exposes them to the elements as an integral part of her process.) It is in this second use of the word that Quaytman stretches the term "landscape" a bit too far by making an equivalence between sign and signifier that constitutes a logical mistake: the paintings might capture some of nature's features—even indexically—but that doesn't mean they are actual "landscapes" (nature itself). They might pick up or register natural events in the same way that thermometers or seismographs do, but such an action does not lead them to become the event they record.

So how to interpret Suter's works? Or, more precisely: how might they be considered "landscapes" in the traditional sense, and in what way do they represent a departure from tradition?

At this juncture it is important to notice that most of the crucial decisions that Suter has taken in her process—the ones that allow us to detect why her works might be considered "actual landscapes"—have meant a considerable reduction of her "selfhood" in them. In other words: like many artists since at least the 1960s or even much earlier, Suter has long been interested in deescalating her agency in her own work.[7] This is, of course, not new. From Robert Rauschenberg to Chuck Close, from Richard Estes to Roxy Paine—to mention just a few—artists around the world have made the restriction of their expressive capacities an integral part of their program and process, in order to produce works of art with the least possible exposure to their choices, their subjectivity.

The idea, in short, is to reduce to a minimum the subjective input required in an artwork. If, with Marcel Duchamp's foundational gesture, the work of art abandoned its "retinal" dependency and the artistic craft shifted from hand and eye to mind and concept, the introduction of chance into the artistic process by John Cage and others, starting in the 1950s, pushed the issue to a new limit.[8] The introduction of indeterminacy effectively opened the possibility of removing agency from art (and thus, paradoxically, of removing artistry from art), as it questioned to the core a basic assumption of modern subjectivity: its unique capacity to produce art. Now it was no longer a matter just of craft but of ontology: you could potentially have art that had not been "made" by artists, that had not been exposed to their subjectivity.

In that sense, what Suter does by relinquishing control is hardly new. What is nonetheless novel and surprising is the way in which her relinquishment of control has been read—and how such reading has led to new decisions (especially in terms of how the pieces are conceived in the exhibition space, which we will discuss later). The crucial difference between all the aforementioned (male) artists and Suter is that whenever the other artists rejected agency, when they chose to control less, they were always equated with a process of automatization proper to machines.[9] In the case of Suter, by contrast, the release of control is understood as a process not of automatization but of being closer to the natural environment. If, when "automatizing," most other artists chose to interface with an industrial process—mimicking "machine production" (think Estes or Chuck Close painstakingly following a process proper to machines to produce a photographic effect in pictorial terms)—Suter's automatization seems to go the opposite way. She seems to wander away, to return to nature.[10] It might just be that nature is the biggest machinery of all—and it seems that she is *mimicking* nature.

LANDSCAPE SIMULACRA

What might it mean to read Suter's work as a "return" to nature? In what ways might any work of art "return to nature"?

This is not the place to discuss the history of the opposition between art and nature, as this opposition is a concept that is as foundational to art as it is possible to be—art is art only insofar as it is artifice and not nature. Suffice it to say that the peaks of their antagonism (as in the late nineteenth century), or the moments in which they have seemed to coexist with almost no opposition (as within certain strands of Romanticism, which are even visible in more recent movements such as the Land art movement of the 1960s), might articulate important precedents for the way Suter's pieces have become unique.

Nonetheless, what is important to keep in mind here is how a "return to nature" as enacted by Suter needs to be linked to a series of displacements at the very center of her work—as it seems that she is always in a constant journey toward an *elsewhere*. And it is the content of this *elsewhere* that matters here.

It matters because this "elsewhere" conflates at least two crucial displacements in Suter's work: her continuous displacement toward a practice involving less conscious control, and her coincidental return to the American continent from Europe. (Born in Buenos Aires in 1949, Suter lived in Switzerland from the time she was twelve until she relocated to Guatemala in 1982.) In other words, there is a direct correlation between moving from the studio to the wilderness (in terms of process) and moving from the metropolis to the periphery—resettling with her mother (the extraordinary artist Elisabeth Wild) in a quiet place in Panajachel, on the shores of Lake Atitlán. Both trajectories are movements in the same direction, which complement each other.

In that sense, the movement away from control and agency toward "natural freedom" is analogous with Suter's move from Europe to the shores of Atitlán. Thus, her "return to nature" effectively implies her surrender of aspects of control in her painting—as we can see in the excellent curatorial text by Ruth Erickson from the show of Suter's work that she organized at the ICA Boston in 2019. Notice that Erickson describes such surrender of control as Suter moving toward a type of "pictorial gesturing," as she works "in concert with rainfall and mudpuddles, with the light that passes between branches and the animals in the forest." According to Erickson, Suter's painting is "gestural," as it makes room for the randomness of nature: "Her method often involves moving her canvases between the indoors and outdoors and exposing them to the climate in order to allow nature to commingle with her broadly painted swaths of vivid color. Inspired by the surrounding vegetation and landscape, Suter's gestural compositions work in concert with rainfall and mud puddles, with the light that passes between branches and the animals in the forest" (fig. 2).[11]

Let's focus a bit more on what Erickson might have meant by "gestural," as it is possible that she may have stretched the term to its very limit: it seems here to describe two simultaneous events, but only one of them is traditionally associated with the word as it is usually understood, that is, in relation to Abstract Expressionism, where it was first applied in visual art.[12] In Erickson's passage on Suter, though, the word "gestural" implies not only a perceived retreat from complete stylistic control but also the encroachment of randomness associated with the exposure of the canvases to "nature." One might be part of the other, but they are not necessarily the same, and in Suter's case they imply two complementary ways in which she is "returning to nature": in the release of her control over the canvas and, at the same time, her surrender of the canvas to the natural environs. To put it in art-historical terms, what we encounter in Suter's surrenders to nature are strands of *art informel*, gestural painting, and automatism—all at the same time. In other words, becoming "natural" implies surrendering on multiple levels, and such surrenders achieve the most crucial aspect here: aligning the mental state of the artist with the exposure to the natural elements of the canvas. Then, and only then, can the *effect* of the return to nature be achieved.

The formal consequences of such surrenders are multiple. As soon as randomness starts to gain territory in Suter's process, the canvases start to lose independent standing and relevance, to the point of appearing to function only as a group. At some point, in fact, Suter's paintings began to be presented (and probably conceived, although that might not be completely clear or necessary for us to discuss here) in groupings that require (and acquire) a spatial dimension in their presentation. There is undoubtedly a direct correlation between the works' loss of independent standing and their acquisition of spatial significance as a group. Obviously this doesn't mean that the unique formal qualities of any of the individual works cease to matter, but it implies that their significance is now entangled in a larger canopy of meanings, one that is (literally) stretched and displayed across the many exhibition spaces in which her work has been presented lately. Take the aforementioned exhibition at the ICA Boston, her show in Greece for documenta 14 in 2017, or her show *A Stone in the Lake*, at the Secession, Vienna, in 2023: in all of these exhibitions the works were displayed as a sort of maze extending across the gallery floor, giving the sense that together they constituted a living organism.[13] In each show, we all of a sudden encountered ourselves beneath a canopy, within a dense forest, or confronting a living organism spreading around the spectator—an organism both living and leaving the gallery space (as was the case at the ICA, in which the show felt as if it was leaving the confines of the exhibition and spreading into the whole building). As Italian curator and art historian Lorenzo Giusti observed recently (the stresses are mine):

> Her recent exhibitions have all presented works dating from periods that were not precisely specified. The different manners in which the works were installed have also resembled one another closely. The approach has always been to

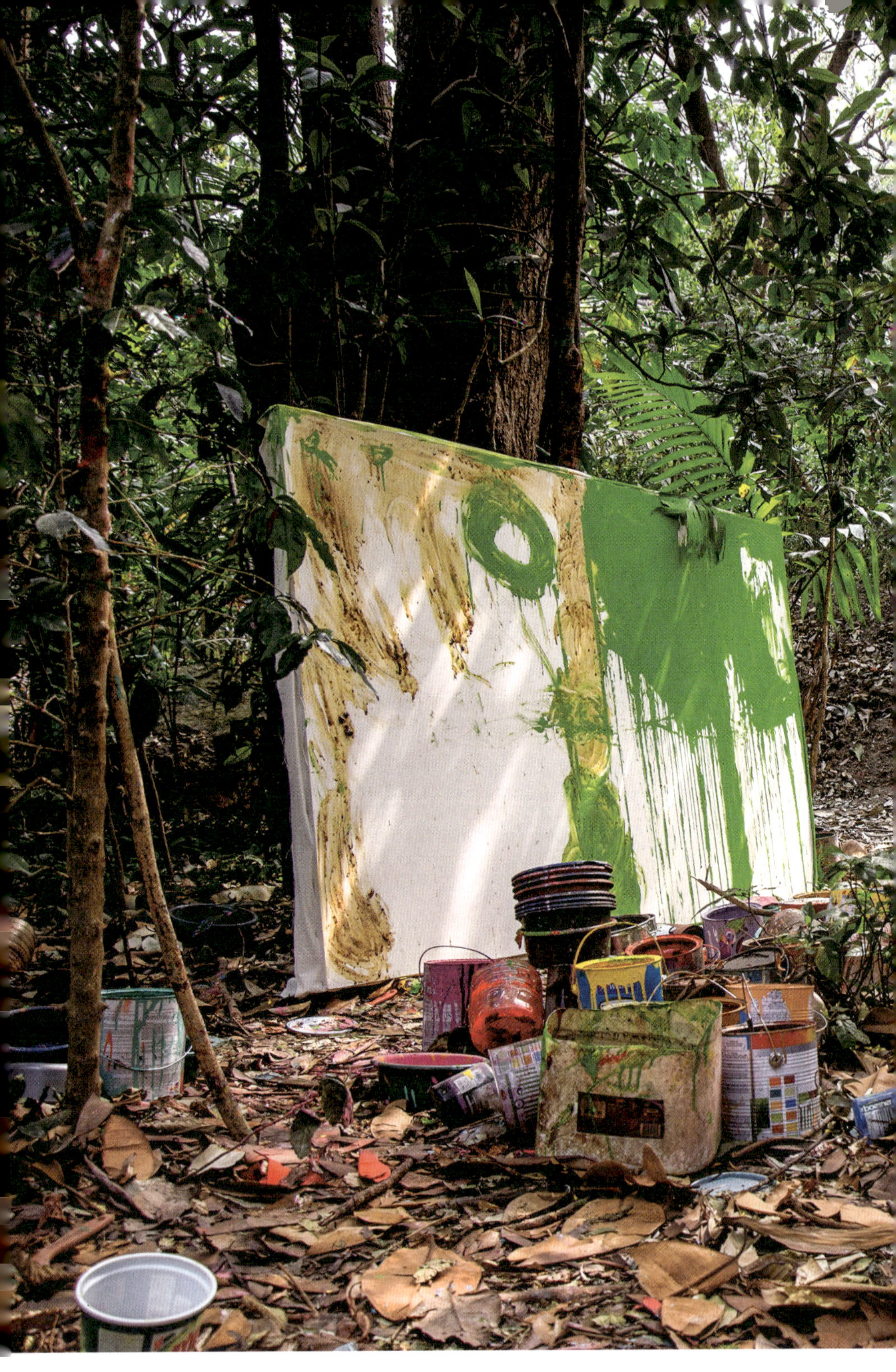

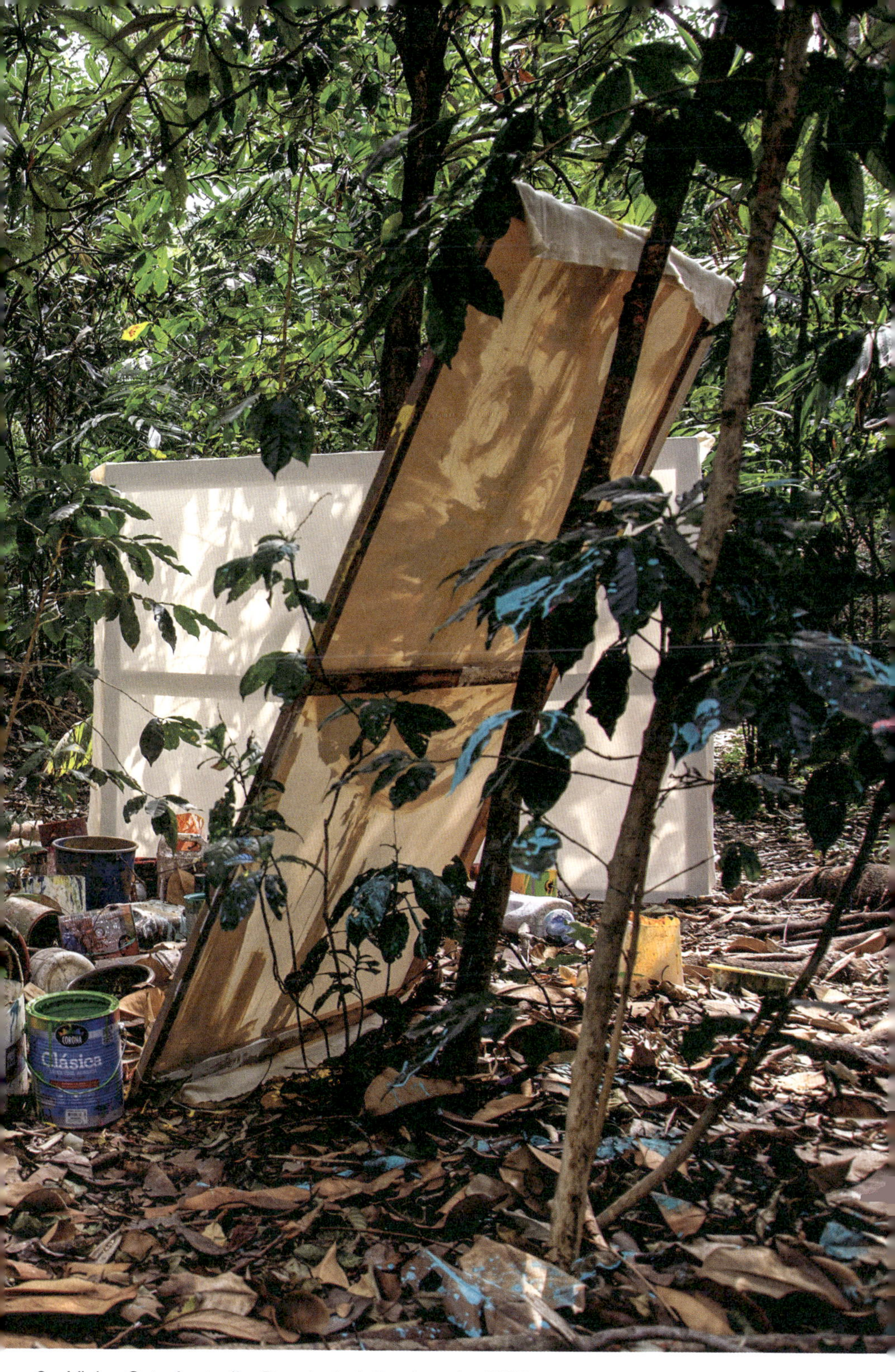

2 Vivian Suter's studio, Panajachel, Guatemala, 2023

> cram the space excessively with materials, with canvases hung pretty much everywhere, to give the idea of a spontaneous, relentless, and constantly evolving creative dimension, but also the idea of a *necessary process of osmosis between painting and the environment*. Viewers are left to wander among the paintings and forms without ever finding a real focus of attention. *It is the same as walking through the depths of a dense wood*, where everything seems to be intertwined, fused together, in motion.[14]

As can be seen in these pieces of writing, and in most texts written on Suter's process, the overriding metaphors used to compare and describe her shows' choreographies are biological ones (*natural* ones), metaphors that are accentuated (in direct correlation) by the coarseness of the pictorial trace *and* the texture of canvas—sensations stressed in turn by the occasional debris of the rainforest that some of the works happen to contain. Every little bit of dirt, every faded watermark, matters here because it is, in itself, matter. In other words, Suter's exhibitions display their "naturalness" by showing not only their reduction of authorship but also the indexical quality that they bear materially. One aspect is not enough—because it is this indexical quality that serves as a simulacrum of what we strictly speaking don't have here: a landscape.

What is ultimately conjured by the installation across the gallery space is the *specter of landscape*. It is clear, as we have said, that the paintings themselves are not landscapes, but through their distinctive accumulation and particular display they activate a unique process, creating a singular effect: to make viewers feel as if they are confronting nature. Suter's works might not be landscapes per se, but they amount to them: the spectator, or inhabiter, of a Suter show is not in a show of landscape paintings, but is experiencing something that acts as a landscape in itself—a simulacrum of it.

CODA: TERRITORIAL SAMPLING

One could argue, with reason, that the "return to nature" underwriting Suter's work might be yet another way of returning to old tropes that have plagued the region, as it harps on and stresses perhaps the most pernicious idea about the American continent: that it is still an untouched Garden of Eden. In other words: that despite all this time, the territory invoked—Panajachel—is elsewhere, existing outside history.

Furthermore, in the case of Suter, this suspicion of operating outside history is compounded by the fact that the history of the terrain—of what has happened in Guatemala—is also not considered at all. Whether out of obliviousness to or disdain for its history, the fact that the terrain is approached as if nothing had happened there—a claim also recently leveled at the Land art artists, who acted when producing their work as if the American West were devoid of history—is a glaring omission, to say the least.[15] It even begs the question of whether it is at all possible to make art in, of, or about those territories without alluding to the constant marks of history there, where they are everywhere inscribed.

These have been, not virginal gardens, but contentious spaces, hosting some of the most complicated and traumatizing events in the history of their local Tzutujil population, going back probably even before the Spanish colonization and reaching as late as the 1990s. This history it is impossible to elude. As is well known, although Panajachel itself lived in relative peace during the Guatemalan Civil War of 1960–96, the whole lake region suffered continuous violations of human rights due to the strict control of the military, which was placed in the area because the guerrillas openly recruited "*atitecos*" (people from Santiago Atitlán, the main city on the lake, right across from Panajachel). Occupying Santiago Atitlán from 1980 on, the army routinely committed offenses against the Indigenous population for years, until, tired of their abuses, a group of inhabitants protested against the garrison on December 2, 1990. The army opened fire, killing fourteen of them, a massacre still remembered every year.[16] It is necessary, then, to question or at least ponder the significant silence on this issue, and even questionable whether one can look beyond it. How to fold (and unfold) the history inscribed in the territory and its people that easily?

The late Fernando Coronil, in one of his most memorable insights, suggested that one possible reason for the sense of immobilism that impregnates the Garden-of-Eden view of the Americas might be found in the concept of nature that is common in the

Western view of the periphery (and particularly of the tropics)—one that is understood as devoid of history (and therefore of progress). In other words, the critical difference between the metropolis and the periphery has been the availability of modern time: history and progress cannot exist outside its confines.[17] According to Coronil, the dialectic between space and time upon which capitalism is based needed vast zones to be understood as mere empty space, beyond culture and in which time does not belong. Precisely because the system required the constant expansion and appropriation of natural resources from areas outside the metropolis, the Americas, ever since their "discovery," have been and continue to be only space.

I find it hard to argue with Coronil. Whatever one could say about his proposed solution to this conundrum—adding "nature" into the Marxist dialectic between labor and capital—it is also obvious that he exposed a foundational aspect of the position in which Latin America has been "placed" since the very beginning: that of being nothing more than a place. Furthermore, this idea also reveals the way in which the nature of "nature" has been perpetually conceived in the region. In fact, the belief, entrenched in every single country in Latin America, that one lives amid untapped and infinite natural riches can be read as the most obvious symptom of the peripheral position that the region has assumed in the world. We might "own" the space (filled with riches) but we don't have access to time (culture, history, progress) in which to exploit it, which means we are perpetually relegated to remain outside of progress.

In this sense, it could be argued that Suter's work plays into the perpetuation of an immobilism that is the defining feature of any sort of exoticism of place. Sure enough, that might be a plausible reading. But there is another one—one that starts by understanding that the genre of landscape has been dynamic across time, and as such has mutated in many latitudes, from landscape into Land art and from there into "eco art." At the center of this mutation is the fact that the very notion of landscape has started to falter, and for decades has proven insufficient, because the "experience of nature" is no longer as coherent as it used to be. This dissatisfaction was already palpable in classic texts such as Kenneth Clark's seminal book *Landscape into Art* (1949):

> Throughout this book, when I have used the word "nature," I have meant that part of the world not created by man which we can see with our unaided senses. Up to fifty years ago this anthropocentric definition did well enough. But since then, the microscope and telescope have so greatly enlarged the range of our vision that the snug, sensible nature which we can see with our own eyes has ceased to satisfy our imaginations. . . . And in the last few years nature has not only seemed too large and too small for imagination: it has also seemed lacking in unity. To anyone but a higher mathematician, nature no longer seems to act consistently in all her operations. In the last few years, we have even lost faith in the stability of what we used hopefully to call "the natural order"; and what is worse, we know that we have ourselves acquired the means of bringing that order to an end.[18]

The insufficiency of landscape as a genre lies in its failure, owing to its static dimension and mere "observational" stance, to make "nature" palpable and visible. This failure propelled new ways of addressing and recalling nature in art, such as Land art, environmental art, Earthworks, and more recently eco art. As Mark A. Cheetham has traced in his *Landscape into Eco Art: Articulations of Nature since the '60s*, the insufficiency of traditional forms of landscape, in which nature is experienced as contained, organized, and rendered by the artist, led to a series of new articulations from the 1960s onward in which the author's mediation has waned.

The ascent of landscape as a genre occurred in collaboration with the industrial revolution of the eighteenth century—a favored starting point for the Anthropocene—and the imperialisms of the nineteenth century. Earthworks and Land art developed at the same time and in the same cultural milieu as mid-twentieth-century environmentalism in the United States and Europe.[19] So modes of apprehending and presenting nature in art have shifted dramatically. And it is clear that Suter's work, in its transformation from representing insufficient landscapes on individual canvases to presenting, by accumulation, a landscape that makes the viewer/inhabiter *experience* a territory, might be a symptom of such a change. This shift in form is crucial for us here because it implies a potential new beginning; it is at this juncture that a novel possibility might open for us.

Let's remember that traditional landscape painting was intrinsically linked to the modern project. As with the novel as a literary form, landscape as a genre of painting is a product of modernity, as it requires a subject who can confront the natural environment, apprehend it, and organize it on the canvas. Thus it is not surprising that the height of the genre coincided with the height of the modern project, during the period of Romanticism—which could be considered the first reaction and criticism toward the rationalism, industrial progress, and city life that modernity had forged, especially in northern Europe and the United States. The climax of landscape as a genre was driven by a strong reaction against the central tenets of modernity.[20]

The problem is that precisely due to Latin America's perceived belatedness and idleness in relationship to Europe, Romanticism didn't develop here, because it did not have a modern background to react against. During the colonial period and even right after the independence of the new Latin American nation-states in the early nineteenth century, no landscapes were made.[21] Our poster boys for natural explorations are overwhelmingly foreigners, such as Alexander von Humboldt and the European painters who followed him, such as J. M. Rugendas.[22] Their figures have simply made the passive position of our local culture in relation to our natural environs more pronounced—a position established since the first contact with Europe. Despite all the efforts of Andrés Bello in Caracas and Santiago, of the members of the 1837 Salón Literario in Buenos Aires, and of other conspicuous liberals during the first part of the nineteenth century in Rio de Janeiro, the cult of Humboldt in Latin America is perhaps the best reminder that Romanticism never grew strong roots in our torrid region—with possibly the exception of certain tendencies in what is now Argentina and Uruguay.[23]

Humboldt and his friends ended up reinforcing an idea among the fathers of the new republics forming at the turn of the nineteenth century that our new nations were (and are) well endowed in terms of natural resources, but outside of history. In other words, instead of introducing a differentiating factor for our understanding of nature, their views ended up reinforcing ideas that go back to Columbus himself, who first installed in the European imagination the belief that America—due to its size and sheer beauty—was "outside" any rational schema.

But it is precisely here that Suter's works might offer a possible new path. As we have seen, the subjective space of the artist has been reduced to its minimum, which means that we are not confronted with an all-powerful explorer sizing himself up against the territory, someone who could capture and render a version of nature. Instead we are confronted with a sampling of nature itself—an ecosystem of works produced by an inhabitant who cares deeply about her connection with the terrain, which means that she is conscious of her inability to control or organize it. The best way of conjuring it in the exhibition space, then, is precisely through showing the majesty of its indeterminacy, of its lack of control. For this reason, insofar as all of Suter's shows act as simulacra of landscapes, they act too as territorial samples (a token of Atitlán): a space that is now no longer outside history, but one that is being sized up, unfolded, with all its traumas and events, and actively carried and deployed across the world. Latent in each of these iterations lies a second chance for Atitlán (and the American natural environs) to finally make a first impression.

1 Aldous Huxley, *Beyond the Mexique Bay* (London: Chatto and Windus, 1950), 149.

2 Together with Carlos Merida and Carlos Valenti, Agustín Iriarte Castro (1876–1963) is considered one of the preeminent figures of the first generation of "modern" Guatemalan artists. He was part of TRIAMA, a legendary group of landscape painters prominent in the 1920s. The name of the group is an acronym of the first letters of some of the artists' surnames: Antonio Tejeda, Ovidio Rodas, Rigoberto Iglesias, Jaime Arimany, Óscar Murúa, and Rafael Alcain. Iriarte is also remembered as one of the most important teachers at the Escuela Nacional de Bellas Artes, Guatemala City.

3 There is often confusion about what one might mean by "modernism" in Latin America. This stems from the fact that in Spanish-American literature (and in Spanish-American literature only), the term refers to the movement created by perhaps the most important writer of the region in the late nineteenth century, the Nicaraguan poet Rubén Darío, whose *Azul* (1888) supposed a revolution of styles and forms. Influenced by French Symbolism (Charles Baudelaire, Arthur Rimbaud, Paul Verlaine) and Parnassianism (Théophile Gautier, Gérard de Nerval), *Azul* presented a universe steeped in medieval and classical European mythology—filled with nymphs, swans, kings, palaces, etc.—and in a stance of "l'art pour l'art." The influence of *Azul* was extraordinary, helping to modernize Latin American letters beyond the meager romanticism inherited from Spain and to anchor the literary references of the region in France. In 1906, Dario distanced himself from this movement, abandoning an apolitical stance and cosmopolitan themes in favor of political issues. Nonetheless, it is important to understand that when shifting to Portuguese literature, "modernism" refers—just as in English literature and fine arts—not to the modernism of *Azul* but to the works created under the different "avant-gardes" of the early twentieth century—usually starting with the Semana da Arte Moderna in São Paulo in 1922. On Darío see Kathleen O'Connor-Bater, ed. and trans., *A Bilingual Anthology of Poems by Ruben Dario 1867–1916: Annotations and Facing Page Translations from Spanish into English* (Lewiston, NY: Edwin Mellen Press, 2015).

4 Here we are thinking of works that form the backbone of the country's landscape tradition, such as the many takes of "*Lago de Atitlán*" that Humberto Garavito painted through the years, in which the lake always appears to be resting in all its majesty of water and clouds; Roberto Cabrera's informalist homage to Garavito's pieces (2002); or the haunting video works of the Indigenous artist Manuel Chavajay.

5 "Conversation between Vivian Suter and R. H. Quaytman," in Adam Szymczyk and Hendrik Folkerts, *Vivian Suter*, exh. cat. (Toronto: The Power Plant, and Berlin: Hatje Cantz, 2019), 65.

6 This type of "collaboration with nature" or "with landscape" has been a continuous trope in Suter's statements, to the point where it seems to have served as a type of origin myth to her poetics, specially in the second phase of her paintings—since she moved to Guatemala in 1982. As the *New York Times* reported in 2019, "When Guatemala was struck by two hurricanes, Stan and Agatha,

in 2005 and 2010, respectively, her house was flooded and much of her work drenched in water and mud. But the catastrophes also brought restorative discoveries. When she opened up an unpainted, waterlogged manta, she found the earthy residue had expressed itself in a series of delicate Rorschach-like forms that resembled X-rays of plants or exotic insects. 'It was like a miracle, you know, just beautiful,' she says. 'This was very special, like a gift.' She calls the painting the Virgin Rorschach." Tess Thackara, "A Painter Who Left the Art World in Order to Actually Make Art," *New York Times Style Magazine*, April 17, 2019. Available online at www.nytimes.com/2019/04/17/t-magazine/vivian-suter.html (accessed September 27, 2023).

7 In fact, seen from a wide perspective, the history of artistic practice in the twentieth century, in any of its mediums, is a history of the erosion of the artist's agency and therefore of the status of the artist as willful creator. The trend appears in such works as Marcel Duchamp's Readymades of the 1910s, or László Moholy-Nagy's *EM1–EM3* of 1923, in which he "ordered" a painting from a factory over the telephone by giving them instructions on how to apply the enamel. In that sense, by Suter's time such de-escalation of agency was a sedimented trend among artists, even to the point of becoming a tradition.

8 Although the incorporation of chance and randomness in artistic processes can be traced to antiquity, the twentieth century saw an intense revival of its application, first through Surrealism, then Abstract Expressionism. Nonetheless, it is not until John Cage in the early 1950s that we are confronted with an artist who made chance central to his practice. And Cage's influence has been enormous, through his teaching at Black Mountain College and beyond. For more on Cage and chance, see his lecture collection *Silence: Lectures and Writings* (Middletown, CT: Wesleyan University Press, 1961).

9 I lack space here to discuss this topic at length, but it seems particularly conspicuous how certain types of "automatism" in art—especially those in which the artistic process purports to "act as a machine"—are read as male. The type of "automatism" implied in surrendering control of the artistic process to a larger, automated process is never understood as a waning of selfhood or a surrender to irrationality but, on the contrary, as a move toward hyperrationality—as the cases of Chuck Close, Donald Judd, or even Roxy Paine attest. By contrast, in the case of Vivian Suter the waning of conscious selfhood, the release of control, seems always to be read as a *retreat* toward the irrational, or the "natural."

10 As is well-known, the process behind Richard Estes's screenprints was "remarkably complex; Estes worked from gouache and acrylic templates, which he then photographed and used to create multiple stencils." See The Metropolitan Museum of Art, "Subway from the portfolio Urban Landscapes III," n.d. Available online at www.metmuseum.org/art/collection/search/383866 (accessed July 28, 2023). In the same vein, the painstaking processes that Chuck Close employed to produce his photorealist paintings has been documented and discussed in many publications. The booklet dedicated to his portraits by the National Gallery of Art in its "Examining Portraits" project has perhaps one of the most

didactic and clear descriptions of the meticulous lengths that Close went to in order to produce his work, especially in his first period, between 1967 and 1988: "Close typically starts with a photograph. Instead of asking someone to sit in front of him while he paints, a slow process that could take days or months, Close takes several photographs of his subject. He then carefully selects one photo. He uses a grid to divide it into smaller units and to maintain the proportional scale between the photo and the much larger canvas. Often applying a grid to the canvas as well, he transfers the image square by square from photo to canvas. It's an exacting and painstaking process that Close has used throughout his career." See National Gallery of Art, *An Eye for Art: Focusing on Great Artists and Their Work* (Chicago: Chicago Review Press, 2013), 64. Available online at www.nga.gov/content/dam/ngaweb/Education/learning-resources/an-eye-for-art/AnEyeforArt-ChuckClose.pdf (accessed July. 28, 2023).

11 Ruth Erickson, "Vivian Suter," website page for Suter exhibition curated by Erickson at the Institute of Contemporary Art, Boston, August 21–December 31, 2019. Available online at www.icaboston.org/exhibitions/vivian-suter (accessed July 28, 2023).

12 As is well documented, the term "gestural" in relation to painting was first applied in relation to Abstract Expressionism, where it was used to describe the correlation between a mental state (anger, pain, suffering, etc.) and a trace or mark left on the canvas. That mark, whether a brushstroke, a spatter, or a drip, was considered a register of the gesture and action by which it was created. Irving Sandler explains, "The gesture painters refused to preconceive particular meanings, regarding the process of painting as an intense, unpremeditated search for the images of their creative experiences. They believed that if, during the direct process of painting, they followed the dictates of their passions, the content would finally emerge." Sandler, *The Triumph of American Painting, A History of Abstract Expressionism* (New York: Harper & Row, 1970), 93.

13 *A Stone in the Lake*, Suter's show at the Secession, ran from April 28 to June 16, 2023.

14 Lorenzo Giusti, "Vivian Suter: Where Is It Home?" in *Vivian Suter, GAMec, Secession* (Vienna: Secession, and Bergamo: GAMec, 2023), 107.

15 The critique of the "obliviousness" of Land art artists to Indigenous history has been mounting in recent years. See, e.g., Jasmine Liu, "What Do Native Artists Think of Michael Heizer's New Land Art Work?" *Hyperallergic*, September 21, 2022. Available online at https://hyperallergic.com/763203/what-do-native-artists-think-of-michael-heizers-new-land-art-work/ (accessed July 31, 2023).

16 For more on the massacre see Robert Loucky, "Massacre in Santiago Atitlán: A Turning Point in the Maya Struggle?," *Cultural Survival*, March 3, 2010. Available online at www.culturalsurvival.org/publications/cultural-survival-quarterly/massacre-santiago-atitlan-turning-point-maya-struggle (accessed July 31, 2023).

17 Fernando Coronil states, "Post-Enlightenment visions of historical progress typically assert the primacy of time over space and of culture over nature. In terms of these polarities, nature is so deeply associated with space and

geography that these categories often stand as metaphors of each other. In differentiating them, historians and social scientists usually present space or geography as the inert stage on which historical events take place and nature as the passive material with which humans make their world. The separation of history from geography and the dominance of time over space has the effect of producing images of societies cut off from their material environment, as if they were fashioned out of thin air. If nature is included, it typically appears in the likeness of the appearance of the air itself, eternally and readily available. Bathed in this deceptive light, the social appropriation of nature does not seem to require particular analytical attention." Coronil, *The Magical State: Nature, Money, and Modernity in Venezuela* (Chicago and London: The University of Chicago Press, 1997), 23–24.

18 Kenneth Clark, *Landscape into Art* (Boston: Beacon Press, 1963), 140–41.

19 Mark A. Cheetham, *Landscape into Eco Art: Articulations of Nature since the '60s* (University Park, PA: The Pennsylvania State University Press, 2018), 5.

20 Far too much has been written on Romanticism's crucial, necessary reaction to the rationalizing project of modernity to reduce it to one paragraph or one quote, but I think this section of Isaiah Berlin's seminal book *The Roots of Romanticism* summarizes its impact and importance well: "What can be said to owe to romanticism? A great deal. We owe to romanticism the notion of the freedom of the artist, and the fact that neither he nor human beings in general can be explained by oversimplified views such as were prevalent in the eighteenth century and such as are still enunciated by over-rational and over-scientific analysts either of human beings or of groups. We also owe to romanticism the notion that a unified answer in human affairs is likely to be ruinous, that if you really believe there is one single solution to all human ills, and that you must impose this solution at no matter what cost, you are likely to become a violent and despotic tyrant in the name of your solution, because your desire to remove all obstacles to it will end by destroying those creatures for whose benefit you offer the solution. The notion that there are many values, and that they are incompatible; the whole notion of plurality, of inexhaustibility, of the imperfection of all human answers and arrangements; the notion that no single answer which claims to be perfect and true, whether in art or in life, can in principle be perfect or true—all this we owe to the romantics." Berlin, *The Roots of Romanticism*, 1965 (reprint ed. Princeton: Princeton University Press, 1999), 146.

21 For more on the absence of landscapes during the colonial period see Natalia Majluf, "Traces of an Absent Landscape: Photographers in Andean Visual Culture," *History of Photography* 24, no. 2 (2000): 91–100.

22 Among the dozens of studies and biographies of Alexander von Humboldt, perhaps the one most interested in tracing him as a deeply Romantic scientist is Maren Meinhardt's *Alexander von Humboldt: How the Most Famous Scientist of the Romantic Age Found the Soul of Nature* (Katonah, NY: Blue Bridge, 2019). On Johann Moritz Rugendas see Pablo Diener's *Rugendas, 1802–1858* (Augsburg: Wissner, 1997),

as well as César Aira's delightful novella *An Episode in the Life of a Landscape Painter*, 2000, Eng. trans. Chris Andrews (New York: New Directions, 2006).

23 As Tulio Halperín Donghi has elaborated in his indispensable *Una nación para el desierto argentino*, to a large degree the Argentine exceptionality of the nineteenth and early twentieth centuries can be seen as a result of the importance of the liberal ideas of the 1837 Generation. See Donghi, *Una nación para el desierto argentino* (Buenos Aires: Centro Editor de América Latina, 1982).

IN DIALOGUE

Allora & Calzadilla

The duo of Jennifer Allora and Guillermo Calzadilla has produced a remarkable, internationally visible body of work addressing questions of ecology and politics, including reflections on system collapse, loss of biodiversity, and coloniality. Since the mid-1990s, their work has focused on different aspects of the environmental, economic, and colonial history of Puerto Rico, where they base and position themselves—a country profoundly affected by the forces of colonialism and extractivism. Among other topics, they have touched on the consequences of the forty-year, highly damaging presence of a US military testing ground on the Puerto Rican island of Vieques, as well as approaching the schemes of dependency in energy production from an interspecies perspective.

In this conversation with scholar Irene V. Small, Allora & Calzadilla unfold the intertwined histories conveyed by their work, as well as the multiple formal and relational strategies they create for each of their projects.

IRENE V. SMALL
In the past decades, thinkers from distinct disciplinary and philosophical traditions have sought to reconfigure the articulation of the terms "nature" and "culture," and particularly the oppositional relationship they have long held within Western thought. New materialisms and object-oriented ontologies have decentered the human subject as the locus of meaning; anticolonial and decolonial thinkers such as Sylvia Wynter have excavated the racialization at work in the "biocentric" conception of the human; anthropologists such as Eduardo Viveiros de Castro and Philippe Descola have demonstrated that many cosmologies refuse the stability or boundedness of either "nature" or "culture," distributing sociality, agency,

and ontology across both; Indigenous and environmental thinkers such as Davi Kopenawa Yanomami and Donna Haraway have compelled us to shift our scales of analysis beyond the so-called Anthropocene. But I'd like to recognize that works of art—and your practice in particular—have long scrambled such supposed dichotomies. When I think of your work, I often circle back to the term "ecology," not in the narrow sense of the biological relationship between organisms and their environment but in the more expansive sense of a living and interactive system, the components of which include not only bodies and things but also forces as they are activated by history, politics, site. To start off our conversation, I wonder if you might reflect on the way you approach the work of art as an agential entity. How do your works exist within, emerge from, or harness larger ecosystems, however they are defined?

ALLORA & CALZADILLA

For us, the notion of an ecosystem in an expanded sense is fundamentally related to the concept of ecological justice. The first time we put these principles into practice was in the late 1990s, when we were working on a series of site-specific projects informed by the working concept *Land Mark*. These projects unsettled the formal and ecological premises of an earlier generation of Land art by posing the following questions: in whose interest is land marked, and to what ends? Which marks are deemed worthy of preservation and which are subject to obliteration? So already we were thinking of the agential quality, not simply of works of art, but of the systems of visuality, bureaucracy, and control, as well as of the practices of living, that mark a given space. These questions were formulated in response to the "transitional geography" of the Puerto Rican island of Vieques, which for sixty years was used by the US Navy as a training ground, leaving its landscape scarred with bomb craters and its ecosystem severely contaminated. To think the ecosystem in this context required moving across the nature/culture divide: the fishermen in Vieques, for example, considered themselves a species in danger of extinction, since their access to fishing was prohibited by the US military. In 2003, a civil-disobedience campaign was successful in forcing the military out, but the land was then designated a federal wildlife refuge, a zone of so-called "nature" in need of protection from humans after years of violent bombardment. But this designation entailed its own violence, marginalizing the interests of island residents, who demanded that the land in question be fully decontaminated and turned over to municipal management so that its future might be democratically debated.

This conflict was the point of departure for our work *Under Discussion* (2005), an "interrogative design" project based on an unlikely bit of reengineering: an overturned conference table retrofitted with an engine and rudder grafted from a small fishing boat. This hybrid vessel functions as something other than a means of transportation or labor; it questions the principle of functionality itself, especially when *making things work* becomes the be-all and end-all of politics. In liberal thought, the table is a common architectural figure for symmetrical communication and the nonviolent resolution of conflict. Yet it often fails to account for the inequalities that structure spaces of negotiation. In the case of Vieques, for example, poor island residents were treated as a threat to a vulnerable environment rather than as citizens who possess and practice their own forms of ecological knowledge. *Under Discussion* is an experimental device for the transmission of such knowledge, not only as a matter of information but as a testimonial rights-claim for environmental justice that interrupts the space of politics itself. In the video, a local activist uses the motorized table to lead viewers on a kind of Situationist eco-tour around the restricted area of the island. This trip re-marks the antagonisms that haunt the picturesque coastal landscape and bear witness to the memory of the fishermen's movement that first initiated acts of civil disobedience in Vieques.

IS

In many ways, the US military presence in Vieques spatialized a state of exception that is ongoing and permutational in Puerto Rico, and I want to talk a bit more about these conditions of exceptionality and emergency. As many Caribbean scholars have underscored, such states are in fact not exceptional but ongoing crises produced when "natural" disasters such as hurricanes and pandemics run up against structured neglect and governmental dereliction, which in turn provide openings for predatory economic capture. This "economy of catastrophe," as the political economist Miriam Muñiz-Varela puts it, is recursive and ongoing, even as it becomes most visible during extreme occurrences such as storms, blackouts, and bankruptcy.[1] As artists, how do you engage with what Rob Nixon calls the "slow violence" of environmental catastrophe and injustice, which tend to resist representation, as well as with the more spectacular dimensions of disaster and emergency?[2] Put another way, how do your works and broader practice aesthetically engage the multiple temporalities and visualities of crisis?

A&C

This question of the repetitive or permutational character of emergency resonates closely with our work on electricity (or the lack thereof). In Puerto Rico, the intensifying recurrence of blackouts across the entire island led us to an interest in electromagnetism, one of the four fundamental forces of nature, and the idea of using this force as both subject and medium. We have experimented with electromagnetism to create forms that are at once abstract and referential. For example, we have dropped iron filings on a canvas and placed it above an array of copper cables connected to an electrical breaker in our studio in San Juan. When the breaker is turned on, the electrical current forces the particles into an arrangement of shapes and patterns governed by the electromagnetic field. To set them in motion, we tap the taut canvas, which sends the heavy bits airborne and toward the positive and negative poles. So the resulting forms of a given work are governed by attraction and repulsion, strength and weakness, accumulation and dispersal. However, the pulsing forces that condition the appearance of the artwork—from stock market cycles to fossil fuel combustions—continue beyond any formal balance achieved within that work. We signal these excessive and ongoing conditions in the parenthetical component of the works' titles. These lengthy sequences of numbers and letters derive from our studio electric bill, and directly relate to the politics of the generation, ownership, and distribution of electricity.

Our ongoing interest in electricity probes the complexities and geopolitics of energy consumption, which connect the oil-futures market and the transnational holders of the Puerto Rico Electric

Power Authority's bond debt to local consumers who suffer the consequences of fiscal mismanagement. Our artistic experiments with electromagnetism are in equal part an exploration of formal principles and a way of confronting the complex nexus that is the energy grid.

IS

In Puerto Rico, exceptionality describes catastrophes like the electricity crisis, but it is also an ongoing political and juridical condition born of coloniality. This status is notoriously exemplified by the phrase "foreign in a domestic sense," a description used in a 1901 US Supreme Court ruling to refer to the archipelago's separate but nonsovereign relationship to the United States. This phrase was the title of your 2017 exhibition at Lisson Gallery in London, in which you showed works related to the Puerto Rican energy and debt crises as well as to longer histories of imperial expropriation. I've always been animated by the way the philosopher Giorgio Agamben describes the figure of *homo sacer* (and thus the state of exception) as existing in a topological relation to sovereignty. In other words, the spatiality is not one of stable inclusion or exclusion, but of an intensified liminality. This helps us to comprehend how Puerto Rico's legal status as an unincorporated territory simultaneously produces its asymmetrical incorporation within US economic and geopolitical interests. How do you go about activating these kinds of paradoxical relationalities and dependencies in your work?

A&C

We did *Foreign in a Domestic Sense* in 2017, just before Hurricane Maria landed on the island. Of course we could not have known the hurricane was coming nor anticipated the destruction it caused, but the exhibition already sought to activate the linguistic, biological, and physical forces at play in Puerto Rico's geopolitical reality. As a "nonincorporated" territory of the United States, the island is in the midst of punishing debt and energy crises that have brought to a head the legacies of colonialism and their complicity with global financial capitalism. The oxymoron "foreign in a domestic sense" was first used by US Supreme Court Justice Edward D. White in the 1901 ruling *Downes v. Bidwell*, which gave legal sanction to the US colonization of foreign territories through the ambiguous formulation of territorial nonincorporation.[3] The actual case involved whether oranges from Puerto Rico entering the port of New York should be subjected to foreign taxes. Justice White reasoned that "while in an international sense Porto Rico was not a foreign country, since it was subject to the sovereignty of and was owned by the United States, it was foreign to the United States in a domestic sense, because the island had not been incorporated into the United States, but was merely appurtenant thereto as a possession."[4] For more than a century, this equivocation has been the basis for Puerto Rico's uneven access to constitutional rights and sovereignty.

In the 2017 exhibition, we explored how this paradox effectively constructs social and political relationships that legitimate authority while obscuring conflict and struggle. The aim was to give formal and conceptual expression to this contradiction. Substitution, analogy, metaphor, and displacement were thus mobilized to redirect the flow of power that runs between the incorporated and the nonincorporated, the foreign and the domestic. The exhibition also invoked seemingly antagonistic art-historical models, from the technological experiments of Russian Constructivism and the base materialism of [Georges] Bataillean Surrealism to the cybernetic systems aesthetics of Jack Burnham and the entropic geographies of Robert Smithson. In this sense we sought to situate contemporary discussions of the power of things and materials to operate beyond the scope and frame of the human, within a far broader historical trajectory, while insisting on an urgent geopolitical awareness about the conditions of Puerto Rico and kindred territories in the global South. One of the works in the exhibition was *Blackout* (2017; fig. 1), which incorporates one of the burnt-out electrical transformers that caused an island-wide blackout in Puerto Rico in September 2016. The deep hum of reverberating electricity buried within that relic of electrically charged copper, ceramic fragments, and transformer coils served as a tuning device for a live vocal performance of a score composed by David Lang. The volatility of the power grid was thus transformed into a distinct materiality and sensorial substrate.

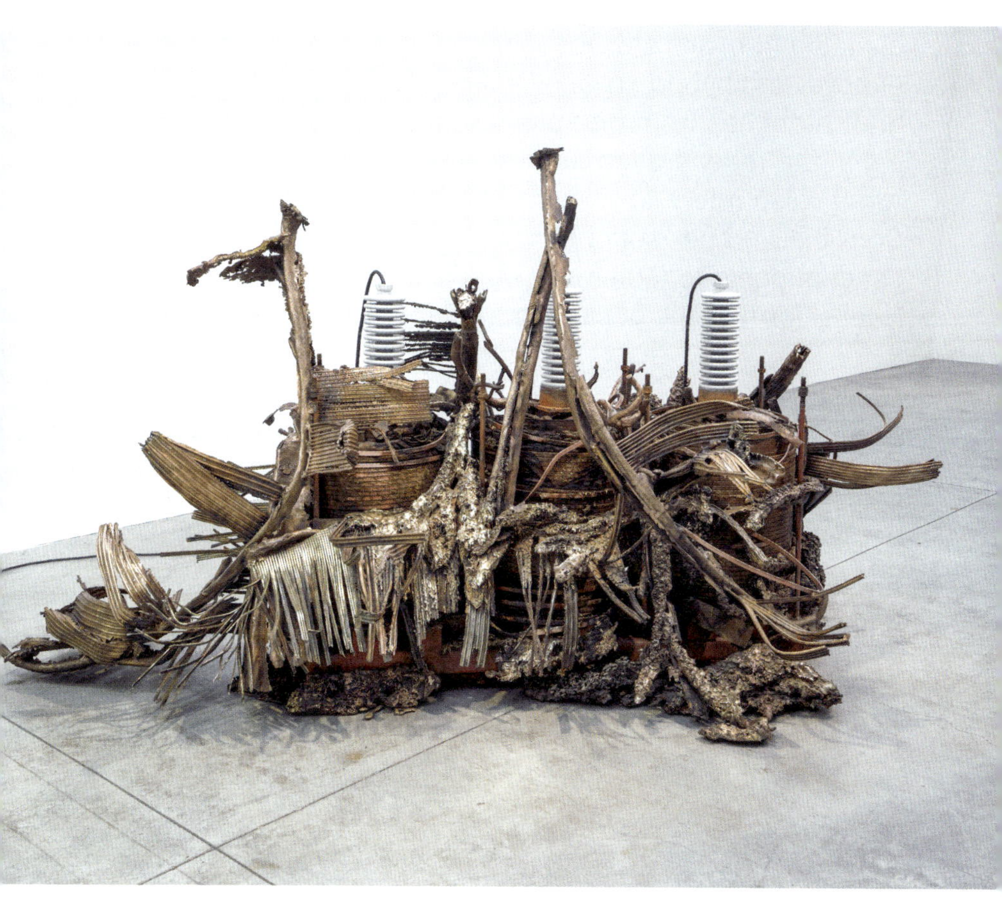

1 Allora & Calzadilla. *Blackout*. 2020. Copper, ceramic, iron, steel, oscillator, speaker, vocal performance, 54 ¾ in. × 8 ft. 7 ⅛ in. × 50 ¾ in. (139 × 262 × 129 cm). Thyssen-Bornemisza Art Collection

IS

In a wonderful interview with Yates McKee, you discuss what you call "the monstrous dimension of art," by which a work of art "exceeds the plans and purposes of its creators."[5] This "monstrous dimension" also suggests an assemblage of parts that take on a new, anomalous life, as in the exploded transformer you describe that emits the sound of electricity. These assemblages evoke a kind of biocollage, speaking both to displacement (the movement of matter from one area, or being, to another) and to the emergence of novel and hybrid entities—I think here of the way horticulturists graft branches from one plant onto the stems of another, resulting in the regeneration and transformation of tissues. Of course, "graft" also refers to political corruption, the redirection of funds from their proper source. How do the multiple inflections of this concept come together in a work like *Graft* (2019), not only at the level of form or site but in terms of labor, capital, and production?

A&C

Some of the primary legacies of colonialism are deforestation, transplantation, and extinction. We made *Graft* with these interlocking effects of colonial exploitation and environmental change in mind. In this sculpture, thousands of yellow blossoms are cast from the flowers of roble trees (*Tabebuia chrysantha*), an oak species native to the Caribbean. The hand-painted petals are reproduced in seven variations or degrees of decomposition, from the freshly fallen to the wilted and brown. These replicas are then installed as if wind had swept them across the floor. The systemic depletion of Caribbean flora and fauna is one of the primary legacies of colonial rule. Nonetheless, the region remains one of thirty-six biodiversity hot spots, areas that support nearly 60 percent of the world's plant, bird, mammal, reptile, and amphibian species but that amount to just 2.4 percent of the earth's land surface. In their plastic and unnatural stillness, the flowers in *Graft* reflect this fragile ecological predicament.

IS

The "monstrous dimension of art" can result in a particular formal or material configuration, as in the hand-painted flowers of *Graft*.

But the kinds of volatile energies you are interested in can also have "the force of an event, a manifestation that interrupts and alters its context," as you have observed.[6] The 2019 "Ricky Renuncia" protests in San Juan had this galvanizing and revelatory potential: those protests led to the resignation of Puerto Rico's governor, Ricardo Rossello, as a symbol of ongoing corruption, but along the way, they opened up countless unexpected, joyous, and even marvelously deviant possibilities—a cavalcade of horses and motorcycles, yoga protests, a *perreo combativo* session in front of the cathedral, unlikely coalitions formed in the street in real time. Certain of your works unmistakably carry this eventlike character; I'm thinking of pieces like *Chalk* (1998), *Returning a Sound* (2004), and *Gloria* (2011). But it's also true that our current realities more often approximate the tense of "a situation," as Lauren Berlant has written, that is closer to an impasse, the exhausting uncertainty that disorganizes and unravels quotidian life and is indicative of the cognitive and affective difficulty in grasping or responding to the scale of climate change, systemic racism, the utter violence of coloniality.[7] Perhaps Berlant's "situation" is closer to the inertia and apathy of acedia, the "noonday demon" that figured prominently in your recent body of work *Specters of Noon*. Can the work of art still have the power of an event? What are its other modalities? Or is the singular nature of the event itself something we need to reconsider?

A&C

We became interested in the notion of acedia after reading about the Christian ascetic Evagrius Ponticus. In the fourth century, Ponticus laid out the seven deadly sins, but described the "most oppressive" of all temptations as acedia, a spiritual dryness and lack of care toward the world and others. For Ponticus, acedia plagues the hot midday hours and is characterized by a feeling of psychic exhaustion and listlessness. Writing in the harsh conditions of the Egyptian desert, he personified this terrible mood as the workings of the "noonday demon" or "Meridian Demon," who "makes the sun appear sluggish and immobile as if the day had fifty hours."[8] In a curious way, this affliction seems to summarize the contemporary moment, in which one finds oneself feeling supremely awake, animated, immersed in very strong sensations and feelings, and yet paradoxically *not alive*.

Our 2020 exhibition at the Menil Collection in Houston, *Specters of Noon*, sought to illuminate elements of this condition, which makes the present intolerable and the future impossible to imagine. We could think of *Entelechy*, a coal sculpture cast from a pine tree felled by lightning, which is to say a massive condensation of sunlight converted into potential energy, and simultaneously inert and dead. At the same time, *Specters of Noon* was a deeply historical undertaking in that it sought to enliven distinct moments in which coloniality was enacted as well as creatively resisted. *Penumbra*, for example, is a shadow projection that re-creates the effect of light in the Absalon forest in Martinique, where a group of dissidents and refugees, including Aimé and Suzanne Césaire, Wifredo Lam, and André Breton, hiked in 1941, in the midst of various states of internment, exile, and resistance. The intellectual and political connections they nurtured during those walks speak to the insurgent potential of aesthetics.

IS

One of the most fascinating elements of your practice is your deep investment in collaboration, not simply with composers and writers but with land activists in Vieques, elementary-school-age choralists, Olympic athletes, experts in prehistoric flutes, and virtuosic whistlers, to name just a few. I see these collaborations as extending specific formal and conceptual elements of your practice; in this sense, *Puerto Rican Light (Cueva Vientos)* (2015–17; fig. 2) is a collaboration—perhaps an unwitting one—with a Dan Flavin sculpture. In your explicitly performative collaborations, however, it seems important that you reconfigure, or *detourn*, to use the Situationist term, the unique and extraordinary capacities of very particular individuals or groups of people.[9] If, in an earlier moment of art history, artists intentionally deskilled their own production, your work often delegates performance, but in a way that willfully redirects the skills of your performers. A clear example is *Stop, Repair, Prepare: Variations of Ode to Joy for a Prepared Piano* (2008), in which trained classical musicians reverse their right and left hands to play Beethoven's *Ninth Symphony* while leaning over a piano keyboard from a hole carved out from inside the instrument, which they also move in space. So redirecting expertise also comes with retraining or reskilling. Of course, most of us lack the virtuosic and exceptional abilities of your collaborators, but it strikes me that the ongoing character of crisis continually requires us to reskill and adjust as mechanisms of survival. This speaks to the instrumentalization of ordinary resilience by institutions and governments, but it also reveals that such "ordinary" resilience is in fact extraordinary and creative. As artists who have, let's say, a polymorphously perverse relationship to media, materials, and subject matter, how do you inhabit and navigate questions of expertise in your practice? Do certain bodies of works oblige you to develop particular skill sets?

A&C
For us, collaboration is about making circuits of connections, and these usually involve both human and nonhuman actors (or actants, as Bruno Latour put it). *Puerto Rican Light (Cueva Vientos)*, for example, concerns questions of travel and displacement, metaphor and actuality. Technically, the work entails harvesting radiant light from the sun outside a cave located on the border between the municipalities of Guayanilla and Peñuelas in Puerto Rico, and then converting it into electric power, which in turn is used to energize three mercury-vapor gas-discharge lamps that produce fluorescent light inside the cave. But this circuit also involves, describes, and manifests an object of art history (Dan Flavin's *Puerto Rican Light [to Jeanie Blake]*), made in 1965; a relationship between autonomy and referentiality (this historic artwork's phenomenological specificity versus its metaphoric dependence on an "Other" that is Puerto Rico); a contemporary artwork we made with the Dia Art Foundation informed by its collection (of which Flavin's sculpture is a part); an archaeological depository (containing artifacts of the ancient American cave-dwelling Saladoid peoples); an Indigenous cosmology (the Columbus-era Taino peoples' myth that they originally emerged from a mountain cave called *Cacibajagua*, or "Cave of the Jagua"); a failed economic policy (the postwar US government program Operation Bootstrap, which gave tax incentives for US corporations to relocate to the island while encouraging Puerto Ricans to emigrate stateside); a host of expressions of

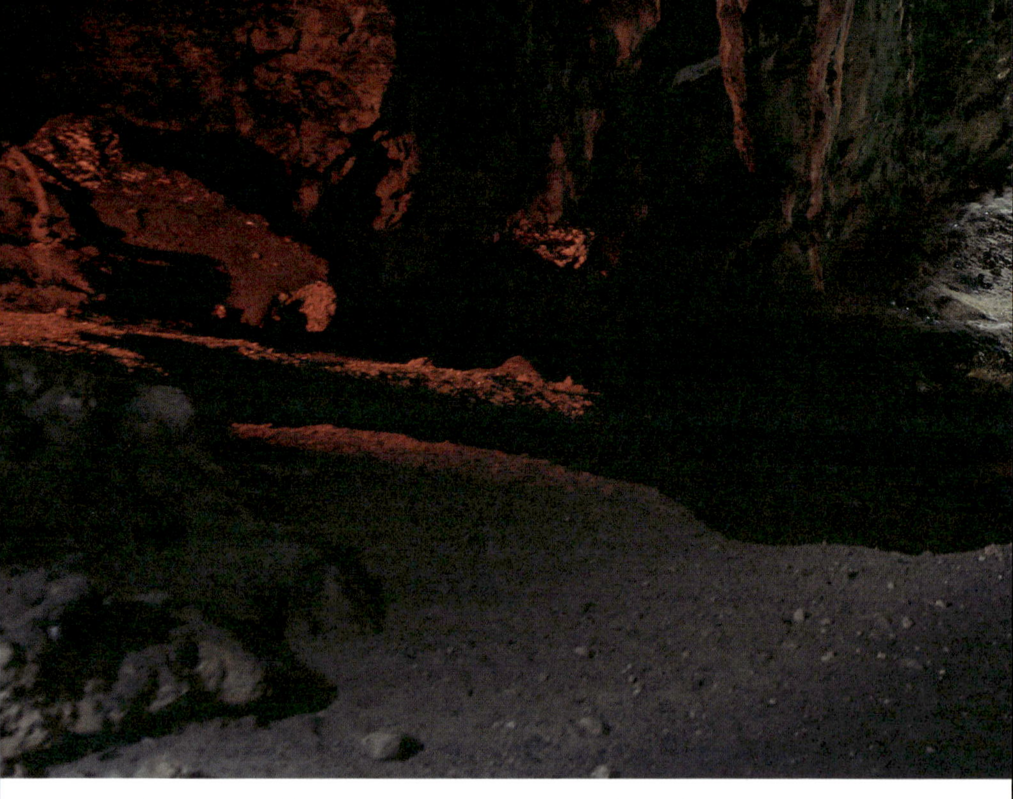

2 Allora & Calzadilla. *Puerto Rican Light (Cueva Vientos)*. 2015–17. Solar-powered batteries, charger, and Dan Flavin, *Puerto Rican Light (to Jeannie Blake)*, 1965, red, pink, and yellow fluorescent lights, 96 in. high. Installation, El Convento Natural Protected Area, Guayanilla-Peñuelas, Puerto Rico. Commissioned by Dia Art Foundation, New York

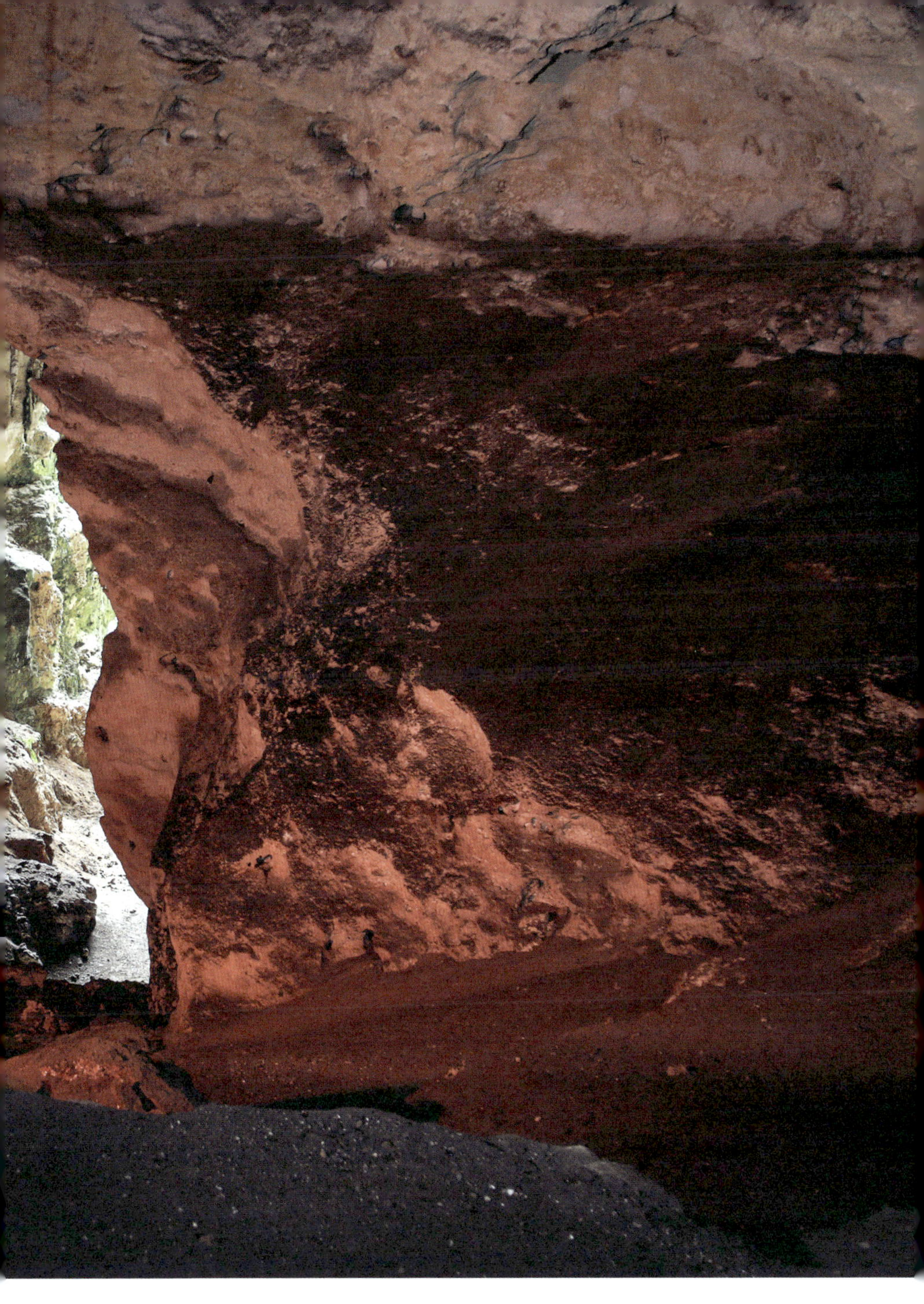

transnational culture (such as the annual Puerto Rican Day Parade in New York City, which began in 1958 and prompted Jeanie Blake, an assistant at the gallery where Flavin exhibited, to tell the artist that the color schemes of his light sculpture reminded her of "Puerto Rican lights"); a blackout in the northeast United States in 1965 (which plunged 30 million people into darkness save for a full moon, inspiring the New York–based novelist José Luis González to write of Puerto Rican migrant workers in *La noche en que volvimos a ser gente* [The night we became people again]); the contemporary wave of stateside migration spurred by the island's current crushing debt obligations (illustrated by the crisis of the island's main electric utility, which passes on impossibly high energy costs to ordinary citizens); an audible sound wave or hum in B-flat emitted by *Puerto Rican Light (to Jeanie Blake)* (as a result of the bulb's continual calibration of its magnetic field); an acoustic environment conditioned by deposits of guano (bat or bird excrement, which is also a sound-absorbing material); a pair of geologically formed oculi in the seventy-five-meter-tall terminal vault of Cueva Vientos (which bring sunlight into the darkened interior of the cave, and at midday, function as a sundial).... In other words, to describe *Puerto Rican Light (Cueva Vientos)* necessarily involves a complex and multilayered network. It is the informational relationships, dependencies, collaborations, and exchanges that occur across this network that produce and sustain the meaning of the project. So perhaps our skill set as artists is weaving the web, or, better, tracing its entanglements.

IS

Speaking of entanglements, many of your works think critically and carefully about the knotted geopolitics and histories of resource extraction—for example, the way farming methods introduced in the nineteenth century depleted soil, which led to a lucrative transnational trade in guano to be used as fertilizer, which in turn spurred colonization and territorial seizure, and so on. Your monumental sculpture *Manifest* (2020) registers and entombs such vertiginous historical and material connections: it takes the form of the engine of an early-twentieth-century US cargo ship cast in guano and then mounted on the wall. There's a double pun, of course, because just as the demand for guano turned shit into gold, the work of art is a site of transformation, wherein the base materiality of an object acquires symbolic value and thus, in turn, exchange value. How do you think about the work of art as an entity that, in sedimenting meaning, also sediments financial value? Are there ways of building in loopholes or escape hatches within a work of art or practice that run counter to these dynamics of capital?

A&C
We think that a good artwork can't be reduced to one thing. It's important to us, for example, that *Manifest* refers to multiple but specific things. At the most basic level, the word refers to a document listing the cargo, passengers, and crew of a ship. These aspects carry their own geopolitics, however. For example, the US Jones Act of 1916 restricted the carriage of goods and passengers between US ports to vessels built and flagged by the United States and crewed predominantly by Americans. That law resulted in an enormous economic burden on Puerto Rico. *Manifest* also refers to the historical idea of "manifest destiny" and the aggressive and interventionist policies that accompanied US imperialism in the nineteenth century. In this context, and that of a worldwide demand for nutrient-rich guano, the US Congress passed the Guano Islands Act (1856), which authorized the expropriation of unoccupied islands in the Pacific and Caribbean containing guano deposits. This act created the legal framework for the kind of territorial expansion that would later be applied to Puerto Rico.

Finally, at a semantic and conceptual level, *Manifest* also means "to make something apparent." As a sculpture that is a massive wall relief derived from the engine of a cargo ship, our work performs this meaning by indexing the geopolitics of the first two meanings. The ship in question belonged to the Crowley Maritime Corporation, which since its founding, in 1892, has been one of the primary maritime transportation companies operating in Puerto Rico. Historically, the corporation has thus benefited from trade laws such as the Jones Act, as well as from the island's severely imbalanced import economy. Our sculpture is splayed open, separated in two, and cast in bat guano. The intestines of the ship engine are thus repurposed as sound paneling that absorbs the room's acoustic energy, creating an eerie stillness. The embodied metaphor sets into motion a circuit of analogies between mechanical and biological technologies. While cargo-ship engines enable the circulation of goods from one territory to another (thus concretely manifesting what is abstractly traded and procured in financial markets), the exceptionally high mineral content of bat manure in turn delivers essential nutrients to soil when used as fertilizer, resulting in an increase of agricultural productivity. Significantly, agriculture accounts for only 1 percent of Puerto Rico's economy, contributing to its dependence on imports. In the wall reliefs of *Manifest*, the guano's potential as fertilizer is left unrealized, just as the defunct engines are unable to move cargo across the sea. So potential remains trapped in a state of latency. The nonproductive, artistic end to which this unlikely combination is now mobilized suggests a radical reimagining of these trenchant and ossified relationships.

IS

Perhaps we might close our dialogue by discussing the multiple temporalities staged by your works: sometimes as ephemeral trace, sometimes as real-time duration, often as a vertiginous clash of timespans or frames. In your film *Raptor's Rapture* (2012), for example, a live griffon vulture becomes a peculiar witness to a flautist's attempt to play an ancient flute made from the bone of a griffon vulture 35,000 years ago. In a beautiful text on this piece, Emily Eliza Scott writes of the "intense presentness" of these protagonists, who coexist in a kind of "adjacency" that presses at "the (indistinct) edges where human and nonhuman meet."[10] As artists particularly invested in deep time, how do you think about the temporality and historicity of your works, or of contemporary art more generally? As contemporary works of art, how do they sit alongside a set of related terms such as "artifact," "document," "myth," "trace"?

A&C

Sonic traces from the deep past: what could early-modern humans have heard as music, and is it possible to listen to those sounds today? These were the starting questions for *Raptor's Rapture*. We read that in 2009, Nicholas J. Conard of the Universität Tübingen,

Germany, unearthed a flute that was carved by *Homo sapiens* 35,000 years ago from the wing bone of a griffon vulture. Found in a cave in southern Germany, it is the oldest-known existing musical instrument. When we met with Nicholas in Germany, he told us that the discovery brings further evidence of the role of music in early humans' social-network development, demographic and territorial expansion, and, ultimately, their evolutionary survival. In other words, the process of procuring this vulture bone, making apertures in it at specific intervals, and blowing air through it to create meaningful sound for others to hear (in other words, music) shows a level of intelligence and brain development that was not previously known. In collaboration with Nicholas, we invited Bernadette Käfer, a flautist specializing in prehistoric instruments, to try to play the flute in the presence of a living griffon vulture. These vultures are evolutionary descendants of some of the oldest creatures that have inhabited the earth, and are currently threatened with extinction. Of course the flute didn't come with a manual, so Bernadette had to play its form from all possible angles; this experimentation determined the duration of the film. The work constituted a moment of sharing the musical remains of a prehistoric human culture with this scavenging bird of prey, while the acoustic trace emitted from the flute offered up a time capsule of sound embedded within the incremental evolutionary emergence of musical capacities in humans. Our video now exists as both a work of art and a document of archaeological and evolutionary biological research.

1 See Miriam Muñiz-Varela, *Adiós a la economía* (San Juan: Ediciones Callejón, 2013), and Rocío Zambrana, *Colonial Debts: The Case of Colonial Puerto Rico* (Durham, NC: Duke University Press, 2021).

2 Rob Nixon, *Slow Violence and the Environmentalism of the Poor* (Cambridge, MA: Harvard University Press, 2013).

3 *Downes v. Bidwell* is one of the Insular Cases (1901–4), a group of fourteen decisions involving the application of the Constitution and the Bill of Rights to overseas territories. The cases arose after the United States acquired island territories after the Spanish–American War (1898). The nation's determination to become a world power, as evidenced by the war and the acquisition of foreign territories, received overwhelming popular endorsement in the presidential election of 1900. The Insular Cases translated the political dispute into the vocabulary of the Constitution, with the Supreme Court eventually echoing popular sentiment. See Kermit Hall, ed., *The Oxford Guide to United States Supreme Court Decisions* (New York: Oxford University Press, 1999). See also Christina Duffy Burnett and Burke Marshall, eds., *Foreign in a Domestic Sense: American Expansion and the Constitution* (Durham, NC, and London: Duke University Press, 2000).

4 *Downes v. Bidwell*, 182 U.S. at 341-42 (White, J., concurring). Downes created the doctrine of "territorial incorporation" to justify keeping the then newly acquired territories without the intention of incorporating them into the Union as states. See Christina D. Ponsa-Kraus, "Recent Publications: Puerto Rico," *Yale Journal of International Law* 23 (1998): 561.

5 Allora & Calzadilla, in Yates McKee, "The Monstrous Dimension of Art," *Flash Art* no. 240 (January–February 2005): 96–99.

6 Ibid.

7 Lauren Berlant, *Cruel Optimism* (Durham, NC: Duke University Press, 2011).

8 Evagrius Ponticus, quoted in Siegfried Wenzel, *The Sin of Sloth: Acedia in Medieval Thought and Literature* (Chapel Hill: University of North Carolina Press, 1967), 5.

9 See Tom McDonough, "Use What Sinks," *Art in America* 96, no. 1 (January 2008): 82–86.

10 Emily Eliza Scott, "Feeling in the Dark," *American Art* 28, no. 3 (Fall 2014): 14–20.

CARLA ACEVEDO-YATES

THE TERRITORIAL BODY

Regina José Galindo, María Evelia Marmolejo, and Beatriz Santiago Muñoz

The imminent global climate catastrophe signals the advent of an alarming geological and historical era, already in the making since "the long sixteenth century."[1] In Latin America and the Caribbean, the devastating effects of the Androcene—the era resulting from the acceleration of capital accumulation, resource extraction, and human exploitation for profit under patriarchy—have deepened the colonial wound, shaping and manipulating nature, displacing Indigenous and other ethnic populations, and generating social and political instabilities.[2] If colonialism, in all of its manifestations, was the condition for modernity, the patriarchal order and its violent forms of masculinity and domination are its main instruments. Anthropologist Rita Laura Segato defines the era that inaugurated modernity as "the stationary time" of what she calls "the patriarchal prehistory of humanity," a historical epoch of ownership that can only be transcended with plural, communal, and collective forms of politics in the feminine. Since colonialism required "the appropriation of bodies and their annexation as territories, but also their damnation, their moral and physical destruction to secure the expropriator's triumph," women's bodies have become a site of contestation and the territory where power is exerted.[3]

The patriarchal prehistory of humanity is most evident in the colonial and neocolonial contexts of Latin America and the Caribbean, where settler occupation and resource extraction for the production of global commodities, often by foreign multinational corporations and the tourist industrial complex, have resulted in population displacement, gendered violence, and death.

Artists in the region have long been engaged with and informed by the social and political contexts produced by the logics of development and the extractive violence of colonial power. Woman-identified artists in Latin America and the Caribbean are particularly well positioned to address the intersection of gender violence and environmental destruction resulting from the centuries-long disconnect between nature and culture produced by modernity, where nature is seen as the feminine wild that necessitates domestication, control, and encroachment. This essay approaches the work of three woman-identified artists who span Central American, South American, and Caribbean territorial narratives: Regina José Galindo, María Evelia Marmolejo, and Beatriz Santiago Muñoz. Employing a wide range of media, including performance, public actions, and video, these artists lay bare the histories of imperial domination, pillage, and resource extraction, offering counternarratives to generate empathy, memory, and cultural transformation toward the future.

Across Latin America and the Caribbean, women, and in particular impoverished Indigenous women in rural areas, are leading the struggle for environmental justice and against extractive violence. Sociologist Maristella Svampa uses the term "territorial ecofeminisms" to describe these struggles, which she defines as "a defense of the land and the territory, showing that the sustainability of life and of the planet is based on another bond with the body and with nature, simultaneously material and spiritual, within the framework of an epistemology of emotions and affects."[4] Svampa references Maya Q'eqchi'-Xinka activist and healer Lorena Cabnal, whose thinking and approach to the "territorial body" and "the body as territory" have shaped thinking and community activism across Latin America. Her slogan "My body, my first territory of defense" makes visible the interrelatedness between extractive violence and gender violence, which have roots in both patriarchy and colonialism. This has little to do with geography but rather with reclaiming ancestral and Indigenous ways of thinking, sensing, and being with the land from a feminist, communal, and antipatriarchal perspective. For artists living and working in Latin America and the Caribbean, the body has also been the territory from which on the one hand to condemn and denounce, and on the other to construct sensibilities in the plural. The epistemological shifts brought on by the work of the artists approached in this essay create spaces of resistance and visibility, offering ways to see and imagine new forms of life that would subvert the normative paradigm of binary thinking and capital accumulation.

Galindo has been using her body as a medium to tell the everyday stories of displaced, murdered, and raped women in her native Guatemala for over twenty years. Making public the worldwide epidemic of misogynist violence, her durational performances deploy pure emotion and energy, with the artist both embodying empathy and silently communicating it to the public. Galindo comes from a generation of artists who came of age after the Guatemalan Civil War of 1960–96, and her visual work has been informed by that decades-long conflict. Focused on the stories of Indigenous women in rural areas of the country, Galindo's work sheds light on the genocide of Indigenous peoples at the hands of the Guatemalan army led by dictator José Efraín Ríos Montt, who came to power in 1982 through a military coup d'état. His regime, although short-lived, was extremely violent. The massacre of Indigenous peoples, including women, children, and the elderly, was part of a larger strategy to completely eradicate Maya-Ixil populations from Guatemalan rural areas.[5] Segato, who served as an expert witness in a case where Indigenous women were subjected to slavery, further shows that "the destruction of the social through the profanation of the female body played an important role in the genocidal war waged by the authoritarian state in the 1980s."[6] This gendered violence, planned and focused on women's bodies, eventually led to the dispossession of Indigenous lands, given away in suspicious contracts to multinational North American, Canadian, and Spanish corporations dedicated to the cultivation of African palm, mining, and other extractive activities.

In March 2013, Ríos Montt and his military intelligence chief, José Mauricio Rodríguez Sánchez, were brought to trial in Guatemala for genocide and other human rights violations. The trial was open to the public and Galindo attended personally, hearing firsthand over 100 victims testify painfully to the atrocities committed by the Guatemalan army, including the torture and rape of women and children and the genocide and forced removal of Mayan-Ixil communities from their lands. Galindo has spoken of how she broke down and cried while listening to these horrific stories. At that moment, during the trial, someone said to her, "We didn't come here to cry, we came here to resist."[7] To this day, these words remain an important mantra for the artist, who finds strength and inspiration through the courage of Indigenous women and men who risk their lives to speak up and demand justice.

From these oral testimonies Galindo made a series of performances. *Tierra* (Earth; fig. 1), performed and filmed in Les Moulins, France, in 2013, references a testimony in which a victim described how an excavator dug mass graves, people were brought in in trucks, many of them then stabbed with bayonets, and all were thrown into the graves; the excavator fell "into the pit on top of the bodies."[8] During the performance, a nude and solemn Galindo stood still on a verdant field as an excavator removed the soil around her, creating a form suggesting the trenches of a mass grave. The excavator continued to remove parts of the land around Galindo until she was left standing in resistance on a small plot. In the film, the camera moves between distant and closer shots of Galindo's body, visibly small in proportion to the excavator but powerful in its unrelenting stance. Her deep breathing often becomes a focus of the viewer's attention, perhaps in remembrance of the multitude of bodies often buried alive in mass graves, but also a reminder of our own bodies in relation to hers. Of her emotional state during the performance, Galindo says that she was thinking of the victims and of their testimonies, words that she transformed into gestures building empathy, compassion, and solidarity in both herself and the work's viewers.[9]

Besides its direct reference to a specific event, *Tierra* also calls attention to the ongoing histories of extractivism and ecological destruction in Guatemala. The military campaigns led by Ríos Montt dispossessed Indigenous populations of their lands, allowing the Guatemalan government free rein to contract transnational corporations to develop extractive projects in rural parts of the country. But communities continue to resist. In one incident, in fact, in a physical gesture similar to Galindo's, a woman stood

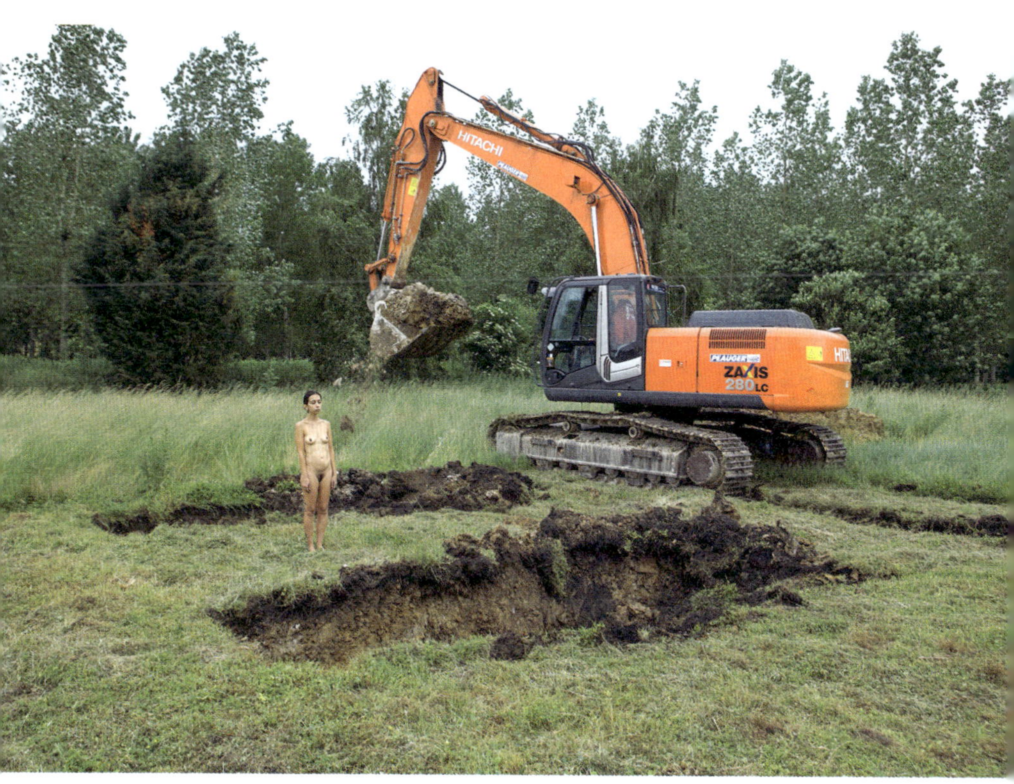

1 Regina José Galindo. *Tierra* (Earth). 2013. High-definition video, color, sound, 33:30 min. Commissioned and produced by Orta Studio, Les Moulins Residence 2013, Paris. The Museum of Modern Art, New York. Gift of Mario Cader-Frech through the Latin American and Caribbean Fund

still in the path of an excavator to protest a gold-mining project in her community.[10] This action inaugurated an activist movement, Resistencia Pacífica La Puya, led mainly by women committed to nonviolent struggle, using strategies such as praying, singing, and human barricades to protect the land and water rights of rural communities in San Pedro de Ayampuc and San José del Golfo. *Tierra*, then, also pays homage to the men and women who, although persecuted and murdered, continue to resist the destruction of their territory in defense of human and nonhuman life for present and future generations.

Galindo took a similar stance in *Mazorca* (Cob), a performance and two-channel video made in 2014 in Totonicapán, a rural department in the western highlands of Guatemala with a predominantly Indigenous population. There Galindo rented a field of corn (called *milpa* in the Nahuatl language) and purchased two consecutive years of harvest from its owner. Then she waited until the color of the crops resembled her skin color in order to position her nude body deep within the field. In the video, two men slash a field of cornstalks with machetes until Galindo's nude body is revealed, standing in resistance amid the slashed stalks of corn. Throughout, the viewer can only hear the sounds of the environment and of the actions performed—the wind and the slashes of the machetes, giving the video a sense of solemn dignity, starkly different from the Guatemalan army's violent attacks, often sexual in nature, against women. As in *Tierra*, Galindo's simple gestures are conceptualized according to place, which provides a contextual framework for her standing nude body. A racialized body, it offers both vulnerability and strength in relation to the historical attempts to destroy the land and, in consequence, a community's existence.

Like *Tierra*, *Mazorca* originates in the testimony of men and women who witnessed the atrocities committed by the country's army, including the rape and murder of women and children. During the war, the military slashed and burned fields of corn as a way to destroy Indigenous communities, which were thought to be involved in guerilla activities against the government.[11] The destruction of corn and other food crops was significant for a number of reasons: not only did it reflect an attempt to strip communities of their main food supply, it also affected their social and spiritual lives.

In Guatemala, the cultivation of *milpa* through sustainable ancestral ways is a relational practice sacred for Mayan communities.[12] *Milpa* is simultaneously a crop, a space, and a way to build relationships. Traditionally, it is harvested through sustainable practices that respect the cycles of nature, where each corn stalk is torn by hand without the violent intervention of a machete or knife. This detail is important, because the cutting of the corn stalk with a sharp object affects the crops' reproducibility. When soldiers cut the corn fields with machetes, they destroyed them, since it takes the crop years to recuperate from this violent assault. *Mazorca* also references the *Ley para la Protección de Obtenciones Vegetales* (Law for the protection of plant varieties), commonly called the *Ley Monsanto*, passed by the Guatemalan congress in 2014 and quickly overturned after public pressure in the form of protests and demonstrations. The law would have privatized seeds such as corn and given way to genetically modified variants, adversely affecting the livelihoods of Indigenous populations through the control and manipulation of one of its main food sources.

Seeing the end of *Mazorca* begs the question: what happened after the performance, after the camera stopped filming? According to Galindo, the fact that she stood nude for several minutes caused outrage.[13] Although she had told the owner of the land what would happen during the performance, she had not asked the permission of the highest Mayan authority. Her nudity threatened the community, whose fundamentalist, evangelical fanaticism is a legacy of US interventionism and the dictatorship of Ríos Montt, who used evangelical and Pentecostal churches as a way to implement counterinsurgency operations against guerrillas during the armed conflict.[14] Galindo was attacked, called the devil, and her right to be there was questioned. These verbal attacks became so frenzied that she was forced to flee in fear of lynching. While two of her colleagues were caught and beaten, Galindo was able to escape, but eventually had to return to the community to ask for forgiveness. Violent attacks of this kind are unfortunately common in communities such as Totonicapán; foreigners and even Mayan people have been lynched and burned alive by Christian fundamentalists, as in the murder of Maya q'eqchí priest, herbalist, and spiritual healer Domingo Choc in 2020.[15] The demonization of

Galindo's nude body at the conclusion of *Mazorca* harkens back to the witch-hunt eras of Protestant Europe and, later, the United States, where women's bodies were "attacked as the source of all evils," as Silvia Federici writes, and the concept of nature as a living, inspirited organism was rejected.[16] As powerful occupations of the land, *Tierra* and *Mazorca* position the body, which takes a solemn, confrontational stance, as a territory of resistance against landscapes of extraction and as the contested site where racial, gendered, and political questions are brought to the fore.

During the COVID-19 pandemic, two environmental defenders, Abelino Chub and Andrés Cano Sierra, reached out to Galindo and invited her to work with communities in defense of land and water. The result was the public action and three-channel video installation *Ríos de Gente* (Rivers of people, 2022). Sparked by the arrest of human and land rights defender Bernardo Caal Xol, the action consisted of a protest across eight different communities where rivers had been diverted, dried up, or contaminated by extractive industries. During the performance, men, women, and children gathered under a long stretch of light-blue fabric to walk, sing, and scream justice for Caal and for the life and livelihood of rivers. Vocalizing the slogans "Freedom for Bernardo. Freedom for water" and "There where there was a river, there, we shall sing," communities walked through the former trajectories of impacted rivers, making visible that which is no longer. In the video, overhead drone footage of the actions shows the flow of the river through lines traced by the fabric. One of the most aesthetically assertive of Galindo's performances, *Ríos de Gente* uses line and color to contend with the impact of extractive economies on the physical landscape. It is also quite different from works such as *Tierra* and *Mazorca* in that it is a collective action, not an individual one, and muddles the difference between activism and performance, art and life.

Galindo's direct, confrontational performance work can be understood in relation to the work of Marmolejo, a pioneer of performance art in Colombia, where the medium did not start up in earnest until the early 1980s.[17] Both artists make their bodies their main medium and adopt a confrontational stance, often involving actions that include self-harm. Marmolejo, however, is older than Galindo, belonging to a generation of artists who came of age in the 1960s and '70s, at the height of the consolidation of the environmental movement focusing on contamination of land, air, and water. After two years in law school at the Universidad de Santiago de Cali (1976–78), she studied art at the Instituto Departamental de Bellas Artes de Cali (1978–80), coming of age as an artist during the presidency of Julio César Turbay Ayala (1978–82). Marmolejo remembers this period as a time of artistic effervescence in Cali. Her art-history professor was the artist Miguel González, who introduced her, through books and magazines, to artists from outside Colombia who worked in performance.[18] Personal experiences from her own upbringing as a young woman growing up in a violent male-chauvinist society have been important to her. Raised in a Catholic household, where gender roles remained quite traditional, she was politically conscious from an early age, and critical of the oppression and marginalization of women. Because Marmolejo's father belonged to the Partido Liberal Colombiano—Colombia's liberal party, whose members were persecuted, tortured, and murdered during the period known as "La Violencia" in the late 1940s and the '50s—her family moved to Cali from Vijes, a rural town whose livelihood depended on limestone mining.[19] Her father, who worked in the small-scale limestone industry, kept copious archives of newspaper clippings of politically motivated massacres and murders. Marmolejo recounts how he used to show these images to her, vividly remembering images of two student massacres in Bogotá and Cali, where bodies were piled on top of each other.

Over the years, Marmolejo has developed a subversive practice that focuses on the body as a personal and political terrain. Her performances are often difficult to watch, including the healing of self-inflicted wounds and the use of unconventional materials such as menstrual blood, bandages, and sanitary napkins. Here, in keeping with Cabnal's theorization of the "body-territory," the body is simultaneously the subject of violence, a means to denounce violence, and also a means by which to process and heal from the psychological and emotional wounds of violence. Marmolejo's work is fueled by her emotions: rage and indignation against violence and injustice, and a deep capacity to sense and feel with others.

She describes her performances as a "rebellious cry, one of protest," with her body a tool to express her frustration and impotence and to generate public awareness of the increasingly dire political situation in her country.[20]

In 1981, Marmolejo initiated a suite of performances she called *Anónimos* (Anonymous), rituals of healing and emotional catharsis in which she used materials associated with the medical industry, such as medical tape, bandages, and gauze. *Anónimo 1* (1981; fig. 2) was performed at the Plazoleta del Centro Administrativo Municipal, a public plaza in Cali. In this twenty-minute performance, Marmolejo, wearing a white cloak and cap and her face covered with Band-Aids, walked down a seventy-foot-long walkway of white paper with the tick-tock sounds of a clock as a backdrop. At one moment during the performance, she sat, made incisions in the tips of her toes with a scalpel, stood up, and continued to walk. Sitting down once again, she cleaned her wounds with a tincture of the antiseptic Merthiolate, which is often very painful, and then bandaged her wounds.[21] She then continued to walk toward the end of the walkway, where the sounds of an alarm "wake her from the nightmare."[22] Of this, Marmolejo says that "the self-harm was realized in homage to the tortured and disappeared during the regime of Turbay Ayala. Calling attention to the violence that surrounded my country, Colombia."[23] Here, Marmolejo's body becomes a surrogate for all bodies tortured and repressed, and a metaphor for the Colombian national body itself as wounded and in need of mending, care, and healing.[24] The work's title, *Anónimo*, refers to the mass graves where thousands of bodies were buried—bodies that remain anonymous but demand to be seen and acknowledged.

Ensuing actions and installations focused on environmental concerns but also span feminism and politics. *Anónimo 3* (1982) was performed without a public on the banks of the Cauca River, outside Cali. Mining companies, particularly those extracting gold, coal, limestone, and bauxite, have had a significant environmental impact on the Cauca; these companies use immense amounts of water for their operations and often dispose their waste directly in the river. The Cauca Valley is one of the most heavily mined regions in Colombia, and defenders of land and water rights have been criminalized and murdered there, as they have in most countries

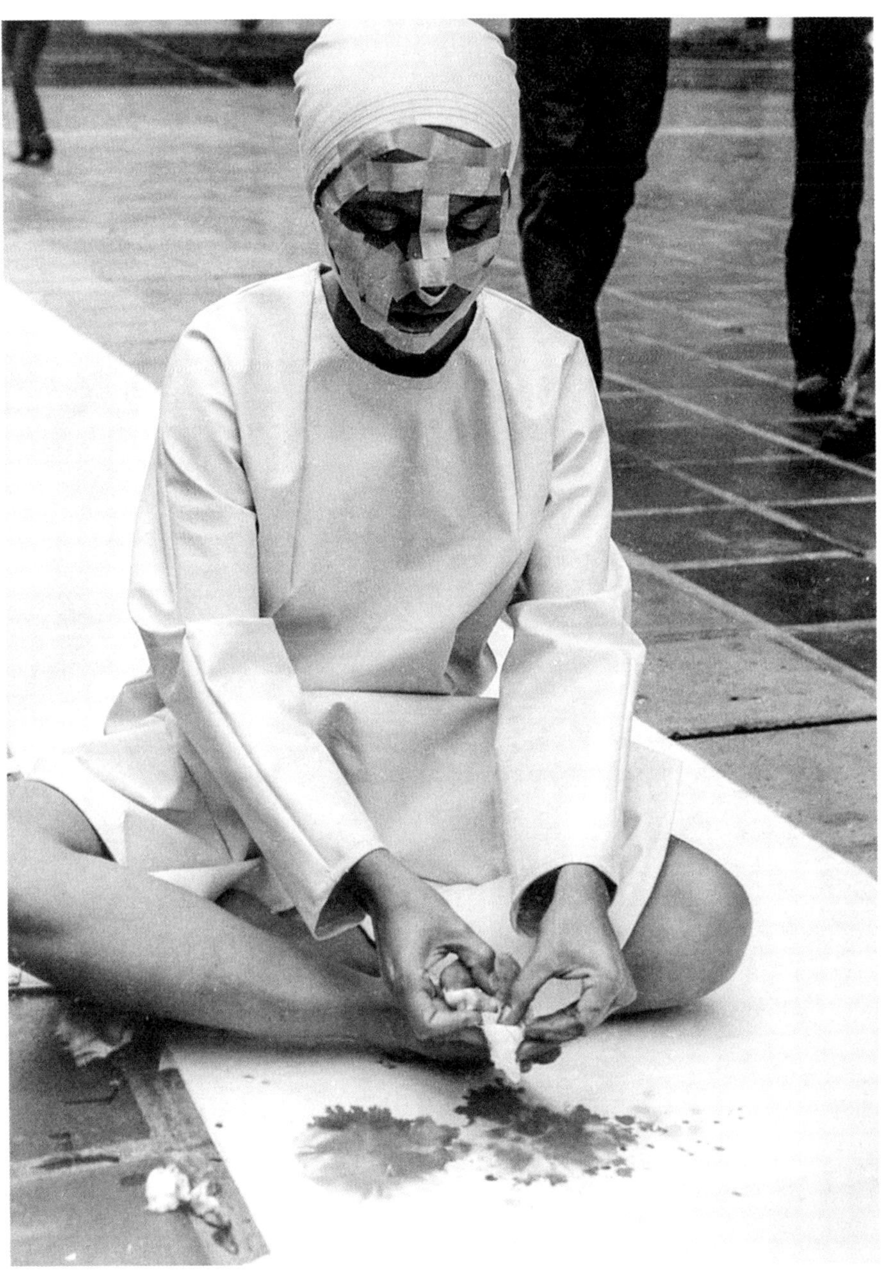

2 María Evelia Marmolejo. *Anónimo 1* (Anonymous 1). 1981. Performance view, Plazoleta del Centro Administrativo Municipal, Cali, Colombia, 20 min. Collection Banco de la República | Colombia

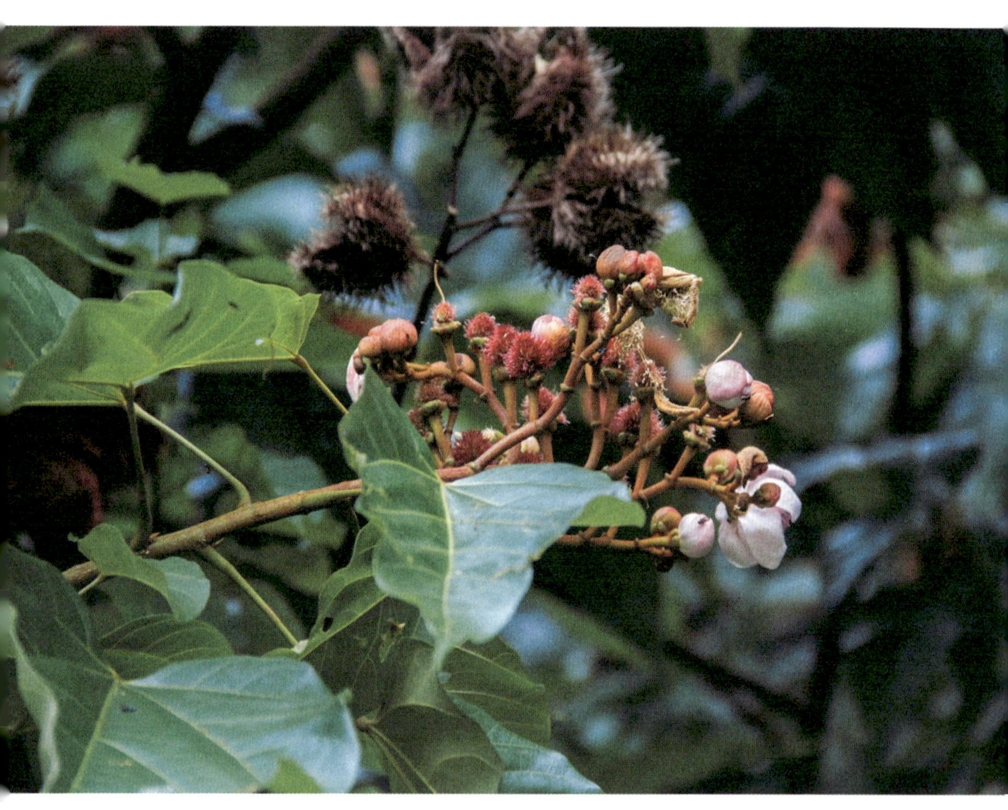

3 Beatriz Santiago Muñoz. *Farmacopea* (Pharmacopoeia). 2013. 16mm film, color, 6 min. Collection the artist

in Latin America. In 2019, Colombia became the most dangerous country on Earth for human rights activists, with over 250 people killed, and it is the Cauca Valley area that has the highest numbers of disappeared.[25] This was the location that Marmolejo chose for *Anónimo 3*. On the riverbank, having bandaged her face with gauze and stuck pieces of medical tape on her body, she drew a large spiral in limestone dust and placed a toilet at the center. With a vaginal douche, she proceeded to cleanse her vagina, letting the fluids fall on the earth while she walked along the spiral. Then, removing the gauze and medical tape, now covered with body hair, from her skin, she laid them on the land as an offering. Marmolejo describes the action as a healing ritual to fertilize the flora and fauna that had been contaminated, an offering to Mother Earth for the destruction and trauma caused by humans and extractive economies.

Although Galindo and Marmolejo come from different generations, their public actions and performances take on related forms of protest and activism, engaging in direct action with the land to denounce the sociopolitical conditions of their time. Santiago Muñoz takes another, more sensory and speculative path. An artist living in San Juan, Puerto Rico, and working mainly in film, she grounds her approach in the colonial histories of pillage, violence, and environmental destruction, as they do, but her work is more process based and is set up as a way to imagine possible futures. Thinking through complex questions on the connections between time, nature, history, and place, films such as *Farmacopea* (Pharmacopoeia, 2013; fig. 3) and *La Cueva Negra* (The black cave, 2013) show the sensorial and affective histories that connect the landscape with the body. Muñoz's work takes on the observational style of cinema verité, combining improvisation with experimental ethnography and feminist thought. She often works with nonactors to make films about the sensorial dimension of coloniality on the island of Puerto Rico, which some consider the oldest colony in the world.[26]

Process is extremely important in Santiago Muñoz's films, shaping their structure and content. The viewer often sees an assemblage of different sources that gives the illusion of a singular, cohesive narrative, but that narrative and the images that carry it are the results of relationships built over time, and of a slow process of informal interviews, conversations, collaborations, and serendipitous encounters. At the onset of the thinking that led to *Farmacopea* and *La Cueva Negra*, Santiago Muñoz was interested in making films about "how the natural is also the historical, how nature and the human coproduce historical events."[27] Both films originate in research conducted at the Paso del Indio archaeological site, a Taíno Indigenous settlement and archaeological site in Vega Baja that was rediscovered in 1992, when the elevations of Puerto Rico Highway 22 were being excavated in the Río Indio Valley.[28] Archaeological excavations conducted between 1993 and 1995 uncovered burial grounds, paved roads, ceramics, organics, and other materials in over thirty stratified cultural layers dating from the Archaic period (2500 BC) to the Late Ostionoid, Taíno period (AD 1200–1500).[29] Santiago Muñoz encountered the site while she was tracing, by foot, the territory that would be crossed by the proposed Gasoducto del Sur (Southern gas pipeline). She uses walking as a research method and sees it as a relational form of learning about place that is "anchored in the specific geography, local knowledge, emerging art practices, and social and political conditions of Puerto Rico."[30] Under the title "Sessions," she has led experimental, pedagogical walking seminars with other artists, thinkers, and creatives through the rural landscape of the island, a landscape fraught with the contradictions of coloniality in the tropics, where military spaces and abandoned postmilitary sites coexist with ecological destruction amid lush, tropical vegetation. These walks often lead to films, collaborations, and future projects, both individual and collective.

Shot in 16mm, *Farmacopea* features a silent narrator—speaking through subtitles rather than vocally—who recounts interconnected stories that unravel through the film, including the description of a catalogue of plants native to Puerto Rico that are considered poisonous or dangerous to humans. The film begins with a self-referential written statement—"This is a film about a disappearing landscape"—followed by a visual enumeration of different species of native plants, such as *guayacán* (soaptree), *manzanilla de la muerte* (manchineel), and *chicharrón* (poison ash). It also features a farmer named Pablo and the land surrounding his house in Orocovis, a town at the heart of the island's mountain range. In one story, the narrator recalls how he was able to time-travel and see his future house by taking tea made from the flowers of *Brugmansia suaveolens* (angel's trumpet), which when ingested provokes strong hallucinations. Another story documents one of the last surviving *manzanilla* trees, almost completely eradicated from Puerto Rico because they are extremely poisonous. Toward the end of the film, a story involving angel's trumpet foregrounds geological time and nonhuman agency, as the narrator describes how the trees started speaking, saying that the human and the nonhuman had been at war for centuries. The disappearing landscape cited at the beginning of the film is the disappearance not only of plants deemed poisonous, deliberately eliminated by settlers, but also of native species that when ingested facilitate cosmic and interspecies communication, crossing the boundary lines between time and space, human and nonhuman worlds. Indeed, for most Indigenous cultures the ingestion of hallucinogenic plants is a medicinal and spiritual ancestral practice, used in ceremonies and rituals for healing purposes and to access other states of consciousness. Moving away from linear, human time and focusing the viewer's attention instead on the time of the land—geological or nonhuman time—the film reveals the porous boundaries and entanglements between beings both human and nonhuman. At the end, the narrator describes how every single tree on the land was cut down. This unexpected outcome suggests the violence that often results when nature is perceived as a menacing force.

La Cueva Negra brings Indigenous histories and the material layers of geological time to the surface. It focuses on Paso del Indio, which, since the construction of the PR-22 highway and the unearthing of the land as an Indigenous site, has become a dumping ground for all sorts of things—a site of death and disposal for objects, animals, and bodies alike. During Santiago Muñoz's visits to the site, she told herself that "the landscape will speak to me." In setting the site's contemporary forms against its excavated histories, *La Cueva Negra* provides a new cosmology, a new way to imagine a history and legacy removed from extraction, development, and death.

The film narrates a creation story as seen through the forms, shapes, and objects of the Paso del Indio site and its surrounding territory. Santiago Muñoz combines that story with elements of the Taíno creation myth, which involves a set of cosmic twins and a black cave where all human and animal life originated. One day on the site she encountered two local boys, Heniel and Keniel Camacho, who were looking for a horse. These boys gave her a way to enter Paso del Indio as a historical site of human life relations and tenderness. Santiago Muñoz met them at Paso del Indio on weekends over a period of months, telling them some of the stories she had heard and asking them to act them out. The film follows these two boys as they run through the landscape, see, touch, and play with the trees, leaves, and objects found there until they reach the dark underpass of a bridge. The film continues to follow them as they interact with their surroundings, at times in staged scenarios that are very formal, intimate, and haptic, such as a scene where one of the boys is lying on his back on top of his horse. For Santiago Muñoz, the film is also about capturing moments of tenderness and affection between these two adolescent boys and collaborating with them on the film's aesthetic decisions. The collapsing of time and space in *La Cueva Negra* can be approached as a narrative based on geological time, where accrued layers of materials, objects, and bodies create new modes of seeing, sensing, and thinking with nature into the future. That narrative offers access to aspects of our collective historical memory that remain elusive or that lack a tangible form.

To build pathways toward the future, we might turn to ancestral ways of thinking and sensing life that find the common humanity in all creatures and things of the natural order, including humans, animals, rivers, and plants.[31] Segato proposes a relational politics of intimacy, rootedness, and nearness as a way to transcend patriarchal prehistory. The historical experiences of women and other sexual, ethnic, and racial groups who have been marginalized and oppressed can provide ways to move forward with a sense of ethics that is plural, transversal, and affective. Feeling plays a fundamental role in the work of the three artists discussed here, and healing is for them a vital need to transcend the structures set up to keep humans separate from nature and unfeeling of the pain and suffering of others. It is not surprising to find that healing, as an aesthetic and political strategy, is a central concern among recent and contemporary Latin American and Caribbean artists, showing the ways in which affect can become an intangible artistic medium. Approaching the sensing body as territory, a body intimately connected with the land, as Cabnal suggests, creates a framework for doing that is grounded on a spiritual and cosmic understanding of life, one in which emancipation rises from the body, from sentience, and from the vital energies of life. That emancipation can be shared by all beings in relations of empathy and close proximity.

1 Historian Fernand Braudel defines the "long sixteenth century" as the period between 1452 and 1630, when capitalism was established as a world economy.

2 The Androcene differentiates itself from the Anthropocene as it considers capital accumulation and destruction under patriarchial structures from an ecofeminist perspective. On the Androcene see Jean-Baptiste Vuillerod, "L'Anthropocène est un Androcène: trois perspectives écoféministes," *Nouvelles Questions Féministes* 40, no. 2 (2021): 18–34.

3 Rita Laura Segato, "A Manifesto in Four Themes," trans. Ramsey McGlazer, *Critical Times* 1, no. 1 (2018): 198–211.

4 Maristella Svampa, "Feminismos ecoterritoriales en América Latina. Entre la violencia patriarcal y extractivista y la interconexión con la naturaleza," *Documentos de Trabajo* 59 (2021). Available online at www.fundacioncarolina.es/wp-content/uploads/2021/11/DT_FC_59.pdf (accessed July 5, 2023).

5 Of the 200,000 people disappeared or dead during the dictatorship of José Efraín Ríos Montt, 80 percent were Mayan.

6 Segato, "A Manifesto in Four Themes," 202–03.

7 Regina José Galindo, telephone conversation with the author, December 2, 2022.

8 See the entry for *Tierra* on Galindo's website, www.reginajosegalindo.com/ (accessed July 5, 2023). *Tierra* was commissioned and produced by Studio Orta Les Moulins when Galindo was participating in the studio's residency program.

9 Galindo, telephone conversation with the author.

10 See "La Puya: una comunidad en resistencia, una empresa insistente," July 1, 2016. Available online at www.plazapublica.com.gt/content/la-puya-una-comunidad-en-resistencia-una-empresa-insistente (accessed July 5, 2023). The mine was the Progreso VII Derivada mine in La Puya, part of a larger project, El Tambor, initially led by the Canadian company Radius Gold, the American company Kappes, Cassiday, and Associates, and their Guatemalan subsidiary Exmingua.

11 Galindo, telephone conversation with the author.

12 According to Mayan cosmology and the ancient sacred book the *Popol Vuh*, humans were first created from maize or corn, after two failed attempts with other materials. In Mayan culture, the four colors of corn (white, yellow, black, red) represent both the four directions of the earth (north, south, east, west) and the four elements (earth, water, air, fire).

13 Galindo, telephone conversation with the author, January 3, 2023.

14 According to a Gallup poll of July 2022, around 40 percent of Guatemalans today identify as evangelical Christians. On the history of evangelical churches in Guatemala and the rise of right-wing evangelical forces in the government, see Jeff Abbott, "The Other Americans: Guatemala Is Constructing a Religious Narco-State," *Progressive Magazine*, September 14, 2022. Available online at https://progressive.org/latest/other-americans-guatemala-religious-narco-state-abbott-091422/ (accessed July 5, 2023).

15 On June 6, 2020, Maya q'eqchí priest and healer Domingo Choc was covered in gasoline and burned alive by a mob in the Mayan village of Chimay, in San Luis, Petén, by evangelical fundamentalists on

accusations of witchcraft. See Rodrigo Rey Rosa, "La oscura condena de Domingo Choc," *El País*, August 11, 2021. Available online at https://elpais.com/revista-de-verano/2021-08-12/la-oscura-condena-de-domingo-choc.html (accessed July 5, 2023).

16 Silvia Federici, *Caliban and the Witch* (New York: Autonomedia, 2004), 137.

17 See Cecilia Fajardo Hill, "El cuerpo politico de María Evelia Marmolejo," *Artnexus* 85 (June–August 2012). Available online at www.artnexus.com/es/magazines/article-magazine/5d64034190c-c21cf7c0a342e/85/maria-evelia-marmolejo-s-political-body (accessed July 5, 2023).

18 Through the material provided by Miguel González, Marmolejo became aware of international movements in performance, including Aktionismus in Vienna, Fluxus in the United States, and the work of the Brazilian artist Lygia Clark. Among other influences, Marmolejo often mentions her experiences with popular theater as a child, as well as a visit to the Coloquio Latinoamericano sobre Arte No-Objetual y Arte Urbano in Medellín in 1981, where she saw performances by Venezuelan artists Yeni & Nan and Carlos Zerpa.

19 "La Violencia," the violent conflict between liberals and conservatives in 1948–58, was further stoked by the murder of Jorge Eliécer Gaitán Ayala, presidential candidate and leader of the Partido Liberal Colombiano, in Bogotá in April 1948.

20 Marmolejo, Zoom conversation with the author, October 3, 2022.

21 Merthiolate is not only painful but can be toxic on sustained exposure.

22 Marmolejo, Zoom conversation with the author.

23 Marmolejo, "Recuento histórico MEM," 2015. A document in the artist's personal archive.

24 *Anónimo 1* was informed by Marmolejo's personal experiences. Apart from having relatives imprisoned, the bodies of two of her childhood friends were found tortured and dismembered in a sugarcane field. According to Fajardo, her biological father was killed as a conservative and an assassination was attempted on her adoptive father as a liberal. See Fajardo, "El cuerpo politico de María Evelia Marmolejo," n. 3.

25 See Cumbre Agraria Campesina Étnica y Popular, "Reseña," July 25, 2019. Available online at www.indepaz.org.co/wp-content/uploads/2019/08/Informe-parcial-Julio-26-2019-Indepaz-Marcha-Cumbre.pdf (accessed July 5, 2023).

26 A series of US Supreme Court decisions known as the "Insular Cases," beginning in 1901, concludes that Puerto Rico "belongs to but . . . is not a part of the United States." On the Insular Cases and the legal arguments around Puerto Rico's continued colonial status see José Trías Monge, *Puerto Rico: The Trials of the Oldest Colony in the World* (New Haven: Yale University Press, 1997).

27 Beatriz Santiago Muñoz, telephone conversation with the author, January 6, 2023.

28 Paso del Indio is the largest and most deeply stratified (approximately five meters deep) multicomponent prehistoric occupation site discovered to date in Puerto Rico and possibly throughout the Caribbean islands. See US Department of the Interior National Park Service, "National Register of Historic Places Registration Form," August 2002. Available

online at https://oech.pr.gov/Propiedades%20en%20el%20Registro%20Nacional/Vega%20Baja/Sitio%20de%20Paso%20del%20indio.pdf (accessed July 6, 2023). See also Mark R. Barnes, "The Paso del Indio Site (VB-4), Puerto Rico: A Site for Change," *North American Archaeologist* 43, no. 3 (2022). Available online at https://www.academia.edu/86813684/The_paso_del_indio_site_VB_4_Puerto_Rico_A_site_for_change?uc-sb-sw=102377835 (accessed June 19, 2024).

29 Ibid.

30 Santiago Muñoz, "Sessions," n.d. Available online at http://fabric-ainutil.com/index.php/projects/sessions--seminar/ (accessed July 5, 2023).

31 See Eduardo Viveiros de Castro, "Cosmological Deixis and Amerindian Perspectivism," *Journal of the Royal Anthropological Institute* 4, no. 3 (September 1998): 269–488.

PART III

Proposals for the Future

This third section of the book highlights the innovative strategies of artists to remedy the capitalist, extractive, and patriarchal ontologies that oppress humans and nonhumans alike. The proposals addressed in these essays share a foundation in concepts of care, collectivity, community, reciprocity, and autonomy. At the same time, in reckoning with social constraints on imagining possible worlds, they break rules, including those governing the understanding of the art object. Rather than taking the form of a discrete thing, the aesthetic can now be found in phantasmagorical hauntings or in multispecies relations.

In many ways, this last section of *Momentum* poetically circles back to the first, which examines artists on relations between human and nonhuman entities, robots included. But it also interrogates the notion of time, suggesting that reorienting ourselves to being "out of time" is necessary to our survival. The texts that follow address various forms and ideas for confronting the environmental crisis. Despite their differences, what unites these ideas is their address of the deep socioeconomic, historical, and ontological roots of our current ecological condition. Seemingly Sisyphean, the tasks they propose nevertheless put into practice modes of living and thinking that may provide us—humans and non—with the tools to live together creatively.

REWRITING THE LANDSCAPE

How the Valparaíso School's *Travesías* Seek to Transform a Continent

It was a long, arduous hike to this faraway location in the Cordillera de los Andes, and the air was clear. High above the tree line there was nothing but rocks, wind, and sky. This was an inhospitable land, architect Juan Ignacio Baixas would recall.[1] The arrival, in 1986, at this desolate site in the shadow of a mountain called Cerro Curimáhuida was an event. After all, the group of forty-five students and five faculty from the Facultad de Arquitectura y Diseño de la Universidad Católica de Valparaíso had made it safely to nowhere. It was time to set up camp and get to work.

The expedition to Curimáhuida, in Chile's central Coquimbo region, was part of the *talleres en travesía*, an ongoing program of study journeys with a design-build component, established in 1984 as part of the curricular and pedagogical activity of the Facultad de Arquitectura y Diseño, which is known as the Valparaíso School. These design-build travel studios, simply known as *travesías*, produce works of architecture and design "*nacidos de la poesía, en lugares lejanos y desconocidos, y con medios precarios*" (born from poetry, in far and unknown places, with precarious means).[2] They are part of the Valparaíso School's foundational mandate to produce an architecture cogenerated with poetry, extending this creative relationship to territory with actual built works and poetic acts across the continent. The inclusion of poetry as one of the disciplines—like building structures or environmental systems—with which architecture must engage is the unique and exceptional contribution of the Valparaíso School to architecture.

The *travesías* have a multipronged relationship with poetry and the poetic word: they literally place poetry across the American land. Pressed on copper strips and wrapped around a boulder,

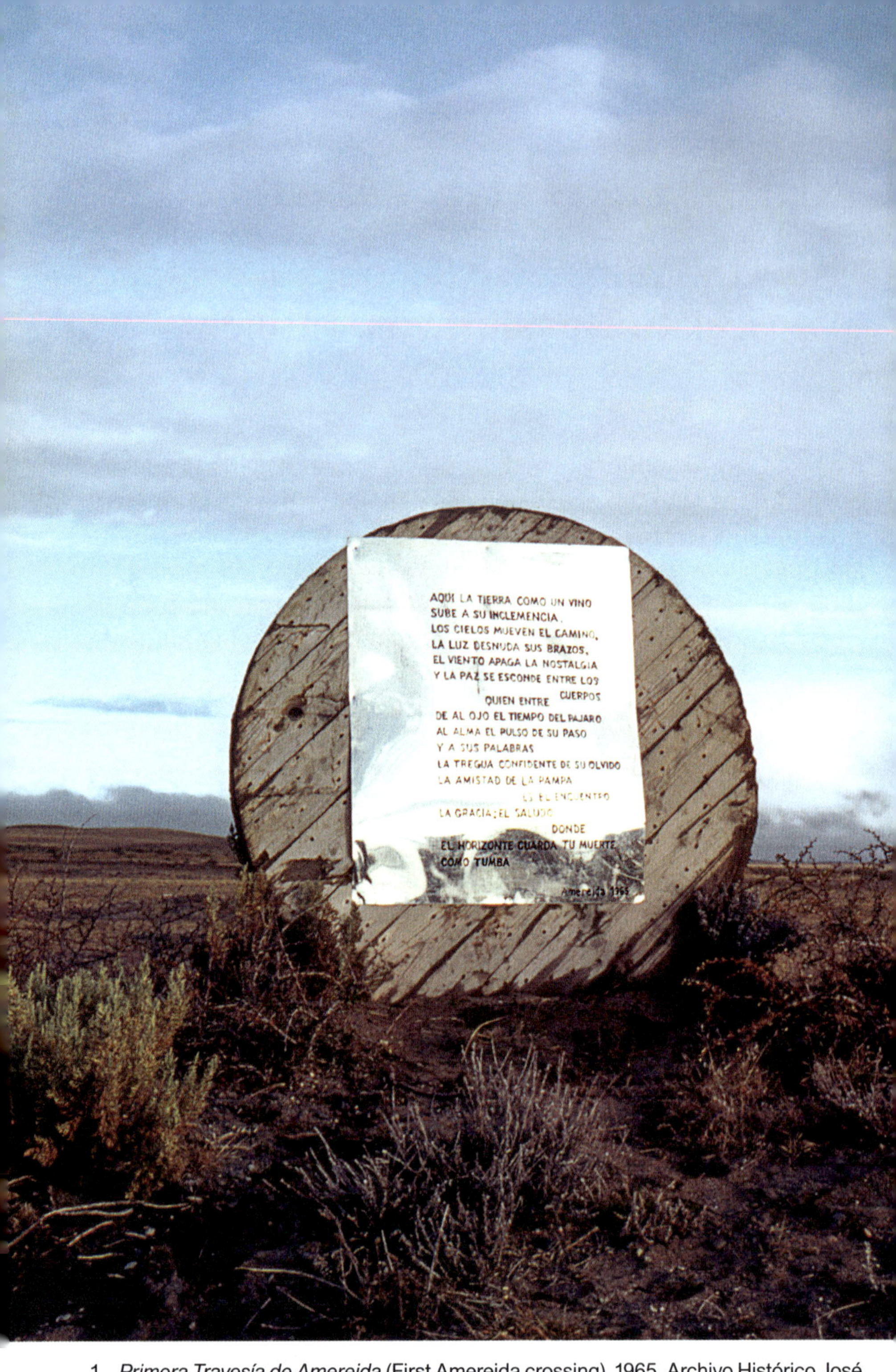

1 *Primera Travesía de Amereida* (First Amereida crossing). 1965. Archivo Histórico José Vial Armstrong, Escuela de Arquitectura y Diseño PUCV, Viña del Mar, Chile

the poem *amereida* (1967)—a result of the very first *travesía*, in 1965 (figs. 1, 2)—remains as a testimony to the 1995 *travesía* to Cerro Chena, on the outskirts of metropolitan Santiago. The journey involved the terracing of the hillside to recall ancient agricultural *andenes* (terraces) and a nearby Inca *pukará*, or outpost—abandoned ruins that, as embodied memories, continue to press on the present. In 1989, at El Juncal, on the road to Argentina and near the old trans-Andean railroad station, verses cast on red tiles were placed high on what looks like a large, haphazard, rickety structure made from discarded railroad tracks and other found and carried metal "sticks," reminding travelers of the vastness of the American territory, which opens up to them "like a sea."[3] These and other *travesías* have disseminated the *amereida* poem by creating a poetic landscape to be encountered by travelers or locals. Also like all the others, they reenact the foundational *travesía* in 1965.

Travesías bring about a multifaceted and dynamic engagement with the American land. The spatialization of a poetic praxis demands a constant and living approximation to the real, an acceptance of the changing yet always present insecurity of human existence. Each journey becomes a series of mappings revealing places that remain elusive. Rekindling the relationship between *la palabra* (the spoken word) and territory, the travesías teach that "*las palabras son como extrañas a las cosas que nombran*" (words are extraneous to the things they name).[4] The disparity between words and places is a key proposition advanced by *amereida* and embodied by the *travesías*. In the philosophy of the Valparaíso School, architecture can breach the gap between words and places, spatializing both poetry and the poetic word. Because the design practice followed in the *travesías* embraces the specific qualities of each site, each expedition produces an intensive and extensive exploration, and therefore solicits and configures a distinct subjectivity—a traveler who inhabits the world with fullness and intensity. Placing oneself in far and unknown places brings to light the precarious condition of inhabitation. This is a fundamental lesson for both architects and nonarchitects: most of the structures that the *travesías* have scattered across the continent look haphazard, impermanent, incomplete, perhaps even on the verge of collapse. Their materials and construction methods seem rudimentary, their techniques craftlike, not those of a refined building practice or sophisticated technology. These structures embody the immediacy and intimacy of local conditions, and also their interconnectedness. They are full of joy and wonder. Playing with scales of inhabitation—local and continental, actual and conceptual, without fixing them—these works advance a unique terraforming.

The Facultad de Arquitectura y Diseño was refounded in 1952 with the premise of producing an architectural practice in the service of poetry. The Chilean architect Alberto Cruz and the Argentine poet Godofredo Iommi established a collective of architects, poets, designers, artists, and educators who advanced a transformative understanding of the way to teach and produce architecture and its allied disciplines. Their pedagogical and professional experiments have developed over the years into a multifaceted set of practices and techniques—impossible to fully cover in this brief essay—that help to plot the development of the school.

In 1965, Cruz and Iommi organized and carried out a journey from Punta Arenas, Chile, in the continent's extreme south, to Santa Cruz de la Sierra, Bolivia, which they declared the "poetic capital" of the American continent.[5] This was the foundational journey of the *travesías*. Known as the *travesía amereida*, it announced a call to traverse the American land: "*mañana comenzaremos a recorrer américa*" (tomorrow we will start to travel America), runs a line in the poem.[6] Since 1984, the *talleres en travesía* have answered this call every year with a series of intense explorations and travels that reveal how one inhabits this land poetically. They carry forth the research and practices developed at and by the Valparaíso School, first under the auspices of the Instituto de Arquitectura y Urbanismo (an early, quasi-independent professional architecture office composed of select members of the school), later, as of 1970, out of the celebrated *ciudad abierta* or "open city," an experimental community officially zoned as a "*Parque Costero Cultural y de Recreación*" (Coastal cultural and recreation park). The city is run by the Cooperativa Amereida, an autonomous association, independent of the university, that superseded the Instituto and serves as an outpost of the school, developing student and faculty projects. The school proper is in the city of Viña del Mar, a resort town on the Pacific Ocean. These different components,

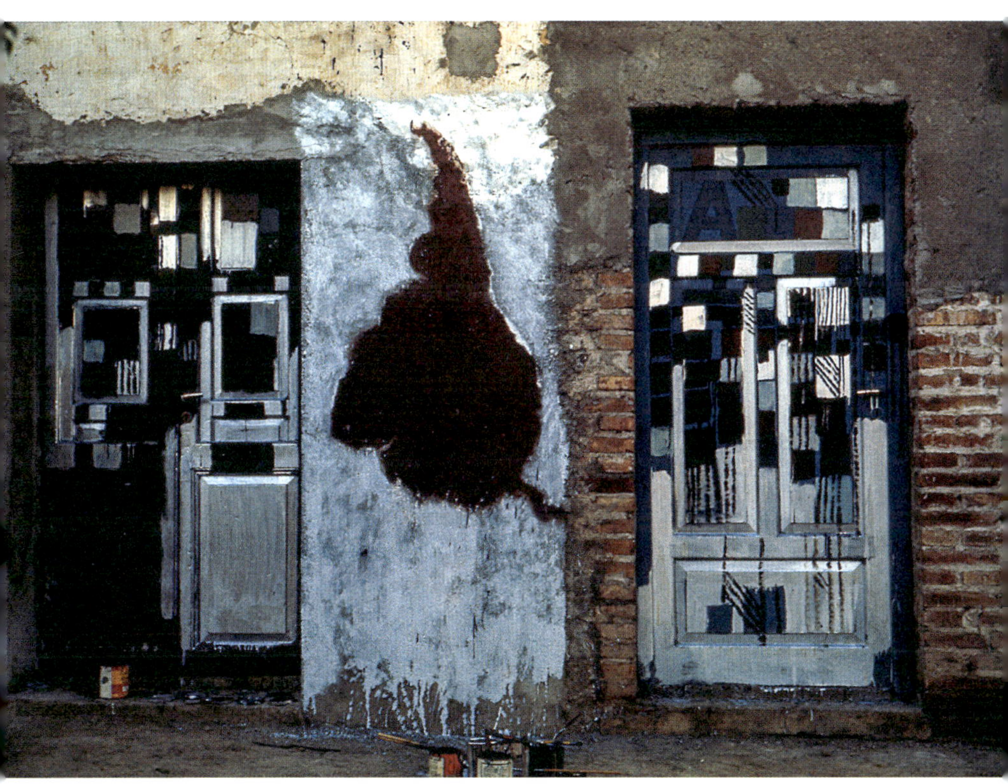

2 *Primera Travesía de Amereida* (First Amereida crossing). 1965. Archivo Histórico José Vial Armstrong, Escuela de Arquitectura y Diseño PUCV, Viña del Mar, Chile

not to mention the consistent engagement with the industrial port-city of Valparaíso as object and subject of study, have provided a theoretical density and institutional porosity that have ensured a consistent and coherent yet lively and dynamic pedagogy. In 1984, the *travesías* marked a new phase in the school's mandate to "live poetically" by expanding their activities beyond the open city to an actual and conceptual territory called *América*.[7]

FARAWAY PLACES

Unlike contemporary forms of travel that make wanderings into holidays and travelers into cultural consumers armed with selfie sticks, the *travesías* take the form of expeditions. Like any expedition, the *travesía de la piedra*—the trip to the foot of Cerro Curimáhuida, in 1986—required extensive, detailed preparation, since traveling to unknown places has risks. *Travesías* are generally planned throughout a semester and usually executed in two weeks during the academic year by the students and faculty of three of the school's departments (architecture, industrial design, and graphic design). They have a defined purpose, but unlike traditional prospecting expeditions, charged with instrumental extractive intent, these start with a poetic charge, a kind of poetic assignment given by a poet or by poetry. With Curimáhuida, the assignment was to make two "oceans" meet: the Pacific Ocean and the *Mar interior* (Interior sea), the vast—actual and conceptual—interior of the South American continent, identified in the 1965 *amereida* journey and later inscribed in the epic poem of the same name.

The challenge of locating the geographic threshold where the Andes establish the edge between the Pacific and the *Mar interior* began with the unknown: to go to the "meeting place" of these "oceans" and make it present—in short, to give it architectural form. Detailed geographical surveys had to be procured and prepared before and during the journey. Cartographic and climatologic charts, historical maps, measurements and observations obtained from previous incursions through several rivers and affluents, and informal conversations with local inhabitants became the basis for the organization of the journey. Every bit of knowledge, every scrap of information and obscure resource, was fair game. Through these multifaceted explorations, a site that manifested the poetic charge, or program, of the journey—to discover the boundary or threshold of the two "seas"—was selected: a concave plain on the edge of a cliff that formed a twenty-kilometer wall in the Andes.[8]

To start an architectural project with no well-defined site or program is a peculiar pedagogical approach; to start it with a charge born from a poem is unheard of. In other words, to start a design-build architecture studio by going somewhere to build something in an unfamiliar place is a daunting task—but radical defamiliarization seems to be the point. Armed with a poetic charge, these travelers were not forlorn wanderers in the landscape. Pragmatically speaking, the poetic charge oriented the actions to be taken. Students and faculty went both to discover and make present the site manifested by poetry and to give it architectural form. Not all journeys, though, are geared to produce an architectural intervention; some produce ephemeral poetic acts—a form of performance or "happening" with neo-avant-garde roots—that builds community and spatializes poetry.

In Curimáhuida, the architectural resolution to the poetic charge—to make present the threshold of the oceans—was negotiated in the field. Built high in the Andes, at almost 11,000 feet above sea level, the work of architecture was a 230-foot-long processional atrium, oriented slightly eastward of the north-south axis. Subdivided in five segments, it was open on one side and bounded on the other by triangular walls that emerged from the land, protecting from the strong winds. Rocks, the only material available on site, were cut, assembled, and fitted to form the distended atrium and its serrated wall. The atrium created an architectural extension, an embodied procession that marked the expansive territory and led to an intersection composed of three distinct spaces, one of these being a small, completely enclosed shelter built out of sheet metal. The decision had been made early on to build a permanent structure, so various materials and preassembled elements, as well as tools, including a specially designed collapsible device capable of bending and joining sheet metal, were brought to the site. The shelter, a fractured rectangular space split in both plan and section, formed two overlapping cubes accessible at their intersection. This enclosed space was fitted with horizontal windows at each end and inscribed with graphics, verses, and colors that manifested the symbolic character of architecture and mediated the "architectural act" and the "*naturaleza americana*" (American nature).[9]

Two additional spaces of different sizes, bound by stone walls and open to the sky, accompanied the shelter and completed the exploration of inhabiting this inhospitable place. Each of these semi-enclosed spaces was equipped with square windows made of thin steel members and glass, inserted at odd angles on the stone walls. Each window oriented the space, charging it with east and west directions and breaking away from the processional axis of the atrium.

Elaborated in the three sheltering spaces, the act of framing a view and orienting space transformed nature into a landscape. The architectural intervention mapped and shaped the multiple scales of the land: the extensive territory holding the atrium, the oriented landscape of the semiprotected open-air spaces that enabled the group to gather outside, and the framed horizontal views from the protective shelter. The architectural act marked the connection between the two seas as the embodiment of the poetic charge: the manifestation of the land in light of the connection of the two seas, a conceptual operation that took actual architectural form as an embodied experience. The intervention was left as a gift in the landscape for any traveler who might encounter it, a moment of shelter and refuge, of orientation and measured habitation in a remote and mysterious place.[10]

AMEREIDA

In the view of the participants in the original *travesía* of 1965, the journey revealed the destiny of the continent and, more pragmatically, established "instructions," penned in a poem, for future *travesías*. The text, like the journey, was a collective undertaking and has no manifest author. It is not a log of the journey; rather, it is an epic poem, composed in free verse and prose poetry and printed for internal use at the school in 1967, after the *travesía* was over.[11] It covers a wide range of topics, from historical events to observations on everyday moments. Considerations of the nature of language, the demarcation lines of the Law of the Indies (the set of laws issued by Spain to govern its colonial possessions), and a local marketplace all reflect on the condition of inhabiting the American continent. What at first seems a series of disconnected thoughts and observations coalesces in the poetic logic of the journey, reenacted by the printed text itself. The poem's form—the compositional play between words and voids on the space of the page—borrows from formalist avant-garde experiments such as Stéphane Mallarmé's poem "*Un coup de dés jamais n'abolira le hasard*" (A throw of the dice will never abolish chance, 1897), revealing *amereida*'s debt to a particular understanding of modernity. Its abstract, formalist character, however, does not trump its performative origins in poetic acts staged in the landscape during the journey, acts that established a bridge between spoken words and places. A key aspect of its poetic program is the spatialization of poetry—the placement of poetry in the land.

Several maps accompany the text of *amereida*. These illustrate four main theses, or observations, that guide the endeavors of the *talleres en travesía*: *Bordes* (Edges), *Mar interior*, *Cuatro estrellas* (Four stars), and *Norte propio* (Proper north, or Our north). These theses refer to the physical occupation of the South American continent unleashed by the wars of conquest and settler colonialism. They capture a historical macro-morphology in which cities were established primarily along the coast, along the *Bordes*, creating vast experiential and conceptual hinterlands that emerge, in the vision of *amereida*, as a *Mar interior*. These conceptual maps, expansive in their power to represent the condition of the land and intense in their ability to prefigure future journeys, such as the *travesía de la piedra* to Curimáhuida, encapsulate the quest for an architecture cogenerated with poetry in the territory.

At the same time, these maps reveal an ongoing preoccupation with geography—that is, with a scientific approach to territory—and with the ideology of *desarrollismo* (Developmentalism), a productivist paradigm of unfettered and unlimited economic growth embraced by governments across Latin America in the postwar period. *Desarrollismo* renegotiated the contract between society and the land. Its unfaltering belief in technology and science demanded a spatial reorganization based on industrial development, resource extraction, and a new scale of infrastructure with which to imagine the region's future.[12] Continent-wide connectivity was key to imagining the creation of this future Latin America. Brazil's Transamazônica—the Trans-Amazonian Highway—and the Carretera Marginal de la Selva, a highway connecting Venezuela, Colombia, Ecuador, Peru, and Bolivia, were two infrastructural projects of the 1960s and '70s that radically changed the continent and germinated more recent undertakings, such as the 2017 highway through the Territorio Indígena y Parque Nacional Isiboro-Sécure (TIPNIS), in the Bolivian Amazon.[13] From the viewpoint espoused in *amereida*, such projects of continental connection were prefigured by what it conceived as historical *travesías*, including the sixteenth-century expeditions of the Spaniards Ñuflo Chavez and Francisco de Orellana and the German Nikolaus Federmann, among others. The conjunction of history, geography, and an economic imperative that places technological innovation at the center of the transformation of the territory resonates in *amereida*, and in the Valparaíso Group's recurring preoccupation with continental connections.

Do the *travesías* escape the destructive flows of development and the telos of modernity? The Valparaíso School practices a dangerous creative praxis that reinscribes the formative role of the conquest and colonization of the Americas within industrial modernity and its capitalist world-system. The aim was not to mitigate the impact of *desarrollismo* but rather to change the subjectivity that shaped it. The notion of the *Mar interior* emerged from a colonial text, the *Historia general y natural de las indias* (1535), in which the Spanish colonist and historian Gonzalo Fernández de Oviedo y Valdés identified the South American hinterland as a "*mare magno e oculto*," a "vast and hidden sea," that needed to be opened—needed an "*abrir camino*" (open road), to quote the *amereida* poem.[14] This set a treacherous path that led and continues to lead to the violence of capitalist accumulation, with its repeated conquest of the land. As instruments of colonization, such colonial texts underpin destructive and extractive practices that continue today, for example in recent railway projects that aim to connect the Pacific and the Atlantic oceans. Early works of the Valparaíso School, such as the unsuccessful 1969 coastal highway or Avenida del Mar project in Valparaíso—a failure that prompted the "retreat" to the *ciudad abierta* in 1970—established uneasy alliances between postwar strategies of development and those of the *Amereida* group. The continued presence and preeminence of the 1965 text in the endeavors of the school cast questions about the changing conditions of production and inhabitation in the Americas. Can the Valparaíso School surmount the inheritance of colonization and modernization?

Poetic hermeneutics, with their openness and assemblages, seem to offer an exit from the destructive flows of "progress." So, too, does the scale of the territorial interventions in the *travesías*. The 2008 journey to Queilén, a fishing village on the island of Chiloé, gave architectural form to the site by reviving an abandoned tradition: *curanto*, the cooking of mollusks in an earthen pit, rather than in the metal pots used today. This *travesía* involved creating a small "plaza" equipped with low platforms that surrounded an open hearth or cooking pit and served as benches and tables for communal gatherings. The *taller en travesía* served as the final exam for first-year students.[15] Such sophisticated interpretations of sites and programs, and their subsequent spatialization, demand measured and scaled interventions. This is more than the consequence of limits of time and money: the *travesías* reveal and teach a distinct attitude toward production, with small, fragmentary, and never totalizing projects. They work through accretion and accumulation, through persistence and insistence, establishing a dialogue between the territory and the landscape—between conceptual and embodied experiences—guided by poetry.

WHAT DOES POETRY MEAN?

Every school, every institution, has its own parlance, one that to a lesser or greater degree creates a distinct space of enunciation, and the Valparaíso School is no different. When visiting the school, one is profoundly struck by its distinct language, one inflected by "poetry." But what does the term "poetry" mean in this context? For the Valparaíso School, poetry is the activity and craft of the poet(s) working with/in the school. Poetry is the work and the body of work that the school embraces. Most fundamentally, poetry is the poem *amereida*, which demands a sustained hermeneutic activity—the interpretation of this and other texts—and serves as the foundation for the *travesías*.

The craft of the poet can be summarized as a *cálculo del lenguaje* (language calculation), a measuring of the world through and with language that produces a distinct embodied hermeneutics, one that the school's graduates carry beyond its institutional confines. (Yet another form of the dissemination of poetry.) The mandate to produce an architecture cogenerated with poetry is but the bringing together of two crafts and two traditions. As Alberto Cruz once explained to me, if architects use structural engineers to calculate their buildings, why not use a poet to calculate their language?[16] The approach is not technologically naive, and does not entail a rejection of science and industry. Rather, it questions whether science and its analytical methods have undisputed claim over the modern understanding of the world, and forwards a philosophical poetry as a tool—and an approach to technology—that both challenges and complements scientific research. If, in modernity, the domain of scientific analysis is empirical observation, the Valparaíso School adds the *cálculo del lenguaje* to this endeavor. This methodology challenges the reductive technorational understanding that fastens empirical methodologies to industrial and technocratic processes of calculation, standardization, and reproduction.

Poetry is far from the subjective escapism or romantic contemplation that the modern industrial world considers it and as such condemns it. It is a probing tool, an alternate system of observing the world that breaks binary thinking by establishing assemblies of concepts, images, and desires. The *cálculo del lenguaje* challenges the value system that springs from modernity's technorational reductions, restoring a qualitative measure to the experience of the everyday.[17] It does this through a particular methodology that involves observing, sketching, and writing, and guides students to ground their work in aesthetic and poetic judgment. The intent is not to make poets out of architects, or architects out of poets, but rather to understand language as an essential tool of experience. In a world such as ours, beyond the brink of environmental catastrophe, this approach may seem like a fanciful indulgence, yet language has real-world implications that can change the way the world functions. This can be seen most recently in the so-called "ecological"

Chilean constitution of 2022—rejected by national plebiscite—which sought to redefine the legal notion of water by freeing it from the web of neoliberal values left from the time of Augusto Pinochet's dictatorship and by terminating its commodity form.[18] One must not forget that language is a political battleground.

WHY FOCUS ON THE LAND?

The *talleres en travesía* begun in 1984 opened a "new" period in the research and activities of the Valparaíso School, a third act that reawakened a particular construction of the American landscape.[19] This period was and continues to be marked by a deliberate return to the first, 1965 *travesía* and by a turn away—albeit not completely—from the *ciudad abierta* as the main site of the school's experimentation.[20] Understood as a return to a foundational moment in the school's history, these new *travesías* speak to a concern with institutional ossification and to the ethos of reflection and self-assessment that shaped the school, and continues to shape it. They also speak to a changing sociopolitical landscape that altered the conditions of inhabitation of both city and countryside—not to mention an educational reform that produced for-profit universities. The school's turn to the *travesías* as a pedagogical tool coincided with the economic and political bankruptcy of the Pinochet regime, a crisis that, beginning in the early 1980s, birthed the national referendum of October 5, 1988, which prepared a transition to democracy still being worked out in 2023. This turn to the land emerged after a decade of economic transformation, including the unfettered expansion of mining activities and the intensification of the agricultural-export industry that would underpin the "successful" transformation of the years following the dictatorship.[21] In short, the *territorio americano* has been sacrificed on the altar of so-called "progress" for a long time—an immolation that continues to this day.

The ongoing process of environmental and ecological wrecking reframes the call of the *amereida* poem to travel a land crisscrossed by the maelstrom of a global extraction economy.[22] This concern with travel was concomitant with the transformation of the Chilean territory into a landscape of neoliberal development, preparing the way, both conceptually and economically, for the entry of Chilean architecture into the celebratory paeans of globalization in the early 1990s, as Jorge Francisco Liernur observes.[23] The turn to the *travesías* prefigured this blissful recent past of Chilean architecture, and cast both doubts and hopes on the expeditions' ability to reinscribe the American territory by placing it beyond the grasp of instrumental technoeconomic reason and its concomitant forms of consumption. The sophisticated and measured interventions deployed since 1984 have produced sensitive approaches to the land, yet do these experiments constitute an ecological thinking and affirmative praxis capable of recoding the fragmenting instruments of modernity, as well as the homogenous abstract spaces these instruments produce? Placing architectural and design creativity in the land, the *travesías* effectively rewrite the landscape and reopen a frontier for human endeavors. Cogenerated with poetry, these modern practices break the promise of modernization to deliver territory to the forces of industrial development and its homogenizing goals (fig. 3).[24]

A glance at the overall trajectories of the *travesías* reveals that these travels have been and continue to be journeys to faraway places, that is, to uninhabited places such as Curimáhuida, as well as to urban sites of diverse magnitude, from capital cities to small towns. The *travesías* build a vast relational and layered map that promotes a distinct spatiotemporal form of experiencing—perhaps of inhabiting—the American territory. These are not rhetorical operations left on paper or shut inside books tucked neatly onto shelves. The poetic calculation of *amereida* seeks to keep open the processes of modernity, to ensure the constant examination of what it is to inhabit the American or any other land. The ever accelerating tempo of technology, the uncontrolled urbanization that plagues the planet, and the insertion of extraction economies in our lives foreclose the future for humans and nonhumans alike. This pedagogy cannot counter the tempo of postindustrial capitalism, but it creates a distinct consciousness and subjectivity with which to inhabit it, offering an unyielding terror of being in the world that can translate into action.

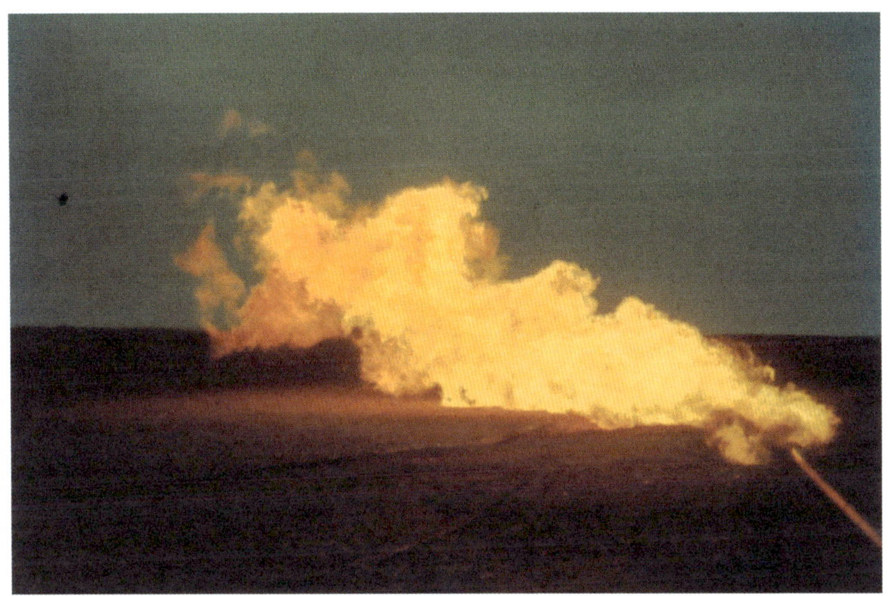

3 *Primera Travesía de Amereida* (First Amereida crossing). 1965. Archivo Histórico José Vial Armstrong, Escuela de Arquitectura y Diseño PUCV, Viña del Mar, Chile

1 From conversations with architect and professor Juan Ignacio Baixas. Along with Baixas, this 1986 expedition was led by professors Fabio Cruz Prieto, Boris Ivelic Kusanovic, Bruno Barla Hidalgo, and Francisco Méndez and included third- and fourth-year architecture students and fourth-year graphic design students. See https://wiki.ead.pucv.cl/Traves%C3%Ada_de_la_Piedra_-_Curimahuida (accessed July 16, 2023).

2 "Los talleres en travesía por América," *CA: revista oficial del Colegio de Arquitectos de Chile* no. 47 (March 1987): 56. All translations from Spanish are the author's.

3 The verses read: "Esta tierra americana/se abre como un mar/que la luz de sus abismos/alumbre/más allá de los/dominios" (This American land/opens up like a sea/may the light of its abysses/shine/beyond the/domains). In 2001, I visited the sites of several *travesías* including Cerro Chena, which I visited with Cruz and Barla, the latter of whom had organized that *travesía* with Alberto Cruz. On El Juncal see https://wiki.ead.pucv.cl/Traves%C3%ADa_Juncal_1989 (accessed July 23, 2023).

4 *Amereida* (Valparaíso: Universidad Católica de Valparaíso, 1986), 77.

5 The participants were the English poet Jonathan Boulting, the Chilean architect Alberto Cruz, the Chilean architect Fabio Cruz, the French poet Cichel Deguy, the French philosopher François Fédier, the Argentine sculptor Claudio Girola, the Argentine poet Godofredo Iommi, the Argentine painter Jorge Pérez Román, the Panamanian poet Edison Simons, and the French industrial designer Henry Tronquoy. See Juan José Ugarte, Alejandro Aravena, Andrea Ortega, and José Quintanilla, eds., *Amereida: poesía y arquitectura* (Santiago: PUC Chile, 1992).

6 *Amereida*, 111.

7 This conceptual and historical research is developed through the school's ongoing pedagogical endeavors, such as the Seminario de América, established in 1969.

8 See "Travesías y obra en la Cordillera de los Andes," *CA: revista oficial del Colegio de Arquitectos de Chile* no. 48 (1987): 48–51, and "Travesía a Llanos de Curimáhuida, travesía de la piedra," in *Amereida, travesías: 1984–1988* (Valparaíso: UCV, 1991), n.p. This research on the two seas began with the Valparaíso School's first *travesía*, to Cape Froward in 1984; the *travesía* to Curimáhuida continued this line of research.

9 *Amereida, travesías: 1984–1988*, n.p.

10 I draw here from a number of the sources previously cited. Many layers of this and other *travesías* lie beyond the scope of this text. For example, the *travesía* to Curimáhuida was galvanized by the phrase "*imágen-palabra: nave*" (image-word: ship), which jump-started the architectural imagination before the actual journey. See "Travesía a Llanos de Curimáhuida, travesía de la piedra," in *Amereida, travesías: 1984–1988*.

11 The Universidad Católica de Valparaíso published a two-volume version of *amereida* for student use in 1986. A log was inserted as an addendum to the second volume. The appearance of parts of the poem early on in journals such as *Orfeo. Revista de Poesía y Teoría Poética* no. 33–38 (1968) demonstrates the group's interest in social participation beyond the confines of the university and their own members.

12 See Maristella Svampa, *Debates Latinoamericanos: Indianismo, desarrollo, dependencia, populismo* (Buenos Aires; Edhasa, 2016), 137–38, and Diego Martín Giller, "Los años dependentistas. Algunas cuestiones en torno de dialéctica de la dependencia," *CLACSO*, 2016, 8.

13 See David E. Snyder, "The 'Carretera Marginal de La Selva': A Geographical Review and Appraisal," *Revista Geográfica—Instituto Panamericano*

de Geografía e Historia no. 67 (1967): 87. On TIPNIS see Dan Collyns, "Bolivia approves highway through Amazon biodiversity hotspot," *Guardian*, August 15, 2017. Available online at www.theguardian.com/environment/2017/aug/15/bolivia-approves-highway-in-amazon-biodiversity-hotspot-as-big-as-jamaica (accessed July 18, 2023).

14 *Amereida*, 28.

15 See "Travesía a Queilen, en la isla grande de Chiloé," March 11, 2009. All references to this journey are taken from this site. Available online at www.ead.pucv.cl/2009/travesia-a-queilen-en-la-isla-grande-de-chiloe/ (accessed August 5, 2023).

16 Alberto Cruz, conversation with the author, *ciudad abierta*, 1995.

17 As Theodor Adorno argued, "Modernity is a qualitative, not a chronological, category." Adorno, quoted in Peter Osborne, *The Politics of Time: Modernity and Avant-Garde* (London and New York: Verso, 1995), 9.

18 See Nick Burns, "Chile's Proposed Constitution: 7 Key Points," *Americas Quarterly*, July 7, 2022, available online at https://www.americasquarterly.org/article/chiles-proposed-constitution-7-key-points/; Michelle Langrand, "Chile's Battle to Reclaim Its Water," *Geneva Solutions*, n.d., available online at https://genevasolutions.news/explorations/the-water-we-share/chile-s-battle-to-reclaim-its-water; and Victor Hugo Moreno Soza, "Los significados y alcances del nuevo estatudo del agua propuesto en el texto constitucional," *Noticias*, Universidad de Chile, July 26, 2022, available online at www.uchile.cl/noticias/188538/sala-constituyente-el-agua-en-la-propuesta-constitucional- (all accessed July 18, 2023).

19 The school's endeavors fall into three distinct episodes: the Instituto de Arquitectura y Urbanismo, the *ciudad abierta*, and, lately, the *travesías*. This periodization, however, is but a backward glance, attempting a normative understanding based on an internal and autonomous developmental logic, and obscuring a conflictive history.

20 Structures continued to be built in the *ciudad abierta* after 1984, and the *torneos*, organized school competitions, make clear that it remained a key site of activities. For a survey of works see Rodrigo Pérez de Arce and Fernando Pérez Oyarzún, *Escuela de Valparaíso: Grupo Ciudad Abierta*, ed. Raúl Rispa (Santiago: Contrapunto, 2003).

21 On the economic conditions of the period see Tulio Halperín Donghi, *The Contemporary History of Latin America*, trans. John Charles Chasteen (Durham, NC: Duke University Press, 1993), 349. On the country's agrarian reform see Heidi Tinsman, *Partners in Conflict: The Politics of Gender, Sexuality, and Labor in the Chilean Agrarian Reform, 1950–1973* (Durham, NC: Duke University Press, 2002). And on its educational reform see Cristián Cox, ed., *Políticas educacionales en el cambio de siglo: la reforma del sistema escolar de Chile* (Santiago de Chile: Editorial Universitaria, 2003).

22 On the concept of environment and ecology see Etienne Benson, *Surroundings: A History of Environments and Environmentalisms* (Chicago: University of Chicago Press, 2020), particularly the introduction, and Murray Bookchin, *The Ecology of Freedom: The Emergence and Dissolution of Hierarchy* (Oakland and Edinburgh: AK Press, 2005).

23 Jorge Francisco Liernur, "1977–1992, El silencio y los susurros," in Liernur and Pedro Bannen Lanata, eds., *Portales del laberinto: arquitectura y ciudad en Chile, 1977–2009* (Santiago: Universidad Andrés Bello, 2009), 30.

24 On the abstract space of modernity see Henri Lefebvre, *The Production of Space*, 1974, Eng trans. Donald Nicholson-Smith (Oxford and Cambridge, MA: Blackwell, 1991).

IMAGINING A POSTEXTRACTIVIST ONGOINGNESS

Reflecting on the landscapes of his native Minas Gerais, the Brazilian poet Carlos Drummond de Andrade voiced critiques and laments for the material and affective deterritorializations generated by the mining industries, which had been installed in the state by Portuguese colonizers and were controlled by national and transnational capital. Minas Gerais—the name means "general mines"—stood for Drummond as a metonym for the global logic of extractivism, and through its landscapes he interrogated extractivism's destructive force. In 1984, presenting mining as an emptying out of the sedimentations of human bonds to territory, he conjured as the synecdoche of extractivism the figure of "*O maior trem do mundo*" (The biggest train in the world), which is "*puxado por cinco locomotivas a óleo diesel/engatadas geminadas desembestadas/leva meu tempo, minha infância, minha vida/triturada em 163 vagões de minério e destruição*" (pulled by five diesel-powered locomotives/twin-hitched runaways/It takes my time, my childhood, my life/crushed into 163 wagons of ore and destruction).[1]

Drummond's words resonate as urgently today as they did when he wrote them. That same earth where the poet felt rooted is also home to the Krenak Indigenous people, who today watch the trains go by, bearing witness to the ongoing dismemberment and extraction from the reserve that contains what remains of their ancestral land. In 2015, the collapse of the Fundão tailings dam at an upstream mining complex owned by a transnational group of corporations, BHP, Vale, and Samarco, unleashed a toxic wave of mud on the Doce River, which runs through Krenak territory, causing Brazil's worst environmental disaster on record. Pondering the fallout of the catastrophe and the persistence of mining "fever," the Indigenous leader and thinker Ailton Krenak, in his book *A vida não é útil* (Life is not useful, 2020), revives Drummond's poetry as a compass with which to navigate the ongoing impacts of extractivism. Krenak reiterates the poet's prediction that once man has colonized all worlds, from Earth to Mars, he will have to "*pôr o pé no chão/do seu coração... descobrindo em suas próprias inexploradas entranhas/a perene, insuspeitada alegria/de con-viver*" (put his feet on the ground/of his heart... discovering in his own unexplored interior/the eternal, unsuspected joy of living-together)."[2] Glossing this idea, Krenak asserts that poetry, dance, and being together are ways of shifting the focus away from the machinery of extractivism and onto collective action and solidarity. "*Temos que parar de nos desenvolver e começar a nos envolver*," he writes, in a clever wordplay: we need to stop developing and start getting involved.[3]

Understanding extractivism is a vital step toward imagining alternative horizons of coexistence and collaboration. Since the 1970s, the political ecology of Latin America has developed critiques of the "one-world world" of modernity, which installed extractivism as part of a global capitalist order, and has instead identified routes toward "pluriverses" that hold multiple ways of worlding.[4] Eduardo Gudynas, for example, has set out a conceptual vocabulary to define extractivism as an outgrowth of colonial exploitation that permeates the territorial, sociocultural, political, and economic fabric of South America (and other parts of the Global South) through the intensive extraction of hydrocarbons, minerals, and crops and their transportation to global markets. As well as causing environmental damage to specific industrial and agroindustrial sites, he shows, extractivism produces "spill effects" such as the usurpation of Indigenous and peasant lands, the displacement of communities, and the violation of the rights of people and nature. Gudynas further signals how states and corporations equate industrial activities with national identity and development, and in doing so entrench an "extractivist theology."[5] Extractivism is thus as much a material phenomenon as a mindset whose "spill effects" create a "stained geography," catalyzing "changes in public policies, in the functioning of the economy, in the understanding of justice and democracy or in the conception of Nature."[6]

By changing conceptions of justice, democracy, and nature, extractivism alters much more than physical landscapes: it modifies the cognitive terrains of how humans relate to nonhumans, imposing a utilitarian commodification of nature over other knowledges and practices of relating to "earth beings."[7]

The question of how to cultivate pathways to postextractive worlds is complex. As Gudynas writes, it entails "very pragmatic proposals of changes to move in that direction, for example in taxation, environmental assessment, a territorial ordering and citizen participation," but it also demands sociocultural paradigm shifts capable of installing ontoepistemologies that would reclaim and (re) instate ecoethical relations for planetary stewardship.[8] The emergence and mainstreaming of the term "Abya Yala" (an Indigenous term for South America) in the 1980s signal the importance of cultural work in energizing alternate horizons of collective life along principles of intercultural coexistence and decoloniality.[9] Similarly, political ecologist Enrique Leff argues that Latin American environmental thought is key to countering the Western objectification of nature and fragmentation of knowledge that have enabled the entrenchment of extractivism and its related socioenvironmental injustices and calamities. Proposing dialogues between ancestral knowledges, emergent territorial praxes, and critical Western thinking to ferment an alternate "environmental rationality," Leff also advocates for learning from the generative work of grassroots social and conservation movements and from their principles of sustainability, solidarity, and the reclamation of biocultural rights.[10]

In Latin America, artistic, curatorial, and collaborative projects are today doing important critical and imaginative work both to expose the destructive logics of extractivism and to nurture relations and actions for postextractive futures. In this essay I explore art projects that delink from the commodification of nature, confront socioenvironmental and cognitive injustice, and cultivate ecologies of knowledges and structures for biocultural conservation, asking: if extractivism perpetuates the logics of *terra nullius* and "expansion of the one world by rendering empty the places it occupies and making absent the worlds that make those places," how do artists work to render present and perceptible those worlds, and the lives affected by such violent negation?[11] What generative worldings emerge from intersections of artistic practice, fieldwork in fragile ecosystems, collaboration, and political resistance that help imagine postextractive worlds?

COGNITIVE JUSTICE AND THE FOREST AS EPISTEMOLOGY

Boaventura de Sousa Santos describes European colonization as an undoing of worlds enacted through not only genocide but also "epistemicide," a coinage referring to the "abyssal thinking" that relegated Indigenous knowledges and ontologies to a radical otherness, one denigrated as a form of barbarism opposed to Western civilization.[12] Abyssal thinking installed scientific knowledge as the enduring dominant epistemology, but in "the strong political presence of peoples and worldviews . . . as partners in the global resistance to capitalism" de Sousa Santos identifies an emergent ecology of knowledges that cannot be brought under a single totalizing epistemology but refracts many "radically different conceptions of alternative society," including those of Latin American peasants and Indigenous peoples' resistance to land-grabbing and megaprojects.[13] The ecology of knowledges shows that "there are not only very diverse forms of knowledge of matter, society, life, and spirit but also many and very diverse concepts of what counts as knowledge and the criteria that may be used to validate it."[14]

In Latin America, this epistemodiversity involves recognizing the environment as a thinking, living phenomenon, a multispecies web of sentient beings. The rise of plant thinking and of more-than-human ethnographies is a recent "turn" in the Western academy, but Indigenous peoples have honored and worked with the inherent intelligence of life for millennia, developing autochthonous knowledge out of intimate relationality with their ecosystems.[15] An ongoing collaboration between leaders and educators of the Inga Indigenous people in Colombia and a team of researchers, designers, and filmmakers convened by the Swiss artist Ursula Biemann realizes an interweaving and negotiation of the ancestral and the academic threads of environmental thought.

Biemann was originally commissioned by curator María Belén Sáez de Ibarra, of the Museo de Arte de la Universidad Nacional de Colombia, to make a film in the country in 2018. After a month of fieldwork in the forest with Inga leader Hernando Chindoy Chindoy, she agreed to his proposal that she support the creation of an Indigenous university crafted to fit the process of knowledge cultivation *in and from* the living forest. In the territories of the Quechua-speaking Inga (the southern Colombian regions of Nariño, Cauca, Caquetá, and Putumayo), some 300 students a year graduate from high school without the option of further study—hence the need for additional educational structures providing certificates of higher education.[16] More like what Marian von Osten terms a "translocal organization" than a conventionally conceived artwork,[17] the collaborative process that ensued has entailed forging transnational connections among knowledge communities, assembling what Biemann calls a "consortium" of academic and creative advisors to collaborate with the Inga on architecture, curriculum design, funding bids, communications strategies, and audiovisual materials that document the university's emergent forms through a bilingual digital platform.[18] Called *Devenir Universidad*—Becoming university—the project, as its name suggests, reflects a process of intercultural epistemological becoming. It is based on the work of a "social assemblage" comprising Inga leaders and educators, European and Colombian academics from the fields of architecture and anthropology, and filmmakers and designers.[19] The aim is to achieve the foundation, consolidation, and accreditation of the university within Colombia's formal educational institutional framework.

For Biemann, on an artistic level the project "shifts the focus from bringing ecology into art to bringing art to ecology by directly intervening in, and co-creating, material and epistemic realities on the ground."[20] It thus marks an inflection in her praxis, moving from counterextractivist initiatives such as *World of Matter*, an art research group launched in 2013 to study global resource entanglements, and *Forest Law*, a 2014 project with Brazilian architect Paulo Tavares researching the Sarayaku Indigenous people's successful case against Chevron, toward a propositional, postextractive imagination of knowledges in which she is just one contributor. *Devenir Universidad*'s audiovisual materials (fig. 1) include films that document assemblies held in Mocoa in 2019 to open the

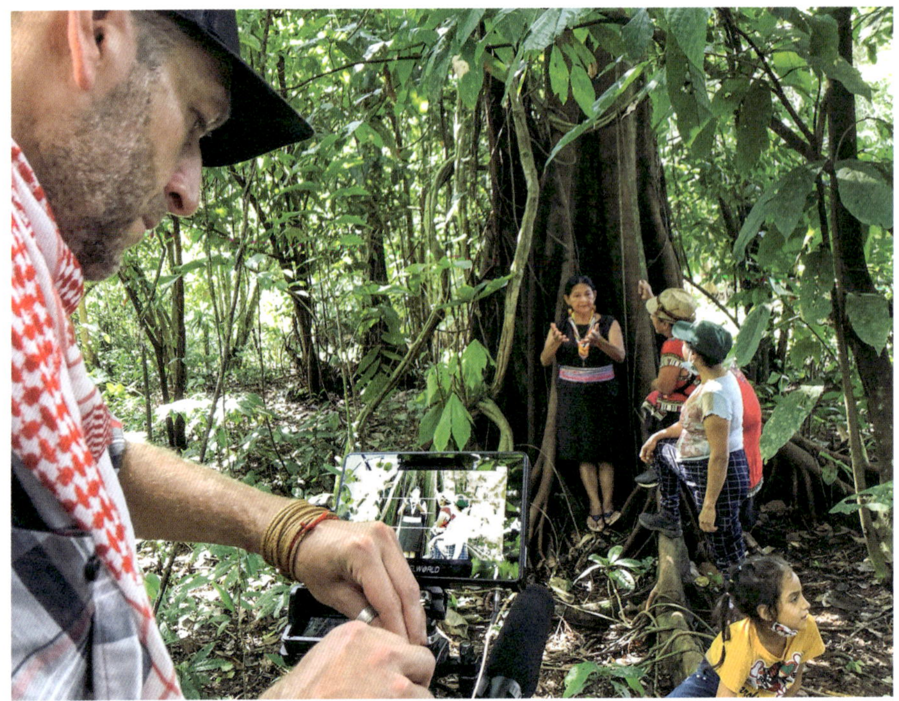

1 An interview with Inga elder Rubiela Mojomboy, filmed for Ursula Biemann and the Inga people of Colombia's *Devenir Universidad* (Becoming university) project in 2021

university project to discussions between local communities and the advisory group. They also include *Vocal Cognitive Territory* (2021), featuring film interviews with Inga leaders and elders, and *Forest Mind* (2021), an artwork by Biemann, presented in Bogotá in 2022, that brings into dialogue Indigenous visionary experiences with sacred plant medicines and images of the plants made with genomic sequencing by scientists at the ETH university in Zurich.[21] The role of the artist in *Devenir Universidad* is articulated as an "inquirer," the term that Isabelle Stengers uses to describe agents who are "experimenters, actively intervening like all who perform experiments, but not in a laboratory, not in order to learn how to obtain reliable knowledge from what they deal with. The aim of today's inquirers should be to learn how to transform the relationship between those who experience and *what they experience*, in such a way that it reactivates the feeling of interdependence."[22]

For all the collaborative spirit of *Devenir Universidad*, Biemann's role treads on the fractious historical grounds of encounter between white European knowledge and Indigenous peoples, terrains, and ontoepistemologies. As decolonial Maōri scholar Linda Tuhiwai Smith argues, research is "inextricably linked to European imperialism and colonialism," and the decolonial "turn" in art must also confront the risks involved in artistic extractivism, which can structure contemporary encounters between artists, researchers, and Indigenous communities.[23] Cognizant of her own background and of the extent of her involvement in a project that exceeds her aesthetic praxis, Biemann acknowledges that her role is limited to that of supporting institutional links between Chindoy and the Inga on the one hand and universities in Bogotá and Switzerland on the other, as well as coordinating film interviews of Inga elders and community members to counter ongoing epistemicidal loss of knowledges.[24] To be an artist-as-inquirer thus entails energizing processes and connections to support a long-standing and ongoing "practical and political imagination" of interdependence that will outlive her collaboration with the Inga's emergent institutional platform.[25]

Devenir Universidad must be understood as an Inga-led endeavor in which Biemann is a temporary instigator and collaborator, while *Forest Mind* is an auteur film that reflects the structure of this "inquiring" role through its narrative voice. Together they situate the overall endeavor as a commitment to articulating the living forest as epistemology. The experimental processes deployed in these works can be articulated as the social practices of an art that confronts the legacies of colonial epistemicide and seeds the ground for reclamation struggles to endure and flourish.

EXPOSING EXTRACTIVISM, BEING RIVER

For the past ten years, the Los Angeles–based Colombian artist Carolina Caycedo has been weaving threads of environmental thinking through a multidisciplinary practice focused on bodies of water and on communities affected by and resisting extractivism. Her practice is predicated on collaboration and solidarity with socioenvironmental activists across the Americas, as well as with "developmental refugees"—that is, groups displaced and dispossessed by large-scale industrial and infrastructure projects.[26] Addressing the challenges to the postextractive imagination, Caycedo's work both reveals the logics of the mining and hydropower industries and supports alternative economies. By appropriating the top-down optics and archival documents of institutions and corporations that enclose territories for profit, she exposes the machinery of extractivism, and by developing aesthetic forms grounded in processes of fieldwork, listening, movement, and performance, she supports river- and water-protector communities in their envisioning of ecologies and economies for just transitions.

Caycedo's ongoing project *Be Dammed*, begun in 2012, has ramified into a delta of film, installation, and performance works whose unfolding reflects an emphatic "liquid turn" in art not only from Latin America and the Caribbean but worldwide: artists are applying their creative practices to exploring water stresses and alternative water cultures. This turn has inspired the curatorial frameworks of major art events such as *Bodies of Water*, the thirteenth Shanghai Biennial, in 2021, and *rīvus*, the Sydney Biennial, in 2022, exhibitions that showcased artistic responses to industrial hydraulic regimes, uneven hydrosocial relations, and ancestral practices of relating through water.[27] Caycedo's work is timely in resonance, since it emerges alongside debates in the social sciences that speculate on future histories of water that will take it beyond dammed, channeled, and polluted rivers, subordinated to human development, toward cultures of care and cogovernance of an emergent hydrocommons.[28] The realization of this hydrocommons will hinge on achieving a broad-based paradigm shift whereby water is no longer objectified as a mere commodity or resource but recognized as a common good and connective tissue with deep, even sacred sociocultural value.

The urgency of reshaping global water cultures underpins Caycedo's representations of bodies of water, which critique the oppositional thinking that separates humans from nature and effect a shift from a view of nature as resource toward environmental relationality.[29] Her artworks and exhibitions, such as *A gente rio* (We river, 2016) and *The Collapse of a Model* (2019), regularly feature large-scale wall-mounted digital montages in which aerial photography of rivers presents compelling visual evidence of the devastating impacts of open-pit mining, megadams, and the collapse of hydraulic structures. By appropriating a technoscientific optics—the satellite view—that surveys and commodifies land and water, rendering them static, these montages strategically stage, on a monumental scale, the hegemonic global logic that converts nature into resource. They mix the Faustian high-modernist tradition, which, as Rob Nixon notes, works visually to create "unimagined communities," with material evidence of the creation of "sacrifice zones" of peoples and territories located in the way of development projects.[30] Contrary to this fixed and depopulated perspective, Caycedo's *Water Portraits* make rivers and waterfalls into subjects of portraiture, amplifying their mutability and vitality through mirror-image photographs of moving water taken by the artist, then printed on long swathes of fabric that may be hung from the ceiling in waves to cascade onto the floor, or stuffed to become cushions to sit on, or, in a recent collaboration with the Brazilian choreographer Marina Magalhães, wrapped around bodies and submerged in water. The latter performance and video work, *Thanks for Hosting Us. We Are Healing Our Broken Bodies/ Gracias por hospedarnos. Estamos sanando nuestros cuerpos rotos* (2019), was filmed on the San Gabriel River and the mouth of the Wanaawna (Santa Ana) River in the unceded Tongva and Acjachemen territories (Orange County, California), demonstrating the hemispheric (rather than Latin American) scope of Caycedo's collaborative work on equitable and sustainable water use.

The performative actions staged in Caycedo's *Geochoreographies* series (2014–18) emphasize the presence of the bodies that have been emptied out of the visual regimes of development produced by nation-states and corporations. Her ongoing series *Cosmoatarrayas* (Cosmonets, 2016–), in which she weaves objects found in her fieldwork into circular artisanal fishing nets, materializes profound relations to land and water and honors the efforts of groups engaged in resisting the commodification of life. This meshing of bodies, bodies of water, and territory lies at the heart of her exploration of the dismemberment enacted by extractivism. Major hydraulic infrastructures change "river rhythmicity"—that is, the relational coexistence of the flows and patterns of bodies of water with the routines and habits of human lifeworlds.[31] The film *A gente rio*—part of the eponymous project Caycedo made for *Incerteza Viva* (Live uncertainty), the 2016 Bienal de São Paulo—gathers the testimonies of artisanal gold-panners, fisherpeople, victims of criminal environmental disaster, and antidam activists. No longer able to pan for gold or fish in polluted rivers, Caycedo's collaborators reenact the actions of panning, fishing, and rowing in a disjointed present, their routines having been ruptured, stilted, and curtailed. In one scene a fisherwoman enacts the intimate relations between body and waterworld by describing and performing the myriad uses that fisherpeople find for their oars. Through the meaning of its title, "We river," *A gente rio* articulates a constant gesture in Caycedo's practice, conjugating humans and rivers as active subjects whose lives are interwoven through both the loss of modes of relationality ruptured by extractivism and alternate ontoepistemologies of care and coexistence in which there is still room for solidarity and hope.

BIOCULTURAL RIGHTS AND PEATLAND PROTECTION

Postextractive futures and sustainable living rely on the cocreation of ecologies of care for the fragile ecosystems affected by environmental degradation. Amid the global heating caused by emissions from fossil fuels, scientific attention has turned to the role that wetlands play as carbon sinks—ecosystems that sequester carbon, mitigating its release into the environment. As temperatures are rising, however, wetlands are shrinking worldwide; since AD 1700, as much as 87 percent of them may have been lost through drainage and reclamation projects, with this process accelerating threefold in the twentieth and twenty-first centuries.[32] Reversing this phenomenon requires deep-rooted attitudinal shifts from a focus on the mere "use value" of wetlands—in water purification, soil substitution, and land reclamation—toward their active role in enabling planetary "ongoingness" (the term that Donna Haraway uses to avoid the developmentalist baggage associated with the term "future").[33] It means moving from property-based ownership to commons-based stewardship.

The care, custodianship, and conservation of wetlands were the concern, local and global, of *Turba Tol Hol-Hol Tol*, an immersive, transdisciplinary, multisensorial installation presented at the Venice Biennale in 2022. Part of an ongoing collaborative project in Karokynka (Tierra del Fuego), *Turba Tol Hol-Hol Tol* was initiated by the Chilean curator Camila Marambio and the Ensayos collective in 2011 (fig. 2). The Selk'nam people have lived on Karokynka's Isla Grande for over 10,000 years; since 2015, through the Corporación Selk'nam Chile, they have called for legal recognition as an Indigenous nation. Today, Chile's Indigenous Law recognizes nine Indigenous ethnicities, together constituting 12.8 percent, or more than 2 million, of the population.[34] The law does not cover the Selk'nam people because they are considered to be extinct, which they are not. After colonizing Chile in the sixteenth century, the Spanish crown claimed land titles and rights to all resources in the country. Even in the late twentieth century, assimilationist policies denied ethnic difference and sustained the notion that Indigenous peoples were almost entirely extinct: a century earlier, concessions of ancestral lands in Tierra del Fuego had been handed over to European, Chilean, and Argentine sheep-breeders, and these settlers had all but wiped out the Selk'nam, with estimates that some 5,000 were murdered and only 100 left alive.[35] On September 5, 2023, after a long fight, the National Congress finally recognized the 1,144 people who identify Selk'nam as one of the eleven original peoples of Chile.[36]

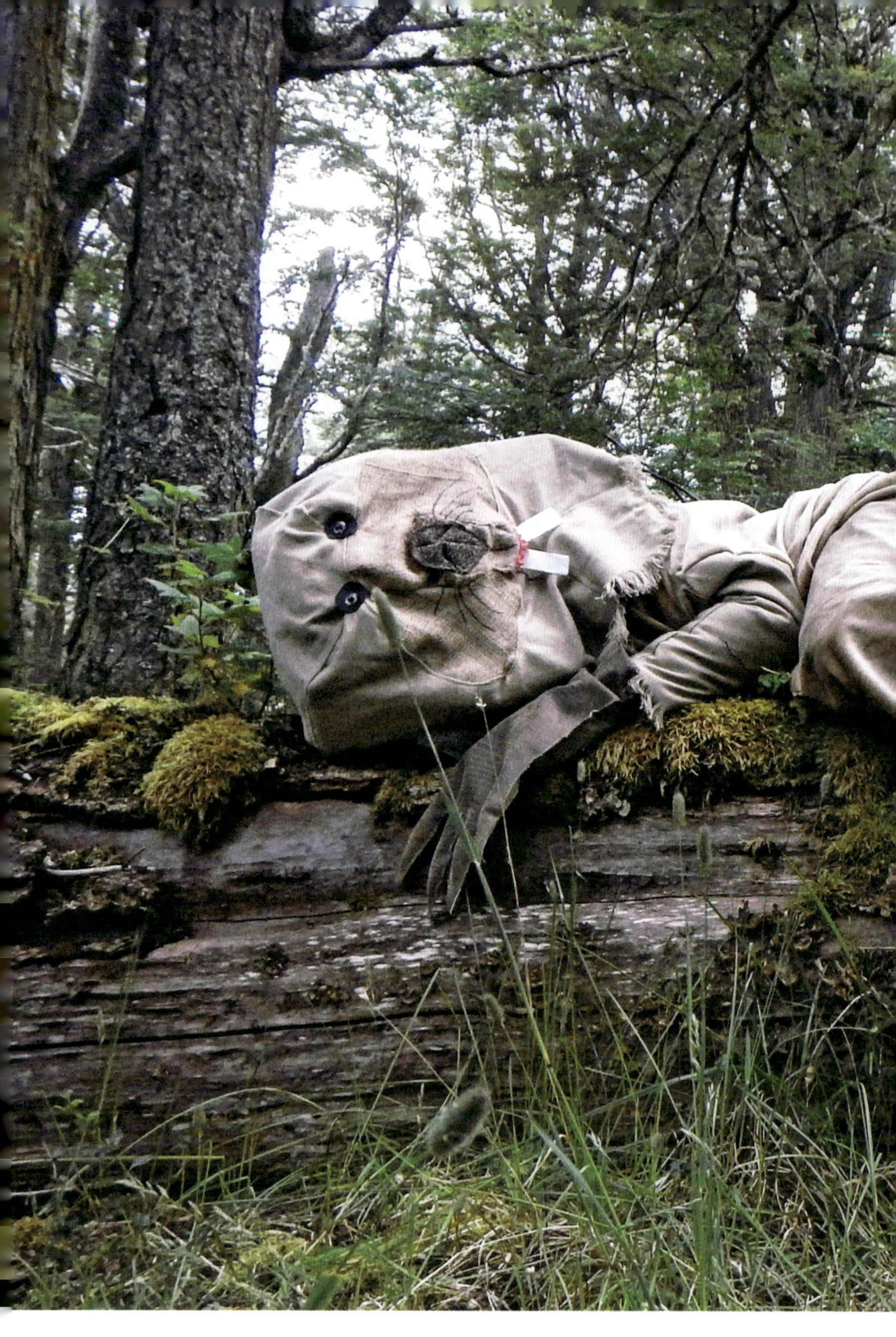

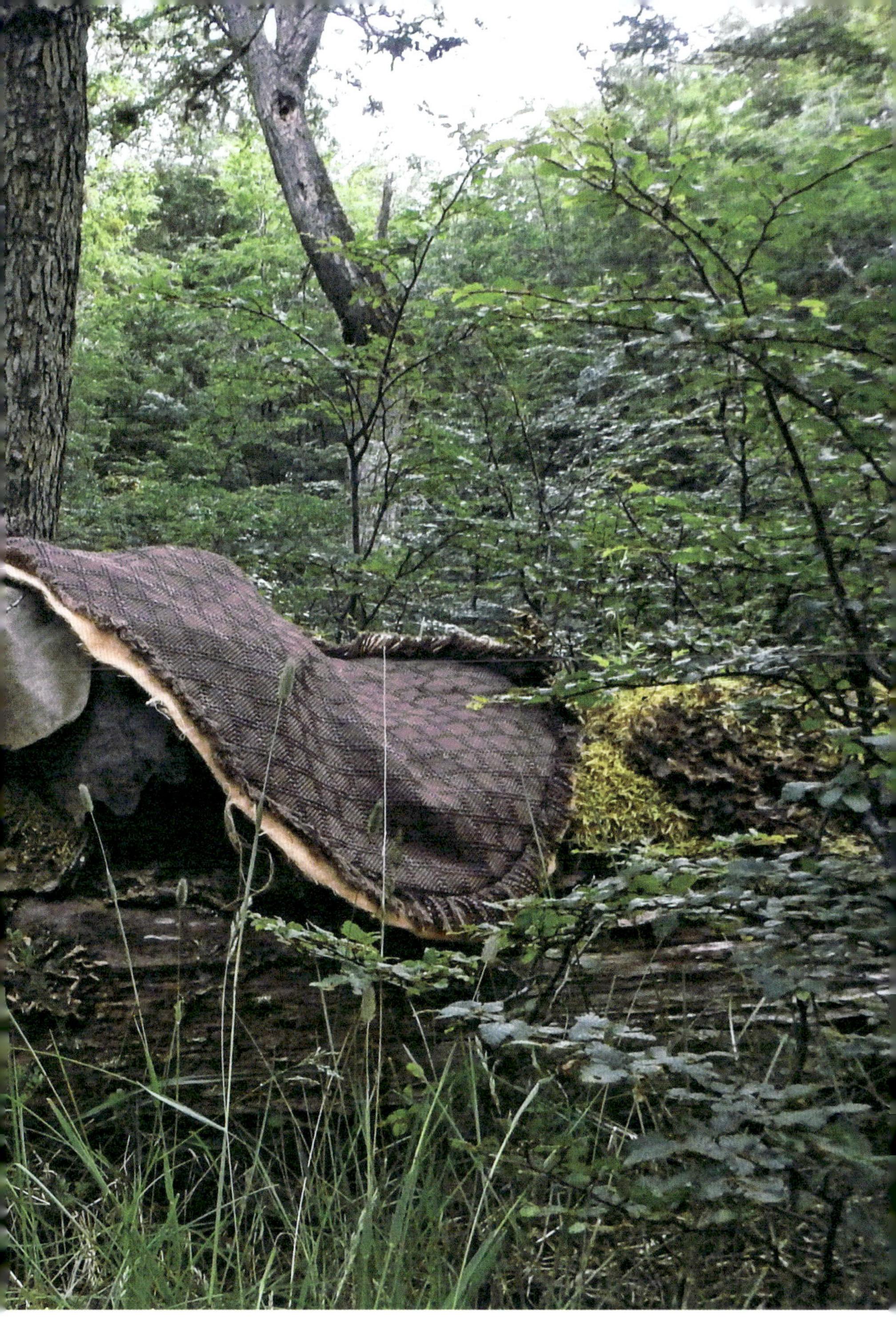

2 Ensayos (Camila Marambio and Christy Gast). *Becoming Beaver*. Performative action, Isla Navarino, Chile, 2014, with performance by Marambio and costume and photo by Gast

Turba Tol addresses the conservation of Karokynka's peat bogs and the cultural rights of the Selk'nam people as interconnected causes and entangled biocultural histories. In an article that weaves together curating, Indigenous leadership, and conservation science, Marambio, Hema'ny Molina, and Bárbara Saavedra make this linkage of historical and contemporary common ground explicit. "The accumulation of decaying peat moss, or *Sphagnum*," they write, "a genus of the approximately 380 nonvascular plants or mosses that grow on the bog, is a process that occurs over thousands of years. In this slow, invisible dance between life and death, dead *Sphagnum* become a carbon sink, maintaining biodiversity and also storing large quantities of freshwater."[37] The writers' transdisciplinary work incubates "emergent forms of place-based eco-cultural ethics" and acts of "co-caring" deriving from human actions that mimic the material onto-logic of peat bogs.[38] Through its material (de)composition, the bog ecosystem resists the acts of separation and isolation that characterize both the epistemological and the physical violence of colonialism. Uniting around (and like) peat bogs means creating an ecocentric structure for resistance and more-than-human solidarity; as the authors put it, "By being deliberately boggy, no one can be stepped on alone."[39]

Visitors could not step on boggy peat in the installation *Turba Tol*. They walked instead across a short platform that stood above a dense, brightly lit bed of mossy life, grown in Germany, then transplanted to Venice. The platform gave access to a circular metal structure hosting an immersive audiovisual space where sounds circled overhead between four speakers. As the projection began, a landscape scene appeared on the circular screen and the camera quickly began to pan downward, sinking into the peat bogs. The soundtrack meanwhile merged fragments of laughter, guttural noises, and squelching sounds evoking the water-clogged land. These deep bass sounds grew strong enough to make the platform underfoot vibrate as the camera shifted back to a recognizable landscape, this time nocturnal, capturing images of people circling the structure and chanting an incantation to the peat bogs. The chant rose in crescendo until flashes of light on-screen gave way to a sudden illumination from LED panels beyond the circular structure, simulating sunlight for the moss. The focus suddenly shifted to the here and now, to the plants growing beneath visitors' feet—an invitation to engage with them, leaning over the bed of moss to observe its tangled forms. The installation generated a deeply physical resonance among bodies, following from the *Rumores* (Rumors)—a series of recordings, released through the *Turba Tol* website, in which human and nonhuman sounds, verbal and nonverbal meanings, create a multispecies cacophony grounded in "arts of attentiveness" to the material lives of the peat bogs and in the affective attachments to it cultivated by different project members.[40] This entanglement of voices and squelches finds in sound (rather than in language) an aesthetic channel to experiment with forms for the multispecies co-becoming of an "ecology of selves"[41]—human, watery, earthly—that would move beyond the taxonomic separation of species, and beyond the idea of vision as the dominant and most important sense, a notion inherent in (Western) science.

This does not amount to a rejection of science, however. As part of a global movement of peat bog conservation, *Turba Tol* directly engages with science, and with scientists, as intrinsic to (not separated from) affective relations to earth beings. In this sense it engages science with a small "s" rather than with a capital "S," a usage that Stengers equates to a hegemonic knowledge system.[42] The installation itself performed this act through the daily cycles of stewardship of the pavilion. At the start of each day, access to the space was restricted (even though the Biennale was open) as peat custodians spent ninety minutes watering the peat and taking samples to monitor pH levels. The imposition of these conditions on public access performed the *eco*-centric (not *anthropo*-centric) grounds of planetary stewardship.

After the Biennale, the waterlogged mass of peat, weighing almost six tons, was transferred to three Venetian agricultural and slow-food producers. This act of sustainability reflects the merging of aesthetic/scientific worlds with sociolegal ones in the Venice Agreement, a pact on peatlands produced during the Biennale through a workshop at TB21—Academy's Ocean Space in the city. Cocreated by participants from global organizations, this digital declaration of care drew on contributions from groups that a month earlier, instigated by *Turba Tol*, had held peatland-conservation workshops around the world. Appropriating its format from policymaking, the Venice Agreement establishes an ecoethical pact of care of and belonging to peatlands, identifying best practices for conservation policy, education, and information sharing and incentivizing others to take action: "I AGREE/WE AGREE TO PROTECT GLOBAL PEATLANDS LOCALLY/AND YOU?"[43] Saavedra—head of the Wildlife Conservation Society, Chile, and a collaborator in the Venice Agreement project—aims for the establishment of a Patagonian Peatland Center, which will implement biodiversity conservation at the nexus of different knowledges and peoples. Beyond Chile, the Agreement contributes to ongoing discussions and collaborations instigated and hosted by the Global Peatland Initiative, linking the spheres of artistic research to possibilities for sociolegal change.

ART TOWARD THE PLANETARY COMMONS

By commodifying the biosphere, extractivism produces nothing but extracts everything, causing a "net loss of natural heritage."[44] As translocal initiatives, each of the projects discussed here demonstrates the important contributions that Abya Yala, as a region, makes to delinking from "extractivist theology" and contributing through creative practice to global environmental thinking. The projects resonate with Leff's claim that "it is from the radical epistemological concept of the environment, which emerges from the environmental crisis as revealing the limitations of the dominant rationality and the potentialities to construct alternative sustainable societies, that an emancipatory environmental knowledge emerges, rooted in the ecological productivity and cultural creativity of the Southern regions."[45] Caycedo's work, and such projects as *Devenir Universidad* and *Turba Tol*, demonstrate how artistic practice can expose the mechanisms of that system and its earlier colonial forms, raising awareness of biocultural loss and cultivating alternative, postextractive worldings based on epistemodiversity and the ecoethics of justice, solidarity, and care. Amid the diversity of contemporary art engaged in reemergent discussions of the commons, some consensus exists, as Amy Elias writes, that "reconceiving the commons requires rethinking the relations between ethics, art, and politics in ways that take into account the hegemony of the world capitalist system and the marketization of everyday life as well as the ongoing degradation of the planetary ecosystem."[46]

In sum, through its entanglement of Indigenous territorial belonging and knowledges, strategies of resistance in marginalized communities, and dialogue with Western sciences, the collaborative environmental work emerging from the arts in Abya Yala is generating responses to the pressing challenge of how to care for the fragile ecosystems so fundamental for regional and planetary ongoingness. Through their participatory and public dimensions and their values of "inclusivity, the exchange of ideas, and play and creativity between human (and nonhuman) entities," Elias writes, these projects can contribute to an art for the planetary commons.[47] The locomotives described by Drummond may continue to move blasted landscapes along the tracks of global capitalism, but in the face of that unworlding, ecocentric practices like those described here have taken up Ailton Krenak's call to shift the worldwide fixation away from extractivist machinery and toward finding ways to get involved in nurturing alternative futures.

1 Carlos Drummond de Andrade published "O maior trem do mundo" in the magazine *O Cometa Itabirano* in 1984. My translation.

2 Ailton Krenak, *A vida não é útil* (Rio de Janeiro: Companhia das Letras, 2020), 25. Krenak is quoting from Drummond de Andrade's poem "O homen; as viagens" (The man; the travels).

3 Ibid., 24.

4 See John Law, "What's wrong with a one-world world?," *Distinktion: Journal of Social Theory* 16, no. 1 (2015): 126–39.

5 See Eduardo Gudynas, "Extractivisms: Tendencies and Consequences," in Ronaldo Munck and Raúl Delgado Wise, eds., *Reframing Latin American Development* (Abingdon and New York: Routledge, 2018), 74. Gudynas notes that while the profits of extractivism mainly flow abroad to transnational companies, under "progressive" South American governments of the early 2000s the commodity boom also bolstered neoextractivist economies, which charged those companies rents that could be redistributed to citizens in the form of social welfare.

6 Ibid., 61–76, 66.

7 On the way Andean Indigenous communities consider relatives, not resources, see Marisol de la Cadena, *Earth Beings: Ecologies of Practices across Andean Worlds* (Durham, NC: Duke University Press, 2015).

8 Gudynas, "Extractivisms: Tendencies and Consequences," 75.

9 See Catherine Walsh, *Interculturalidad, crítica y (de)colonialidad. Ensayos desde Abya Yala* (Quito: Ediciones Abya Yala, 2012).

10 Enrique Leff, "Latin American Environmental Thought: A Heritage of Knowledge for Sustainability," South American Environmental Philosophy Section, *ISEE Publicación Ocasional* no. 9 (2010): 1–16.

11 Mario Blaser and de la Cadena, "Introduction. Pluriverse: Proposals for a World of Many Worlds," in Blaser and de la Cadena, eds., *World of Many Worlds* (Durham, NC: Duke University Press, 2018), 3.

12 Boaventura de Sousa Santos, *Epistemologies of the South: Justice against Epistemicide* (New York: Routledge, 2014).

13 Ibid., 192–93.

14 Ibid., 245.

15 See, for instance, Michael Marder, *Plant Thinking* (New York: Columbia University Press, 2013), and Eduardo Kohn, *How Forests Think: Toward an Anthropology beyond the Human* (Berkeley, Los Angeles, and London: University of California Press, 2013).

16 Ursula Biemann, in an interview with the author, July 21, 2022.

17 See Binna Choi and Marion von Osten, "Trans-Local, Post-Disciplinary Organizational Practice: A Conversation between Binna Choi and Marion von Osten," in Choi, ed., *Cluster: Dialectionary* (Berlin: Sternberg Press, 2014), 274.

18 Biemann, in the interview with the author.

19 See Choi and von Osten, "Trans-Local, Post-Disciplinary Organizational Practice."

20 Devenir Universidad, "Art & Audiovisuals," 2023. Available online at https://deveniruniversidad.org/en/art-and-audiovisual-media/ (accessed July 9, 2023).

21 The exhibition *Forest Mind: sobre la interconexión de la vida* opened on November 9, 2022, at the Claustro de San Agustín, Universidad Nacional de Colombia, Bogotá. The show was presented as a collaboration between Biemann and the

Panamazonian Inga people, and the opening included traditional music by three Inga *taitas* (elders).

22 Isabelle Stengers, "We Are Divided," *e-flux Journal* no. 114 (December 2020). Available online at www.e-flux.com/journal/114/366189/we-are-divided/ (accessed July 9, 2023). Derived from the thinking of the philosopher John Dewey, the figure of the "inquirer" contrasts with that of the "diplomat," who represents entrenched institutionality and the failure to confront ecological crisis.

23 Linda Tuhiwai Smith, *Decolonial Methodologies: Research and Indigenous Peoples* (New York: Zed Books, 2012).

24 Biemann, in the interview with the author.

25 Stengers, "We Are Divided."

26 See Rob Nixon, *Slow Violence and the Environmentalism of the Poor* (Cambridge, MA: Harvard University Press, 2011).

27 See Lisa Blackmore and Liliana Gómez, "Beyond the Blue: Notes on the Liquid Turn," in Blackmore and Gómez, eds., *Liquid Ecologies in Latin American and Caribbean Art* (New York and London: Routledge, 2020), 1–10.

28 See ethnographic accounts of Costa Rican water committees' collaborative administration of their common water sources in Andrea Ballestero, *A Future History of Water* (Durham, NC: Duke University Press, 2019). For a posthuman, interdisciplinary perspective see Astrida Neimanis, "Bodies of Water, Human Rights and the Hydrocommons," *Topia* 21 (2017): 161–82.

29 See Blackmore, "Turbulent River Times: Art and Hydropower in Latin America's Extractive Zones," in Blackmore and Gómez, eds., *Liquid Ecologies*, 12–34.

30 Nixon, *Slow Violence*, 150.

31 See Sue Jackson, Elizabeth P. Anderson, Natalia C. Piland, Solomon Carriere, Lilia Java, and Timothy D. Jardine, "River Rhythmicity: A Conceptual Means of Understanding and Leveraging the Relational Values of Rivers," *People and Nature* 4 (2022): 949–62.

32 See Nick C. Davidson, "How much wetland has the world lost? Long-term and recent trends in global wetland area," *Marine and Freshwater Research* 65, no. 10 (January 2014).

33 See Donna Haraway, *Staying with the Trouble: Making Kin in the Chthulucene* (Durham, NC: Duke University Press, 2016).

34 See Elizabeth Jane Macpherson, "Recognising and Allocating Indigenous Water Rights in Chile," in Macpherson, *Indigenous Water Rights in Law and Regulation: Lessons from Comparative Experience* (Cambridge: Cambridge University Press, 2019), 163. Census numbers from 2017.

35 See Marcio Pimenta, Nina Radovic Fanta, and Caio Barretto Briso, "'We are alive and we are here': Chile's lost tribe celebrates long-awaited recognition," *Guardian*, October 3, 2023. Available online at www.theguardian.com/global-development/2023/oct/03/we-are-alive-and-we-are-here-chiles-lost-tribe-celebrates-long-awaited-recognition?CMP=Share_iOSApp_Other (accessed October 14, 2023).

36 Ibid.

37 Camila Marambio, Hema'ny Molina, and Bárbara Saavedra, "Careful Thinking: Pensar Cuidando—Henvupen Yaconso," in Samuel Alexander, Sangeetha Chandrashekeran, and Brenda Gleeson, eds., *Post-Capitalist*

Futures: Paradigms, Politics, and Prospects (London: Palgrave Macmillan, 2022), 166.

38 Ibid., 165.

39 Ibid., 170.

40 See Thom van Dooren, Eben Kirksey, and Ursula Münster, “Multispecies Studies: Cultivating Arts of Attentiveness,” *Environmental Humanities* 8, no. 1 (May 1, 2016).

41 The phrase, quoted in ibid., 4, is originally from Kohn, *How Forests Think*, 134.

42 Stengers, “Introductory Notes on an Ecologies of Practices,” *Cultural Studies Review* 11, no. 1 (2005): 183–96.

43 Venice Agreement poster. Available online at https://www.turbatol.org/venice-agreement.html (accessed July 15, 2023).

44 Gudynas, “Extractivisms: Tendencies and Consequences,” 62.

45 Leff, “Latin American Environmental Thought,” 2.

46 Amy Elias, “Art and the Commons,” *ASAP/Journal* 1, no. 1 (January 2016): 4.

47 Ibid., 3.

JENS ANDERMANN

ALLIANCES OF SURVIVAL

Ala Plástica, thislandyourland, and the Arts of Entanglement

A beach by the brackish brown waters of the River Plate estuary in Argentina. A creek on the outskirts of Belo Horizonte, Brazil, one of its margins bulldozed over by roadworks. In both locations, small groups are busy planting seedlings. What could at first glance appear as acts of eco-activist protest, government-sponsored community outreach, or a school or college lesson are, in fact, art actions—though whether or not primarily so is very much open to question. *Junco/Especies emergentes* (Rushes/Emergent species, 1995), devised for the Plate estuary by artists, activists, and curators Silvina Babich, Alejandro Meitin, and Rafael Santos, working together as "Ala Plástica," was a project for planting seedlings of California bulrush, a native freshwater marsh plant characterized by its rhizomatic root structure and self-propagating capacity. Bulrushes encourage soil sedimentation and contribute to the removal of excess nutrients from contaminated water, while their dense, upright foliage provides a nesting habitat and food shelter for birds, reptiles, and small mammals. In time, the sedge lands expanding from Ala Plástica's small plantation would do their part (if on a very small scale) in halting shore erosion and stabilizing the native biome from environmental stress caused by the extractive industries that line the Paraná River basin. The Belo Horizonte work, *Clareira* (Clearing, 2012–13; fig. 1), a collaboration between Louise Ganz and Ines Linke, who sign their work as "thislandyourland," invited kids from a nearby school to collect and transplant samples of vegetation from the intact to the damaged margin of a small river, to revisit the site later and observe the plants' success or failure in adapting to their new ground, and to make drawings of leaves, stamens, and pistils. Later, during an

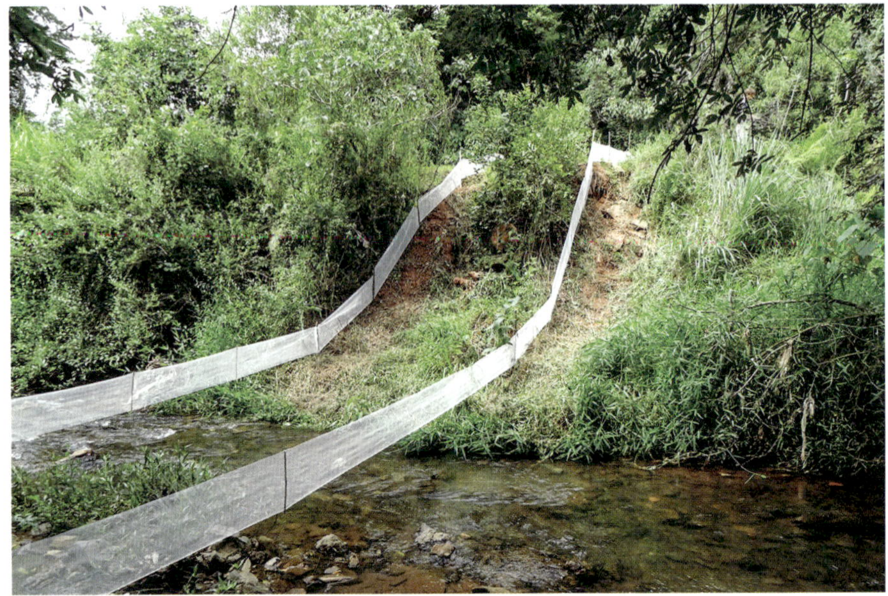

1 thislandyourland. *Clareira* (Clearing). 2012–13. Collaboration with schoolchildren to plant a degraded riverbank, a plot of c. 164 × 16 ft. 5 in. (50 × 5 m), Nova Lima, Minas Gerais, Brazil

on-site exhibition of their drawings for parents and relatives, the kids would conduct a guided tour, with the idea of incentivizing other forms of rescuing local flora in family backyards or on neighborhood streets and squares.

Ala Plástica and thislandyourland can be seen as contributing to the "new regime of the arts" that critic Reinaldo Laddaga has seen emerging since the millennium, in which the institutional conduits of the modern art system have made way for "local spaces with which [artworks] connect through multiple transitions."[1] At the same time, these novel, often collectively authored propositions have benefited from wider, translocal networks facilitated by the expansion of communications technologies (indeed, the rise of the Internet and of social media parallels and in a sense facilitates the dissemination of what Ala Plástica calls "an artistic way of thinking and acting").[2] They have also manifested within such constellations of resistance as Argentina's social uprising of 2001 and Brazil's wave of urban protests surrounding the 2014 soccer World Cup and the 2016 Rio Olympics.

Art theorists including Nicolas Bourriaud, Grant H. Kester, and Florencia Garramuño have put forth the monikers of "relational art," "collaborative art," or "unspecific art" to characterize this dimension of indiscernibility both within and beyond the artistic field, with artists relinquishing the "autonomy" on which the modern notion of art had been predicated. Instead, unspecificity becomes the condition that allows them to enter into horizontal and open-ended networks of cross-pollination with social collectives, which often come together only in response to the questions that the artists have posed, in "an exchange that is not wholly subsumable to conventional, pragmatic demands, but is consciously marked as a form of artistic practice. In fact, it is in part the lack of categorical fixity around art that makes this openness possible."[3] Indeed, even with the proliferation of medialities and genres in the expanded field of art actions, it is often difficult to circumscribe what exactly is at stake and where and when an individual "work" starts and ends. As Ines Linke of thislandyourland puts it, "When cataloguing a piece, we always need to think what kind of work this is, where to place it: is it a performance, a process, an installation? Is it site-specific or not? Is it something that continues to be active or is it a process that has run its course?"[4]

Moreover, the work of Ala Plástica and thislandyourland also reconnects with, and rereads, the "environmental turn" that Latin American theorists such as Mário Pedrosa and Nelly Richard had already detected in the experimental merger of art and politics heralded by actions such as *Apocalipopótese*, the public happening organized by Rogério Duarte, Lygia Pape, and Hélio Oiticica in the environs of Rio de Janeiro's Museu de Arte Moderna in 1968, or Argentina's legendary *Tucumán Arde* (Tucumán is burning) intervention in media and politics of the same year. If collaborative modes of production, collective authorship, formal and medium unspecificity, and the incorporation of "the dimension of social exteriority in the production of art," in Richard's expression,[5] were already present as defining aspects of these art actions, the work of Ala Plástica and thislandyourland adds to these qualities an ecological dimension that has only come to the fore with the more recent, catastrophic acceleration of anthropogenic global warming and with the critique of extractivism in the Global South. For even though the exposure of ecological hazards was already at issue in works such as Frans Krajcberg's *Queimadas* (Fires), a series of installations of charred wood produced in the 1970s juxtaposed with photographs of slash-and-burn deforestation in the Amazon, and Roberto Evangelista's *Mater Dolorosa—In Memoriam I* (1976), a heap of charcoal and ash collected from burned rainforest, first displayed at an industrial fair in Manaus, Brazil, this critique remained predicated on the autonomy and object quality of the individual artwork, and continued to draw on well-established rhetorics of allegory and denunciation.

By contrast, the kind of art practice that Ala Plástica and thislandyourland have made their hallmark eschews representation and objecthood to create spatiotemporally fluid, processlike entanglements, at the same time that it expands the "social exteriority" incorporated in previous iterations of collaborative art toward the more-than-human. Rather than just the instability of mediums and expressive languages, "unspecific art" also implies here an extension of the aesthetic faculty beyond the species boundary: an "experimental practice of world-making," in T. J. Demos's expression, that opens up "sites of expanded creativity."[6] In thislandyourland's and Ala Plástica's actions, rather than accruing figurative value on account of their status as objects of ruinous inertia, as in Krajcberg's and Evangelista's installations, plants, liquids, and earth participate as gathering agents of collectives, which they prompt to imagine what I call "alliances of survival": modes of togetherness that thrive only in the measure that they are capable of including humans and nonhumans alike.

How, I will ask, following Demos's lead, "might the world-generating activity of aesthetics become a multinatural, multispecies affair, and not simply the reserve of human exceptionalism?"[7] And I will suggest that it does through weaving together alliances of survival, by which I mean, on the one hand, the exemplary, or prototypical, character of the artworks—their propositions of miniature models of experimental community—and on the other, the way they are designed to dissolve into the very processes of ramification and entanglement that they unleash. Both *Junco* and *Clareira*, as we have seen, draw on the resilient, literally ground-breaking capacities of plants at the same time that they draw attention to their precarious survival under conditions of intensifying extractivism—which is also, as the participants of these actions may come to realize, the condition they share with other existents in their living environment. At the same time, if successful, both kinds of vegetation will also outlive their iteration as participants in an art action and enter into new assemblages (as will the human participants, who may or may not use these entanglements as incentives to invent new and different forms of community). But this "afterlife" in diverse and unforeseeable forms of socioecological practice is very much part of the setup of these works: their ultimate destination is not the museum but a kind of *art de vivre*, in which art and life merge to herald the emergence of a new commons.

BEYOND LANDSCAPE: MAKING WORLDS FROM SPACES OF ABANDONMENT

To open up new spaces and times of commonality means daring to imagine alternative modes of use for sites ravaged by extractivism, the "world-ecological" foundation of global capitalism.[8] The work of thislandyourland consequently "centers on one issue: the modes of occupation and use of the land."[9] In Latin America, the key apparatuses of subjugating places to the designs of extractivism have been twofold: on the one hand, the genre tradition of landscape as a mode of objectifying land and its (mineral, vegetable, animal, human) "contents" into resources at the behest of a beholder placed outside the frame—a disembodied, extractive eye; and on the other, the tabula rasa design of cities as "civilizational" bastions *against the land*, from the Spanish colonial "*damero*" (checkerboard grid) to its postcolonial sequels, including the planned cities of La Plata, Belo Horizonte, and, later, Brasília.[10] Working mainly though not exclusively in and around Belo Horizonte, the state capital of Minas Gerais—home for over three centuries to one of the most destructive mining economies in the world—thislandyourland has consistently engaged in a deconstructive critique of the forms and gestures of modern-colonial landscaping and urbanism, at the same time that it has also devised alternative modes of dwelling in and interacting with both our ecosocial and our built surroundings.

In a spatial context comprehensively ravaged by environmental destruction and speculative urban sprawl, the work of thislandyourland suggests, to imagine other spaces and commonalities demands an embodied critical practice that moves to counter the reifying, disembodied detachment of the landscape view and of the urban master plan alike. Instead of adopting the all-seeing gaze of colonial landscaping, the pair's work delves into the "anthropological, poetic, and mythical experience of space," in Michel de Certeau's expression, that is being teased out by "blindly, opaquely moving through the inhabited city" and land.[11] This move from representation toward a "sensorial approach to the territorial and the public," as Alejandro Meitin of Ala Plástica puts it, triggers cascading affections, encounters, and aftereffects in both the artists/proponents and the communities that gather around them.[12] It involves exploration of the resilient modes of world-making put into practice by those surviving amid the rubble that extractivism leaves behind. It is also a doubly critical practice, at once deconstructing the formal frameworks of landscape and urbanism and taking stock of their ruinous consequences, of the *unworldings* they unleash. At the same time, it proposes and tries out alternative modes of reciprocity, of the "livable collaboration" that, as anthropologist Anna

Tsing argues, is a requisite for "staying alive" even, or especially, amid extractivism's ruins: "The diversity that allows us to enter collaborations emerges from histories of extermination, imperialism, and all the rest. Contamination makes diversity."[13]

This embodied critique of the landscape-form is already in evidence in some of thislandyourland's *Percursos* (Hikes, 2007), in which the pair walked along large infrastructures, such as aqueducts and high-voltage cables, running through Belo Horizonte's sprawling periurban belt and into the heavily mined mountain ranges around it, across traffic arteries as well as dirt tracks and meadows. In the course of their journey, "following the logic of infrastructure, of distribution, and contrasting this industrial logic with that of nature," Louise and Ines performed minimal interventions, isolating plants with paper disks that converted them into singular objects, for example, or setting up ephemeral outdoor receptions, bedrooms, and dining tables to be shared with neighbors in the area.[14] Continuing in subsequent years with more far-ranging actions such as *Expedição na Bahia* (Expedition in Bahia, 2012), an exploration of modes of land use in the northeastern *sertão* (arid backlands), these hikes also critically reperformed the naturalists' journey into tropical nature, complete with the collection of trophies—the floral displays of herbs and weeds, picked from roadsides and brownfields, that thislandyourland included in their 2012 show *Outros lugares* (Other places, 2012), at the Museu de Arte da Pampulha, in Belo Horizonte.

Meanwhile, *Paisagem de águas e monstros* (Landscape of waters and monsters, 2016) was part of a series of interventions in Torre H, an uninhabited apartment tower—a speculative ruin—designed in the early 1970s by Oscar Niemeyer in the center of the wealthy Barra da Tijuca neighborhood in southwestern Rio de Janeiro. Here, on the building's roof terrace, thislandyourland installed a viewing platform from which visitors could take in southside Rio's neo-Venetian panorama of channels and lagoons, all sprinkled with motorways and high-rise clusters. Through maps, audiovisual registers, and liquid samples, the installation also invited them to emulate a ground-level boat trip the pair took through the toxic, nausea-inducing seascape of what is, in fact, a large-scale periurban open sewer. Taking their cue from Robert

Smithson's notion of the "nonsite," and from his legendary forays into the "monumental" timelessness of the postindustrial edgelands around Passaic, New Jersey, thislandyourland's hiking works (indeed, the duo's name was originally intended for an artistic "travel agency" offering guided tours through the outskirts of Belo Horizonte) force the modern-colonial landscape into confrontation with the ruinous afterlives of extractivism. As a result, Louise argues, landscape's "approach towards nature through its plasticity and visuality from a fixed point of view turns more complex the moment when, beyond the dualism of nature and culture . . . , new environmental and political problems are being incorporated, amplifying the artistic field."[15] The juxtaposition of rural spaces with industrial materialities and large infrastructures complicates the oppositions between country and city, nature and culture, body and mind, on which landscape's perspectivist relation was predicated, inviting other, immersive and nonfinalistic approaches to the land.

The encounter, on the sprawling edge of the city, with figments of rural modes of life informs a series of urban interventions by thislandyourland that both draw on and deconstruct the metropolitan counterpart of the traveler's landscape: the garden. Just as the pair's engagement with the artistic journey has an ironic edge at the same time that it reasserts the immersive dimension of displacement, their deployment of the garden form is twofold. First, it performs a critical archaeology of colonial gardening, which disavowed the violence of plantation monoculture toward the human and more-than-human through what art historian Jill H. Casid calls "a prodigious variety of introduced flora held in place by an ordering system of careful segmentation."[16] At the same time, their gardening practice also appeals to the "almost organic connection between gardens and forms of conviviality," in literary critic Robert Pogue Harrison's phrase.[17]

This is perhaps most evident in *Museu Campestre* (Countryside museum, 2012), the centerpiece of the 2012 Pampulha show. In this year-long intervention in a vacant lot opposite the Museu (the site, incidentally, of an as-yet-unbuilt annex designed by Niemeyer), the pair—in collaboration with Cláudio Ribeiro, a gardener from the nearby community of Vila Aeroporto, and five local trainees—planted and grew a variety of herbs, vegetables, corn, and sugarcane as well as native shrubs and grasses. A wood stove and picnic tables invited visitors to prepare snacks and teas from on-site produce. In marked contrast with the abstract shapes of the Museu building and its surrounding park and artificial lagoon (designed in 1940 by Niemeyer, Cândido Portinari, and Roberto Burle Marx), this temporary orchard brought into the very heart of the modernist city the kinds of vegetable patches that Louise and Ines had stumbled upon in their hikes through the suburbs and settlements: "We had the idea of replicating that—what if those lots were nearby, mixing city and nature? It would create a kind of geology, an original geology, an archaeology of this land. Is it possible to create neighborly relations? Here, whoever lives nearby starts to use these kinds of spaces. So it has to do with inventing a mode of life, with inventing a mode of leisure, a relation nature-city."[18]

The orchard/garden also offered a practice-based, experimental response to the Museu de Arte da Pampulha's quest to establish a different, less vertical relation with Vila Aeroporto, a community that had settled over the years on the Ribeirão Pampulha, a creek that was dammed to create an artificial lake and that decades of abandonment had turned into a sewage drain. Not only are Seu Cláudio's trainees now themselves relandscaping the creek into a community-led urban park, but their design decisions are probably informed by their previous collaboration in *Museu Campestre*, and by visiting thislandyourland's show in the Museu next door with friends and relatives, a space they might otherwise have seen as forbidding and exclusive. As a collaborative process, "encountering common elements among the vocabularies [of artistic landscape and community orchards], the intervention functions as a communicative tool," as the biologist Marcio Gibram, who collaborated in the project, describes it. "The environmental and landscaping intervention translated this dialogue into a shared language: the garden."[19]

Gardening, then—understood in the widest sense as a performance of alliance-making unfolding over time, a process of playful learning from one another that involves diverse existents, human and more-than-human—underwrites much of thislandyourland's art practice, which is born out of their shared pleasure, as Ines explains, “in working with the earth, in weeding, planting: a manual kind of labor that involves the earth.”[20] In a series of urban interventions, some more ephemeral than others, they interrupt the city’s spatiotemporal continuum by staging other modes of use, taking advantage of Ines’s background in stage design and Louise’s in architecture and urban planning. Their juxtaposition, in unforeseen ways, of elements associated with the public and domestic domain, with work and leisure, teases out the emergence of *places* from spaces of abandonment.

The series *Lotes vagos* (Idle lots), started by Louise in 2004 and subsequently revisited by thislandyourland between 2006 and 2010, with iterations in Belo Horizonte and Fortaleza, proposed temporary modes of occupying urban locations held in suspension: islands of greenery in the midst of traffic intersections, post-demolition sites or lots left vacant by real estate speculation. Vaguely inspired by actions including Gordon Matta-Clark’s *Odd Lots* (initiated in 1973) or Alan Sonfist’s *Time Landscape* (initiated in 1965), and using only the slightest material props—lawn mats, construction sand, beach chairs, umbrellas—the pair, in collaboration with friends, neighbors, and occasional passersby, created, among other things, a 100-foot open-air communal table near a housing estate (*Banquete*); lounging cushions fitted to the slope of a hillside backyard, where people could take a nap or chill out overlooking the city below (*Topografia*); a beach on a downtown street corner, complete with its own minifridge for refreshments and coconuts (*Praia*); a unisex hair-and-beauty parlor (*Cabelereiro*), a playpen (*Brinquedo*), and a roadside beach resort (*Praia Atlântico Clube*). In *Muro Jardim* (Wall-garden, 2011) they intervened in an empty backyard, opening holes in the street-facing brick wall to fit planter bags accessible from both sides, growing herbs and vegetables including parsley, lettuce, and laurel. A bench and table, complete with a gas stove, washbasins, and kitchen utensils, likewise broke through the wall, inviting neighbors to gather to prepare a meal, or just chat, while moving interchangeably between backyard and sidewalk. Similarly, *Cozinha temporária* (Temporary kitchen, 2012), an intervention in the nearby town of Nova Lima, invited residents to forage for edibles in their own backyards. These spaces would then be mapped and cross-referenced through recipes for dishes and refreshments, to be cooked, mixed, and enjoyed together at temporary street kitchens and food stands arranged by thislandyourland, with neighbors lending plastic furniture or opening empty yards and garages.

In the course of these interventions, as Louise explains, "We began to understand what is 'public' and what is 'collective' as well as the differences between these two. . . . Public space, within a logic of collective use, is the square or the sidewalk, the street, and we began to see that there could be other kinds of collective space."[21] Indeed, thislandyourland's "actions" are always part of a longer process of sharing ideas and experiences, and it is only from these conversations and gatherings—rather as in Augusto Boal's participative-theater exercises in the 1970s—that the "script" of the event emerges. As Ines puts it, "There is never a project before we arrive at a place. Instead, there is conversation and listening. . . . we like to sit down and chat with people. And in this negotiation, we try to downplay this notion of the artist and rather encourage an idea of the collective event, of something that is being created out of these encounters, which we don't control."[22]

The event, which transpires in a limited period of time, continues to resonate beyond its performative "core," sometimes in subtle, almost imperceptible fashion—through new friendships, for example—sometimes in more tangible ones, such as neighborhood orchards, food shares, and communal leisure spaces. Feminist science-scholar Karen Barad's notion of "intra-action"—"the mutual constitution of entangled agencies" that only emerge as such through this very entanglement with one another, and that therefore "are only distinct in relation to their mutual entanglement"—well describes the kinds of assemblages, at once "real" and "imaginary," that stem from and are performed by the artists, participants, and onlookers alike in thislandyourland's performances (roles that may well shift in the course of the action).[23] As the pair put it, "We are provoking this action which becomes a reality, at the same time as it's completely surreal, fictional, and it creates a juncture of both, which somehow adds to our notion of reality."[24]

We could think of these events, then, as what Laddaga calls "gathering acts," which, despite their nature as prototypes for larger, more long-term networks and practices, lay no claim to a "universal" significance, as the modernist work of art did.[25] Rather, they remain focused on the specificity of their local context of occurrence, which they at the same time intervene in and transform by triggering unspecified and open-ended aftereffects. These aftereffects make many of their works difficult to confine and periodize within a given time and place; in fact, it is precisely these zones of transition, where the properly artistic or aesthetic touches on, and branches out into, other modes of practice, that constitute the ex-centric core of their work. Thislandyourland's projects zero in on "the construction of borderline objects that facilitate communication among the parts of the community," in Laddaga's poignant expression, on the condition "that at least some of those participating in these projects will make a non-artistic use of the resources they provide."[26]

THE RHIZOME AND THE WEAVE

Whereas the core of thislandyourland's work lies in critical engagements with, and departures from, the legacies and afterlives of landscape as a mode of picturing, and designing, spaces and places, including earlier iterations of land and site-specific art, the art of Ala Plástica intervenes on a different level of spatial imagination: the territory and its visual correlate, the map. "Who designs territories? For whom are they designed?"[27] These questions are a starting point of many of their interventions, which often open up or reclaim "local community centers as bases of resistance to the mega–construction projects being imposed on the region (dams, bridges, etc.)," and as initial triggers for "the creation of space and platforms for sharing information and generating ideas."[28]

Emerging from La Plata's alternative art scene in the late 1980s, Ala Plástica (initially comprising artist/educator Babich, lawyer/activist Meitin, and artist/agronomist Santos) has disputed the cartographic imaginaries of the kind of regional development that overwrites the entangled materialities of places in favor of an abstract, docile space awaiting the corrective interventions of bio- and geopower.[29] They did this by "hacking" the tools of governmental productions of space, putting satellite imaging, short-wave-radio broadcasting, and alternative energy grids to community-based forms of use, and by creating their own, "bio-regional" countercartographies that track environmental damage as well as the activist networks mobilizing to oppose it: from the city of La Plata, including its industrial coastline scattered with shipyards and oil refineries, to the Paraná-Paraguay river basins, under immense environmental stress from cellulose factories, agroindustrial toxins, and hydropower megadams, and to a still larger, subcontinental dimension, including soil erosion and the impact of glacial melting in the Andean uplands. "Estuary, Basin, and Hydroamerica, the subcontinent . . . all of this, simultaneously, was the territory of work, of conceptualization and expansion," as Alejandro recounts.[30]

At the same time, Ala Plástica's countercartographic practice has also been developed in a bottom-up rather than a top-down fashion, deriving as it did from on-site actions that gather "communities of interaction," in Alejandro's phrase.[31] It also involves close observation of, and alliance-making with, the "soft resistance of the natural order as it challenges the walls that try to contain it and . . . generates new territories of life."[32] This more-than-human territorializing practice, as observed in the modes of self-propagation and the artisanal use of California bulrush—an "emergent species" that takes root in polluted shorelines and marshlands, where it provides fibers for baskets and fishing utensils—is deployed in Ala Plástica's work in both material and metaphorical fashion, manifest in two elementary forms of expression: the rhizome and the weave.

In critical theory, the figure of the rhizome has enjoyed widespread success thanks to Gilles Deleuze's and Félix Guattari's famous introduction to their book *A Thousand Plateaus* (1980), where it features as an alternative to "arborescent," cause-and-effect relations. In botany, "running rhizomes" are the modified stems sustaining the asexual reproduction of plants such as California bulrush and other reed species, many of them native to the Plate River. These plants are able to extend both shoots and nodes at the same time as they absorb nutrients such as proteins and glucose, maximizing their resilience in unfavorable conditions. Rhizomes, crucially, are not a specific part of a singular plant organism (as roots are in nonrhizomatic plants) but are rather the connective/regenerative tissue common to entire populations that cannot be broken up into "individuals." Harvested and sun-dried, the shoots of California bulrushes also provide the natural fibers used in basket-weaving, a craft introduced to the Plate estuary in the late nineteenth century by Italian immigrants from the Veneto and Piedmont but also practiced by many river-dwelling Amerindian communities such as the Wichí and Tobá (Qom), originally from the Pilcomayo River in the Chaco region but displaced to the environs of La Plata through deforestation and enclosures for sugar and soy farming. As different but interconnected modes of networking, the rhizome and the weave provide Ala Plástica with a repertoire of gathering strategies that are at once tangible, deceptively simple, and abysmally complex.

In 1997, for instance, in response to a plan for a bridge across the Plate River to connect Colonia del Sacramento, Uruguay, to Punta Lara, Argentina—a trade-driven megaproject intended to create a new "Meta-Polis" that would have incrementally increased traffic and warehousing, causing long-term environmental damage to one of the last remaining native forests on the estuary—Ala Plástica occupied Punta Lara's abandoned train station, not far from where, a few years earlier, they had started their first bulrush plantings. Bringing to the space the weaving workshops it had been holding with local basket-makers since 1995, the group invited neighbors and allies from nearby La Plata to meet and discuss their hopes and anxieties about the megaproject and to pass on their experiences with local plants and earths for medicinal and culinary uses. Participants also learned about alternative approaches to land, such as permaculture, and about the hydrodynamic impact of the planned bridge on estuarine currents. These conversations sometimes led to temporary exhibitions and installations using fibers and recycled plastics. The squatters also edited a monthly newsletter, distributed locally and through activist networks along the coastline.

To the abstract space of transnational infrastructure-planning, then, Ala Plástica opposed the recovery of locality through multiple forms of learning and alliance-making. It took the resilient materiality of its vegetal allies—the *junco*, the reeds—not just as a "raw material" but as a co-custodian of land and water, and as a figurative model for the weaving together and amplification of networks of solidarity (fig. 2). Only two years later, this alliance-making proved urgent and effective when a massive oil spill, caused by a tanker accident close to the nearby town of Magdalena, devastated a coastal biosphere reserve and impacted the health and sanitary conditions of some 17 million inhabitants of the wider region. Thanks to the artist/activist networks forged through the Punta Lara occupation, Ala Plástica and its allies were able to quickly assemble an exhibition documenting the contamination and its impact on artisans and fishermen. The exhibition was shown locally and in the Netherlands, home to Royal Dutch Shell, the oil company responsible for the accident. Ala Plástica also coordinated a report that a coalition of NGOs later included in an indictment of Shell's environmental crimes that was submitted to its annual shareholders' meeting in 2002, the same year the company was ordered to pay $35 million in remediation. This was the highest penalty for environmental damages ever handed down in Latin America up to that point.

The starting point, then, was relatively straightforward: countering the capital-driven production of a smooth, geography- and ecology-defying megaspace with the occupation and refunctionalizing of an earlier abandoned infrastructure (the train station) for the purpose of creating community. Yet from this beginning a series of more and more complex and far-ranging actions ensued. The Punta Lara train station—like the bulrush plantings, previously—defined the initial, local departure point where an "aesthetic way of thinking and action" could intervene in sociopolitical and ecological realities, with ripple effects that could be public and explicit or intimate and subtle. As Alejandro reflects,

> Theories of complexity tell us that emergent complexity is the result of subjacent simplicity. And that's what happened: it was very simple, everything was very simple. Out of this simplicity, a path for action would open up. . . . many times, when we met up with people working in art theory, they would tell us, no, what you're trying to say is too complicated. But actually, in the way we associated with the reed pickers, the weavers, the basket-makers, communication was always very straightforward. It was always very easy to

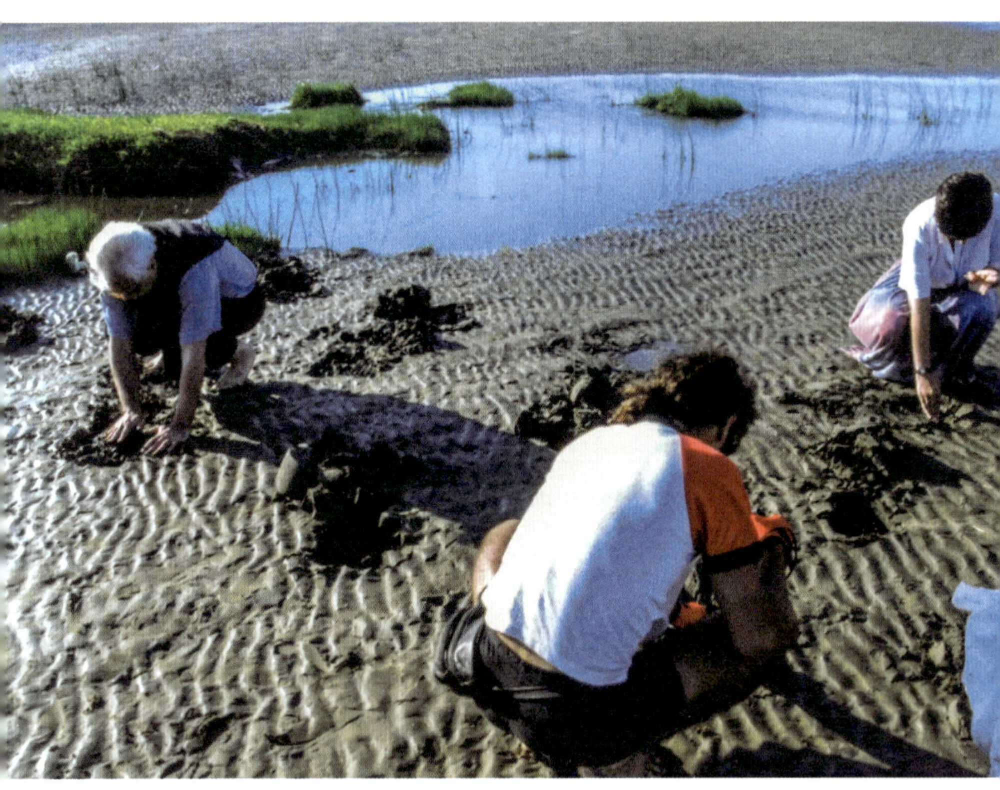

2 Ala Plástica. *Plantación de rizomas* (Planting rhizomes). 1995. Planting of bulrushes at the mouth of the Arroyo La Guardia in the Plate estuary, Punta Lara, Argentina

> convey what we were doing. And, in this simplicity, thought became transposed into action very quickly.[33]

At the same time, the fact that this "departure point" took the form of a network rather than a single origin—the rhizomatic organism of the bulrushes, the weave of the basket fibers, the train station as an (abandoned) connector between localities—also allowed for, and anticipated, subsequent processes of amplification onto a "bioregional" or "translocal" scale. Over the next decade and a half, Ala Plástica would "bring to its art innumerable practices, sometimes without them being fully enounced, in an uninterrupted continuity of constant production."[34] These would include, among others, the design of a solar-powered energy grid with inhabitants of a threatened island community near the La Plata oil refinery (*Ejercicio de Desplazamiento* [Displacement exercise], 2002); the construction of a "nomadic archive" in collaboration with artist and activist groups farther upriver, allied under the umbrella of the Red Delta del Paraná (Paraná delta network), to which Ala Plástica, together with the Ecuadorian sound-art collective Centro Experimental Oído Salvaje, contributed a low-frequency radio transmitter for the live broadcasting of on-site actions, interviews, and habitat recordings (*Territorio y radialidad* [Territory and radiality], 2009–12);[35] and the use of satellite imagery to document both the active, intense earth stewardship of the Ayoreo—one of the last nomadic Indigenous communities in the Chaco—and the encroachment of monocrop agriculture on their territory, a project developed in collaboration with the Indigenous-rights organization Iniciativa Amotocodie.[36]

At the same time, Ala Plástica used shows and residences abroad—among others, *Groundworks* (Miller Gallery, Pittsburgh, 2005), *Citizen Culture* (Santa Monica Museum of Art, 2014), and *Rights of Nature* (Nottingham Contemporary, 2015)—for site-specific projects. In 2008, for example, it intervened in a customs fence separating the working-class neighborhood of Wilhelmsburg, Hamburg, from the river Elbe, setting up a stepladder that supplied views across this urban border and a notebook in which residents and passersby could record what they had seen. In time this action sparked a residents' movement that eventually forced the city to remove parts of the boundary.

EXTENDED COMMUNITY AND MULTISPECIES AESTHETICS

Over the years, both thislandyourland and Ala Plástica have established strategic relationships with art institutions—museums, galleries, festivals. In addition to producing "archives" of their work (it is often only through museum exhibitions and the accompanying literature that their more ephemeral, site-specific interventions gain a documentary afterlife), these institutions have provided the support necessary to develop further actions, given the relative marginality of both groups to mainstream art and activism. As Alejandro remembers, "What we were doing was completely obscure to all our interlocutors. On the one hand, folks working within environmental dynamics couldn't conceive of the question of art as integral to their discussions. On the other hand, people in the field of art thought what we were doing was in Chinese, something impossible to understand."[37]

Rather than, in the words of art historians Sophie Halart and Mara Polgovsky Ezcurra, "assault those systems of representation that constitute the canon," as in older iterations of neo-avant-gardist "sabotage art,"[38] Ala Plástica and thislandyourland draw on the institutional circuits of art as "resources, which can be employed less for the purpose of individual appropriation than for extending collective networks."[39] And their relationship to the art institution can be less conflictive precisely because "art" is no longer the sole target of their interventions: once established (with or without formal support) through an action that shares characteristics with ambient or performance art, their "practices... unfold as social facts," in Alejandro's succinct expression. "Nowadays," he continues, "there's a number of processes that were seeded in these initial actions, which completely ignore these origins. For us, that's the best possible outcome, the fact that it's been transformed into a form of life, that a representation and an action should have encountered a resonance so that others, consciously or unconsciously, reproduce, amplify, and improve it.... they have become part of a network of territorializations."[40]

For this process of "decomposition" and "sedimentation" of the aesthetic into shared forms of life and conviviality to succeed, the work of Ala Plástica and thislandyourland relies on the organic materialities that their interventions bring into play—quite literally so, since a certain ludic, playful (or even "theatrical") element is central to their imaginings of different modes of togetherness in inhabiting city and country. Their works usher in a "what if" space and time that includes human and nonhuman actors—and the uncertain, indeterminable nature of their "actorship" is very much at the core of many of their actions—as well as networking and communications technologies proper to an itself already "posthuman" age. Pushing the aesthetic toward, and beyond, the species barrier doesn't mean that reeds, vegetables, or insects become subject to a shared experience of beauty—though can we be so sure?—but rather that they participate in a form of imaginative ecosocial practice that has the capacity for world-making (if on an infinitely small scale). But these worlds—or "experimental microspheres," in Laddaga's expression—don't simply restore damaged environments to their "natural order"; on the contrary, they forge "artificial forms of social life" beyond the landscape-form, mobilizing, through a mode of playful, theatricalized togetherness, an uprooting of the territorial belonging that is the shared condition of multiple identities (social, ethnic, species) under conditions of extractive capitalism.[41] And this "entanglement of boundaries and this wager on the unspecific," as Garramuño argues, also "offer figures of nonbelonging that conjure up images of expanded community."[42] As such, they also put forth an intriguing answer to the timeworn argument about art's capacity for absorbing even the most extreme acts of dissidence: what if this very permeability was, in the end, a form of resilience, showing us a way to survive?

1 Reinaldo Laddaga, *Estética de la emergencia. La formación de otra cultura de las artes* (Buenos Aires: Adriana Hidalgo, 2006), 261. All translations into English are the author's.

2 Alejandro Meitin, "Iniciativas artístico-ambientales de gestión comunitaria en el estuario del Río de la Plata," in José Esteban Castro et al., eds., *Territorialidades del agua. Conocimiento y acción para construir el futuro que queremos* (Buenos Aires: Fundación CICCUS, 2019), 230.

3 Grant H. Kester, *The One and the Many: Contemporary Collaborative Art in a Global Context* (Durham, NC: Duke University Press, 2011), 28.

4 Thislandyourland (Louise Ganz and Ines Linke), conversation with the author, February 14, 2022. Author's transcript, 10.

5 Nelly Richard, *Márgenes e instituciónes. Arte en Chile desde 1973/Margins and Institutions: Art in Chile since 1973* (Melbourne: Experimental Art Foundation, 1986), 53.

6 T. J. Demos, *Beyond the World's End: Arts of Living at the Crossing* (Durham, NC: Duke University Press, 2020), 18.

7 Ibid., 18–19.

8 Jason W. Moore, *Capitalism in the Web of Life: Ecology and the Accumulation of Capital* (London: Verso, 2015), 14.

9 Ganz, *Imaginários da terra. Ensaios sobre natureza e arte na contemporaneidade* (Rio de Janeiro: Faperj/Quartet, 2015), 11.

10 See Jens Andermann, *Tierras en trance. Arte y naturaleza después del paisaje* (Santiago de Chile: Metales Pesados, 2018), and Ángel Rama, *The Lettered City* (Durham, NC: Duke University Press, 1996).

11 Michel de Certeau, "Pratiques d'espace," in *L'Invention du quotidien. Arts de faire* (Paris: Gallimard, 1990), 141–42.

12 Meitin, conversation with the author, March 8, 2022. Author's transcript, 2.

13 Anna Lowenhaupt Tsing, *The Mushroom at the End of the World: On the Possibility of Life in Capitalist Ruins* (Princeton: Princeton University Press, 2015), 29.

14 Thislandyourland, conversation with the author, 1.

15 Ganz, *Imaginários da terra*, 13–14.

16 Jill H. Casid, *Sowing Empire: Landscape and Colonization* (Minneapolis: University of Minnesota Press, 2005), 4.

17 Robert Pogue Harrison, *Gardens: An Essay on the Human Condition* (Chicago: The University of Chicago Press, 2008), 45.

18 Thislandyourland, conversation with the author, 2.

19 Marcio Gibram, "Vila São Tomaz e Aeroporto," in *Outros Lugares. Ines Linke e Louise Ganz/Mónica Nador* (Belo Horizonte: Museu de Arte de Pampulha, 2012), 57.

20 Thislandyourland, conversation with the author, 4.

21 Ibid., 3.

22 Ibid., 4.

23 Karen Barad, *Meeting the Universe Halfway: Quantum Physics and the Entanglement of Matter and Meaning* (Durham, NC: Duke University Press, 2007), 33.

24 Ibid., 4.

25 Laddaga, *Estética de la emergencia*, 147.

26 Ibid., 279, 285.

27 Meitin, "Las pedagogías de lo anegado y el poder de lo performativo," in Meitin, Graciela Carnevale, Brian Holmes, and Colectivo Materia, *La tierra ~~no~~ resistirá* (La Plata: Casa Río, 2020), 26.

28 Meitin, quoted in Kester, "Otros-nosotros: An Interview with Ala Plástica," in Bill Kelley and

Kester, eds., *Collective Situations: Readings in Contemporary Latin American Art* (Durham, NC: Duke University Press, 2017), 264, 266.

29 According to Meitin, Ala Plástica was active between 1991 and 2016 and included a variety of other, location-specific collaborators and alliances with partner organizations. Its name (loosely translatable as "the visual arts wing") stems from the group's origins in a performance space they had occupied together with an underground theater group (the "Ala Teatral," or theater wing). Meitin, *Escape hacia la desespecialización. Las formas que adopta el arte de los bordes* (La Plata: Casa Río, 2021), 1; Meitin, conversation with the author, 1.

30 Meitin, conversation with the author, 8.

31 Meitin, *Escape hacia la desespecialización*, 2.

32 Meitin, quoted in Kester, "Otros-nosotros," 265.

33 Meitin, conversation with the author, 6.

34 Ibid.

35 See Vera Coleman, "Emergent Rhizomes: Posthumanist Environmental Ethics in the Participatory Art of Ala Plástica," *Confluencia* 31, no. 2 (Spring 2016): 92.

36 See Jennifer Flores Sternad, "Interview with Ala Plástica," *LatinArt.com*, July 1, 2007. Available online at http://www.latinart.com/transcript.cfm?id=88 (accessed July 20, 2022).

37 Meitin, conversation with the author, 6.

38 Sophie Halart and Mara Polgovsky Ezcurra, "Introduction," in *Sabotage Art: Politics and Iconoclasm in Contemporary Latin America* (London: I. B. Tauris, 2016), 3.

39 Laddaga, *Estética de la emergencia*, 201.

40 Meitin, conversation with the author, 6, 9.

41 Laddaga, *Estética de la emergencia*, 22.

42 Florencia Garramuño, *Mundos en común. Ensayos sobre la inespecificidad en el arte* (Buenos Aires: Fondo de Cultura Económica, 2015), 39.

CAMILA MARAMBIO

CREATURES OF THE FUTURE

Listening to the Song of the Bichas

"I felt a shiver of emotion run up and down my spine, and I don't know whether or not I realized then that the Amerindian flame was lighting up my heart."[1] In 1994, the Colombian artist Bárbara Santos, then seventeen, was traveling by bus from Arequipa, Peru, to Arica, Chile. On crossing the border between the two countries, the driver turned up the music, and as if in ecstasy, the passengers stood up and danced in the aisles to the sound of *cuecas* by the Chilean singer/songwriter Violeta Parra. Most were families of exiles returning to Chile, Bárbara tells me, and their combination of song and political passion touched her deeply.

Bárbara began her training as an artist when she returned from that trip to Chile. In her hometown of Bogotá, Colombia, she dedicated herself to studying the technologies of the emission, flow, and circulation of matter, light, and empty space. During her years as a student, she acquired the theoretical and practical tools for what she would soon discover to be her art: electronics. Far from the European circuits and the modernist modes of exhibiting art in galleries, she has worked with excombatants, forest dwellers, and the wise men and women of the Pirá Paraná, a tributary of the Apaporis River in the Amazon Basin.

There, in the Colombian jungle, under the tutelage of those with an intimate knowledge of the river, she received her real training, finding herself immersed in a creative vocation that "seeps into the pores as an indivisible collective force." This force, "contained in the land, makes the individual disappear." It showed her the path toward the transmission and protection of traditional knowledge, while she herself became essentially invisible. Dedicated to fostering connections and codesigning vehicles of this ancestral knowledge "under friendly protocols," Bárbara works to create systems, networks, and relationships that ensure the sovereign participation of the river peoples in the dissemination of their own knowledge, within and beyond their communities.

The first of these processes lasted ten years and resulted in *Hee yaia godo-bakari: El territorio de los jaguares del Yuruparí* (Hee Yaia Godo-Bakari: territory of the jaguars of the Yuruparí, 2015), a book written as a collective effort of the Tatuyo, Eduria, Barasana, Itana, and Makuna peoples and published by the Asociación de Capitanes y Autoridades Tradicionales Indígenas del Pirá Paraná (ACAIPI; Association of Indigenous captains and traditional authorities of the Pirá Paraná). Bárbara is credited as a coeditor. From 2005 to 2015, ACAIPI developed a plan to administer its territory and environment, to provide organization on a collective and regional scale, and to establish relations with state officials and other representatives of the outside world.

When Bárbara and I discussed the dangers faced by those fighting extractivism both material and immaterial in Amazonia, she stressed the importance of ACAIPI's legal structure. She is no stranger to violence and has been harassed and stalked because she rejects the logic of "ecocapitalism." Under this free-market economy, natural resources are regarded as capital, and Indigenous people's knowledge of the world is both suppressed and co-opted in order to administer their territory as a carbon bank, a mere repository of "credits" to be exchanged in the international carbon markets. "The falsehoods they use to greenwash in Amazonia—like the fake business of carbon credits and the dynamics of hegemonic power that prettify it, even as they stifle ancestral knowledge—are another instance of colonialism and violence against collective processes," processes that by their very nature are shaped completely differently from Western rhythms and technologies. While carbon is the building block of life, carbon credits are a financial instrument, a derivative of carbon as an underlying commodity that allows businesses producing carbon emissions to pay into certified climate-action projects (land trusts, conservation initiatives, and so on) in compensation for polluting. This instrumentalization of nature "greenwashes" organizations by making them appear more environmentally sound than they really are.

Bárbara and I wept together over the abuse hurled against collective processes we both have followed, something Violeta Parra also endured, as she tirelessly traveled the length of Chile to document rural life and its ecological and spiritual values. The pain we share is evident in Parra's final album, *Las últimas composiciones* (Last compositions, 1966), when marginalization already threatened to tarnish her splendid dream of opening a national university of folklore in her circus tent in La Reina, a neighborhood in the foothills of the Andes in Santiago, Chile, where she had been granted a plot of land.

Toward the end of 1964, Violeta abandoned the grand stages of Paris—spaces that she had boldly braved, exposing herself to the world in order to advocate for popular song—in the conviction that the struggle for *el buen vivir*, a dignified life, in Latin America had to be accomplished through both education and pleasure, and that it should be done at the foot of the Andes.[2] The price she paid for her effort was her own suicide.

As we talked together, Bárbara asked me, "How do we care for ourselves?" I thought of the epigenetic trauma inherited by all of Violeta's adopted children, all of us who aspire to a greater admixture of high and low, beautiful and ugly, rich and poor, human and nonhuman, and who suffer her disillusion because the class, race, and gender binaries set forth by heteropatriarchal colonialism are inescapable without a tremendous amount of pain, violence, and death. We fell silent, thinking/feeling through Violeta's daring, her placement of her body on the front line, about how she offered herself up almost pornographically to represent the forgotten, the obscure, the denigrated. After a while, it was Violeta herself who offered us the gift of her failure through her song:

El huma-
El humano está formado
De un espí-
De un espíritu y un cuerpo,
De un cora-
De un corazón que palpita
Al son de
Al son de los sentimientos. . . .

No entiendo los amores
Ai a yai
Del alma sola,

Cuando el cuerpo es un río
Ai a yai
De bellas olas.
De bellas olas, sí,
Que le dan vida;
Sí falta un elemento,
Ai a yai
Negra es la herida.

¡Comprende que te quiero
De cuerpo entero!

A huma-
A human is formed
From a spir-
From a spirit and a body,
From a bea-
From a beating heart
To the sound of
To the sound of feelings.

I don't understand the loves
Ai a yai
Of the solitary soul,
When the body is a river
Ai a yai
Of lovely waves.
Lovely waves, yes,
That give it life;
If an element's missing,
Ai a yai
Black is the wound.

Understand that I love you
With my entire body!

—Violeta Parra, "*De cuerpo entero,*" *Las últimas composiciones*, 1966

Unlearning the separation of body and spirit, I kept coming back to this issue with Bárbara and the other prophetic artists to whom I dedicate this essay—*bichas del futuro*, creatures of the future, who predict that the cure to relieve human pain exists and at the same time is not enough. Because the illusion that the world could be better is our weakness. We don't need to improve the world but to learn from our mistakes, so that in the future we will only make new ones.[3]

THE EARTH COVERS YOU

El Cuenco de Cera (The wax basin) came into being on Bárbara's initiative, serving as an independent platform, based in Bogotá, for organizing meetings, conversations, and exchanges in order to build bridges between the traditional knowledge of the Amazon and Western science. Working with a network of friends (artists, anthropologists such as Stephen Hugh-Jones, and scientists) and guided by traditional wisdom-bearers such as *Heegu* Reynel Ortega *Yai Hoa* and other representatives of the Tatuyo, Barasano, Macuna, Ticuna, and Yanomami peoples, El Cuenco de Cera meets the urgent need to tackle the crushing distance between—and hierarchies of—the so-called "real sciences" and traditional knowledge.

Creating a space that invites "dialogue around the uncertain, the complex, what isn't subject to control, the unmeasurable and unquantifiable," is a project to which Bárbara devotes herself with all her heart/soul/reason, like Violeta with her tent. Listening to her speak is as nourishing and clarifying as listening to Violeta's lyrics: both generate words as if a nonhuman power were whispering in their ear. Like transmitters, they know how to hear or sympathize with an original sound, achieving in their sentences the resonance of a primal call that renounces the progressivist idea of artistic autonomy.

El Cuenco de Cera asks such questions as: is there a relationship between the patterns woven into caraná-palm hut roofs and digital data processors? This provocative contemplation of the ancestral, the rural, and the working class as a matrix of sophisticated and complex knowledge must be understood as an artistic practice that connects and rechannels energy to those forms of knowledge that sustain the world and that are the source of any possible future well-being.

Guardians of remembrance and rebels against forgetting, Violeta and Bárbara mediate with death. In her song "*Gracias a la Vida*" (Thanks to life, 1966), Violeta thanked life for giving her "an ear that night and day records crickets, canaries, hammers, turbines, barking, cloudbursts, in all their range," as well as for the "sound and alphabet" with which she dedicated herself to preserving and sharing Chilean folklore through the hundreds of songs that she recorded in her journey through the countryside. Unlike Violeta, Bárbara received lessons in ethical recording practices, faithful archiving, and "healing as technology" from the wise women and men of Amazonia, and she transmits their hidden ways of knowledge at their request.[4]

"For the sake of the land, which needs protection," the women of Pirá, led by Rosa Marin, Hilda Marin, and Rosalía León, seek to be recognized and guarded in a network that Bárbara cultivates along with the anthropologist Carolina Duque and the photographer Sergio Bartelsman. The mending of relationships that Bárbara practices requires rigorous training and specific tools. Communing with the sources of life is no joke and much less a holistic whim; it is a struggle that demands a radical attitude, a sweet gentleness, and subtle science. These are the raw materials of Bárbara's work.

In *~Kii RɨKɨ Bare Hako. El Conocimiento ANCESTRAL de las Madres de Semilla (~Kii RɨKɨ Bare Hako*: ancestral knowledge of the seed mothers),[5] the knowledge of the women of the Pirá is preserved and protected by using virtual reality to transmit information to younger generations of girls. Moving through Organizmo, the center for research and training in regeneration and intercultural knowledge on the outskirts of Bogotá, where I visited Bárbara in 2022, I hear the songs of the women of the river and see their drawings. They emerge from my cell phone screen when I scan a QR code printed in booklets inserted in *~Kii RɨKɨ Bare Hako*. I see the cassava and breast milk and I surrender to their power. Cassava is here, alive, because there are women who grow it in their plots, who pray to it, and who, in so doing, care for us tangentially as well. I recognize this kind of service: it is that of the doula, the midwife who cares for the mother, not the baby. Caring for the mother is an art, as is valuing the slut.

LA PUTA, THE SLUT

> For me, the place of *la puta* is that of the lover of life, as we called her during the October workshops in Bolivia.
>
> The whore is the one who holds the keys and mysteries for disabling the violent body of rapists, blackmailers, and hypocrites. Because just as the housewife can gather all her knowledge of life and give it back as something fundamental to humanity, just as the lesbian can collect all her knowledge of her body and give it back to all women, so can the whore collect all her knowledge of the violent, prostituting other and give it back to women. In her and in her rebellious self, many things can become clear. If she can disable the mechanisms of reification that trap her body and her pleasure, her work will rain down on all of us and soak us with fresh water.[6]

These words of the writer, creator, graffiti artist, radio producer, videographer, activist, and out queer María Galindo sound to me like a public-health manifesto that I would happily sign. Based in La Paz, Bolivia, María studied psychology in Italy, but her multiple artistic languages reflect above all her identity as a Latin American agitator who challenges authority. Her stage is usually the street, where she invades the social imaginary through murals, monumental sculptures, dramatic actions—transvestism, song, dance, theater, performance—and poetic graffiti (fig. 1). All of these creations are attributed to Mujeres Creando (Women creating), the Bolivian anarcha-feminist movement of which María is a member.[7]

On Monday September 4, 2022, I stand at the door of La Virgen de los Deseos (The virgin of desires), the home of Mujeres Creando. María isn't there, she's off in Berlin, but another woman invites me in. They're not serving lunch today but there are women chatting in the kitchen; I see them through the stairwell. The woman working at the little store outside, which sells office supplies and toiletries, lets me look at the books on the shelves in the lobby. Among T-shirts and banners are books published by Mujeres Creando: *Machos, varones y maricones. Manual para conocer tu sexualidad por ti mismo* (Machos, men, and queers: manual for knowing your

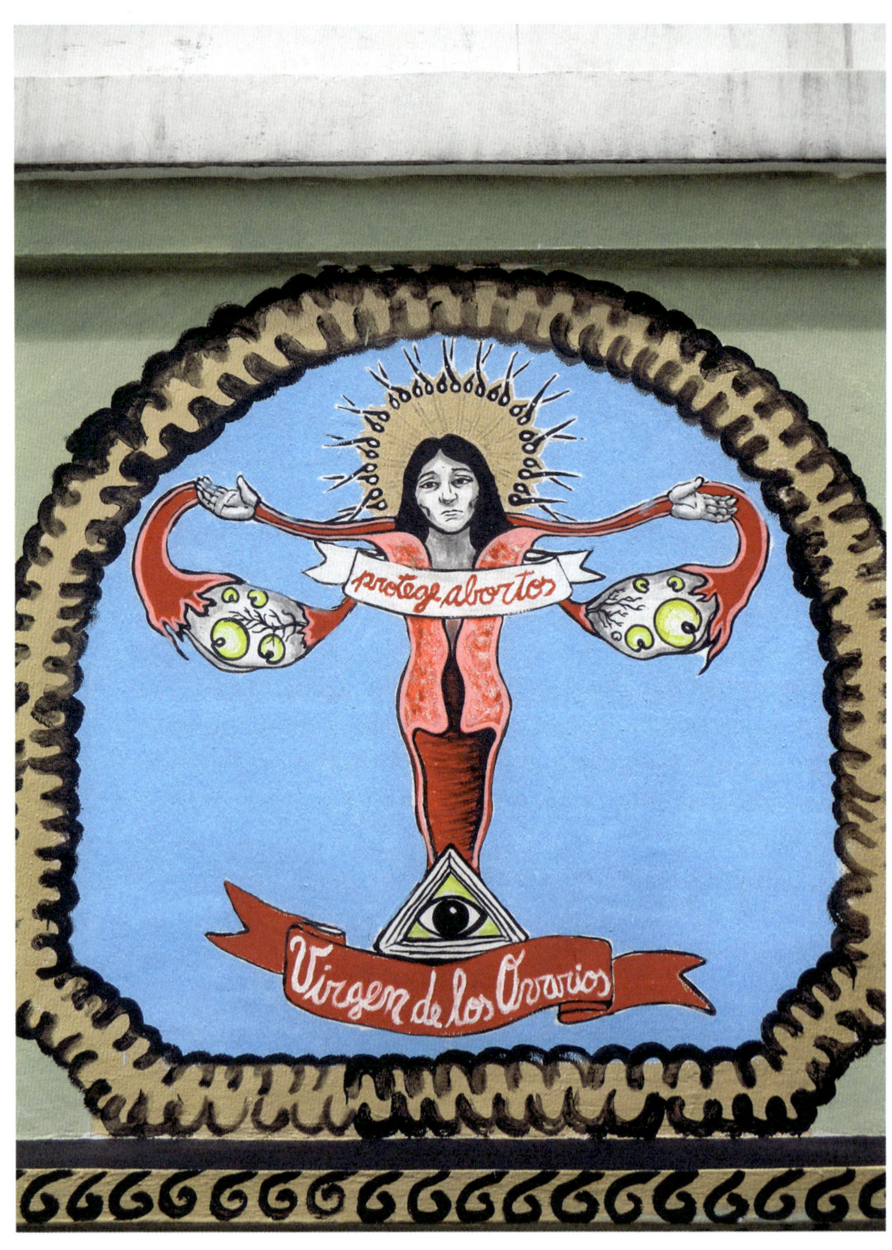

1 Mujeres Creando (María Galindo, Esther Argollo, and Danitza Luna). *Virgen de los Ovarios* (Virgin of the ovaries), detail of *Milagroso Altar Blasfemo* (Miraculous blasphemous altar). Mural, c. 16 ft. 5 in. × 45 feet 11 in. (5 × 14 m). Part of the installation *La Intimidad es política* (Intimacy is political), shown at the Centro Cultural Metropolitano, Quito, Ecuador, July–August 2017, where it was subject to censorship, and on other occasions in Bolivia (where it was also censored) and Chile

sexuality for yourself, 2001), *Así como tú me quieres, yo no quiero ser de ti* (The way you want me, I don't want to be yours, 2006), *No se puede descolonizar sin despatriarcalizar. Teoría y propuesta de la despatriarcalización* (You can't decolonize if you don't depatriarchalize: theory and proposal of depatriarchalization, 2013), *No hay libertad política si no hay libertad sexual* (There is no political freedom without sexual freedom, 2018), *Soy lo prohibido* (I am the forbidden, 2019), and *Feminismo bastardo* (Bastard feminism, 2021). Most of these books have "a concrete, direct, simple style," writes the Peruvian curator Viola Varotto. "While maintaining a style of prose accessible to a diverse readership, they still have considerable philosophical and intellectual content. They include texts on the history of the Mujeres Creando movement, pedagogical manuals, manifestos, and syntheses of their research."[8] Once I had chosen a book to take with me, my eyes drifted toward the walls, where I recognized the work of the Colombian muralist and graffitist Bastardilla, whose sinuous lines shape the bodies of voluptuous women, sex workers and sluts, free, liberated, and liberating: "The place of the whore is as the host of social change."[9]

"I'm 'downwardly mobile' because even though I was born into an upper-middle-class family, the direction I took in my life was not to move up in the world but to socially move down. I'm going down, and best of luck to those going up."[10] I savor these sentences, feeling/thinking how boldly they defy the notion of progress that we Latin American women have so thoroughly swallowed. Having grown up in the United States until I was sixteen, I know all too well the practices and consequences of the drive to climb the social ladder. That's why I was so moved to read Viola Varotto's argument that María Galindo's audiovisual work demonstrates "her capacity and desire for a certain conceptual, ideological, and class 'mobility,'" and that this desire must be understood as reinforced "upon her return to Bolivia following her experience in Europe as a student and worker. This inaugurated a new moment in what she describes as a 'birth without parents.'"[11] Divesting oneself of Europeanness, even when one lives or works in Europe, always seems to inaugurate *cuir* (queer) movements, changes of skin that give rise to a demand for multiplicity as an aesthetic and political act, nurturing the natural right to be many and to rename oneself how and when one wishes.

Like María Galindo, the late Guatemalan artist Margarita Azurdia was self-engendered (fig. 2). She called herself by many names: Margot Fanjul, Anastasia Margarita, Margarita Rita Rica Dinamita, and others. In 1982 in Guatemala City, Margarita Rita Rica Dinamita cofounded the Laboratorio de Creatividad (Creativity laboratory), a group that improvised performances in cafés, plazas, and other public spaces, shunning the solemnity of established art spaces such as museums and galleries. The group provoked extreme situations in response to the climate of cultural censorship in Guatemala after the country had endured massacres, military counterinsurgencies, and coups d'état. Repudiating her parents by dropping her surname, Margarita left behind the colonial heritage of social climbing. The scholar and curator Rosina Cazali writes, "After her time in the Laboratorio de Creatividad, Azurdia began to combine corporeal experimentation with idiosyncratic practices of a spiritual and healing nature. She gave the name 'rituals' to a whole series of practices and reflections centered on the search for processes that affected the freedom of the feminine. They also expressed her constant preoccupation with the deterioration of the environment as a problem both local and global. These rituals were performed in exhibition spaces or in her home, alone or with small groups of mostly women."[12]

Unable to ask Galindo whether she would consider Azurdia a proto-anarcha-feminist (certainly less anarcha and more feminist),[13] I am left with the title of the book I chose at La Virgen de los Deseos: *No hay libertad política si no hay libertad sexual*. The book is the result of an investigation sponsored by the Bolivian vice-president's office after the homophobic slurs against Mujeres Creando issued by former legislator Daniel Rojas, who was the head of the political party Movimiento al Socialismo (MAS) at the time. The investigation, carried out in 2015–16 through extensive interviews within the Bolivian Plurinational Legislative Assembly, was

> a subversive mechanism that turned a public offense against [Mujeres Creando] to our advantage: operating not from the usual stance of victimization, but from our ability to challenge the entire Parliament to broadcast its homophobic justifications through interviews that we used in exactly

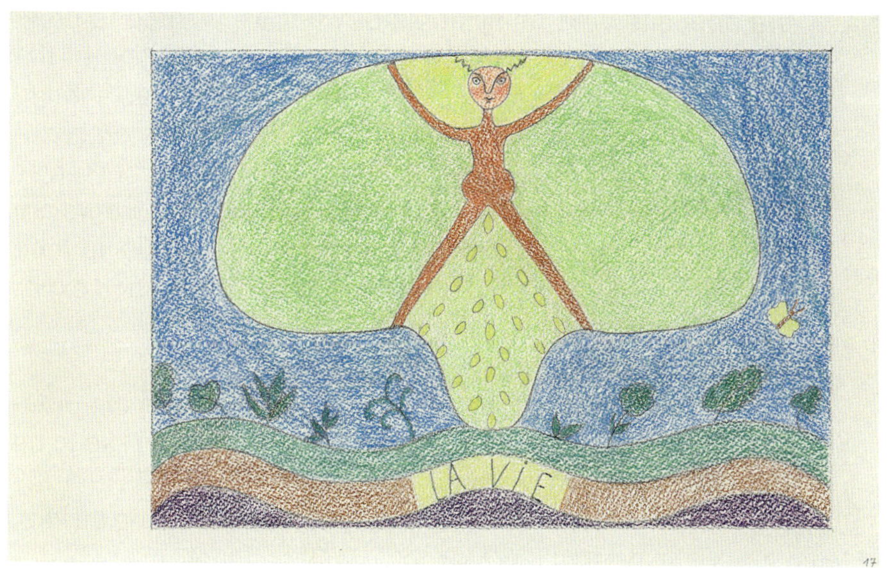

2 Margarita Azurdia. *La Vie* (The life), page from the artist's book *Rencontres par Margarita Azurdia*. 1980. Graphite, wax crayon, and ink on paper, 8 ⅝ × 14 ⅛ in. (22 cm × 36 cm). Collection Milagro de Amor

> the opposite way, as symbols of what we wanted to challenge and change. The result was a demonstration of the legislators' lack of reflection. This investigation was nothing more nor less than a politicization of the debate on sexuality, homosexuality, and the body. We allowed ourselves the possibility of moving away from the bench where we are constantly questioned to act instead as people using their sexualities and bodies to question the sexualities and experiences of our bodies in our society as a whole.[14]

Proposing a connection, however tenuous, between the practices of Margarita and María is a provocation that I would like to be understood as a seedpod I'm hurling into the middle of this text. My Latin America is not identity based. Here, I am constructing an idea of Latin American art that has little to do with nationalisms or historicity; rather, I am trying to trace anomalous lineages that delineate a field where there is room for artistic practices that combine corporeality, sensuality, and political emancipation. Whoever we are, have been, or were engendered as, we know very well that we are sick. Within and without, in our territories and our diasporas, we suffer the triumvirate of colonialism, alienation from the Earth, and neoliberalism, a triple patriarchal conjunction that has gestated what is now known as Latin American art.

Margarita Azurdia has a graphic poem called *Todo es una* (All is one, 1992). As I examine it, I think of how women's knowledge—intrinsically related to menstruation, to fertility, to being able to birth and unbirth things from and with our bodies—has been held in contempt until now, insistently, by calling us witches, hysterics, separatists, madwomen.[15] And then I write my own poem:

> Seated on the egg body
> I hear the critters sing.
> They traverse me.
> I am they, and I give them my voice.
> All this making and unmaking
> exhausts us.
> On the verge of being killed,
> supported by the vibration,

we continue to create life forms.
It's coming,
from the rot
rises
the ceiba tree,
the seed,
the bitch,
the aquatic phenomenal.

Written in cursive and without ornamentation, one of Mujeres Creando's street poems reads in part, "*Ni la tierra ni las mujeres somos territorio de conquista*" (Neither the Earth nor women are territory for conquest). This opposition to imperialism and patriarchy, this active resistance to the ravaging of the soil and the control exercised over women's bodies, is fundamental to the ecofeminist movement. And although María does not call herself an ecofeminist, I see her work as a recipe for escaping the jaws of ecocide and femicide plaguing the continent.

Since the imperialist patriarchy attacks anything that threatens its regulating hegemony, it is no surprise to discover that the gestation of (eco)systems that enable and protect the survival of (bio)diversity lay at the origin of the founding of Mujeres Creando, in 1992. As an activist in the Bolivian political party Organización Política Militar (OPM), María demanded greater recognition of women as part of the party's project.[16] When her petition was denied, she left OPM. María is skilled at refusing to be intimidated.

Since 2007, Mujeres Creando has been broadcasting critical discussions twenty-four hours a day, seven days a week, over its radio station Radio Deseo (Radio desire).[17] Listening to María on her program *Barricada* is dramatic and inspiring; her character is indomitable and she asks spontaneous and uncomfortable questions, patiently insisting on the healing potential that can emerge from the most basic encounter of bodies confronting one another without a script.

POETHICAL MEDICINE

"Bring a stone," I text Valentina. "I was looking at the ground on my way over here and I didn't see one. Maybe you'll have more luck."

I am waiting for her in front of New York's Museum of Modern Art (MoMA), on West 53rd Street in Manhattan. The museum is not yet open this morning, but the guards have been told to expect us: they know we will go to the gallery where the Brazilian artist Lygia Clark's *Casulo no. 2* (Cocoon no. 2, 1959) is on view. This work of art is hinged both physically and figuratively: the little piece not only folds but marks a pivot in Lygia's work, from the wall to the relational field, from geometrical abstraction to participatory experimentation, from the work to be contemplated to the work to be performed. In the same year that she made *Casulo no. 2*, Lygia wrote, "If I work . . . my reason is more than anything to achieve myself in the highest ethical-religious sense"—this in a time-traveling letter addressed to the artist Piet Mondrian (who had died a decade and a half earlier).[18]

Looking at the work, sitting on a blanket we laid on the floor in front of it, Denise Ferreira da Silva, Valentina Desideri, and I—accompanied by Inés Katzenstein, MoMA curator and director of the Patricia Phelps de Cisneros Research Institute—begin to read it as if it were a tarot card, heavy with symbolism, a messenger of the future. Denise and Valentina have collaborated as artists for more than a decade; Denise is Afro-Brazilian, Valentina is Italian, and together we embody a neologism of ecofuturist Latin American art. In 2011, Denise was already reading the tarot and Valentina was reading palms. Both used tools of the so-called "occult arts" in expanded forms, directing them toward social action, for example by asking the tarot how to support the cause of Haitian emancipation.

The first major arcana card of the tarot, the Fool (*el loco* in Spanish, *il matto* in Italian), is numbered 0. Zero suggests an embryo, an egg, on the verge of giving birth. The cards that follow, from 1 to 21, represent the forces, virtues, vices, and archetypes unfolding in the Fool's journey through the world. He is usually portrayed on the card as a pageboy on the edge of a precipice, at the point of leaping into the void, into the future of a preexisting world. His appearance in a traditional tarot reading is generally seen as implying a new beginning, an infinite potential, or the resting point before a new journey or undertaking.

In the deck that Denise and Valentina are creating together, the Echo Tarot, they read the Fool as metamorphosis, a process of material and elemental transmutation that holds the potential for a new way of thinking in tune with what Denise and Arjuna Neuman call "deep implicancy."[19] By applying Reiki, astrology, and tarot reading to sociopolitical problems on a global scale, Denise and Valentina integrate bodies, emotions, inclinations, and spirituality into the processes of producing knowledge and vision.

Not long before they met and began studying together, in Naples, Italy, using a method they would later call "Poethical Readings," Valentina had created two decks of cards containing "made-up" instructions to be used when one person wants to help another to heal. These instructional cards, which today circulate freely online, for anyone to download and adapt, emerged from two modes of relational art that she practiced under the names "Fake Therapy" and "Political Therapy," in which she explored how a group of people (or, in the case of Political Therapy, a couple) can heal each other without necessarily having any specialized knowledge.[20]

From this position of deliberate nonmastery, Denise and Valentina continued to explore ways of collectivizing their subaltern practices, inaugurating another format they called the Sensing Salon. For years they held this salon—a kind of dialogic laboratory—whenever they exhibited their work, but in 2018, tired of having to adapt the practice to the institutionality of the art world, they decided to turn it into an annual event held in the former convent occupied by the Performing Arts Forum (PAF) in the small French town of Saint-Erme-Outre-et-Ramecourt, two hours from Paris.

PAF is a communal experiment that is difficult to summarize and describe, but for the purposes of this essay, let's say that it is a place for meetings and creative production where there are no slaves; whoever creates, decides; and desire and responsibility are self-administered. The only rules of coexistence in the building, which holds up to 120 people, are that those who spend time there pay a small fee to become members for the year, and that under this "contract" they have the right to live there, again for a small fee, and to use the spaces to develop their practices, trying to leave no trace, to respect asymmetries among the people there at any given moment, and to facilitate experiences for others.

In the 1970s and '80s, Lygia Clark produced a series called *Corpo coletivo* (Collective body), a step toward the deconstruction of the autonomous artwork and the creation of works made communally by groups of participants. Through these explorations and her experience of psychotherapy, she delved into issues of public health and developed an artistic/therapeutic practice, applying a system she called *Estruturação do Self* (Structuring the self) that involved the use of "*objetos relacionáis*" (relational objects) activated on the body of the patient to recover exiled sensory capacities and stimulate healing. A relational object might be a pen, a plastic bag full of water, a wrinkled mat, or pillows filled with sand, seeds, or air, but Lygia always asked the patient to hold a stone in their hand so as not to lose themselves completely and to reconnect after the therapeutic journey.[21]

On her path from geometric abstraction to healing, Lygia left exhibitions behind and began to treat her patients in her office, a little room she set up in her Rio de Janeiro home. This change of direction, which distanced her from the art world, resembles the shifts of Violeta Parra and Margarita Azurdia: bored and fed up with art institutions, all three women left Paris (no more than four years apart; could they have met?) and came home, to heal their bodies, their souls, and their ancestral memories.

Let us linger a moment on the houses of collective healing that unite not only Violeta, Lygia, and Margarita but all the artists brought together in this essay. I began writing without knowing where I would find their point of convergence, and here it is: the collective house of healing, the house of welcome, the educational center for cultivating the care of the body, the soil, diversity, pleasure, orality, reciprocity. Removing themselves from the idea of the right to one's own house—a foundation of capitalism—Bárbara, Violeta, María, Margarita, Denise, Valentina, and Lygia aspire to the regeneration or, better yet, the autonomous reconfiguration of the world from a shared house. House, body, and healing are the major technologies with which to liberate the embodied knowledge of the flux between death and life and thus recover stolen dignity.

Reestablishing the communal house as a place for practicing the healing arts is a radically futurist gesture because it proposes to invest not in "progress" but in curing the sicknesses produced by racism, homophobia, transphobia, xenophobia, exile, the addiction to consumption, and the need to be in charge and dominate, to name a few. The tent house, the *Organizmo* house, the forum house, the basin house, the study house, the process house, the kitchen house, the discotheque house, the whorehouse, the song house, the retirement home—all rebel against museumification and propose a better house for art.

Latin American art is hybridization and service; it changes and adapts itself to the needs of healing. The system of Western art violates this principle, trying to dictate the quality of artworks, demanding visuality, visibility, specific identity, objectification, and above all subjectivity. These women, *bichas del futuro*, are doing something different: they become mediums, transmute energy, translate languages, lend each other clothes, disappear into the text and reappear on the walls. They broadcast on the radio, disrupt the senate, dance without stopping, go crazy, heal.

The Echo Tarot deck created by Denise and Valentina was intended for shared readings and inspired by the poems of Ai Ogawa. The first card (fig. 3) was designed by the Peruvian artist Arely Amaut. In the hands of Denise and Valentina, each reading decomposes and recomposes a situation, reconfiguring any notion of the subject in question. I was invited to play the Fool. I liked the role—I always identified with that androgynous page and their adventurous attitude—but this figure is in crisis. The crisis is poetic: it challenges the idea that the Fool is human. What is yet to be born isn't interested in recovering an earlier form; it is a vibrational body on the point of exploding into an infinity of differences, but it senses and ponders an ethical criterion in this expansion.

This is why I invited Denise and Valentina to look at *Casulo no. 2*. Detaching itself timidly from the wall, the work is in the beginning stages of a transformative displacement, a metamorphosis, an erasure of binaries. White and black blend into one another; the shadow appears. This larval state presages the birth of the *Bichos* (Critters), the series of artworks/creatures that would consume Lygia for the next five years. *Bichos* screech and squeal at birth.[22] The threatening screech of the unknown, of the "other," is the sound of something new awakening and suffering pain in order to be. The single ego doubles itself and becomes other, another, many.

In the Aymara, Quechua, and Toba cultures, it is said that the past is ahead of us while the future is at our backs.[23] To think of time in this way makes me look backward, to what has already been said. I see the artistic duo of Denise Ferreira da Silva and Valentina Desideri, then farther back, the anarcha-feminist María Galindo, and in the distance, Bárbara Santos and the women of Pirá. Following the threads leading back from them, my perceptual field and I, myself, expand, and I see the bodies of three women now deceased: the great Chilean folklorist Violeta Parra; the extravagant Guatemalan artist Margarita Azurdia; and the Brazilian healer Lygia Clark.[24] Their practices differ greatly, both among themselves and from those of the women living today, yet like their precursors, they exist at the margin of what is commonly considered (Latin American) art. It may be because they do not participate in museum circuits, because their works are understood as activism rather than art, or because, in Denise's and Valentina's case, they do not live in Latin America and are seen as "global artists." True, they are; but if there is an argument in this essay/journal/narrative of my trips to Bogotá, La Paz, and New York to conspire with them, it is that their practices are genuinely Latin American in the deepest sense: rooted in a cosmological conception of time, close to the transcestors, conscious of the sensuality of bodies and of their erotic and dissident powers, powerful adversaries of historical and contemporary abuses.[25] Toughened by struggle, they fiercely persist, metabolizing the dream of an inclusive, biodiverse collectivity, beyond politics, beyond the human.

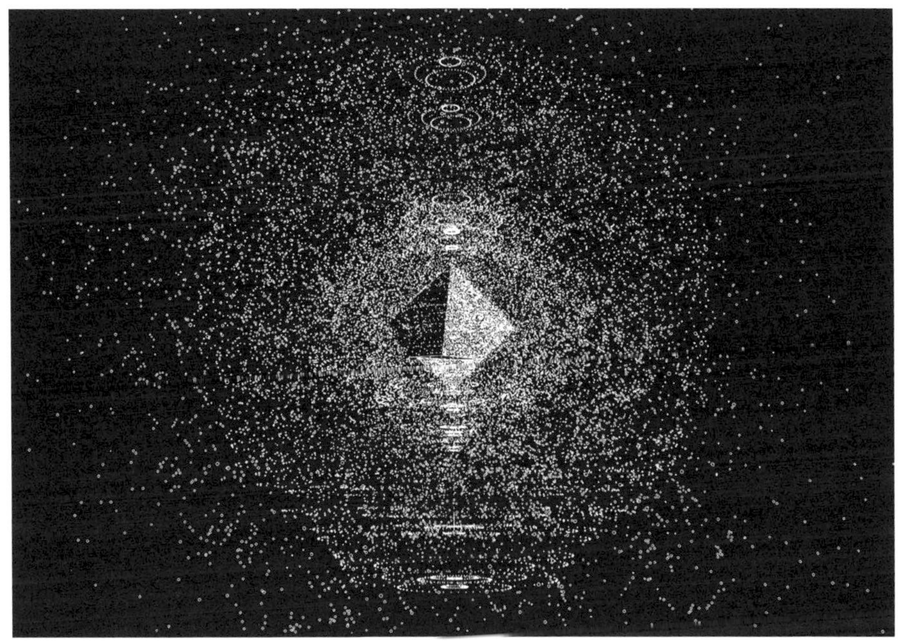

3 Valentina Desideri and Denise Ferreira da Silva. *Reading with Echo: Study for the Fool*. 2023. Digital image designed by Arely Amaut

1 Bárbara Santos and I talked for days when I visited her in Bogotá in August and September 2021. All of her quotations that follow come from notes I took while we cooked, weeded, laughed, cried, and danced.

2 See Catherine Boyle, "Violeta Parra and the Empty Space of La Carpa de la Reina," in Lorna Dillon, ed., *Violeta Parra: Life and Work* (Martlesham, Suffolk: Boydell & Brewer, 2017), 173–88.

3 See Vanessa Machado de Oliveira, *Hospicing Modernity: Facing Humanity's Wrongs and the Implications for Social Activism* (Berkeley: North Atlantic Books, 2021).

4 "Healing as technology" is the title of Bárbara Santos's own first book, *Curación Como Tecnología: Basado en entrevistas a sabedores de la Amazonia* (Bogotá: Idartes, 2019), a research project into technologies and ancestry sponsored by the mayoralty of Bogotá.

5 The book Bárbara was working on at the time of this writing in 2023, commissioned by the women of the river.

6 María Galindo and Sonia Sánchez, *Ninguna mujer nace para puta* (Buenos Aires: Cooperativa de Trabajo Lavaca, 2007), 192–95.

7 On Mujeres Creando see *La Virgen de los deseos* (La Paz: Mujeres Creando, 2005) and "Territorial Re-Connections: How Can We Generate New Utopias from the South?," a conversation between Catarina Duncan and Mujeres Creando members Julieta Ojeda and Danitza Luna. A project of the Patricia Phelps de Cisneros Research Institute for the Study of Art from Latin America, The Museum of Modern Art, New York. Available online at https://www.youtube.com/watch?v=dR1GxQzzoM4&t=7s (accessed January 7, 2023).

8 Viola Varotto, "Ay de mí que ardiendo . . . *¡puedo!* Notas extensas sobre el cine bastardo de María Galindo," 2021, in Diana Coryat, Christian León, and Noah Zweig, eds., *Small Cinemas of the Andes: New Aesthetics, Practices and Platforms* (Cham, Switzerland: Palgrave Macmillan, 2023), 239.

9 Galindo and Sánchez, *Ninguna mujer nace para puta*, 192–95.

10 Galindo, in Pamela Valdéz, "Feminismo Bastardo. Entrevista a María Galindo," *re-Vista* (La Paz) no. 2 (May 17, 2020). Available online at https://mujerescreando.org/wp-content/uploads/2020/05/re-VISTA-No2.pdf (accessed January 7, 2024).

11 Varotto, "Ay de mí que ardiendo . . . *¡puedo!*," 240.

12 Rosina Cazali, "Margarita Azurdia: Vida y Obra," *ArtNexus* no. 110 (September–November 2018). Available online at www.artnexus.com/es/magazines/article-magazine-artnexus/6042bb38d7dac6104e1b9ab2/110/margarita-azurdia (accessed January 7, 2024).

13 See Galindo and Sánchez, *Ninguna mujer nace para puta*, 192–95.

14 "'No hay libertad política si no hay libertad sexual.' María Galindo, Mujeres Creando," *Traficantes*, February 27, 2020. A statement on the occasion of a book launch for *No hay libertad política si no hay libertad sexual*. Available online at https://traficantes.net/actividad/%C2%ABno-hay-libertad-pol%C3%ADtica-si-no-hay-libertad-sexual%C2%BB-mar%C3%ADa-galindo-mujeres-creando-bolivia (accessed January 7, 2023).

15 *Todo es una* (All is one) is from Azurdia's book *Iluminaciones*

(Guatemala: Las ediciones de Margarita, 1992).

16 I thank Varotto for stressing the importance of this political context in the essay cited above. Her curatorial and academic work has helped me to see more deeply into the social nuances of María's work, which I was unable to recognize on my own when I visited La Paz.

17 Available online at https://radiodeseo.com/ (accessed February 1, 2024).

18 Lygia Clark, "Letter to Piet Mondrian," 1959. Quoted here from Cornelia H. Butler and Luis Pérez-Oramas, *Lygia Clark: The Abandonment of Art, 1948–1988*, exh. cat. (New York: The Museum of Modern Art, 2014), 59.

19 Denise Ferreira da Silva and Arjuna Neuman introduced the term "deep implicancy" in their film *4 Waters—Deep Implicancy* (2019). It refers to a way of thinking that takes responsibility for global issues such as migration, displacement, colonial inheritances, and ecological devastation. Deep implicancy is an embodied ethics that confronts the violence of abstraction.

20 The cards can be downloaded at https://faketherapy.wordpress.com/2011/10/30/cards-deck/ (accessed February 1, 2024).

21 I thank the critic Santiago García Navarro for meeting virtually with me, Denise, and Valentina right after our reading of *Casulo no. 2*. Among many things we learned, he confirmed the use of the stone as a "reality test," a key part of the therapy's process of integration.

22 In "El híbrido de Lygia Clark" (Lygia Clark's hybrid, 1996), the Brazilian psychoanalyst Suely Rolnik writes of the relationship between the *Bichos* and screeching, an idea she develops on the basis of a letter Clark wrote to the critic Mario Pedrosa: "How many beings am I in order always to go seeking in the other being that dwells in me the realities of contradictions? How many joys and sufferings has my body offered to the other being that is secretly within my Self, opening itself like a gigantic cauliflower? In my belly dwells a bird, in my breast, a lion. The latter doesn't stop walking from here to there. The bird screeches, grim reaper, and is sacrificed. The egg continues to envelop it, like a shroud, but it is already the beginning of another bird that is born immediately after death. There's no Interval. It's the feast of life and death interwoven." Clark, quoted in Rolnik, "El híbrido de Lygia Clark," 1996, repr. in *Mundo Performance: Plataforma de Investigación y Creación Alrededor del Arte de Performance*, May 28, 2021. Available online at https://mundoperformance.net/2021/05/28/el-hibrido-de-lygia-clark/#_ftn2 (accessed January 10, 2024).

23 See Gabriel Luis Bourdin, "En los tiempos de Ñaupa. El cuerpo y la deixis temporal en lenguas originarias de Sudamérica," *Península* 9, no. 1 (Mexico City: Instituto de Investigaciones Antropológicas, Universidad Nacional Autónoma de México, January–June 2014), 33–58.

24 As I was finishing this essay, the sky collapsed around me and everything became rain here in the heights of the Sierra de Cayey, on the island of Borikén (the Taino name for Puerto Rico), where I now live and where I first met the Cuban curator Tamara Díaz Bringas, to whom I dedicate this essay. A couple of years before her death, Tamara told me, "Latin America is in the future." Or did she predict: "Latin America is in your future"? The storm is drowning

my memory, and now I will never know for sure whether she knew Bárbara Santos's books on "occult culture," whether she had the good fortune to participate in the divinatory practices of Denise Ferreira da Silva and Valentina Desideri, whether she heard one of María Galindo's *Barricada* programs on Radio Deseo. No matter—I'm happy to feel/think that, like me, she would have admired them, and that we wrote this text together.

25 In the essay "How to Tame a Wild Tongue," Gloria Anzaldúa writes, "Deep in our hearts we believe that being Mexican has nothing to do with which country one lives in. Being Mexican is a state of soul not one of mind, not one of citizenship. Neither eagle nor serpent, but both. And like the ocean, neither animal respects borders." In Anzaldúa, *Borderlands/La Frontera: The New Mestiza* (San Francisco: Aunt Lute Books, 1987), 42. As a Chicana poet and scholar, Gloria advocated for the decolonization of borders—between nations, classes, genders, languages, and species—and proposed radical new ways to conceive of who is included when we say *nosotras* (third-person plural, feminine) *latinoamericanas*.

COATL TIME

Learning to Live in an Era of Mass Extinction

In the second year of the pandemic caused by the COVID-19 virus, at some slightly decreased peak in contagiousness, I went to a lunch at a friend's house. I knew all the other guests except for one woman, who had a certain intensity that frightened me—which says less about her than it does about my fragile emotional state. I spent the afternoon dodging gazes and questions. I kept my distance. At some point, while I was washing dishes, she and I were left alone in the kitchen. She asked me what I do for work. I responded vaguely, saying that I was a researcher at the Instituto de Investigaciones Estéticas, at the university, and that I taught classes. "What kind of classes?" she asked me. "Aesthetic theory," I said without saying anything. She insisted, "But what is your class about?" Without any way around it, I responded, "It's a seminar about how to think aesthetics in the age of the sixth extinction." She looked at me combatively: "Which extinction? Has it already happened? What is going extinct? Is it the future human extinction?" Taken aback that there could still be someone in our circuits who is not aware that we may be living in an era of mass extinction, I responded in the least emotional way possible: "Some studies indicate that it's likely that we are in an age of mass extinction. It isn't possible to know if it will lead to human extinction, though it's probable. What is certain is that entire species of plants, insects, and animals are dying at a scale and speed that presage a radical change in what we know as life on this planet." Surprised, she scolded me: "What does that have to do with aesthetics? Isn't it too sad to talk about all that? Why do you do it?" Saying that my daughter was calling me, I excused myself and took refuge in another conversation.

At some moment later that night, when there were just a few of us left, the woman was talking with a friend of mine, sitting across from me. I heard her saying, with no qualms whatsoever that I might be listening, "I know why Helena works on what she works on, she's scared and thinks that this will help reduce her fear." I felt ashamed. Indeed, it was fear that had pushed me to pursue an interest in what it might mean to live in an era of mass extinction. Is fear a valid motivation to incite thinking? To set research into motion? To work? If my work is developed in the spaces between aesthetics, art, and politics, isn't it my responsibility to think through time, this time, our time?

Since the Enlightenment, time has been a central element of aesthetics, along with space, since it is these two phenomena that lead—whether transcendentally (Kant) or historically (Michel Foucault)—to the conditions of possibility for sensibility, that is, experience. And it is also through art that times are projected, imagined, figured. Awareness develops of their rhythms and counterrhythms, their intensities and pauses, their continuities and fracturings. So why not think through the aesthetics of this age of mass extinction?

After a couple of days of turning this issue around in my head, I thought that perhaps fear is not bad as a detonator of thinking. It helps us to discern—perhaps not with clarity, but it does unfurl an unusual capacity to create strategies for positioning ourselves, for moving. It allows for speculation about what is to come and what has passed. At the very least, it helps us to overcome night terrors—at least some of them—and go back to sleep at night.

The uncertainty awakened by the pandemic was an initial fracture that made me situate myself in the time we're currently living, making me want to think through it, and also to distrust it. To repudiate its measure and seek other temporalities in which the ending would not necessarily be extinction. In art I have found other ways of counting. It has not been easy but it is perhaps my way of being, my way of learning, if possible, to live in unsettled times. This text is a journey through a time enchanted by ghosts—the ghost of the *coatl*, the serpent.

Perhaps, as Timothy Morton asserts in his small book *All Art Is Ecological*, "you may find yourself living in an age of mass extinction."[1] It's possible, not certain. The possible can't be read as affirmation or negation: it is rather the hole, the hollow. If it's the case that we're living in an age of mass extinction, this would be the sixth. The previous ones, we are told, occurred through a combination of factors external to planetary life, such as collisions of extraterrestrial bodies or high volcanic activity, or through internal ones such as transformations of species that made life for other beings nonviable. Although these extinctions were massive—each caused the disappearance of between 76 and 96 percent of its period's species, whose durations varied but were in some cases millions of years—life persisted and found other modes of organization.

What is the time of what is called the sixth extinction? How to measure it? How to count its time? The crisis we are going through, the sixth extinction, goes hand in hand with the universal project we've called humanity. The Anthropocene—a disputed categorical convention—takes productive human organization as the starting point for transformations in geological and planetary life. However, while it is clear that from the time agriculture began, an alteration of nonhuman life by human life began that would have consequences for other species, it is necessary to recognize and acknowledge the radical shifts in velocity, transformation, and violence that modernity and capitalism entail in relation to death in the era in which we are convened. For this reason, it is perhaps better to think not of the Anthropocene but of the Capitalocene, a term used by Donna Haraway, among others.[2] The age of the sixth extinction, the age we are living in, is the result of a logic of mass industrial production on a global scale that has generated such phenomena as the pollution of the oceans, the loss of habitats, changes in the soil, and climate change.[3] This last, specifically, is linked to the time of capitalism—to its development and its passage, that is its history, but also to the productivist logic of time within capitalism. Time is undoubtedly a convention that does not just situate and mark material realities but also produces them. How was this time produced, and what did it generate?

It is perhaps in the apparatus of a universalist imaginary, as Jaime Vindel has signaled, that life—human and nonhuman—was articulated in the interlocking links between energy, work, and industrialism. Time became productive thanks to regulations that allowed its passage to be situated as the duration between occurrence *a* and occurrence *b*. Time became a line, an arrow, a clock for measuring the duration between one event and another. Following Vindel, then, we might think of this era from the perspective of a "fossil aesthetic," understanding the latter term in the philosophical sense toward which we gestured earlier, and the former as a characterization derived from the substances produced through the accumulation of the fossilized remains of once living beings, now used as fuel. Fossil aesthetics, Vindel tells us, correspond "to the gestation of a cosmic imaginary (a cosmopolitics, if you will) and of a sensibility at the confluence of the colonial matrix of power, the cultural hegemony of the dominant classes, fossil fuels, and capital."[4]

Alteration of their surroundings is something species do. Humans, with their needs and their social and material projects, have been taking over resources and modifying their environment forever. But the radical transformation we are now experiencing is linked, in its scale and force, on the one hand to capitalism and its ways of organizing bodies to be exploited for labor (through gender, race, and class), and on the other to an ideology and rationality that created a separation between nature and culture. That separation, foundational to and constitutive of the time of capitalism, allowed for plunder and what Marx called primitive accumulation. From here things started to go wrong, and got worse and worse.

Although any transformation of or use of force on a body can be conceived as work on various scales beyond just the human, bodies—and tools as extended devices—were for a long time the limit of the energy that could be spent; the energies available for the kinds of mass displacement that produce and transmute things were limited to bodies. So the appearance of machines that used steam, coal, gas, and fuel to function represented a radical transformation that changed speeds and ways of being in the world. Through fossil fuels that exceeded the limits of living bodies, it became possible to produce and expend an enormous amount of force. Production exploded in speed (the rate of manufacture), in quantity (the vast numbers of consumer objects that appeared), and in extension (of resources and territories).

This model began to spread, and the time of thermodynamics was imposed, a time that, as Vindel signals, is both detonated and detonator, in a scientific construct that is inevitably linked to capitalist forms of production. A duration from *a* to *b*, from *b* to *c*... in this way the idea of progress was constituted, a progress that might be defined as the historico-political temporality of capitalism—a continuum of production, of expenditure, of energy. The argument could be condensed, Vindel maintains, in the idea that capitalism understood and perfectly utilized the first law of thermodynamics, which points to the conservation of energy. That is to say, if one system exchanges heat with another, the energy will change, will transform, but will not be destroyed; and therein lies the productivist logic of machines. Nonetheless, what does not seem to have been considered here is the role of entropy. All the work produced by bodies and machines generates energy, but its residue dissipates as heat, resulting in a rise in temperatures on the planet. All work produces energy—plants performing photosynthesis, my breathing—but the energy of capitalist production, still dependent on fossil fuels, exacts a brutal and excessive cost.[5] To offer just one example that might situate us within this problem: how much energy was needed in order for me to write this essay? How much work—not just my own, but that of the people producing the devices and everything else presupposed by the material existence of books, libraries, computers, electricity, Internet (raw material, production, distribution, point of sale, waste)—is needed to write an essay? Where does this work dissipate? Into the air.

> Something more than smoke metaphors condenses in the air: we breathe the result of a modernity that, to accelerate its pace toward the final ramp of history, in just two centuries has consumed the fossil resources that had been deposited in the bowels of Earth for millions of years. Fossil aesthetics no longer refer to the object imprint that facilitates the return of a previous temporality or access to the traces of human work. The geological spiral gasifies on the surface of the planet like a toxic cloud that threatens the very survival of the idea of the future. More than an index of the past, it represents an indication of the collapse underway. We must resist this disastrous

> destiny by clinging to life with a weak messianic force, the kind of hope that has left all optimism behind.[6]

If toxicity and heat threaten the very idea of a future, from what time can we summon that fragile messianic force?

Enter the Ghost/Exit the Ghost/Reenter the Ghost.

Time is out of joint, crazed and turned upside down, the philosopher Jacques Derrida warns us in his reading of Shakespeare's *Hamlet* in *Specters of Marx: The State of the Debt, the Work of Mourning, and the New International* (1994), a fundamental work for confronting the neoliberal chants about the end of history and of politics in the 1990s, after the fall of the Berlin Wall.[7] The ghost of Hamlet's father appears at the heart of deconstruction to dislocate any possible historical continuity. He lurks, perforating the present time to announce injustice. Here there are no invocations of the paranormal; the concept traces the wound of Western rational thought, which promised to break the spell cast on the world and declare its right to destruction. In the face of catastrophe, we needed a concept that would reveal other possible relationships, hence the ghost: a figure both present and absent at the same time, a figure that harasses, demands, reclaims, that links to a place, a history, a debt. The Derridean narrative of *hantologie* ("hauntology"), an ontology besieged by phantoms, involves the rupture of any possible metaphysics of presence—a logic of sameness—and the possibility of the appearance of alterity in the West. The ghost is the political eruption without which there is no yet-to-come. With Derrida this being becomes not a speculative figure of the supernatural but a political concept that would undo the knot of time and justice, a foundational entanglement present in Western rationality since the time of Anaximander. The specter appears and we know that something is out of joint.

The "time out of joint" that Hamlet speaks of signals that time is disarticulated, disjointed, displaced, dislocated, disrupted, harassed, disturbed, deranged, disarranged, and crazed. This disjunction of "time out of joint" speaks of the dislocation of time; an injustice looms in it, because something that has "gone wrong" has disrupted the direction of time and of the world. Because the exhaustion caused by imbalance also affects the world, the disjunction of being out of joint affects both time and history.

Maybe the only way to measure time in this extinction is through injunction. Perhaps in order to situate ourselves in the dimension of what is happening to us, we need to renounce the time of progress, which moves like an arrow toward the teleological destiny of history, with no regard for what is living or for life. To stop thinking of the future and instead summon the yet-to-come, we need to open ourselves to the ghost. Time is constructed by the uncertainty of alterity, a concept that no longer measures the duration of events, nor expects their continuity, but rather positions itself within time's leaps and misfires, its *glitch*. Derrida writes,

> The time of the "learning to live," a time without tutelary present, would amount to this, to which the exordium is leading us: to learn to live *with* ghosts, in the upkeep, the conversation, the company, or the companionship, in the commerce without commerce of ghosts. To live otherwise, and better. No, not better, but more justly. But *with them*. No *being-with* the other, no *socius* without this *with* that makes *being-with* in general more enigmatic than ever for us. And this being-with specters would also be, not only but also, a *politics* of memory, of inheritance, and of generations.[8]

The problem with ghosts is that they are elusive—perhaps because they are invisible, at least to the human eye. Their materiality languishes and dissipates. Appearing indirectly and obliquely, they disturb us. As Elaine Gan, Anna Tsing, Heather Swanson, and Nils Bubandt write,

> Haunting is quite properly eerie: the presence of the past often can be felt only indirectly, and so we extend our senses beyond their comfort zones. Human-made radiocesium has this uncanny quality: it travels in water and soil; it gets inside plants and animals; we cannot see it even as we learn to find its traces. It disturbs us in its indeterminacy; this is a quality of ghosts.[9]

In the face of this disturbance we sometimes prefer to negate the existence of ghosts, and if we manage to open up the possibility of believing in them, we demand proof. We want to be certain they are there, as if that would mitigate our fear. But no matter how much we try to flee the indeterminate, if we are to deal with ghosts we have to seek ways to see the invisible, to intuit it, listen to it, savor it, make it intelligible.

The work of noticeability, a concept Tsing gifts us, allows us to become aware of ghosts[10]—ghosts that aren't ours, that don't necessarily result from trauma or from lineage.[11] They are specters of what was, of what at one time was alive. Life is what we have in common, the responsibility that unites us and for which we take on a debt. There is no methodology for making ghosts appear. We don't know if it's a science or an art; it is certainly a game of speculation that attaches to what it can. Its aim is to intervene, almost as a medium, in order to glimpse beings to whom the world was once open and without whom our world would not exist—to make their absences appear, be magnified, and, in that vortex, to open ourselves to their passage through us and be enchanted by them, and by the times and rhythms they carry with them.

Enter the Ghost/Exit the Ghost/Reenter the Ghost.

Amotla otlacualacac oncan tlanahuatiz quename ye huitz quiahuitl...*mocualnezcayotl*...The lightning did not strike to announce that it would rain...your beauty. In 2013, the Museo del Chopo in Mexico City was taken over by ghosts, coyote machines, invocations of animal gods, whose movements did not respond to the will or the presence of the museum's visitors but rather to the air, the rain, or the wind in San Pedro Atocpan, a town twenty-eight miles or so away. The gestures of the coyote-robots demanded that the audience look toward a place where they were not, but that appeared on the arm, head, or body that was moving about. The machines danced to announce the invisible, an existence that we could intuit only through their movements. The ghost moving them was atmospheric change. Their actions in no way responded to the people watching them, who hoped something would happen, tried to activate them with their movements, their voices, their minds, or their desires, but the coyote-machines remained static. Hours might pass in this way, until suddenly something in the atmosphere of San Pedro Atocpan—a gust of wind, or rain from a storm absent in Mexico City—made the robots start to dance, sabotaging the audience's certainties.

This exhibition, by the Nahua artist Fernando Palma Rodríguez, was part of a project that intersected with art from the field of mechatronics (fig. 1).[12] Palma's production of machine-deities began in 1994, with works such as *Tumba la casa co's is for the raza, Greetings, Zapata Moles*, and *Xipetotec*.[13] Here he explored another type of robotic logic through which the machine is not created to resemble the human but rather, through its very differences, invites us to contemplate other ways of being and becoming in the world. Educated as an engineer at the Instituto Politécnico Nacional in Mexico City, and later through a master's program at Goldsmiths College, London, Palma uses mechanics, electronics, and coding to subvert Western epistemologies that conceive machines as ontologically different from and inferior to the human since even in their mimetic state they have no soul or spirit. Palma's machines to the contrary respond to other logics and worldviews, and thus presuppose a break in modern aesthetics and politics. With no pretensions to access or testify to a lost origin, these beings replicate and amplify a broken time: an Indigenous time, an imagined, dreamt, supposed time, to be found in prints, in plants, in words, in images in which the relationship between nature and culture implies not a gap but a collaboration. Animals, gods, *nahuales*, beings with an abundance of attributes that recuperate relationships with the living from within a Nahuatl worldview—even given the certainty that return is impossible.

Although Nahuatl was the language of Palma's mother, he did not learn it as a child, but felt the need to speak it and think from inside it with his politicization later on. What was opened for him in the language was the world out of which his poetic/machinic figurations emerged. In Nahuatl as in other Indigenous languages, the term "person" is extended to objects, organic elements, work processes. These are, Palma tells us, *téotl*, energies, that are also persons, a conception that he maintains "brings consequences with it": "When an [object] is a person for you, a relationship is

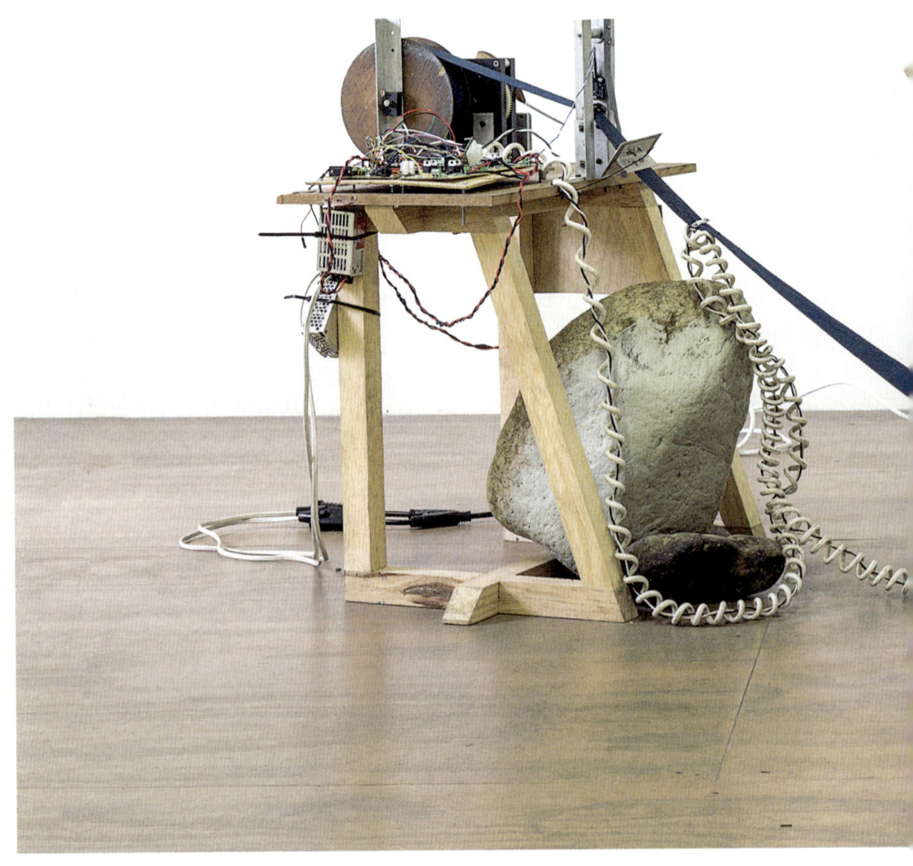

1 Fernando Palma Rodríguez. *Soldado* (Soldier). 2001. Wooden structure, electronic circuits, sensors, and software, dimensions variable. Collection the artist

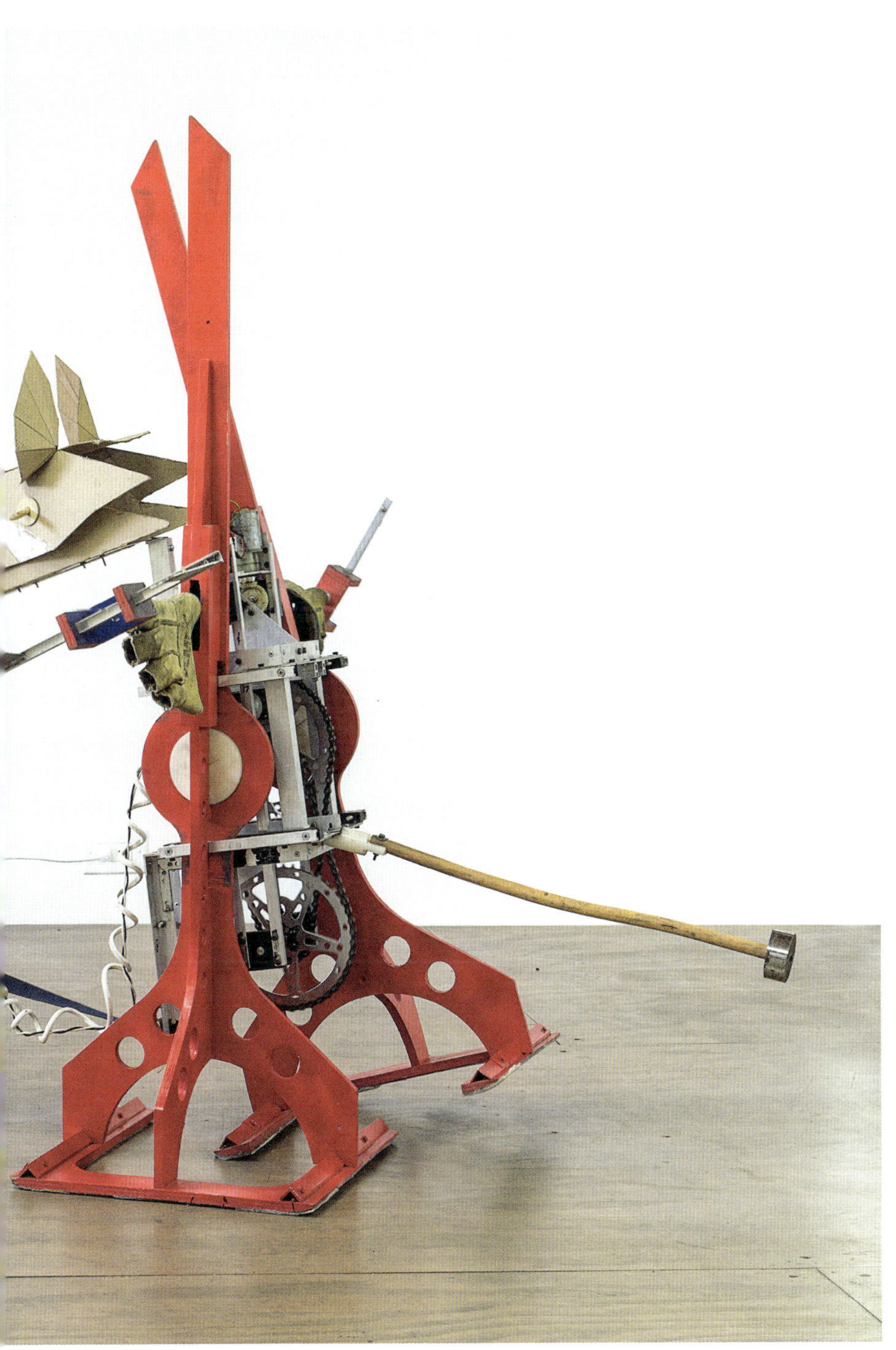

established according to another ethic, in which this person can become a friend, and this is therefore extended to the natural world, or what is considered as such."[14] His machines come alive and communicate in another, more malleable language.

Nahuatl is agglutinative in principle, which allows it to take in words from different sources in order to build concepts; it was "deconstructive" well before deconstruction. There are no differences in Nahuatl between nouns and verbs or between objects and subjects. We, the air, cans: we are all persons. Palma's coyote-machines, then, are not objects but something else. They are not mimetic, not useful productive devices; they are monsters inhabited by the breath of failed endeavors. Although sophisticated, they insist on a *low-tech* formalization. They don't hide their systems but expose them in guts and internal organs made of visible wires and connections. Their presence is constructed out of materials considered "poor": cardboard, cloth, dirt, sheet metal, wire, strings. They are "robots from the South" that present themselves as nonutilitarian and appeal to a "recharged animism," in the words of the Argentine artist Paula Gaetano Adi, who works in the same artistic and political realm.[15] The use of technology in Palma's work points to contemporary messianic paradoxes, but here futurist daydreams of salvation are sabotaged. Technology and its logic—how it is produced and what is produced—are part of the problem, and Palma does not avoid that question but rather approaches it through other forms of figuration. Western technology is transformed in his work through the question of how to avoid the patterns of aleatory systems and frustrate expectations of productivity. The counterrhythms of these machines resist functionality to the benefit of poetics.

On the other hand, Palma's work breaks with any fantasy—almost always a white fantasy—of an Indigenous purity. It seeks not to satisfy imaginaries of the autochthonous but to open the maw of a world that is coming alive today, in the conditions and with the objects we have, through devices that resemble certain Western technologies but are crazed, saturated with anima. His "temperamental robotic machines," as Jean Fisher called them, tell "new stories of the urban landscape and the fragile relationship between humans and technology."[16] This is an aesthetic problem, and therefore also a political one. It's here that our relationship with the world comes into play, our experience of it and of our very selves.

If in the tele-technical there is a ghost effect that harasses the contemporary image, as we see something that isn't there (here we can trace a genealogy that takes us to Roland Barthes's *punctum*), it's also true that the apparatuses themselves, and their frequencies, seem to keep away those ghosts that open us to other, less rational, less contemporary, less human worlds. Palma maintains that contamination—from waste but also from radio frequencies—means that there is no longer any space for ghosts; perhaps this is why he evokes and produces them in his work.[17] From within his Nahua machinations, he summons them so that they might appear, might harass us and stalk us. The vortex his machines open—their noticeability—makes ghosts appear that belong to no fixed time but to worlds that have perished, because in the silence of speech they have stopped thinking and activating their powers.

Along with these machinic constructions, Palma's work has a performative dimension in which the artist himself, wearing a mask of cardboard, hair, and lights, transforms into Huehuecoyotl-Fer, a coyote thief traversing a devastated landscape. The artist enacts these *dérives* regularly in Mexico City's partly rural borough of Milpa Alta, or on walks to the extinct volcano Teuhtli (which lies on the border between Milpa Alta and the boroughs of Tláhuac and Xochimilco), and he has recorded them in videos such as *Si no fuera por estos momentos* (If it weren't for these moments, 2000), or the video included in the installation *Huehuetlazotemoani (Old Beloved Searcher), Xico* (2015), made with the artist Nuria Montiel. In this sort of *nahualismo*—the transformation of an individual into some animal or natural phenomenon with a shared destiny—the artist traverses peripheral areas of the city and uses his magic powers to intervene in landscapes devastated by poverty, plunder, and violence. On these journeys, Huehuecoyotl-Fer shows those who walk with him the city under its own spell.

What appears, according to someone who accompanied Palma in 2022, is "a kind of negative sunrise in which, strictly speaking, the city is not brought to light by dawn but rather disappears under a chemical haze as its electric lights darken."[18]

Magic, Palma shows us, does not promise salvation; instead, *nahualismo* involves mediation, connection, and intervention. Huehuecoyotl-Fer communicates from other cosmic planes and juxtaposes times along his journey. On his *dérives*, place is shot through with many different temporalities: that of a city in collapse, and that of a nature that buzzes like serpents tangled in a net. The moment contains the time of the volcano, the time of the sun, the time of waste, the time of the robot, the time of the coyote, the time of ourselves, the time of telluric forces. These are not the same times, nor are they even similar, but they coincide, touch for a moment, and then dislocate again. In that spiral the continuum collapses and the yet-to-come appears—or more precisely, the yet-to-comes.

This performative invocation takes further shape out of Palma's work with Calpulli Tecalco, an educational and environmental association, based in Milpa Alta, where he works with his family. Calpulli Tecalco mounts a clear and powerful defense of the land, and thus of nature and language. Both issues are foundational to any process of confronting the challenges facing Palma's community: the uncontrolled growth of the city, the destruction of the terraces in the *milpa* (an ancient technique of cultivation that relieves stress on the land and sustains its mineral richness), the emigration of the Indigenous population and the immigration of people dispossessed by the cost of living in the city (making the area a bedroom community), the loss of language and customs, and the violence that batters and devastates Milpa Alta, as it does the rest of the country.

Although Palma's work takes shape in a variety of ways, it centers on a broad notion of repair. According to the artist himself, his work comes from grief and the need to heal the social body in the earth and in memory as in speech. Resistance is activist and poetic at the same time, an ongoing effort with no beginning and no end. When something seems to end, it does so only to begin again, at another point, from another opening.

Beginning at the end of the 1990s, Palma has made a series of works called *Tetzahuitl* (Biennale de Lyon, 2019; Mexico Pavilion, Venice Biennale, 2022) in which dresses—sometimes women's dresses, sometimes dresses for young girls—move up and down on mechanical structures in a sort of choreography that seems to invoke usurped bodies. These works reference femicide, forced disappearance, and death as lurking constantly, and they uncover systemic violence through figurations that bring ghosts close, materializing them and presenting them on a poetic plane. Rather than offer melancholic consolation for dark events, Palma's interventions gather forces to confront ruin and demand compensation. His operation doesn't assume a return to being "in joint," much less a serving of justice in a legal sense. It is a work in the present, made in the awareness that there will be neither resolution nor happy ending. We never finish repairing what is broken; just as when we mend something torn, we can reunite what has been ripped apart, but the patch always leaves a gap—a scar, a mark, a cavity, a hole where darkness might seep through, but light as well. The repair offers no foreseeable future, but its fissures signal a time yet-to-come in which we can glimpse a more just life, a better life, a way of living well.

Palma's work invokes a *coatl* time, a time of serpents, a time that repeats synodal movements—a series of movements that at each turn, again and again and again, announces the return of a new cycle, a new time count, with neither beginning nor end, in incommensurable temporalities that overlap, fluttering all at once. His work parades beings who move in response to entities that we do not see, that are not there, and that nonetheless accompany us, embodied in *nahual* machines, revealing to us their beauty and, perhaps, our own as well.

In a telephone conversation a little while ago, Fernando told me that light was born from darkness. I'm not quite sure what he was trying to tell me: perhaps that gloom is part of being, that we have to trust that life will continue, that it will persist in reappearing. That what is opaque, though it might be invisible to our eyes, contains light itself, that light is in its innards. That perhaps they are the same thing.

There are no certainties, and that frightens me, but perhaps, Fernando reminds me, that's what it is to be in the presence of the mystery of life. What remains for us is to learn to live with that mystery and with its ghosts, in their company and in our glimpses of them. To listen to the cosmic echo of *coatl*, the serpent, and to let the times unfold: forward and backward, up and down, in a diagonal, in a spiral, in death and in life.

To live in an era of mass extinction. To respond to it, to confront it, we will have to make space for ghosts, so that they can lead us to times that are not the times of capitalism, to times that aren't that of an arrow heading toward catastrophe but rather of what is yet-to-come, with all the uncertainty and all the darkness it carries. In that darkness, says Huehuecoyotl-Fer, is also the light.

1 Timothy Morton, *All Art is Ecological* (London: Penguin Books, 2018), I.

2 See Donna Haraway, *Staying with the Trouble: Making Kin in the Chthulucene* (Durham, NC: Duke University Press, 2016).

3 Some studies note the possibility of identifying at least three moments within this extinction: an initial stage with the expansion of *Homo sapiens*, a second stage with the development of agriculture, and a third with the climate change that resulted from industrial development.

4 Jaime Vindel, *Estética fósil. Imaginarios de la energía y crisis ecosocial* (Barcelona: Arcadia and MACBA, 2020), 21.

5 Here it's important to note the warning of Emilio Santiago Muíño: "The Gordian knot of our time can be summarized by the following paradox: renewable energies, which are the only energies capable of insuring the ecological viability of humanity, can't be responsible for the social model of our present if that model does not undergo a profound shift." Muíño, "De nuevo estamos todos en peligro. El petróleo como el eslabón más débil de la cadena neoliberal," in Muíño, Yayo Herrera, and Jorge Reichmann, *Petróleo* (Barcelona: Arcadia and MACBA, 2018), 46–47. In this same sense, Muíño posits that a society whose production system was based totally on renewables could achieve only one fifth of the production achieved by a society using cheap oil. Any shift in energy use on a global level, then, would have to presume a change in the models of production, circulation, and consumption.

6 Vindel, *Estética fósil*, 188.

7 Jacques Derrida, *Specters of Marx: The State of the Debt, the Work of Mourning, and the New International*, trans. Peggy Kamuf (New York and London: Routledge, 1994).

8 Ibid., xviii–xix.

9 Elaine Gan, Anna Tsing, Heather Swanson, and Nils Bubandt, Introduction, in Gan, Tsing, Swanson, and Bubandt, eds., *Arts of Living on a Damaged Planet* (Minneapolis: University of Minnesota Press, 2017), 2.

10 Tsing devotes a chapter of her book *The Mushroom at the End of the World* to presenting this concept as a critical methodology for thinking outside the limits of scientific methods. "To listen and to tell a rush of stories is a *method*. And why not make a strong claim and call it science, an addition to knowledge? Its research object is contaminated diversity; its unit of analysis is the indeterminate encounter. To learn anything we must revitalize arts of noticing and include ethnography and natural history." Tsing, *The Mushroom at the End of the World: On the Possibility of Life in Capitalist Ruins* (Princeton: Princeton University Press, 2015), 38. We propose to add art here as an integral part of this new methodology.

11 The leap proposed by ecological ethnography from inheritance to not just human history but other species, and to what is alive in general, may be the radical movement that will allow us to conceive of an ecopolitics.

12 My initial approach to the work of Fernando Palma Rodríguez appears in my essay "Hacer bailar a los espectros," in "*... amotla otlacualacac oncan tlanahuatiz quename ye huitz quiahuitl...mocualnezcayotl" ... no relampagueó para anunciar que llovería...tu hermosura*, exh. cat. (Mexico City: Museo Universitario del Chopo, 2015).

13 It's important to note that the birth

of these machines coincided with the uprising of the Ejercito Zapatista de Liberación Nacional (Zapatista army for national liberation), in January of 1994. This "coincidence" also marks the struggle for the significance of the Indigenous in Mexican politics, a movement that led, among many other things, to the emergence of the Indigenous not as an object of Indigenist policies but as a political subject. This change in grammar signified a questioning that continues to oblige us to consider the current situations of Indigenous peoples.

14 Palma, in an interview with Manuel Guerrero, "In Ixtli in Yollotl," *Código: Arte—Arquitectura—Diseño*, May 25, 2018. Available online at https://revistacodigo.com/entrevista-con-fernando-palma-2/ (accessed January 2, 2024).

15 Paula Gaetano Adi, "Insurrección y Robocalipsis: Poéticas tecnológicas en disidencia," lecture, Materia Abierta, Casa del Lago, Mexico Citiy, August 18, 2022.

16 Jean Fisher, "Fernando Rodríguez Palma. New Horizons of the Past: The Indigenous Landscape," in Simon Read, ed., *Plot,* Confluens 3 (London: Middlesex University Fine Art Research Publications, 2008), 22, 23.

17 Palma, in Tate, "Artist Fernando Palma Rodríguez—'There's no room for ghosts,'" YouTube, n.d. Available online at www.youtube.com/watch?v=-hlw0aL_OGM (accessed January 3, 2024).

18 Guillermo Canek García, "From These Dark Mirrors of the World: 'The Rise of the Coyote,' Materia Abierta Summer School," *e-flux education*, October 4, 2022. Available online at www.e-flux.com/education/features/495433/from-these-dark-mirrors-of-the-world-the-rise-of-the-coyote-materia-abierta-summer-school (accessed January 3, 2024).

PART IV

Interdisciplinary Essays

EDUARDO VIVEIROS DE CASTRO

COSMOLOGICAL DEIXIS AND AMERINDIAN PERSPECTIVISM

This study discusses the meaning of Amerindian "perspectivism": the ideas in Amazonian cosmologies concerning the way in which humans, animals, and spirits see both themselves and one another. Such ideas suggest the possibility of a redefinition of the classical categories of "nature," "culture," and "supernature" based on the concept of perspective or point of view. The study argues in particular that the antinomy between two characterizations of Indigenous thought—on the one hand "ethnocentrism," which would deny the attributes of humanity to humans from other groups, and on the other hand "animism," which would extend such qualities to beings of other species—can be resolved if one considers the difference between the spiritual and corporal aspects of beings.

> *La reciprocité de perspectives où j'ai vu le caractère propre de la pensée mythique. . . .*
>
> The reciprocity of perspectives that I have seen as the specific character of mythic thought. . . .
> —Claude Lévi-Strauss, *La potière jalouse*, 1985

INTRODUCTION

This article deals with that aspect of Amerindian thought which has been called its "perspectival quality": the conception, common to many peoples of the continent, according to which the world is inhabited by different sorts of subjects or persons, human and nonhuman, which apprehend reality from distinct points of view.[1] This idea cannot be reduced to our current concept of relativism, which at first it seems to call to mind.[2] In fact, it is at right angles, so to speak, to the opposition between relativism and universalism.

Such resistance by Amerindian perspectivism to the terms of our epistemological debates casts suspicion on the robustness and transportability of the ontological partitions which they presuppose. In particular, as many anthropologists have already concluded (albeit for other reasons), the classic distinction between Nature and Culture cannot be used to describe domains internal to non-Western cosmologies without first undergoing a rigorous ethnographic critique.

Such a critique, in the present case, implies a redistribution of the predicates subsumed within the two paradigmatic sets that traditionally oppose one another under the headings of "Nature" and "Culture": universal and particular, objective and subjective, physical and social, fact and value, the given and the instituted, necessity and spontaneity, immanence and transcendence, body and mind, animality and humanity, among many more. Such an ethnographically based reshuffling of our conceptual schemes leads me to suggest the expression "multinaturalism" to designate one of the contrastive features of Amerindian thought in relation to Western "multiculturalist" cosmologies. Where the latter are founded on the mutual implication of the unity of nature and the plurality of cultures—the first guaranteed by the objective universality of body and substance, the second generated by the subjective particularity of spirit and meaning—the Amerindian conception would suppose a spiritual unity and a corporeal diversity. Here, culture or the subject would be the form of the universal, whilst nature or the object would be the form of the particular.

This inversion, perhaps too symmetrical to be more than speculative, must be developed by means of a plausible phenomenological interpretation of Amerindian cosmological categories, which determine the constitutive conditions of the relational contexts we can call "nature" and "culture." Clearly, then, I think that the distinction between Nature and Culture must be subjected to critique, but not in order to reach the conclusion that such a thing does not exist (there are already too many things which do not exist). The flourishing industry of criticisms of the Westernizing character of all dualisms has called for the abandonment of our conceptually dichotomous heritage, but to date the alternatives have not gone beyond the stage of wishful unthinking. I would prefer to gain a perspective on our own contrasts, contrasting them with the distinctions actually operating in Amerindian perspectivist cosmologies.

PERSPECTIVISM

The initial stimuli for the present reflections were the numerous references in Amazonian ethnography to an Indigenous theory according to which the way humans perceive animals and other subjectivities that inhabit the world—gods, spirits, the dead, inhabitants of other cosmic levels, meteorological phenomena, plants, occasionally even objects and artefacts—differs profoundly from the way in which these beings see humans and see themselves.

Typically, in normal conditions, humans see humans as humans, animals as animals, and spirits (if they see them) as spirits; however, animals (predators) and spirits see humans as animals (as prey) to the same extent that animals (as prey) see humans as spirits or as animals (predators). By the same token, animals and spirits see themselves as humans: they perceive themselves as (or become) anthropomorphic beings when they are in their own houses or villages and they experience their own habits and characteristics in the form of culture—they see their food as human food (jaguars see blood as manioc beer, vultures see the maggots in rotting meat as grilled fish, etc.), they see their bodily attributes (fur, feathers, claws, beaks, etc.) as body decorations or cultural instruments, they see their social system as organized in the same way as human institutions are (with chiefs, shamans, ceremonies, exogamous moieties, etc.). This "to see as" refers literally to percepts and not analogically to concepts, although in some cases the emphasis is placed more on the categorical rather than on the sensory aspect of the phenomenon.

In sum, animals are people, or see themselves as persons. Such a notion is virtually always associated with the idea that the manifest form of each species is a mere envelope (a "clothing") which conceals an internal human form, usually only visible to the eyes of the particular species or to certain transspecific beings such as shamans. This internal form is the "soul" or "spirit" of the animal: an intentionality or subjectivity formally identical to human consciousness, materializable, let us say, in a human bodily schema concealed behind an animal mask. At first sight, then, we would have a distinction between an anthropomorphic essence of a spiritual type, common to animate beings, and a variable bodily appearance, characteristic of each individual species but which rather than being a fixed attribute is instead a changeable and removable clothing. This notion of "clothing" is one of the privileged expressions of metamorphosis—spirits, the dead, and shamans who assume animal form, beasts that turn into other beasts, humans that are inadvertently turned into animals—an omnipresent process in the "highly transformational world" proposed by Amazonian ontologies.[3]

This perspectivism and cosmological transformism can be seen in various South American ethnographies, but in general it is only the object of short commentaries and seems to be quite unevenly elaborated.[4] It can also be found, and maybe with even greater generative value, in the far north of North America and Asia, as well as amongst hunter-gatherer populations of other parts of the world.[5] In South America, the cosmologies of the Vaupés area are in this respect highly developed,[6] but other Amazonian societies, such as the Wari' of Rondônia[7] and the Juruna of the Middle Xingu,[8] also give equal emphasis to the theme.

Some general observations are necessary. Perspectivism does not usually involve all animal species (besides covering other beings); the emphasis seems to be on those species which perform a key symbolic and practical role such as the great predators and the principal species of prey for humans—one of the central dimensions, possibly even the fundamental dimension, of perspectival inversions refers to the relative and relational statuses of predator and prey.[9] On the other hand, however, it is not always clear whether spirits or subjectivities are being attributed to each individual animal, and there are examples of cosmologies which deny consciousness to postmythical animals[10] or some other spiritual distinctiveness.[11] Nonetheless, as is well known, the notion of animal spirit "masters" ("mothers of the game animals," "masters of the white-lipped peccaries," etc.) is widespread throughout the continent. These spirit masters, clearly endowed with intentionality analogous to that of humans, function as hypostases of the animal species with which they are associated, thereby creating an intersubjective field for human-animal relations even where empirical animals are not spiritualized.

We must remember, above all, that if there is a virtually universal Amerindian notion, it is that of an original state of undifferentiation between humans and animals, described in mythology. Myths are filled with beings whose form, name, and behavior inextricably mix human and animal attributes in a common context of intercommunicability, identical to that which defines the present-day intrahuman world. The differentiation between "culture" and "nature," which Claude Lévi-Strauss showed to be the central theme of Amerindian mythology, is not a process of differentiating the human from the animal, as in our own evolutionist mythology. The original common condition of both humans and animals is not animality but rather humanity. The great mythical separation reveals not so much culture distinguishing itself from nature but rather nature distancing itself from culture: the myths tell how animals lost the qualities inherited or retained by humans.[12] Humans are those who continue as they have always been: animals are exhumans, not humans exanimals. In sum, "the common point of reference for all beings of nature is not humans as a species but rather humanity as a condition."[13]

This is a distinction—between the human species and the human condition—which should be retained. It has an evident connection with the idea of animal clothing hiding a common spiritual "essence" and with the issue of the general meaning of perspectivism. For the moment, we may simply note one of its main corollaries: the past humanity of animals is added to their present-day spirituality hidden by their visible form in order to produce that extended set of food restrictions or precautions which either declare inedible certain animals that were mythically cosubstantial with humans, or demand their desubjectivization by shamanistic means before they can be consumed (neutralizing the spirit, transubstantiating the meat into plant food, semantically reducing it to other animals less proximate to humans), under the threat of illness, conceived of as a cannibal counterpredation undertaken by the spirit of the prey turned predator, in a lethal inversion of perspectives which transforms the human into animal.[14]

It is worth pointing out that Amerindian perspectivism has an essential relation with shamanism and with the valorization of the hunt. The association between shamanism and this "venatic ideology" is a classic question.[15] I stress that this is a matter of symbolic importance, not ecological necessity: horticulturists such as the Tukano or the Juruna (who in any case fish more than they hunt) do not differ much from circumpolar hunters in respect of the cosmological weight conferred on animal predation, spiritual subjectivation of animals, and the theory according to which the universe is populated by extrahuman intentionalities endowed with their own perspectives. In this sense, the spiritualization of plants, meteorological phenomena, or artefacts seems to me to be secondary or derivative in comparison with the spiritualization of animals: the animal is the extrahuman prototype of the Other, maintaining privileged relations with other prototypical figures of alterity, such as affines.[16] This hunting ideology is also and above all an ideology of shamans, insofar as it is shamans who administer the relations between humans and the spiritual component of the extrahumans, since they alone are capable of assuming the point of view of such beings and, in particular, are capable of returning to tell the tale. If Western multiculturalism is relativism as public policy, then Amerindian perspectivist shamanism is multinaturalism as cosmic politics.

ANIMISM

The reader will have noticed that my "perspectivism" is reminiscent of the notion of "animism" recently recuperated by Philippe Descola.[17] Stating that all conceptualizations of nonhumans always refer to the social domain, Descola distinguishes three modes of objectifying nature: totemism, where the differences between natural species are used as a model for social distinctions; that is, where the relationship between nature and culture is metaphorical in character and marked by discontinuity (both within and between series); animism, where the "elementary categories structuring social life" organize the relations between humans and natural species, thus defining a social continuity between nature and culture, founded on the attribution of human dispositions and social characteristics to "natural beings";[18] and naturalism, typical of Western cosmologies, which supposes an ontological duality between nature, the domain of necessity, and culture, the domain of spontaneity, areas separated by metonymic discontinuity. The "animic mode" is characteristic of societies in which animals are the "strategic focus of the objectification of nature and of its socialization,"[19] as is the case amongst Indigenous peoples of America, reigning supreme over those social morphologies lacking in elaborate internal segmentations. But this mode can also be found coexisting or combined with totemism, wherein such segmentations exist, the Bororo and their *aroe/bope* dualism being such a case.[20]

These ideas form part of a theory which I cannot discuss here as fully as it would merit. I merely comment on the contrast between animism and naturalism but from a somewhat different angle from the original one. (Totemism, as defined by Descola, seems to me to be a heterogeneous phenomenon, primarily classificatory rather than cosmological: it is not a system of *relations* between nature and culture as is the case in the other two modes, but rather of purely logical and differential *correlations*.)

Animism could be defined as an ontology which postulates the social character of relations between humans and nonhumans: the space between nature and society is itself social. Naturalism is founded on the inverted axiom: relations between society and nature are themselves natural. Indeed, if in the animic mode the distinction "nature/culture" is internal to the social world, humans and animals being immersed in the same sociocosmic medium (and in this sense "nature" is a part of an encompassing sociality), then in naturalist ontology, the distinction "nature/culture" is internal to nature (and in this sense, human society is one natural phenomenon amongst others). Animism has "society" as the unmarked pole, naturalism has "nature": these poles function, respectively and contrastively, as the universal dimension of each mode. Thus animism and naturalism are hierarchical and metonymical structures (this distinguishes them from totemism, which is based on a metaphoric correlation between equipollent opposites).

In Western naturalist ontology, the nature/society interface is natural: humans are organisms like the rest, body-objects in "ecological" interaction with other bodies and forces, all of them ruled by the necessary laws of biology and physics; "productive forces" harness, and thereby express, natural forces. Social relations, that is, contractual or instituted relations between subjects, can only exist internal to human society. But how alien to nature—this would be the problem of naturalism—are these relations? Given the universality of nature, the status of the human and social world is unstable and, as the history of Western thought shows, it perpetually oscillates between a naturalistic monism ("sociobiology" being one of its current avatars) and an ontological dualism of nature/culture ("culturalism" being its contemporary expression). The assertion of this latter dualism, for all that, only reinforces the final referential character of the notion of nature, by revealing itself to be the direct descendant of the opposition between Nature and Supernature. Culture is the modern name of Spirit—let us recall the distinction between *Naturwissenschaften* and *Geisteswissenschaften*—or at the least it is the name of the compromise between Nature and Grace. Of animism, we would be tempted to say that the instability is located in the opposite pole: there the problem is how to administer the mixture of humanity and animality constituting animals, and not, as is the case amongst ourselves, the combination of culture and nature which characterizes humans; the point is to differentiate a "nature" out of the universal sociality.

However, can animism be defined as a projection of differences and qualities internal to the human world onto nonhuman worlds, as a "sociocentric" model in which categories and social relations are used to map the universe? This interpretation by analogy is explicit in some glosses on the theory: "if totemic systems model society after nature, then animic systems model nature after society."[21] The problem here, obviously, is to avoid any undesirable proximity with the traditional sense of "animism," or with the reduction of "primitive classifications" to emanations of social morphology; but equally the problem is to go beyond other classical characterizations of the relation between society and nature such as Alfred Reginald Radcliffe-Brown's.[22]

Tim Ingold showed how schemes of analogical projection or social modeling of nature escape naturalist reductionism only to fall into a nature/culture dualism which by distinguishing "really natural" nature from "culturally constructed" nature reveals itself to be a typical cosmological antinomy faced with infinite regression.[23] The notion of model or metaphor supposes a previous distinction between a domain wherein social relations are constitutive and literal and another where they are representational and metaphorical. Animism, interpreted as human sociality projected onto the nonhuman world, would be nothing but the metaphor of a metonymy.

Amongst the questions remaining to be resolved, therefore, is that of knowing whether animism can be described as a figurative use of categories pertaining to the human-social domain to conceptualize the domain of nonhumans and their relations with the former. Another question: if animism depends on the attribution of human cognitive and sensory faculties to animals, and the same form of subjectivity, then what in the end is the difference between humans and animals? If animals are people, then why do they not see us as people? Why, to be precise, the perspectivism? Finally, if animism is a way of objectifying nature in which the dualism of nature/culture does not hold, then what is to be done with the abundant indications regarding the centrality of this opposition to South American cosmologies? Are we dealing with just another "totemic illusion," if not with an ingenuous projection of our Western dualism?

ETHNOCENTRISM

In a well-known essay, Lévi-Strauss observed that for "savages" humanity ceases at the boundary of the group, a notion which is exemplified by the widespread auto-ethnonym meaning "real humans," which, in turn, implies a definition of strangers as somehow pertaining to the domain of the extrahuman. Therefore, ethnocentrism would not be the privilege of the West but a natural ideological attitude, inherent to human collective life. Lévi-Strauss illustrates the universal reciprocity of this attitude with an anecdote:

> In the Greater Antilles, some years after the discovery of America, whilst the Spanish were dispatching inquisitional commissions to investigate whether the natives had a soul or not, these very natives were busy drowning the white

> people they had captured in order to find out, after lengthy observation, whether or not the corpses were subject to putrefaction.[24]

The general point of this parable (from which Lévi-Strauss derived the famous moral: "The barbarian is first and foremost the man who believes in barbarism") is quite simple: the Indians, like the European invaders, considered that only the group to which they belong incarnates humanity; strangers are on the other side of the border which separates humans from animals and spirits, culture from nature and supernature. As matrix and condition for the existence of ethnocentrism, the nature/culture opposition appears to be a universal of social apperception.

At the time when Lévi-Strauss was writing these lines, the strategy of vindicating the full humanity of savages was to demonstrate that they made the same distinctions as we do: the proof that they were true humans is that they considered that they alone were the true humans. Like us, they distinguished culture from nature and they too believed that *Naturvölker* are always the others. The universality of the cultural distinction between Nature and Culture bore witness to the universality of culture as human nature. In sum, the answer to the question of the Spanish investigators (which can be read as a sixteenth-century version of the "problem of other minds") was positive: savages do have souls.

Now, everything has changed. The savages are no longer ethnocentric but rather cosmocentric; instead of having to prove that they are humans because they distinguish themselves from animals, we now have to recognize how *in*human *we* are for opposing humans to animals in a way they never did: for them nature and culture are part of the same sociocosmic field. Not only would Amerindians put a wide berth between themselves and the Great Cartesian Divide which separated humanity from animality, but their views anticipate the fundamental lessons of ecology which we are only now in a position to assimilate.[25] Before, the Indians' refusal to concede predicates of humanity to other men was of note; now we stress that they extend such predicates far beyond the frontiers of their own species in a demonstration of "ecosophic" knowledge which we should emulate in as far as limits of our objectivism permit.[26] Formerly, it had been necessary to combat the assimilation of the savage mind to narcissistic animism, the infantile stage of naturalism, showing that totemism affirmed the cognitive distinction between culture and nature; now, neoanimism reveals itself as the recognition of the universal admixture of subjects and objects, humans and nonhumans against modern *hubris*, the primitive and post-modern "hybrids," to borrow a term from Bruno Latour.[27]

Two antinomies then, which are, in fact, only one: either Amerindians are ethnocentrically "stingy" in the extension of their concept of humanity and they "totemically" oppose nature and culture; or they are cosmocentric and "animic" and do not profess to such a distinction, being models of relativist tolerance, postulating a multiplicity of points of view on the world.

I believe that the solution to these antinomies lies not in favoring one branch over the other, sustaining, for example, the argument that the most recent characterization of American attitudes is the correct one and relegating the other to the outer darkness of pre-post-modernity.[28] Rather, the point is to show that the "thesis" as well as the "antithesis" are true (both correspond to solid ethnographic intuitions), but that they apprehend the same phenomena from different angles; and also it is to show that both are false in that they refer to a substantivist conceptualization of the categories of Nature and Culture (whether it be to affirm or negate them) which is not applicable to Amerindian cosmologies.

The first point to be considered is that the Amerindian words which are usually translated as "human being" and which figure in those supposedly ethnocentric self-designations do not denote humanity as a natural species. They refer rather to the social condition of personhood, and they function (pragmatically when not syntactically) less as nouns than as pronouns. They indicate the position of the subject; they are enunciative markers, not names. Far from manifesting a semantic shrinking of a common name to a proper name (taking "people" to be the name of the tribe), these words move in the opposite direction, going from substantive to perspective (using "people" as a collective pronoun "we people/us"). For this very reason, Indigenous categories of identity have that enormous contextual variability of scope that characterizes pronouns, marking contrastively Ego's immediate kin, his/her local group, all humans, or even all beings endowed with subjectivity: their coagulation as "ethnonyms" seems largely to be an artefact of interactions with ethnographers. Nor is it by chance that the majority of Amerindian ethnonyms which enter the literature are not self-designations, but rather names (frequently pejorative) conferred by other groups: ethnonymic objectivation is primordially applied to others, not to the ones in the position of subject. Ethnonyms are names of third parties; they belong to the category of "*they*" not to the category of "*we*." This, by the way, is consistent with a widespread avoidance of self-reference on the level of personal onomastics: names are not spoken by the bearers nor in their presence; to name is to externalize, to separate (from) the subject.

Thus self-references such as "people" mean "person," not "member of the human species," and they are personal pronouns registering the point of view of the subject talking, not proper names. To say, then, that animals and spirits are people is to say that they are persons, and to attribute to nonhumans the capacities of conscious intentionality and agency which define the position of the subject. Such capacities are objectified as the soul or spirit with which these nonhumans are endowed. Whatever possesses a soul is a subject, and whatever has a soul is capable of having a point of view. Amerindian souls, be they human or animal, are thus indexical categories, cosmological deictics whose analysis calls not so much for an animist psychology or substantialist ontology as for a theory of the sign or a perspectival pragmatics.[29]

Thus, every being to whom a point of view is attributed would be a subject; or better, wherever there is a point of view there is a subject position. Whilst our constructionist epistemology can be summed up in the Saussurean formula: *the point of view creates the object*—the subject being the original, fixed condition whence the point of view emanates—Amerindian ontological perspectivism proceeds along the lines that the *point of view creates the subject*; whatever is activated or "agented" by the point of view will be a subject.[30] This is why terms such as *wari'*,[31] *dene*,[32] or *masa*[33] mean "people," but they can be used for—and therefore used by—very different classes of beings: used by humans they denote human beings; but used by peccaries, howler monkeys, or beavers they self-refer to peccaries, howler monkeys, or beavers.

As it happens, however, these nonhumans placed in the subject perspective do not merely "call" themselves "people"; they see themselves anatomically and culturally as *humans*. The symbolic spiritualization of animals would imply their imaginary hominization and culturalization; thus the anthropomorphic-anthropocentric character of Indigenous thought would seem to be unquestionable. However, I believe that something totally different is at issue. Any being which vicariously occupies the point of view of reference, being in the position of subject, sees itself as a member of the human species. The human bodily form and human culture—the schemata of perception and action "embodied" in specific dispositions—are deictics of the same type as the self-designations discussed above. They are reflexive or apperceptive schematisms by which all subjects apprehend themselves, and not literal and constitutive human predicates projected metaphorically (i.e. improperly) onto nonhumans. Such deictic "attributes" are immanent in the viewpoint, and move with it.[34] Human beings—naturally—enjoy the same prerogative and therefore see themselves as such.[35] It is not that animals are subjects because they are humans in disguise, but rather that they are human because they are potential subjects. This is to say *Culture is the Subject's nature*; it is the form in which every subject experiences its own nature. Animism is not a projection of substantive human *qualities* cast onto animals, but rather expresses the logical equivalence of the reflexive *relations* that humans and animals each have to themselves: salmon are to (see) salmon as humans are to (see) humans, namely, (as) human.[36] If, as we have observed, the common condition of humans and animals is humanity not animality, this is because "humanity" is the name for the general form taken by the Subject.

MULTINATURALISM

With this we may have discarded analogical anthropocentrism, but only apparently to adopt relativism.[37] For would this cosmology of multiple viewpoints not imply that "every perspective is equally valid and true" and that "a correct and true representation of the world does not exist"?[38]

But this is exactly the question: is the Amerindian perspectivist theory in fact asserting a multiplicity of representations of the same world? It is sufficient to consider ethnographic evidence to perceive that the opposite applies: all beings see ("represent") the world in the same way—what changes is the world that they see. Animals impose the same categories and values on reality as humans do: their worlds, like ours, revolve around hunting and fishing, cooking and fermented drinks, cross-cousins and war, initiation rituals, shamans, chiefs, spirits. "Everybody is involved in fishing and hunting; everybody is involved in feasts, social hierarchy, chiefs, war, and disease, all the way up and down."[39] If the moon, snakes, and jaguars see humans as tapirs or white-lipped peccaries, it is because they, like us, eat tapirs and peccaries, people's food.[40] It could only be this way, since, being people in their own sphere, nonhumans see things as "people" do. But the things that they see are different: what to us is blood, is maize beer to the jaguar; what to the souls of the dead is a rotting corpse, to us is soaking manioc; what we see as a muddy waterhole, the tapirs see as a great ceremonial house.

(Multi)cultural relativism supposes a diversity of subjective and partial representations, each striving to grasp an external and unified nature, which remains perfectly indifferent to those representations. Amerindian thought proposes the opposite: a representational or phenomenological unity which is purely pronominal or deictic, indifferently applied to a radically objective diversity. One single "culture," multiple "natures"—perspectivism is multinaturalist, for a perspective is not a representation.

A perspective is not a representation because representations are a property of the mind or spirit, whereas the point of view is located in the body.[41] The ability to adopt a point of view is undoubtedly a power of the soul, and nonhumans are subjects in so far as they have (or are) spirit; but the differences between viewpoints (and a viewpoint is nothing if not a difference) lies not in the soul. Since the soul is formally identical in all species, it can only see the same things everywhere—the difference is given in the specificity of bodies. This permits answers to be found for our questions: if nonhumans are persons and have souls, then what distinguishes them from humans? And why, being people, do they not see us as people?

Animals see in the *same* way as we do *different* things because their bodies are different from ours. I am not referring to physiological differences—as far as that is concerned, Amerindians recognize a basic uniformity of bodies—but rather to affects, dispositions or capacities which render the body of every species unique: what it eats, how it communicates, where it lives, whether it is gregarious or solitary, and so forth. The visible shape of the body is a powerful sign of these differences in affect, although it can be deceptive since a human appearance could, for example, be concealing a jaguar-affect. Thus, what I call "body" is not a synonym for distinctive substance or fixed shape; it is an assemblage of affects or ways of being that constitute a *habitus*. Between the formal subjectivity of souls and the substantial materiality of organisms there is an intermediate plane which is occupied by the body as a bundle of affects and capacities and which is the origin of perspectives.

The difference between bodies, however, is only apprehendable from an exterior viewpoint, by an other, since, for itself, every type of being has the same form (the generic form of a human being): bodies are the way in which alterity is apprehended as such. In normal conditions we do not see animals as people, and vice versa, because our respective bodies (and the perspectives which they allow) are different. Thus, if "culture" is a reflexive perspective of the subject, objectified through the concept of soul, it can be said that "nature" is the viewpoint which the subject takes of other body-affects; if Culture is the Subject's nature, then *Nature is the form of the Other as body*, that is, as the object for a subject. Culture takes the self-referential form of the pronoun "I"; nature is the form of the nonperson or the object, indicated by the impersonal pronoun "it."[42]

If, in the eyes of Amerindians, the body makes the difference, then it is easily understood why, in the anecdote told by Lévi-Strauss, the methods of investigation into the humanity of the other, employed by the Spanish and the inhabitants of the Antilles, showed such asymmetry. For the Europeans, the issue was to decide whether the others possessed a soul; for the Indians, the aim was to find out what kind of body the others had. For the Europeans the great diacritic, the marker of difference in perspective, is the soul (are Indians humans or animals?); for the Indians it is the body (are Europeans humans or spirits?). The Europeans never doubted that the Indians had bodies; the Indians never doubted that the Europeans had souls (animals and spirits have them too). What the Indians wanted to know was whether the bodies of those "souls" were capable of the same affects as their own—whether they had the bodies of humans or the bodies of spirits, nonputrescible and protean. In sum: European ethnocentrism consisted in doubting whether other bodies have the same souls as they themselves; Amerindian ethnocentrism in doubting whether other souls had the same bodies.

As Ingold has stressed, the status of humans in Western thought is essentially ambiguous: on the one hand, humankind is an animal species amongst others, and animality is a domain that includes humans; on the other hand, humanity is a moral condition which excludes animals.[43] These two statuses coexist in the problematic and disjunctive notion of "human nature." In other words, our cosmology postulates a physical continuity and a metaphysical discontinuity between humans and animals, the former making of man an object for the natural sciences, the latter an object for the "humanities." Spirit or mind is our great differentiator: it raises us above animals and matter in general, it distinguishes cultures, it makes each person unique before his or her fellow beings. The body, in contrast, is the major integrator: it connects us to the rest of the living, united by a universal substrate (DNA, carbon chemistry) which, in turn, links up with the ultimate nature of all material bodies.[44] In contrast to this, Amerindians postulate a metaphysical continuity and a physical discontinuity between the beings of the cosmos, the former resulting in animism, the latter in perspectivism: the spirit or soul (here not an immaterial substance but rather a reflexive form) integrates, while the body (not a material organism but a system of active affects) differentiates.

THE SPIRIT'S MANY BODIES

The idea that the body appears to be the great differentiator in Amazonian cosmologies—that is, as that which unites beings of the same type, to the extent that it differentiates them from others—allows us to reconsider some of the classic questions of the ethnology of the region in a new light.

Thus, the now old theme of the importance of corporeality in Amazonian societies acquires firmer foundations.[45] For example, it becomes possible to gain a better understanding of why the categories of identity—be they personal, social, or cosmological—are so frequently expressed through bodily idioms, particularly through food practices and body decoration. The universal symbolic importance of food and cooking regimes in Amazonia—from the mythological "raw and the cooked" of Lévi-Strauss, to the Piro idea that what literally (i.e. naturally) makes them different from white people is "real food";[46] from the food avoidances which define "groups of substance" in Central Brazil[47] to the basic classification of beings according to their eating habits;[48] from the ontological productivity of commensality, similarity of diet and relative condition of prey-object and predator-subject[49] to the omnipresence of cannibalism as the "predicative" horizon of all relations with the other, be they matrimonial, alimentary, or bellicose[50]—this universality demonstrates that the set of habits and processes that constitute bodies is precisely the location from which identity and difference emerge.

The same can be said of the intense semiotic use of the body in the definition of personal identities and in the circulation of social values.[51] The connection between this overdetermination of the body (particularly of its visible surface) and the restricted recourse in the Amazonian *socius* to objects capable of supporting relations—that is, a situation wherein social exchange is not mediated by material objectifications such as those characteristic of gift and commodity economies—has been shrewdly pinpointed by Terence Turner, who has shown how the human body therefore must appear as the prototypical social object. However, the Amerindian emphasis on the social construction of the body cannot be taken as the culturalization of a natural substract but rather as the production of a distinctly human body, meaning *naturally* human. Such a process seems to be expressing not so much a wish to "deanimalize" the body through its cultural marking, but rather to particularize a body still too generic, differentiating it from the bodies of other human collectivities as well as from those of other species. The body, as the site of differentiating perspective, must be differentiated to the highest degree in order completely to express it.

The human body can be seen as the locus of the confrontation between humanity and animality, but not because it is essentially animal by nature and needs to be veiled and controlled by culture.[52] The body is the subject's fundamental expressive instrument and at the same time the object par excellence, that which is presented to the sight of the other. It is no coincidence, then, that the maximum social objectification of bodies, their maximal particularization expressed in decoration and ritual exhibition, is at the same time the moment of maximum animalization, when bodies are covered by feathers, colors, designs, masks, and other animal prostheses.[53] Man ritually clothed as an animal is the counterpart to the animal supernaturally naked. The former, transformed into an animal, reveals to himself the "natural" distinctiveness of his body; the latter, free of its exterior form and revealing itself as human, shows the "supernatural" similarity of spirit. The model of spirit is the human spirit, but the model of body is the bodies of animals; and if, from the point of view of the subject, culture takes the generic form of "I" and nature of "it/they," then the objectification of the subject to itself demands a singularization of bodies—which naturalizes culture, i.e. embodies it—whilst the subjectification of the object implies communication at the level of spirit—which culturalizes nature, i.e. supernaturalizes it. Put in these terms, the Amerindian distinction of Nature/Culture, before it is dissolved in the name of a common animic human-animal sociality, must be reread in the light of somatic perspectivism.

It is important to note that these Amerindian bodies are not thought of as given but rather as made. Therefore, an emphasis on the methods for the continuous fabrication of the body;[54] a notion of kinship as a process of active assimilation of individuals[55] through the sharing of bodily substances, sexual and alimentary—and not as a passive inheritance of some substantial essence; the theory of memory which inscribes it in the flesh,[56] and more generally the theory which situates knowledge in the body.[57] The Amerindian *Bildung* happens in the body more than in the spirit: there is no "spiritual" change which is not a bodily transformation, a redefinition of its affects and capacities. Furthermore, while the distinction between body and soul is obviously pertinent to these cosmologies, it cannot be interpreted as an ontological discontinuity.[58]

As bundles of affects and sites of perspective, rather than material organisms, bodies "are" souls, just, incidentally, as souls and spirits "are" bodies. The dual (or plural) conception of the human soul, widespread in Indigenous Amazonia, distinguishes between the soul (or souls) of the body, reified register of an individual's history, site of memory and affect, and a "true soul," pure, formal subjective singularity, the abstract mark of a person.[59] On the other hand, the souls of the dead and the spirits which inhabit the universe are not immaterial entities, but equally types of bodies, endowed with properties—affects—sui generis. Indeed, body and soul, just like nature and culture, do not correspond to substantives, self-subsistent entities, or ontological provinces, but rather to pronouns or phenomenological perspectives.

The performative rather than given character of the body, a conception that requires it to differentiate itself "culturally" in order for it to be "naturally" different, has an obvious connection with interspecific metamorphosis, a possibility suggested by Amerindian cosmologies. We need not be surprised by a way of thinking which posits bodies as the great differentiators yet at the same time states their transformability. Our cosmology supposes a singular distinctiveness of minds, but not even for this reason does it declare communication (albeit solipsism is a constant problem) to be impossible, or deny the mental/spiritual transformations induced by processes such as education and religious conversion; in truth, it is precisely because the spiritual is the locus of difference that conversion becomes necessary (the Europeans wanted to know whether Indians had souls in order to modify them). Bodily metamorphosis is the Amerindian counterpart to the European theme of spiritual conversion.[60] In the same way, if solipsism is the phantom that continuously threatens our cosmology—raising the fear of not recognizing ourselves in our "own kind" because they are not like us, given the potentially absolute singularity of minds—then the possibility of metamorphosis expresses the opposite fear, of no longer being able to differentiate between the human and the animal, and, in particular, the fear of seeing the human who lurks within the body of the animal one eats[61]—hence the importance of food prohibitions and precautions linked to the spiritual potency of animals, mentioned above. The phantom of cannibalism is the

Amerindian equivalent to the problem of solipsism: if the latter derives from the uncertainty as to whether the natural similarity of bodies guarantees a real community of spirit, then the former suspects that the similarity of souls might prevail over the real differences of body and that all animals that are eaten might, despite the shamanistic efforts to desubjectivize them, remain human. This, of course, does not prevent us having amongst ourselves more or less radical solipsists, such as the relativists, nor that various Amerindian societies be purposefully and more or less literally cannibalistic.[62]

The notion of metamorphosis is directly linked to the doctrine of animal "clothing," to which I have referred. How are we to reconcile the idea that the body is the site of differentiating perspectives with the theme of the "appearance" and "essence" which is always evoked to interpret animism and perspectivism?[63] Here seems to me to lie an important mistake, which is that of taking bodily "appearance" to be inert and false, whereas spiritual "essence" is active and real.[64] I argue that nothing could be further from the Indians' minds when they speak of bodies in terms of "clothing." It is not so much that the body is a clothing but rather that clothing is a body. We are dealing with societies which inscribe efficacious meanings onto the skin, and which use animal masks (or at least know their principle) endowed with the power metaphysically to transform the identities of those who wear them, if used in the appropriate ritual context. To put on mask-clothing is not so much to conceal a human essence beneath an animal appearance, but rather to activate the powers of a different body.[65] The animal clothes that shamans use to travel the cosmos are not fantasies but instruments: they are akin to diving equipment, or space suits, and not to carnival masks. The intention when donning a wet suit is to be able to function like a fish, to breathe underwater, not to conceal oneself under a strange covering. In the same way, the "clothing" which, amongst animals, covers an internal "essence" of a human type, is not a mere disguise but their distinctive equipment, endowed with the affects and capacities which define each animal.[66] It is true that appearances can be deceptive;[67] but my impression is that in Amerindian narratives which take as a theme animal "clothing" the interest lies more in what these clothes do rather than what they hide. Besides this, between a being and its appearance is its body, which is more than just that—and the very same narratives relate how appearances are always "unmasked" by bodily behavior which is inconsistent with them. In short: there is no doubt that bodies are discardable and exchangeable and that "behind" them lie subjectivities which are formally identical to humans. But the idea is not similar to our opposition between appearance and essence; it merely manifests the objective permutability of bodies which is based in the subjective equivalence of souls.

Another classic theme in South American ethnology which could be interpreted within this framework is that of the sociological discontinuity between the living and the dead.[68] The fundamental distinction between the living and the dead is made by the body and precisely not by the spirit; death is a bodily catastrophe which prevails as differentiator over the common "animation" of the living and the dead. Amerindian cosmologies dedicate equal or greater interest to the way in which the dead see reality as they do to the vision of animals, and as is the case for the latter, they underline the radical differences vis-à-vis the world of the living. To be precise, being definitively separated from their bodies, the dead are not human. As spirits defined by their disjunction from a human body, the dead are logically attracted to the bodies of animals; this is why to die is to transform into an animal,[69] as it is to transform into other figures of bodily alterity, such as affines and enemies. In this manner, if animism affirms a subjective and social continuity between humans and animals, its somatic complement, perspectivism, establishes an objective discontinuity, equally social, between live humans and dead humans.[70]

Having examined the differentiating component of Amerindian perspectivism, it remains for me to attribute a cosmological "function" to the transspecific unity of the spirit. This is the point at which, I believe, a relational definition could be given for a category, Supernature, which nowadays has fallen into disrepute (actually, ever since Émile Durkheim), but whose pertinence seems to me to be unquestionable. Apart from its use in labeling cosmographic domains of a "hyperuranian" type, or in defining a third type of intentional beings occurring in Indigenous cosmologies, which are neither human nor animal (I refer to "spirits"), the notion of supernature may serve to designate a specific relational context and particular phenomenological quality, which is as distinct from the intersubjective relations that define the social world as from the "interobjective" relations with the bodies of animals.

Following the analogy with the pronominal set[71] we can see that between the reflexive "I" of culture (the generator of the concepts of soul or spirit) and the impersonal "it" of nature (definer of the relation with somatic alterity), there is a position missing, the "you," the *second person*, or the other taken as other subject, whose point of view is the latent echo of that of the "I." I believe that this concept can aid in determining the supernatural context. An abnormal context wherein a subject is captured by another cosmologically dominant point of view, wherein he is the "you" of a nonhuman perspective, *Supernature is the form of the Other as Subject*, implying an objectification of the human I as a "you" for this Other. The typical "supernatural" situation in an Amerindian world is the meeting in the forest between a man—always on his own—and a being which is seen at first merely as an animal or a person, then reveals itself as a spirit or a dead person and speaks to the man.[72] These encounters can be lethal for the interlocutor who, overpowered by the nonhuman subjectivity, passes over to its side, transforming himself into a being of the same species as the speaker: dead, spirit, or animal. He who responds to a "you" spoken by a nonhuman accepts the condition of being its "second person," and when assuming in his turn the position of "I" does so already as a nonhuman. The canonical form of these supernatural encounters, then, consists in suddenly finding out that the other is "human," that is, that it is the human, which automatically dehumanizes and alienates the interlocutor and transforms him into a prey object, that is, an animal. Only shamans, multinatural beings by definition and office, are always capable of transiting the various perspectives, calling and being called "you" by the animal subjectivities and spirits without losing their condition as human subjects.[73]

I would conclude by observing that Amerindian perspectivism has a vanishing point, as it were, where the differences between points of view are at the same time annulled and exacerbated:

myth, which thus takes on the character of an absolute discourse. In myth, every species of being appears to others as it appears to itself (as human), while acting as if already showing its distinctive and definitive nature (as animal, plant, or spirit). In a certain sense, all the beings which people mythology are shamans, which indeed is explicitly affirmed by some Amazonian cultures.[74] Myth speaks of a state of being where bodies and names, souls and affects, the I and the Other interpenetrate, submerged in the same presubjective and preobjective milieu—a milieu whose end is precisely what the mythology sets out to tell.

First published in *The Journal of the Royal Anthropological Institute* 4, no. 3 (1998): 469–88. A shorter version was presented as a Munro Lecture at the University of Edinburgh in 1998. The article is the result of an extended dialogue with Tânia Stolze Lima, who, in parallel with and synchronous to its earlier version (published first in Portuguese), has written a masterful article on perspectivism in Juruna cosmology ("O dois e seu múltiplo: reflexões sobre o perspectivismo em uma cosmologia tupi," *Mana* 2, no. 2 [1996]). Peter Gow (who, together with Elizabeth Ewart, translated most of the article into English), Aparecida Vilaca, Philippe Descola, and Michael Houseman made invaluable suggestions at various stages in the elaboration of the materials I present here. Bruno Latour's *Nous n'avons jamais été modernes. Essai d'anthropologie symétrique* (Paris: Éditions La Découverte, 1991) was an indirect but crucial source of inspiration. After this article had reached its present form, I read an essay by Fritz Krause ("Maske und Ahnenfigur. Das Motiv der Hülle und das Prinzip der Form," *Ethnol. Stud.* 1, 1931; mentioned in Marianne Boelscher, *The Curtain Within: Haida Social and Mythical Discourse* [Vancouver: University of British Columbia Press, 1989], 212 n. 10) that advances ideas strikingly similar to some developed here.

1 Karl Århem, "Ecosofía makuna," in *La selva humanizada: ecología alternativa en el trópico húmedo colombiano*, ed. François Correa (Bogotá: Instituto Colombiano de Antropología, Fondo FEN Colombia, Fondo Editorial CEREC, 1993).

2 Tânia Stolze Lima, "A parte do cauim: etnografia juruna," PhD diss., Museu Nacional, Universidade Federal do Rio de Janeiro, 1995; and Lima, "O dois e seu múltiplo: reflexões sobre o perspectivismo em uma cosmologia tupi," *Mana* 2, no. 2 (1996): 21–47.

3 Peter Rivière, "WYSINWYG in Amazonia," *JASO* 25, no. 3 (1994): 256. This notion of the body as a "clothing" can be found amongst the Makuna (Århem, "Ecosofía makuna"), the Yagua (Jean-Pierre Chaumeil, *Voir, savoir, pouvoir: le chamanisme chez les Yagua du nord-est péruvien* [Paris: École des Hautes Études en Sciences Sociales, 1983], 125–27), the Piro (Peter Gow, personal communication), the Trio (Rivière, "WYSINWYG in Amazonia"), and the Upper Xingu societies (Thomas Gregor, *Mehinaku: The Drama of Daily Life in a Brazilian Indian Village* [Chicago: University of Chicago Press, 1977], 322). The notion is very likely pan-American, having considerable symbolic yield, for example, in Northwest Coast cosmologies (see Irving Goldman, *The Mouth of Heaven: An Introduction to Kwakiutl Religious Thought* [New York: Wiley Interscience, 1975], and Marianne Boelscher, *The Curtain Within: Haida Social and Mythical Discourse* [Vancouver: University of British Columbia Press, 1989]), if not of much wider distribution, a question I cannot consider here.

4 For some examples see amongst many others: Gerald Weiss, *The Cosmology of the Campa Indians of Eastern Peru*, PhD diss., University of Michigan, 1969, 158; Weiss, "Campa Cosmology," *Ethnology* 11 (1972); Gerhard Baer, *Cosmología y shamanismo de los Matsiguenga* (Quito: Abya-Yala. 1994), 102, 119, 224; France-Marie Renard-Casevitz, *Le banquet masqué: une mythologie de l'étranger* (Paris: Lierre & Coudrier, 1991), 24–31 (Matsiguenga); Pierre Grenand, *Introduction à l'étude de l'univers wayãpi: ethno-écologie des Indiens du Haut-Oyapock (Guyane Française)* (Paris: SELAF/CNRS,

1980), 42 (Wayapi); Eduardo Viveiros de Castro, *From the Enemy's Point of View: Humanity and Divinity in an Amazonian Society* (Chicago: University of Chicago Press, 1992), 68 (Arawete); Ann Osborn, "Eat and Be Eaten: Animals in U'wa (Tunebo) Oral Tradition," in Roy Willis, ed., *Signifying Animals: Human Meaning in the Natural World* (London: Unwin Hyman, 1990), 151 (U'wa); and Fabiola Jara, *El camino del Kumu: ecologia y ritual entre los Akuriyó de Surinam* (Quito: Abya-Yala, 1996), 68–73 (Akuriyo).

5 See for example, Bernard Saladin d'Anglure, "Nanook, Super-Male; The Polar Bear in the Imaginary Space and Social Time of the Inuit of the Canadian Arctic," in Willis, ed., *Signifying Animals*; Ann Fienup-Riordan, *Boundaries and Passages: Rule and Ritual in Yup'ik Eskimo Oral Tradition* (Norman: University of Oklahoma Press, 1994; Eskimo); Richard Nelson, *Make Prayers to the Raven* (Chicago: University of Chicago Press, 1983); Roger McDonnell, "Symbolic Orientations and Systematic Turmoil: Centering on the Kaska Symbol of *dene*," *Canadian Journal of Anthropology* 4 (1984; Koyukon, Kaska); Adrian Tanner, *Bringing Home Animals: Religious Ideology and Mode of Production of the Mistassini Cree Hunters* (St. John's: Memorial University of Newfoundland, 1979); Colin Scott, "Knowledge Construction among Cree Hunters: Metaphors and Literal Understanding," *Journal de la société des américanistes* 75 (1989); Robert A. Brightman, *Grateful Prey: Rock Cree Human-Animal Relationships* (Berkeley: University of California Press, 1993; Cree); A. Irving Hallowell, "Ojibwa ontology, Behavior, and World View," in Stanley Diamond, ed., *Culture in History: Essays in Honor of Paul Radin* (New York: Columbia University Press, 1960; Ojibwa); Goldman, *The Mouth of Heaven* (Kwakiutl); Marie-Françoise Guédon, "An Introduction to the Tsimshian World View and Its Practitioners," in Margaret Seguin, ed., *The Tsimshian: Images of the Past, Views for the Present* (Vancouver: University of British Columbia Press, 1984; Tsimshian); and Boelscher, *The Curtain Within* (Haida). See also Signe Howell, *Society and Cosmos: Chewong of Peninsular Malaysia* (Oxford: Oxford University Press, 1984), Howell, "Nature in Culture or Culture in Nature? Chewong Ideas of 'humans' and Other Species," in Philippe Descola and Gísli Pálsson, eds., *Nature and Society: Anthropological Perspectives* (London: Routledge, 1996), and Wazir-Jahan Karim, *Ma'betisek Concepts of Living Things* (London: Athlone Press, 1981), for the Chewong and Ma'Betisek of Malaysia; and for Siberia, Roberte Hamayon, *La Chasse à l'âme. Esquisse d'une théorie du chamanisme sibérien* (Nanterre: Société d'Ethnologie, 1990).

6 Århem, "Ecosofía makuna"; Århem, "The Cosmic Food Web: Human-Nature Relatedness in the Northwest Amazon," in Descola and Pálsson, eds., *Nature and Society*; Stephen Hugh-Jones, "Bonnes raisons ou mauvaise conscience? De l'ambivalence de certains Amazoniens envers la consommation de viande," *Terrain* 26 (1996): 123–48; and Gerardo Reichel-Dolmatoff, "Tapir Avoidance in the Colombian Northwest Amazon," in Gary Urton, ed., *Animal Myths and Metaphors in South America* (Salt Lake City: University of Utah Press, 1985).

7 Aparecida Vilaça, *Comendo como gente: formas do canibalismo Wari'*

(Pakaa-Nova) (Rio de Janeiro: Editora da UFRJ, 1992).

8 Lima, "A parte do cauim," and Lima, "O dois e seu múltiplo."

9 Århem, "Ecosofía makuna," 11–12; Vilaça, *Comendo como gente*, 49–51.

10 Joanna Overing, "There Is No End of Evil: The Guilty Innocents and Their Fallible God," in David Parkin, ed., *The Anthropology of Evil* (London: Basil Blackwell, 1985), 249; Overing, "Images of Cannibalism, Death and Domination in a 'Non-Violent' Society," *Journal de la société des américanistes* 72 (1986): 245–46.

11 Baer, *Cosmología y shamanismo*, 89; Viveiros de Castro, *From the Enemy's Point of View*, 73–74.

12 Brightman, *Grateful Prey*, 40, 160; Claude Lévi-Strauss, *La Potière jalouse* (Paris: Plon, 1985), 14, 190; Weiss, "Campa Cosmology," 157–72.

13 Descola, *La Nature domestique. Symbolisme et praxis dans l'écologie des Achuar* (Paris: Maison des Sciences de l'Homme, 1986), 120.

14 See Århem, "Ecosofía makuna"; J. Christopher Crocker, *Vital Souls: Bororo Cosmology, Natural Symbolism, and Shamanism* (Tucson: University of Arizona Press, 1985); Hugh-Jones, "Bonnes raisons ou mauvaise conscience?"; Overing, "There Is No End of Evil"; and Vilaça, *Comendo como gente*.

15 For Amazonia see Chaumeil, *Voir, savoir, pouvoir*, 231–32, and Crocker, *Vital Souls*, 17–25.

16 Århem, "The Cosmic Food Web"; Descola, *La Nature domestique*, 317–30; Philippe Erikson, "De l'apprivoisement à l'approvisionnement: chasse, alliance et familiarisation en Amazonie amerindienne," *Techniques & Culture* 9 (1984): 110–12.

17 Descola, "Societies of Nature and the Nature of Society," in Adam Kuper, ed., *Conceptualizing Society* (London: Routledge, 1992); Descola, *La Nature domestique*.

18 Descola, *La Nature domestique*, 87–88.

19 Ibid., 115.

20 Or, as we may add, the case of the Ojibwa, where the coexistence of the systems of *totem* and *manido* (Lévi-Strauss, *Le Totémisme aujourd'hui* [Paris: Presses Universitaires de France, 1962], 25–33) served as a matrix for the general opposition between totemism and sacrifice (Lévi-Strauss, *La Pensée sauvage* [Paris: Plon, 1962], 295–302) and can be directly interpreted within the framework of a distinction between totemism and animism.

21 Århem, "The Cosmic Food Web," 185.

22 See Alfred Reginald Radcliffe-Brown, "The Sociological Theory of Totemism," in *Structure and Function in Primitive Society*, 1929 (reprint ed. London: Routledge & Kegan Paul, 1952), 130–31, who, amongst other interesting arguments, distinguishes *processes of personification* of species and natural phenomena (which "permits nature to be thought of as if it were a society of persons, and so makes of it a social or moral order"), like those found amongst the Eskimos and Andaman Islanders, from *systems of classification* of natural species, like those found in Australia and which compose a "system of social solidarities" between man and nature—this obviously calls to mind Descola's distinction of animism/totemism as well as the contrast of *manido/totem* explored by Lévi-Strauss.

23 Tim Ingold, "Becoming Persons: Consciousness and Sociality in Human Evolution," *Cultural Dynamics* 4, no. 3 (1991): 355–78; Ingold, "Hunting and Gathering as Ways of Perceiving the Environment," in Roy F. Ellen and

Katsuyoshi Fukui, eds., *Redefining Nature: Ecology, Culture and Domestication* (London: Berg Publishers, 1996).

24 Lévi-Strauss, "Race et histoire," in *Anthropologie structurale deux* (Paris: Plon, 1973 [1952]), 384.

25 Reichel-Dolmatoff, "Cosmology as Ecological Analysis: A View from the Rain Forest," *Man* 11, no. 3 (September 1976): 307–18.

26 Århem, "Ecosofía makuna."

27 Bruno Latour, *Nous n'avons jamais été modernes. Essai d'anthropologie symétrique* (Paris: Éditions La Découverte, 1991).

28 The uncomfortable tension inherent in such antinomies can be gauged in Howell's article "Nature in Culture or Culture in Nature?," on Chewong cosmology, where the Chewong are described as being both "relativist" and "anthropocentric"—a double mischaracterization, I believe.

29 Anne Christine Taylor, "Des fantômes stupéfiants: Langage et croyance dans la pensée achuar," *L'Homme* 33, no. 126–28 (1993): 429–47; Taylor, "Remembering to Forget: Identity, Mourning and Memory among the Jivaro," *Man* 28, no. 4 (1991): 653–78; Viveiros de Castro, "Apresentaçao to A. Vilaça," in Vilaça, *Comendo como gente*.

30 "Such is the foundation of perspectivism. It does not express a dependency on a predefined subject; on the contrary, whatever accedes to the point of view will be subject." Gilles Deleuze, *Le Pli. Leibniz et le baroque* (Paris: Minuit, 1988), 27.

31 Vilaça, *Comendo como gente*.

32 McDonnell, "Symbolic Orientations and Systematic Turmoil," 39–56.

33 Århem, "Ecosofía makuna."

34 Brightman, *Grateful Prey*, 40.

35 "Human beings see themselves as such; the Moon, the snakes, the jaguars and the Mother of Smallpox, however, see them as tapirs or peccaries, which they kill." Baer, *Cosmología y shamanismo*, 224.

36 If salmon look to salmon as humans to humans—and this is "animism"—salmon do not look human to humans (they look like salmon), and neither do humans to salmon (they look like spirits, or maybe bears; see Guédon, "An Introduction to the Tsimshian World View," 141)—and this is "perspectivism." Ultimately, then, animism and perspectivism may have a deeper relationship to totemism than Descola's model allows for.

37 The attribution of humanlike consciousness and intentionality (to say nothing of human bodily form and cultural habits) to nonhuman beings has been indifferently denominated "anthropocentrism" or "anthropomorphism." However, these two labels can be taken to denote radically opposed cosmological outlooks. Western popular evolutionism is very anthropocentric, but not particularly anthropomorphic. On the other hand, "primitive animism" may be characterized as anthropomorphic, but it is definitely not anthropocentric: if sundry other beings besides humans are "human," then we humans are not a special lot.

38 Århem, "Ecosofía makuna," 124.

39 Guédon, "An Introduction to the Tsimshian World View," 142.

40 Baer, *Cosmología y shamanismo*, 224.

41 "The point of view is located in the body, says Leibniz." Deleuze, *Le Pli*, 16.

42 Émile Benveniste, "La Nature des pronoms," in *Problèmes de linguistique générale* (Paris: Gallimard, 1966), 256.

43 Ingold, "Humanity and Animality," in *Companion Encyclopedia of Anthropology: Humanity, Culture and Social Life* (London: Routledge, 1994); Ingold, "Hunting and

Gathering as Ways of Perceiving the Environment," 1996.

44 The counterproof of the singularity of the spirit in our cosmologies lies in the fact that when we try to universalize it, we are obliged—now that supernature is out of bounds—to identify it with the structure and function of the brain. The spirit can only be universal (natural) if it is (in) the body.

45 A theme that much predates the current "embodiment" craze—see Anthony Seeger, Roberto Da Matta, and Viveiros de Castro, "A construção da pessoa nas sociedades indígenas brasileiras," *Boletim do Museu Nacional. Antropologia* no. 32 (1979): 2–19.

46 Gow, *Of Mixed Blood: Kinship and History in Peruvian Amazonia* (Oxford: Clarendon Press, 1991).

47 Seeger, "Corporação e Corporalidade: Ideologia de Concepção e Descendência," in *Os indios e nós* (Rio de Janeiro: Campus, 1980), 127–32.

48 Baer, *Cosmología y shamanismo*, 88.

49 Vilaça, *Comendo como gente*.

50 Viveiros de Castro, "Alguns aspectos da afinidade no dravidianato amazônico," in Viveiros de Castro and Manuela Carneiro da Cunha, eds., *Amazônia: etnologia e história indígena* (São Paulo: NHII [USP/FAPESP], 1993).

51 George Mentore, "Tempering the Social Self: Body Adornment, Vital Substance, and Knowledge among the Waiwai," *Journal of Archaeology and Anthopology* 9 (1993): 22–34; Terence Turner, "Social Body and Embodied Subject: Bodiliness, Subjectivity, and Sociality among the Kayapo," *Cultural Anthropology* 10, no. 2 (May 1995): 143–70.

52 Rivière, "WYSINWYG in Amazonia," 255–62.

53 Goldman, *The Mouth of Heaven*; Turner, "'We Are Parrots, Twins Are Birds': Play of Tropes as Operational Structure," in James W. Fernandez, ed., *Beyond Metaphor: The Theory of Tropes in Anthropology* (Stanford, CA: Stanford University Press, 1991); Turner, "Social Body and Embodied Subject," 143–70.

54 Viveiros de Castro, "A fabricação do corpo na sociedade Xinguana," *Boletim do Museo Nacional* 32 (1979): 2–19.

55 Gow, "The Perverse Child: Desire in a Native Amazonian Subsistence Economy," *Man* 24 (1989): 567–82; Gow, *Of Mixed Blood*, 1991.

56 Viveiros de Castro, *From the Enemy's Point of View*, 201–7.

57 Kenneth M. Kensinger, *How Real People Ought to Live: The Cashinahua of Eastern Peru* (Prospect Heights, IL: Waveland Press, 1995), chapter 22; Cecilia McCallum, "The Body That Knows: From Cashinahua Epistemology to a Medical Anthropology of Lowland South America," *Medical Anthropology* 10, no. 3 (1996): 1–26.

58 Graham Townsley, "Song Paths: The Ways and Means of Yaminahua Shamanic Knowledge," *L'Homme* 33, no. 126–28 (1993): 454–55.

59 McCallum, "The Body That Knows," 1–26; Viveiros de Castro, *From the Enemy's Point of View*, 201–14.

60 The rarity of unequivocal examples of spirit possession in the complex of Amerindian shamanism may derive from the prevalence of the theme of bodily metamorphosis. The classical problem of the religious conversion of Amerindians could also be further illuminated from this angle; Indigenous conceptions of "acculturation" seem to focus more on the incorporation and embodiment of Western bodily practices (food, clothing, interethnic sex) rather than on spiritual assimilation (language, religion, etc.).

61 The traditional problem of Western mainstream epistemology is how to connect and universalize (individual substances are given, relations have to be made); the problem in Amazonia is how to separate and particularize (relations are given, substances must be defined). On this contrast see Brightman, *Grateful Prey*, 177–85, and Fienup-Riordan, *Boundaries and Passages*, 46–50)—both inspired by the ideas about the "innate" and the "constructed" in Roy Wagner, "Scientific and Indigenous Papuan Conceptualizations of the Innate: A Semiotic Critique of the Ecological Perspective," in Timothy Bayliss-Smith and Richard Feachem, eds., *Subsistence and Survival: Rural Ecology in the Pacific* (London: Academic Press, 1977).

62 In Amazonian cannibalism, what is intended is precisely the incorporation of the subject-aspect of the enemy (who is accordingly hypersubjectivized, in very much the same way as that described for Melanesian warfare in Simon Harrison, *The Mask of War: Violence, Ritual, and the Self in Melanesia* [Manchester: University of Manchester Press, 1993], 121), not its desubjectivization as is the case with game animals. See Viveiros de Castro, *From the Enemy's Point of View*, 290–93; Viveiros de Castro, "Le Meurtrier et son double chez les Araweté: un exemple de fusion rituelle," *Systèmes de pensée en Afrique noire* 14 (1996): 98–102; Carlos Fausto, "A dialética da predação e familiarização entre os Parakanã da Amazônia oriental," PhD thesis, Museu Nacional, Universidade Federal do Rio de Janeiro, 1997.

63 Århem, "Ecosofía makuna," 122; Descola, *La Nature domestique*, 120; Hugh-Jones, "Bonnes raisons ou mauvaise conscience?"; Rivière, "WYSINWYG in Amazonia."

64 See the definitive observations of Goldman, *The Mouth of Heaven*, 63.

65 Gow, in a personal communication, tells me that the Piro conceive of the act of putting on clothes as an animating of clothes. See also Goldman, *The Mouth of Heaven*, 183, on Kwakiutl masks: "Masks get 'excited' during Winter dances."

66 "'Clothing' in this sense does not mean merely a body covering but also refers to the skill and ability to carry out certain tasks." Rivière, in Cees Koelewijn with Rivière, *Oral Literature of the Trio Indians of Surinam* (Dordrecht: Foris, 1987), 306.

67 Hallowell, "Ojibwa Ontology, Behavior, and World View," in Diamond, ed., *Culture in History*; Rivière, "WYSINWYG in Amazonia."

68 Carneiro da Cunha, *Os mortos e os outros. Uma análise do sistema funerário e da noção de pessoa entre os índios Krahó* (São Paulo: Hucitec Editora, 1978).

69 Donald Pollock, "Personhood and Illness among the Culina of Western Brazil," PhD diss., University of Rochester, NY, 1985, 95; Stephan Lane Schwartzman, "The Panara of the Xingu National Park: The Transformation of a Society," PhD diss., The University of Chicago, 1988, 268; Turner, "Social Body and Embodied Subject," 152; Vilaça, *Comendo como gente*, 247–55.

70 Religions based on the cult of the ancestors seem to postulate the inverse: spiritual identity goes beyond the bodily barrier of death, the living and the dead are similar in so far as they manifest the same spirit. We would accordingly have superhuman ancestrality and spiritual possession on one side, animalization of the dead and bodily metamorphosis on the other.

71 Benveniste, "La Nature des

pronoms" and "De la subjectivité dans le langage," in *Problèmes de linguistique générale*.

72 The dynamics of this communication are well analyzed by Taylor, "Des fantômes stupéfiants." This would be the true significance of the "deceptiveness of appearances" theme: appearances deceive because one is never certain whose point of view is dominant, that is, which world is in force when one interacts with other beings. The similarity of this idea to the familiar injunction not to "trust your senses" of Western epistemologies is, I fear, just another deceitful appearance.

73 As we have remarked, a good part of shamanistic work consists in desubjectivizing animals, that is in transforming them into pure, natural bodies capable of being consumed without danger. In contrast, what defines spirits is precisely the fact that they are inedible; this transforms them into eaters par excellence, i.e. into anthropophagous beings. In this way, it is common for the great predators to be the preferred forms in which spirits manifest themselves, and it is understandable that game animals should see humans as spirits, that spirits and predator animals should see us as game animals, and that animals taken to be inedible should be assimilated to spirits (Viveiros de Castro, "Alguns aspectos do pensamento yawalapíti [Alto Xingu]: classificações e transformações," *Boletim do Museu Nacional*, nac. 26 [1978]. The scales of edibility of Indigenous Amazonia (Hugh-Jones, "Bonnes raisons ou mauvaise conscience?") should therefore include spirits at their negative pole.

74 David M. Guss, *To Weave and Sing: Art, Symbol and Narrative in the South American Rainforest* (Berkeley: University of California Press, 1989), 52.

FLORESMILO SIMBAÑA

SUMAK KAWSAY AS A POLITICAL PROJECT

Sumak kawsay is a concept that has aroused broad debate in academic and political circles. This is not only because it forms part of the normative structure of the Bolivian and Ecuadorian constitutions, but because it was one of the fundamental discourses that enabled the Indigenous movement and other social organizations to confront neoliberalism. But if we want to define it, we are obliged to refer to the historical memory of the originary peoples, since it stems from that. It is in the combination of these two processes or time periods, then, that the elements allowing a better understanding of *sumak kawsay* must be sought.

This must be kept in mind in order to avoid falling into the common absurdity of presenting *sumak kawsay* as something of a quantitative notion, in which are amassed, as in an empty container, rights, policies, moral standards, and everything we can think of that could show ourselves as broad-minded and original, thereby ensuring that *sumak kawsay* is "the satisfaction of needs, the achievement of a dignified quality of life and death, loving and being loved, and the healthy flowering of all, in peace and harmony with nature, for the propagation of human cultures and biodiversity."[1]

In Ecuador today, this debate is determined by the political circumstances of the constitutional referendum of 2008 and the process since, characterized by the government's policies of "civic revolution" for the building of the new juridical and institutional framework of the state and its economic model. In that process, the Indigenous movement and the national government confronted each other's arguments and proposals. In no way can this be reduced to a "politically fueled fight," and even less to "defending spaces and privileges," as some governmental functionaries have asserted. What is at stake are different visions of proposals for confronting the capitalist model.

SUMAK KAWSAY AS A RESPONSE TO NEOLIBERALISM

From a historical perspective, *sumak kawsay* has subsisted in the historical memory of the Indigenous communities of the Andean region as a sense of life, an ethic that orders the life of the community. But in the period of the originary pre-Columbian states, it served to organize not only the community but society as a whole, including the state itself. This latter characteristic, of course, did not survive the destruction of the pre-Columbian states by conquest and subsequent colonization. For centuries, *sumak kawsay* was rescued and practiced by families, ayllus, communities; and as such it was precisely what the present Indigenous movements took up and claimed as an ethico-civilizational perspective and for its current elaboration as a political project.

During the structural adjustment of the 1990s, antineoliberal resistance concentrated on the struggle against free-trade agreements. This mobilization questioned the neoliberal discourse that was being presented as a definitive response to Latin America's permanent crisis: namely, through entry into globalization and the world market. For the neoliberals, this was the only possible path to progress and development. This discourse, along with its references to and promises of liberty and democracy, emphasized an economic model of abundance, free access to modernization (in particular to technology), the flow of food, and so on. The Indigenous and peasant movement had to denounce and fight this discourse and deployed alternative proposals around it.

Faced with this discourse, social movements, particularly the Indigenous one, developed elements of an economic model opposed to that projected by neoliberalism. In the case of Ecuador, these proposals spanned the economic and the agrarian, and from there moved toward food sovereignty and agrarian reform as indispensable conditions for an economic model contrary to that proposed by capitalism. But the idea of agrarian reform as a basis for food sovereignty could not repeat the experiences of the reforms of the 1960s and '70s, nor those carried out by older socialist processes. It was in this need for a new sociopolitical and economic project that *sumak kawsay*, that concept underpinning the memory and spirit of the Indigenous peoples, was transformed. And it is in this way that the Ecuadorian Indigenous movement is constructing its proposal for a plurinational state, a proposal in which the demand for an agrarian revolution is vital.

These two proposals, plurinationality and *sumak kawsay*, developed in the heat of resistance to neoliberalism, and their strength and social acceptance were such that in the new period marked by the government of Rafael Correa, and especially in the Asamblea Constituyente (Constituent assembly), the new political class had no alternative but to take it up and enshrine it in the country's new constitution, as well as in government discourses.

ELEMENTS FOR CONCEPTUALIZING SUMAK KAWSAY

The major foundation of Western philosophy is the conception of the human being as a separate entity from nature: the more civilized a society is, the more distanced it is from the natural world; having any perception of or relationship to nature as an active tie is proof of barbarism. Nature is conceived as counterposed to the civilized, the human, to reason, so it must be controlled and subdued as a mere object of domination and the maximum source of wealth.

Outside the Western orbit, and also within it, other people held and hold other conceptions. For them, reaching a high level of civilization is necessarily bound up with nature, because it is impossible to understand oneself as outside of nature: society and nature are a totality. To conceive of oneself as "part of" is not synonymous with barbarism—quite the contrary. This is the case of the originary peoples of America. For these peoples, Abya Yala was not a continent rich in natural resources but a "land of abundant life"; nature was not a resource but Pachamama, the "mother" of everything existing.

Sumak kawsay is a concept historically constructed by the Indigenous peoples of what we know today as the Andean area of South America, and makes reference to the attainment of a full life, a good way of living. For this to be possible, the life of nature and society must be regulated according to the principle of harmony and balance: "In harmony with the cycles of Mother Earth . . . , of life and history, and in balance with all forms of existence."[2] This involves various dimensions—social, cultural, economic, environmental, epistemological, and political—as an interrelated and interdependent whole, each of whose elements depends on the others. Human life cannot survive without nature. Because of this, *sumak kawsay* is undergirded by the concept of Pachamama, which refers to the universe, like the mother who gives and organizes life. Guaranteeing a good life for society, then, implies considering nature as "subject."

From this perspective, *sumak kawsay* does not depend on economic development, as dictated by capitalism, and much less so on the economic growth demanded by neoliberalism; but neither does it depend on extractivism. It depends on the defense of life in general. Thus *sumak kawsay* is not an individual moral reference or an abstract idea, as some government functionaries try to insinuate: "*Sumak kawsay* implies improving the population's quality of life, developing capacities and potentials: relying on an economic system that promotes equality through the social and territorial redistribution of the benefits of development."[3] Yet some are more audacious, attempting to convince us that *sumak kawsay* is an individual moral reference sustained "not only by 'having' but above all in 'being,' 'existing,' 'doing,' and 'feeling': in living well, in living fully."[4] For these rulers, *sumak kawsay* is reduced to a "redistribution of the benefits of development," not necessarily to a change of model or to the destruction of the real structures that sustain it.

As a system founded on the rights of nature, *sumak kawsay* demands reorganization and new approaches in the politico-economic model that would in turn transform not only society but above all the state. It is unthinkable to sustain or, worse, expand mining, petroleum exploitation, and other natural assets under the promise of redistribution and greater state participation without continuing to weaken the social economy of peoples. In the case of Ecuador, the "civic revolution" model is demonstrating that in the last analysis it ends up resting on the overexploitation of nature, maintaining the vigor of unproductive economies (financial and commercial) and empowering other, new ones, such as agribusiness and industrial food production. These were the economies that had the most dynamic growth during the first four years of Correa's government, and were also concentrated in two big monopolies: PRONACA and the Corporación Favorita C.A. As is obvious, this "new reality" has aggravated social conflicts.

COMMUNITY AT THE CENTER OF SUMAK KAWSAY

The central element of the proposal of plurinationality, hence of *sumak kawsay*, is the community. There are at least two understandings of this concept: on one hand, that of the state, and on the other, that of the Indigenous peoples. The state sees the community only as a form of social organization of a reduced, marginal, essentially rural segment of society. This concept is used as a strategy to gain access to assets (land) and/or services (drinking water, means of communication, etc.), but is seen as anachronistic and inefficient in management, administration, and socioeconomic reproduction. Within this perspective, the community as a political, economic, cultural, and juridical system has no place, and thus the state only grants it weak institutional support. In Ecuador, this was the original sin of legislation dating back to the first *Ley de Comunas* (Law of communes) of 1937, which put together an extensive administrative approach to lands, with laws referring to communal property and with no recognition of social self-government. The same logic applied in the agrarian-reform laws of 1964 and 1973; communal property was accorded an abstract, purely administrative mention but in terms of concrete public policies, what was promoted first was cooperativism and, subsequently, the "free association of individual producers." In neoliberal times—especially with the *Ley de Desarrollo Agrario* (Law of agrarian development) of 1994—this left the communes defenseless, obligated to adapt or "transform themselves" into other organizational forms as a means of subsistence.

To its regret, this way of conceiving and "regulating" communities did not bring about the end of their existence and historical relevance, and least of all of Indigeneity. Here is the other vision. According to Luis Macas, one of the most prominent Indigenous leaders, the commune is one of the crucial institutions that "in the process of reconstruction of peoples and ancestral nations . . . have established themselves throughout history and whose primordial function is to secure and provide continuity for the historical and ideological reproduction of the Indian peoples. For us, the commune is the *llacta*, or the ayllu or *jatun ayllu*. The commune is the organizational nucleus of Indigenous society. According to our understanding, the institution of the commune constitutes the fundamental axis that expresses Indigenous society and gives it cohesion."[5] As we can see, from this perspective the commune and/or community cannot be reduced to a prop or circumstantial instrument. Rather, its components and reality go much further: they encompass more areas of life, from the material to the historical and subjective (cultural and spiritual) dimension; "it is the fundamental basis of cultural, political, social, historical, and ideological concentration and processing."[6]

According to Macas, the following principles are re-created in the communitarian space:

> Reciprocity.
> A system of collective property.
> A relationship and coexistence with nature.
> Social responsibility.
> Consensus.[7]

These principles are ethical norms and practices of conviviality and of collective and individual relationships that go well beyond the "local" or the "rural." They are ideological imaginaries, the "must be," "the coordinative center of the Indigenous cosmovision" and identity, with their own cognitive parameters; but they are also concrete and supported models, openly contradicting capitalist neoliberalism and its paradigms of progress and development.

Communitarianism is one of the organizing principles of the political project of the Confederación de Nacionalidades Indígenas del Ecuador (CONAIE), the national organization of the country's Indigenous peoples and nationalities. As part of this project, a document was drawn up in 1994 (and revised in 2007) that includes the following definition:

> Historically, the Indigenous Nationalities and Peoples have constructed and practiced a communal mode of living for millennia. Communitarianism is the principle of life for all Indigenous Nationalities and Peoples, based on reciprocity, solidarity, equality, and self-management. Therefore, communitarianism is for us a regime of property and of systems of economic and sociopolitical organization that is collective

> in nature and that promotes the active participation and well-being of all its members. Our communitarian systems have historically adapted to external economic and political processes; they have been modified but they have not disappeared, they are alive and practiced daily by the Indigenous Nationalities and Peoples within the family and community. The sociopolitical model we advocate is a Communitarian, intercultural Society. In the new Plurinational State, familial, communitarian, and public property will be recognized and strengthened, and its economy will be organized by means of communitarian, collective, and familial forms.

As we can see, *sumak kawsay* is not a concept to be understood in isolation; it is necessarily united with the concept of plurinationality and it is directly tied to communitarian practices, which are the constitutive basis of both.

First published as "El sumak kawsay como proyecto político" in Miriam Lang and Dunia Mokrani, eds., *Más Allá del Desarrollo* (Quito: Abya Yala and Fundación Rosa Luxemburg, 2011), 218–26.

1 René Ramírez Gallegos, "Socialismo del Sumak Kawsay o biosocialismo republicano," in *Socialismo y Sumak Kawsay. Los nuevos retos de América Latina* (Quito: SENPLADES, 2010), 61.
2 Fernando Huanacuni Mamani, *Buen Vivir/Vivir Bien. Filosofía, políticas, estrategias y experiencias regionales andinas* (Lima: Coordinadora Andina de Organizaciones Indígenas, 2010), 34.
3 Ana María Larrea, "La disputa de sentidos por el buen vivir como proceso contrahegemónico," in Secretaría Nacional de Planeación y Desarrollo, ed., *Los nuevos retos de América Latina: Socialismo y Sumak Kawsay* (Quito: SENPLADES, 2010), 22.
4 René Gallegos Ramírez, "Socialismo del Sumak Kawsay o biosocialismo republicano," in *Socialismo y Sumak Kawsay, los nuevos retos de América* (Quito: SENPLADES, 2010), 61.
5 Luis Macas, "Instituciones Indígenas: La comuna como eje," *Boletín ICCI "Rimay"* 2, no. 17 (August 2000). Available online at http://icci.nativeweb.org/boletin/17/macas.html (accessed November 18, 2022).
6 Ibid.
7 Ibid.

THE FALLING SKY

Words of a Yanomami Shaman

THE FIRST SHAMAN

It was *Omama* who created the land and the forest, the wind that shakes its leaves, and the rivers whose water we drink. It is he who gave us life and made us many. Our elders have made us hear his name from the beginning. *Omama* and his brother *Yoasi* first came to existence alone. They did not have a father or a mother. Before them, in the beginning of time, only the people we call *yarori* existed.[1] These ancestors were human beings with animal names. They constantly metamorphosed. Gradually, they became the game we arrow and eat today. Then it was *Omama*'s turn to come into being and to re-create the forest, for the one that existed before was fragile. It constantly became other until finally the sky fell on it. Its inhabitants were pushed underground and became the meat-hungry ancestors we call the *aõpatari*.[2]

This is why *Omama* had to create a new, more solid forest, whose name is *Hutukara*. This is also the name of the ancient sky, which fell long ago. *Omama* set the image of this new land and carefully extended it little by little, like when one spreads clay to make a plate to bake a *mahe* cassava bread. Then he covered it with tight lines traced with annatto dye, like word drawings.[3] He planted immense pieces of metal in its depths so it wouldn't collapse. He also used them to root the sky's feet.[4] Without this, the land would have remained sandy and friable and the sky would not have stayed in place. Later *Omama* turned the remaining metal harmless and used it to make our ancestors' first metal tools.[5] Finally, he put the mountains down on the surface of the land so it would not shake in the storm winds and frighten human beings. He also drew a first sun to give us light. But it burned too hot and he had to get rid of it by destroying its image. Finally, he created the sun we still see in the sky, along with the clouds and the rain, so he could interpose them when it gets too hot. This is what I heard my elders say.

Omama also created the trees and the plants by sowing the pits of their fruit all over the ground. These seeds sprouted in the soil and gave rise to all the vegetation of the forest we have lived in ever since. And so the *hoko si*, *maima si*, and *rioko si* palms, the *apia hi*, *komatima hi*, *makina hi*, and *oruxi hi* trees, and all the other plants from which we get our food grew. At first their branches were bare. Then fruit formed on them. Finally, *Omama* created the bees that came to live on them and drink the nectar of the flowers with which they produce their honey.

At first, there weren't any rivers either; the waters ran deep underground. All you could hear of them was a distant roar, like that of powerful rapids. They formed a great waterway the shamans call *Motu uri u*. One day *Omama* was working in his garden with his son when the boy started to cry because he was thirsty. To quench his son's thirst, *Omama* made a hole in the ground with a metal bar.[6] When he pulled it out, water leaped up to the sky. It pushed back his child, who had come to drink his fill, and shot all the fish, skates, and caimans into the sky. The stream rose so high that another river formed on the sky's back, where the ghosts of our dead live. Then the waters accumulated on the earth and ran off in every direction to form the rivers, streams, and lakes of the forest.

In the beginning, no human beings lived there yet. *Omama* and his brother *Yoasi* lived there alone. There were no women yet. The two brothers met the first woman much later, when *Omama* fished *Tëpërësiki*'s daughter out of a big river.[7] In the beginning, *Omama* copulated in the fold of his brother *Yoasi*'s knee. With time, the latter's calf became pregnant and this is how Omama first had a son.[8] Yet we, the inhabitants of the forest, were not born this way. We came later, from the vagina of *Omama*'s wife, *Thuëyoma*, the woman he pulled out of the water.[9] The shamans have brought her image down since the beginning of time. They also refer to her as *Paonakare*. This was a fish being that let itself be captured in the appearance of a woman. It is so. If *Omama* hadn't fished her out of the river, perhaps human beings would still copulate behind their knees!

Later *Omama* got angry at his brother *Yoasi*, for *Yoasi* had furtively made appear the evil beings of disease we call *në wãri*, as well as those of the *xawara* epidemic, who are also eaters of human flesh.[10] *Yoasi* was bad and his thought full of oblivion. As for *Omama*, he had created the sun being who never dies and whom the shamans call *Mothokari*.[11] I am not speaking here of the sun whose warmth settles on the forest and which ordinary people see, but of the image of the sun.[12] It is so. The sun and the moon possess images that only the shamans can bring down and make dance. They have a human appearance, like us, but the white people cannot know them.

Omama wanted us to be as immortal as the sun. He wanted to do things handsomely and put a truly solid breath of life in us. So he searched the forest to cut a piece of hardwood to stand up and imitate his wife's shape with. He chose a *pore hi* ghost tree whose skin is constantly renewed. He wanted to introduce the image of this tree into our breath of life so it remained long and resistant.[13] This way, when we became old, our skin could have sloughed off and remained smooth and new forever. We could have constantly become young again and never died. This is what *Omama* wanted. But *Yoasi* took advantage of his absence and placed the bark of a tree we call *kotopori usihi* in *Omama*'s wife's hammock. This soft bark folded back on the side and hung from the hammock to the ground. The toucan spirits immediately broke into pained wails of mourning.[14] *Omama* heard them and got angry at his brother. But it was too late, the damage was done. *Yoasi* had taught us how to die once and for all. He had introduced death, that evil being, into our mind and our breath, making them so fragile.[15] Since then, human beings have always been close to death. This is why we also sometimes refer to white people as *Yoasi thëri*, People of Yoasi. We see their merchandise, machines, and the epidemics that constantly bring us death as traces of *Omama*'s bad brother.

It was also *Yoasi* who created the *Poriporiri* moon being. This is why *Poriporiri* also always dies. *Poriporiri* is a man who travels through the immensity of the sky every night, sitting in his pirogue like in a kind of plane. At first he is a young man, but he gets older and older day after day. By the end of his journey, he has become emaciated and his hair has turned white. Then finally he dies.

Then his daughters and the toucan spirits cry for him relentlessly. Their tears turn into heavy rain, which falls on the forest for a long time. Once their father's body has decomposed, they carefully gather his bones. Then they bloom again and *Poriporiri* comes back to life. It is so. The moon being is also a thing of death. *Yoasi* wanted it to be so because he lacked wisdom. Unlike him, *Omama* truly wanted us to be eternal. If he had been alone, we would never perish and our breath of life would always be as vigorous. But it was not so and, alas, *Yoasi* made our ancestors become other.

In the end, *Omama* created the *xapiri* so we could take revenge on disease and protect ourselves from the death with which his evil brother afflicted us.[16] Then he created the *urihinari* spirits of the forest, the *mãu unari* spirits of the waters, and the *yarori* animal spirits. He hid them on the mountain peaks and in the deepest part of the woods until his son became a shaman. At first, I thought the *xapiri* came into being by themselves, but I was wrong. I only fully understood who they were later, once I was able to see them and hear their singing. My wife's father also tells how it was *Omama*'s wife, the woman of the waters, who was the first to ask for the *xapiri* to come into being. We are her children and our elders became many starting from her. This is why after procreating she asked her husband: "What will we do to cure our children if they are sick?" This is what worried her. Her husband *Omama*'s thought remained in oblivion. No matter how hard his mind searched, he could not think what else he could create. Then the woman of the waters told him: "Come out of your perplexity. Create the *xapiri* who will cure our children!" *Omama* approved: "*Awe!* These are wise words. The spirits will chase the evil beings away. They will tear the image of the sick from them and bring it back into their body!" This is how he made the *xapiri* appear, as innumerable and powerful as we know them today.

Later *Omama*'s son became a young man and his father wanted him to know how to call the *xapiri* to heal his people. He found a *yãkoana hi* tree in the forest and told his son: "With this tree, you will prepare the *yãkoana* powder! You will mix it with the odorous *maxara hana* leaves and the bark of the *ama hi* and *amatha hi* trees and drink it! The *yãkoana*'s power reveals the *xapiri*'s voice. By drinking it, you will hear their clamor and you too will become a spirit!" He blew the *yãkoana* into his son's nostrils with a tube made from a *horoma* palm.[17] Then *Omama* called the *xapiri* for the first time and added to his son: "Now it is up to you to make them come down. If you behave well and the spirits really want you, they will come to you to do their presentation dance and remain by your side. You will be their father. When your children are sick, you will follow the path of the evil beings and fight them to bring back their image! You will also bring down the *ayokora* cacique bird spirit to help you to regurgitate their dangerous objects, which you will tear out from inside the sick people.[18] This way you will truly be able to cure human beings!" This was how *Omama* revealed the use of the *yãkoana* to his son—the first shaman—and taught him how to see the spirits he had just created. Our elders have continued to follow the trail of his words to this day. This is why we continue to drink the *yãkoana* to make the *xapiri* dance. We do not do that without reason. We do it because we are inhabitants of the forest, sons and sons-in-law of *Omama*.

Omama's son listened to his father's words carefully and set his mind on the *xapiri*. He entered a ghost state and became other.[19] He began to see the beauty of their presentation dance. He quickly became a shaman, for he was able to be friendly with all the spirits. The *xapiri* had kept their eyes set on him since he was a little child and his father had often spoken to him about them. Now he was grown up and they had finally come to him in great numbers. He could see them come down, dazzling with light, and hear their melodious songs. He exclaimed: "Father! Now I know the spirits and they have come to stand by my side. From now on, human beings will be able to multiply and fight off disease." *Omama* was the only one to know the *xapiri*. He gave them to his son, for if he had died without teaching their words, there would never have been any shamans in the forest. He did not want human beings to be left helpless and pitiful. This is why he made his son into the first shaman. He left him the *xapiri*'s path before he fled far away. This is what he wanted.

He told him: "With these spirits, you will protect human beings and their children, no matter how many there are. Don't let the evil beings and jaguars come devour them. Prevent the snakes from biting them and the scorpions from stinging them. Divert the *xawara* epidemic smoke from them. Also protect the forest.

Prevent the river waters from flooding it and the rains from mercilessly drenching it. Repel the cloudy weather and darkness. Hold up the sky so it does not fall apart. Don't let the lightning strike the earth and soothe the thunder's vociferation. Prevent the giant *Wakari* armadillo spirit from cutting up the roots of the trees and the *Yariporari* storm wind from arrowing them and making them fall!" These were *Omama*'s words to his son. This is why the shamans continue to defend the *Yanomami* and the forest today. They also protect the white people, even if they are a different people, as well as all lands, however vast and distant.

Omama's son first took the *yãkoana* with his father. Then he continued to drink it alone, again and again, to call the spirits in increasing numbers so he could know their songs. He was magnificent to see when he made their images dance. He was a very beautiful young man. His skin was coated with vermilion annatto and covered in shiny black drawings. His curassow crest armbands held a profusion of scarlet macaw tails, toucan tail pendants, and bunches of *paixi* feathers.[20] His eyes were dark and his hair covered in dazzling white *hõromae* down feathers.[21] A thick black saki tail was wrapped around his forehead.[22] He danced slowly with his back arched, filled with joy by contemplating the beauty of the *xapiri*. He called them and brought them down endlessly because he truly carried them in his thought. This was because he came from the sperm of *Omama*, who is their creator.

Today I think that *Omama*'s son is dead. Yet his image still exists, very far from here, downstream of the rivers, to the east, or maybe in the sky. I saw him during the time of dream. His image accompanied the image of our forest in tears. Becoming ghost under the effect of epidemic fumes, the sick forest was asking the *xapiri* to heal the suffering the white people's rage had inflicted upon her. She implored them to clean her trees and to make their leaves shiny; to make their flowers grow and to make her fertility return. She told them: "You are mine, you must avenge me!" I see all that in dreams when the inner part of my body becomes other, after the *yãkoana* has turned me into a ghost by day.[23] Otherwise I would not be able to speak about all this like I do.

Omama's son was the first to become spirit, before anyone else. He was the first to study and to see things with the *yãkoana*.

Following on from him, many of our elders became shamans. He taught them to make the spirits dance. He told them what *Omama* had taught him: "When the evil beings of the forest capture your children's image to devour it, the *xapiri* will take it back and avenge you!"[24] It was by following these words that the elders started drinking the *yãkoana* and contemplating the splendor of the spirits. We continue to do this to the present day. This is why shamans can so often be seen working in our houses.[25] Without their movement, they would be empty and silent. It is so. These words are very old but they will not disappear, for they are very beautiful and their value is very high.

First published in Kopenawa and Albert, *The Falling Sky: Words of a Yanomami Shaman* (Cambridge, MA: The Belknap Press of Harvard University Press, 2013), 27–33.

1 This human/animal ancestor's name comes from *yaro*, game, followed by the suffix *-ri* (plur. *pë*), which denotes the original times, the nonhuman, superlative, monstrous, or something of extreme intensity. These first ancestors transformed (*xi wãri-*) into game while their "images" became *xapiri* spirits (see Davi Kopenawa and Bruce Albert, *The Falling Sky: Words of a Yanomami Shaman*, trans. Nicholas Elliott and Alison Dundy [Cambridge, MA: Belknap Press, 2013], chap. 1, note 26).

2 On the fall of the sky and the pushing of the first ancestors into the underworld, where they were transformed into chthonian voracious predators—*aõpatari* (plur. *pë*)—see M 7 in Johannes Wilbert and Karin Simoneau, eds., *Folk Literature of the Yanomami Indians* (Los Angeles: UCLA Latin American Center Publications, 1990; subsequent numbered myths are also from this volume), and Kopenawa and Albert, *Falling Sky*, chaps. 6 and 7.

3 Vermilion dye obtained from oily seeds found in the fruit of a small cultivated tree (*nara xihi*).

4 The Yanomami think the celestial level (*hutu mosi*) curves at its ends to get closer to the terrestrial level (*warõ patarima mosi*), where it rests upon it with huge "feet" (stakes).

5 On the pathogenic power of the metal *Omama* hid under the ground, see Kopenawa and Albert, *Falling Sky*, chap. 16.

6 On *Omama* and the origin of rivers, see M 202. On *Omama* and the origin of metal, see Kopenawa and Albert, *Falling Sky*, chap. 9.

7 On the aquatic monster *Tëpërësiki* (sometimes associated with the anaconda), his daughter's coupling with *Omama*, and the origin of cultivated plants, see M 197 and 198.

8 On the birth of *Omama*'s son, see M 22. Davi Kopenawa sometimes calls him *Pirimari a*, which is also the name of the star that the Yanomami refer to as "the moon's son-in-law," Venus.

9 This name is based on a feminine redundancy: *thuë*, "woman, wife"; *-yoma*, feminine name suffix (like in *napëyoma*, "white woman"). This emphasizes the fact that this is the first woman. It is also a "fish being–woman," with which Davi Kopenawa freely associates our image of the siren (see Kopenawa and Albert, *Falling Sky*, chap. 20).

10 These evil forest beings (see ibid., chap. 1, note 23) are also described by the expressions *yanomae thë pë rããmomãɨwi*, "those who make humans sick," and *yanomae watima thë pë*, "eaters of humans."

11 The *Watorikɨ* shamans say that the ghostly form (his "value of ghost," *në porepë*) of *Omama* (equivalent to his "image," *utupë*) "has many names" (*thë ã waroho*), such as the sun being *Mothokari*, the jaguar being *Ɨramari*, and the evil being *Omamari*.

12 The "ordinary people," *kuapora thë pë* (literally, "the people who exist simply"), are opposed here to shamans, *xapiri thë pë* (literally, "the spirit people"). The latter consider that the former have "ghost eyes" that can only see the deceptive appearance of beings and phenomena. Conversely, shamans are said to be able to see the image-essence (*utupë*) of existing beings at the time of their mythical creation. This primordial image-form is denoted by the suffix *-ri* (plur. *-ri pë*; see above, note 1).

13 For another version of this origin myth of short life, see M 191. Western

Yanomami mothers tie their newborn's umbilical cord to this tree and carry the baby around in it in order to assure him a long life (Jacques Lizot, *Diccionario enciclopédico de la lengua yãnomami: Morfología* [Puerto Ayacucho, VE: Vicariato Apostólico, 2004], 321).

14 The Yanomami consider the toucan's hoarse and piercing call particularly sad. They associate it with mourning and nostalgia. Hearing the toucans "cry" in the forest announces a death in a distant home; listening to their call in the house at dusk gives rise to amorous longing.

15 Here, Davi Kopenawa refers to *pihi*, a term that denotes conscious thought and volition (but also the gaze), and to the "breath of life" (*wixia*). Death is referred to as *noma a*.

16 The shamanic cure's actions against pathogenic entities that devour the sick person's body image/vital essence (*utupë*) are described by warlike expressions such as *në yuai*, "to take revenge"; *nëhë rëai*, "to ambush"; *nëhë yaxuu*, "to scare away" (see Kopenawa and Albert, *Falling Sky*, chap. 6). The Yanomami conceive all relationships of otherness/hostility in terms of predation and vengeance, be it between humans or between humans and nonhumans.

17 This tube, between fifty and ninety centimeters long, is generally made from the hollowed-out stalk of a small *horoma a* palm or from a cultivated *xaraka si* bitter cane.

18 The greatest Yanomami shamans are considered able to regurgitate (*kahiki ho-*) the pathogenic objects affecting the sick person's body image/vital essence (*utupë*) or animal double (*rixi*). See Kopenawa and Albert, *Falling Sky*, chap. 7.

19 The expressions "act in/enter a ghost state" (*poremuu*) or "to become other" (*në aipëi*) both refer to states of altered consciousness induced by hallucinogens, pain, or sickness in which the "ghost" (pore) enclosed in each living being (similar to our "unconscious") takes precedence over consciousness (*pihi*).

20 These bunches consist of feathers torn horizontally and attached to a small stick. Green *werehe* parrot feathers and black and white *maraxi* piping-guan feathers are most commonly used. They can also be made with white ventral feathers from the *paari* wild turkey and the *wakoa* hawk.

21 An ornament made of down feathers from the *watupa aurima* vulture and the *wakoa* and *kãokãoma* birds of prey.

22 This monkey's tail is covered in thick and shiny black fur. This ornament and those described in the previous notes are worn by shamans during their sessions and, more generally, by all men during *reahu* feasts.

23 *Uuxi* refers to "the inner part" of the person, as opposed to his or her body ("the skin"), *siki*. The expression *xi wãri-* (literally, "turn bad") refers to mythical transformations and any kind of ontological change of form or identity. In this case it is synonymous with *në aipëi*, "to take the value of/become other." It can also mean "to lose one's mind, to be out of one's self," and, more concretely, "to get tangled up, become inextricable, no longer cease (state or action), remain stuck."

It should be noted here that nocturnal shamanism associated with the dream time is a fundamental part of Yanomami shamanism. Shamanistic initiation and work colonize the dreams of shamans with hallucinatory remnants of their daytime sessions (see Kopenawa and Albert, *Falling Sky*, chap. 22). Both *yãkoana* snuff and dreaming allow shamans to access the mythical times of origin which, for them, continue to unfold immutably in an eternal present, as a parallel dimension to historical time (that of migrations and wars).

24 The Yanomami consider any form of lethal aggression, whether human (war, sorcery) or nonhuman (evil beings, enemy shamanic spirits), visible or invisible, as a form of cannibalism (see Bruce Albert, "Temps du sang, temps des cendres: Représentation de la maladie, espace politique et système rituel chez les Yanomami de sud-est [Amazonie brésilienne]" [PhD diss., Université de Paris X Nanterre, 1985]).

25 Shamanic activity is referred to by the verb *kiãɨ*, "to move, to work."

NONHUMAN RIGHTS

> *Our culture abhors the world.*
> —Michel Serres, *The Natural Contract*, 1990

While the crew of technicians worked through the last hours of a daylong operation to plug the drilling hole, a mass of hydrocarbons leaked into the bottom of the bore well, some 13,000 feet below the sea floor. Undetected by the safety sensors, the material rapidly flowed upward inside the giant tubular structure. As it rose closer to the surface, where pressure is much lower, gases present in the mixture quickly expanded in volume, pushing drilling fluids within the riser further above. This sudden influx led to the complete deregulation of the pressure balance within the well, generating a pump effect that sucked large quantities of high-pressure oil and gases from the reservoir deep down in the earth's crust. Emergency alarms and valves designed to shut down the pipes failed to function. At 21:47 on April 20, 2010, an uncontrolled stream of mud—a mixture of seawater, oil, gas, and other components—burst over the deck of the *Deepwater Horizon* floating rig in the Gulf of Mexico. Two minutes later, the hydrocarbons inevitably ignited, causing the first of a sequence of massive explosions.[1]

For the next thirty-six hours, an army of ships tried in vain to halt the fire, until the platform finally collapsed and sank five thousand feet below the surface of the ocean. Then the real ecological catastrophe began. Because the marine riser failed to disconnect from the rig, the tubular structure fractured completely while *Deepwater Horizon* was sinking, leaving a large opening to the giant Macondo oil field in the seabed. A series of systems were engineered to try to stop the spill, all of which either totally failed or were ineffective. Meanwhile, as large-scale contamination loomed from the deep sea and forecasts of a severe hurricane season prompted concerns of further ecological chaos in the gulf, response operations escalated to warlike proportions.[2] Only six months later, after the installation of a cement cap, was the well considered permanently sealed.

One year on, local testimonies and international media reports revealed that oil was still leaking under the ocean.[3]

By any comparison, the "BP oil spill" was one of the largest offshore oil disasters in history and the gravest ever registered within US jurisdictional waters. The official (and highly controversial) quantification of the amount of oil that was discharged into the gulf amounts to a volume more than six times greater than the quantity discharged into the ocean by the *Exxon Valdez* spill in 1989.[4] Apart from the damage caused to coastal communities, scientists paint a grim ecological picture of planetary proportions. Being one of the most important migration corridors in the world, the Gulf of Mexico is the feeding ground for multiple species of fish, seabirds, and marine mammals. The spreading of toxic fluids and gases may significantly alter the rates of mortality of sessile plants and damage coral reef habitats, thus severely affecting the food chain and impacting the population size of certain species of fish. Furthermore, scientific analysis of real-time satellite data concluded that there is enough evidence to show that the biochemical and biophysical effects of the oil spill led to the break of a crucial ocean stream named the Loop Current, which could possibly trigger chain reactions that will alter the thermoregulation functions of the Gulf of Mexico and thus affect the global climate.[5]

"To invent the sailing ship or steamer is to invent the shipwreck," wrote architect Paul Virilio with respect to the philosophical dimensions of technological catastrophes. "It follows that fighting against the damage done by Progress above all means uncovering the truth of our success in this accidental revelation."[6] Perhaps more relevant than all the other historical records broken by the BP oil spill was the fact that it inaugurated a previously unknown form of disaster, for which the best available means of environmental defense were completely inadequate. Up until this point, the science of governing oil spills was informed by what had been learned with the *Exxon Valdez* oil spill on March 24, 1989, in the Gulf of Alaska, a tank vessel disaster. The technologies of containment that were developed were thus prepared only to deal with surface-based leaks, battling waves of crude in order to protect shorelines.[7] The ultra-deep-water spill in the Gulf of Mexico, however, exposed a totally different accidental chemistry and physics. It demonstrated that as the frontiers of resource extraction are expanded toward formerly unreachable geological depths and land surfaces of the earth, the nature of catastrophes to come will be completely other than those already recorded, from the Gulf of Mexico to Fukushima—far more violent and devastating than before.

Through examining the disastrous blowout at the Macondo well and the multiple reactions that ensued, what becomes apparent is that this event challenged not only the currently available knowledge and technologies of environmental remediation, but also the means of representation that render socioecological catastrophes culturally and politically meaningful. The unprecedented scope of the disaster, and the extraordinary qualitative dimensions of the ecological damage it unleashed, exposed the limits of a certain regime of visibility that shape our perceptions and relations to what we have named "nature." By looking to a crucial space of mediation through which our conceptions of the natural world are constructed (the courts, codes, and protocols of law), probing how nature appears within legal forums and texts, and questioning the ways by which violence against forms of life other than human are legally moderated, this text contends that, ultimately, the fundamental conflict is not so much related to the question of whether the existing legal provisions can ensure proper protection to ecosystems, but rather to the very concept of nature itself that is being inscribed by law. As such, I conclude by speculating on the idea that power is concerned as much with relations of cultural hegemony as with those of "natural hegemony," insofar as power seeks to impose a particular view of what nature stands for in such a manner that it appears as a universal condition, thus reducing the diversity of forms of entanglement between society and environment to a univocally utilitarian perspective. The hidden laws and politics that constituted this natural order have functioned and continue to function as a subtle yet powerful instrument of domination of both humans and nonhumans, and, crucially, of the relations established between them.

IN DEFENSE OF THE RIGHTS OF THE SEA

What was even more troubling in the case of the BP oil spill was that the unforeseeable socioecological impact it unleashed was not only a consequence of the spill itself, but in large part the result of the technology that was used to contain it and to "clean up" the sea. Besides setting a record for the quantity of discharged oil, the Macondo well blowout became a landmark event in the history of catastrophes because BP's response strategy released an incalculable volume of dispersant agents into the ocean. Corexit—the carcinogenic chemical compound deployed as a dispersant—is manufactured under the monopoly of a US-based company named Nalco whose managerial board includes executives from powerful petroleum multinationals such as Exxon and, not surprisingly, BP.[8] Designed to break up oil into tiny droplets and thus avoid the formation of large waves of crude, dispersants help to reduce impact on shorelines. At the same time, their toxic chemistry also has the capacity to severely damage and even kill various forms of marine life. As if we were watching another version of the chemical attacks deployed against the sea of tropical forests of Indochina during the Vietnam War, aircraft such as the US Air Force C-130 sprayed millions of liters of Corexit over the Gulf of Mexico and millions of liters more were directly injected deep into the ocean.[9]

"Never before in human history," wrote journalist Julia Whitty after visiting the disaster zone while clean-up operations were taking place, "has the vast food web of the ocean—rooted in the dark, and flowering at the surface—come under so many assaults from below, above, and within the water column: marine warfare masquerading as a cleanup."[10] It is almost as if we were trapped within a vicious, deadly loop: the same means of inflicting violence against nature deployed as instruments for its healing, the industry of production and destruction converging into a global conveyor belt lubricated by heavy crude and its petrochemical derivatives. While economists have long built upon Karl Marx's original insights to demonstrate how capital finds in devastation its most energizing forces, nothing can be compared to the geographical scope achieved by the logic of "creative destruction" today, which now operates at the scale of Earth's climate.

"What action could we take?," activist and writer Esperanza Martínez asked herself when news about the *Deepwater Horizon* catastrophe started leaking through the media. "What could function as a pedagogic instrument to bring public awareness to the limits imposed by extreme disasters such as the BP oil spill?"[11] During an interview conducted in Quito in late 2011, Martínez, member of NGO Acción Ecológica and Oilwatch representative in Ecuador, explained to me why the response came in the somewhat unusual form of a lawsuit filed in the constitutional court of Ecuador, a country which, in principle, had little if any direct relation to the devastation that swept through the currents of the Gulf of Mexico. Signed by the main Indigenous organizations of Ecuador together with an international coalition of NGOs and activists, the preamble of the lawsuit set the dispute in the following terms:

> We hereby submit, in *defense of the rights of the sea*—understood as an integral part of nature which the Ecuadorian Constitution of 2008 recognizes as a subject of rights and which we recognize as a giver of life of which we form part—the present lawsuit filed under the *principle of universal jurisdiction* against the transnational corporation British Petroleum PLC, headquartered in the United Kingdom, as the responsible party for the environmental disaster that struck the Gulf of Mexico on 20 April 2010.[12]

Legal advocacy is of course not a novel strategy in responding to ecological catastrophes. Most notably after the publication in 1962 of *Silent Spring*, Rachel Carson's seminal indictment of chemical pollutants, the alliance between the natural and legal sciences was firmly consolidated as lawyers and ecologists increasingly joined efforts to take action on behalf of the environment. The lawsuit filed against BP in Ecuador draws from this historical tradition, but at the same time projects it anew, appropriating the classic tools of environmental advocacy to expose its own limitations, searching for means to expand the political force of what has been called "environmental justice." More than an action with law as its instrument, the lawsuit was an intervention within the very frames of law itself, which sought to make visible how the existing legal order inevitably legitimizes the ecological violence it should help to restrain.

Although rigorously crafted to respond to juridical protocols, the lawsuit subverted the traditional grammar of legal demands by framing the dispute as a political-cultural conflict of broader implications.

The plaintiffs justified the indictment through a forceful criticism of the "model of growth, overexploitation and plunder" of the global fossil-fuel industry, the "development philosophy antagonistic to nature" upon which that model is based, and the international legal system that regulates it.[13] Besides the clarity with which the lawsuit exposed the hidden economic, political, and juridical mechanisms that lie behind large-scale ecological disasters such as the BP oil spill, what made this legal text a rather exceptional intervention within the larger context of environmental politics was the position that the environment occupies within it. Similarly to traditional class suits, the plaintiffs emphasized the necessity to redress the harm caused to local communities but, in a radically different manner, they situated the demand beyond the arena of human rights and socioeconomic rights, advocating not on behalf of the interests of the users of the Gulf of Mexico but "in *defense of the rights of the sea*." Nature thus appears as the primary subject whose rights had been violated rather than only the medium through which the rights of persons were impaired.

From this juridical gesture unfolded a series of pragmatic consequences that challenged the then current mechanisms of restorative justice with regard to the environment—the most significant being that, while acting as the legal guardians of the rights of nature, the plaintiffs rejected any form of economic reparation, penalty, or award. Instead, following a pioneer initiative launched in Ecuador to keep large reserves of crude in Amazonia *bajo tierra*, they demanded that BP commit to leaving underground the equivalent amount of oil that was discharged into the ocean. There is a dissident conception of environmental damage and compensation operating here which, as opposed to traditional juridical approaches, aims to respond to forms of injury and rights violations that extrapolate the geography of the (collective) human body and the temporality of social history, thus requiring a legal framework that corresponds to the scaleless spatiality of nature and the deep time of the geological.

Starting from the understanding that nature itself should be considered as a subject entitled to rights, the reinstatement of the ecology of the Gulf of Mexico is considered imperative yet not sufficient. By displacing the human-centric foundations of modern/colonial law, the lawsuit projects a legal concept of environment that cannot be bound to regional ecological dynamics or the artificial lines that demarcate national jurisdictions. Therefore leaving oil in the ground, the plaintiffs argued, could be a more just way of compensating nature for the impact that the BP oil spill had on global regulatory thermodynamic cycles. Conceived as a "pedagogic instrument," as Martínez described it, the lawsuit's central message dwelled within the field of the biopolitical, situating environmental crimes as a form of violence against life itself, but only insofar as the inscription of life into the realm of politics is not limited to the philosophical tenets of humanism. Rather, biopolitics is considered to be contingent upon the multispecies agency that makes natural history, before and beyond the human, against state-led forms of territorialization.

There have been several attempts at defining provisions for including severe environmental damage alongside other forms of international crime, such as, for example, establishing "ecocide" as a crime against peace. However, there exists no international or national court that recognizes the rights of nature apart from courts in Ecuador, under the constitutional visions established in 2008, and more recently in Bolivia, under the Law of the Rights of Mother Earth, passed in 2010. Therefore, the lawsuit against BP was filed on the basis of the principle of *universal jurisdiction*, a doctrine of international law which contends that a state can claim jurisdiction over crimes committed outside its territorial boundaries, regardless of whether those crimes were committed against its own citizens, in the context of certain types of international crime such as genocide and crimes against humanity. Because those types of crimes are considered severe attacks on the dignity and consciousness of humanity, the means of legislating them are not ascribed to domestic courts, but belong to a conceptual "universal court" that can potentially be activated anywhere around the globe. By invoking the principle of universal jurisdiction, the lawsuit filed against BP sought to position the oil spill as a matter that concerned humanity as such, calling attention to the absence of an adequate legal forum capable of protecting the vital cycles of global ecosystems.

"We are filing this lawsuit to break with the long-standing colonial logic of positive rights, which closes the doors to us for demanding fulfillment of the rights of nature in formal spaces," the lawsuit concluded.[14] Environmental advocacy has since its early beginnings made use of strategies of exposing wrongs to the "court public," trying to mobilize the attention of society in order to put pressure on both states and private enterprises to change their practices and conducts. By making visible certain facts and findings that had previously remained largely concealed from the public, activists seek to build up new forms of conscience, calling for more transparency in policy-making and tighter scrutiny over governments and transnational corporations. The lawsuit filed against BP operated on the same plane, albeit in a totally different mode. Rather than trying to intervene in the realm of collective sensibility, it sought to question the regime of visibility that shapes the gaze of the law, exposing the partitions between those who count and those who do not count as legal subjects, challenging the economy of distribution of rights, the limitations of existing juridical forums, and the structures of power that this regime helps to sustain.

BLACK SEA, DARK EARTH

The framing of environmental catastrophes as events that concern universal rights and international politics is inevitably conditioned by means and channels of representation and mediatization. To a large extent, globalization and humanity—the objective and subjective conditions of universality—have been forged at the juncture between global communicational systems and what ecologist Wolfgang Sachs has called "massive accidental internationalization," the entanglements between transnational flows of information and pollution that forge a sense of a shared earthly space.[15] The ways in which ecological disasters are rendered visible and narrated therefore have a determining effect on their political and ethical impact. One of the most notable examples in this respect, perhaps paralleled only by Chernobyl, is the accident of the *Exxon Valdez* oil spill in 1989, a paradigmatic case at many different levels. Probably no other marine catastrophe has been so extensively reported by the international media or exerted such a great influence on collective consciousness. The images of dead fishes washing up on shores, seagulls soaked in oil, and the darkened beaches of the Prince William Sound formed an iconography of sea-life destruction of sorts, a set of recognizable signs and symbols which have since been repeatedly mobilized as political-aesthetical instruments within the field of environmental advocacy and by the media at large.

The *Deepwater Horizon* disaster broke with this visual language. Just as the uncontrolled blowout challenged the technologies of environmental defense that had been developed in response to the contamination in the Gulf of Alaska, the devastation in the Gulf of Mexico could not be adequately indexed using the iconographic vocabulary traditionally associated with maritime oil spills. Because the leak occurred in ultradeep water, and most of the oil was fractured by dispersants into hidden plumes before surfacing, there was little corresponding to the visual imagery that emerged from the *Exxon Valdez* disaster. Rather than pictures of darkened beaches and oily birds, the canonical images of the BP oil spill were recorded by a remotely controlled underwater camera whose primordial objective was not to mobilize public empathy but rather to perform strictly technical functions, such as aiding in the assessment of the amount of oil that was leaking and shutting down the hole. Perhaps even more than Prince William Sound, the Gulf of Mexico has been extensively monitored by reporters and activists, but the available visual code could not properly convey the real extent and intensity of the damage. The definitive image of the disaster then became much more dependent on scientific evidences.

Since the *Exxon Valdez* disaster, when scientists developed a technology called "geochemical fingerprinting"—a means of identifying a chemical signature that, similar to human fingerprints, is unique to each type of oil and thus allows precise identification between residues to their sources—environmental forensic investigations became a common feature within oil-spill litigations. This methodology enables the reconstruction of what experts call "release histories" of the leak.[16] By identifying chemical traces of oil compounds in contaminated elements, chiefly in the bodies of wildlife species, scientists can reconstruct dispersion patterns of oil flows within the ocean currents, building up a sort of model of the regional ecology. Introduced as evidentiary material within a court of law, the information extracted from this model is used to corroborate damage assessments and determine the necessary compensation measures for the reinstatement of the environment.

In the case of the BP oil spill, in order to grasp the full picture of destruction, biologists have been archiving carcasses of dead dolphins in nitrogen freezers, as their bones and tissues are considered key evidence of the crime. Being at the top of the food web, once geo-prints are identified and it is proven that the dolphins were killed by BP, the reconstruction of the chain of events that led to their death will serve to render the most accurate view possible of the real scale of the disaster. So far there have been more than 100 similar investigations, but it is likely that the true extent of the environmental impact of the spill will in the future remain an open question, since the more scientists investigate the effects of the disaster the deeper they plunge into the unknown nature of the ocean.[17]

A similar methodology was tested in the Ecuadorian Amazon in the early 2000s during the trial for a lawsuit that was filed by local peasants and Indigenous communities against Texaco/Chevron. During almost thirty years of operating in this region, the US oil-giant corporation deliberately dumped billions of gallons of toxic waste directly into the Amazonian soils, generating massive contamination, endemic disease, and death in one of the most biodiverse regions on the planet. Famously named by activists the "Amazon Chernobyl," this ecological catastrophe was as devastating as the BP and the *Exxon Valdez* oil spills. It also shared the traits of being highly mediatized and that the full scope of the destruction was nearly impossible to represent or visualize without the aid of forensic techniques. In attempting to map the extent of the contamination and its long-term effects, earth samples were taken from hundreds of sites scattered around oil-production facilities and housing settlements, and then brought to makeshift laboratories installed in the middle of the jungle where they were precisely archived according to GPS data and geological taxonomies. Mediated by the language and tools of expert witnesses, the darkened portions of earth became central pieces of evidence within the juridical dispute.

Because of the vast geographic area that was affected and the long timespan of the process of contamination, causal links between sources of pollution and its direct effects upon human and nonhuman bodies are very difficult to trace. Science is thus called into court to interrogate the opaque testimony that was gradually recorded in soil transformations, the earth samples functioning as a model of an entire socioecological dynamic constructed over decades. The mud registers the historical agency of multiple forces and actors—the impact of the technology used to extract oil, the negligence of state institutions and corporations, the misfortune of migrant peasants and Indigenous communities, wildlife refugees, polluted water streams, and contaminated atmospheres—out of which a complex political history can be narrated. As the microchemistry of the hydrological transport chain carries the traces of breaches of rights, the mud serves as a murky prism in which human and natural forces are entangled into a single, relational historical force field, the environment itself appearing as the very medium of violence and, ultimately, its victim.

Insofar as nature has become a fundamental space to which cultural and political rights are bound, with increasing frequency and relevance, ecological systems inhabit the courtrooms of national and transnational forums as potential witnesses of legal violations. Shattering the limits of predefined forums, merging the laboratory and the court, ecological and legal sciences, nature participates in a scaleless political construct that connects the universal and the particular, articulating ethical engagements on behalf of humanity with the particularity of local political struggles.

BEYOND CALCULATION

Although legal advocacy has been a central instrument in forming contemporary environmental ethos, ecology has arguably been much less influential in shaping juridical mechanisms. In the last four decades or so, a series of important international and national protocols directed at guaranteeing environmental protection has been implemented, but at a more fundamental level there has been little significant change, the ethical and philosophical foundations of modern law being rather resistant to incorporating bolder transformations that would ensure the proper protection of natural elements. The legal definition of what constitutes damage to the environment is in most cases conditioned by human-centric concepts of injury, such as damage to personal health, damage to property, or the "loss of profit" caused by the impairment of ecosystems. Therefore reinstatement of nature is often treated as secondary to human compensation. Moreover, even when restoration is directed toward the environment, the available methodologies for assessing the extent of the damage are based on cost-benefit models that reproduce similar anthropocentric rationales. Rather than protecting ecosystems per se, the current legal regime is primarily focused on the damages to the economic interests of natural or juridical persons that occur through the environment. In international maritime law, for example, liability for oil pollution other than that framed under the principles of human-based injury is limited to "costs of reasonable measure," a provision which implies that if the scale of ecological destruction is so massive that financial costs for restitution are prohibitive or that reinstatement is technically impossible, the legal resolution is to leave the environment to recover itself without any external means of remediation. In other words, the worse the damage, the more likely it is that polluters will be absolved from environment-centered retributive and restorative penalties.[18]

A more progressive example is found in the domestic environmental law of the United States, particularly in the Oil Pollution Act of 1990 (OPA), which arguably imposes stronger parameters for determining responses to damage to the environment itself. Elaborated in the aftermath of the *Exxon Valdez* oil spill, the design of OPA's legal codes reflects the public outrage over the devastation caused to the ecology of Prince William Sound and the subsequent environmentalist momentum that gathered within the US policy-making establishment. As opposed to the existing international law, under OPA the responsible party may face unlimited liability and is obliged to restore the affected environment to pre-damage conditions. When reinstatement of a particular ecosystem is considered financially "unreasonable" or technically unfeasible, some of OPA's provisions can be interpreted to request that polluters re-create a comparable biotope elsewhere.[19] In practice, however, the application of these measures has been limited by political compromises, and OPA's level of efficiency in assuring the whole restoration of natural elements seems to be rather feeble in the face of the powerful lobbying apparatus that surrounds the global oil industry. Paradoxical as it may sound, the same event that prompted the creation of OPA, the *Exxon Valdez* oil spill, is usually the example used to demonstrate its limitations, insofar as the settlements of this landmark case gave much more importance to human-based injury claims than to ecological reparation. The State of Alaska and the US government, acting as the legal trustees of *ferae naturae*, i.e., "unowned property," sought only one third of the estimated $3 billion in environmental damages, a number already considered highly conservative; in contrast, awards directed toward the economic harm to users of the Sound were initially set at $5 billion, roughly ten times the amount estimated.[20]

The perverse flaws of this regime have never been more clearly exposed than in the catastrophe at the Macondo well. Immediately after the disaster struck, BP reportedly rented nearly every hotel room in Louisiana, hired local scientists under confidentiality clauses, and employed virtually every worker who was suddenly made jobless by the spill. In parallel, the company allocated unprecedented sums for an aggressive advertising campaign.[21] Promoting a good public image and assuring the public that affected communities are financially compensated may be costly, but in the final balance the political profit is rewarding. On the one hand, this strategy certainly helps to contain popular grievances, co-opt oppositional voices, and avoid public confrontation. On the other hand, when everything comes down to a matter of paying the right price, fundamental questions about the economic and political structures of power that lay behind the disaster are left practically untouched, and little significant change in the logics of the system itself need ensue.

Two months after the blowout, BP set up a $20 billion compensation fund for the victims and says it has so far spent nearly $25 billion in clean-up efforts and restoration measures.[22] The company is en route to face a major litigation for violating US environmental laws such as the Oil Pollution Act and the Clean Water Act, which could lead to fines of up to $17.5 billion. But regardless of the disputes over these numbers, the crucial conflict that emerged out of the *Deepwater Horizon* disaster is not so much related to settling the right compensation formula as precisely the opposite: to what is in excess of the economy of calculations. "Even if we were capable of meaningfully establishing a price for ecological harm," explained law professor David Uhlmann, the former head of the environmental crimes section of the US Department of Justice, "there is so much that we do not know about the harm to the Gulf of Mexico—and will not know for years—that it may never be possible to come up with an accurate natural resource damage assessment."[23]

It is common sense that the fulfillment of justice in cases of ecological damage requires adequate compensation to persons that were directly or indirectly affected. Nevertheless, it is important to identify and understand the consequences of the political-ideological structures that are hidden within the legal differentiation between retributive measures to humans and nonhumans. This is not only because failing to properly recover the environment impairs the rights of future generations to access natural resources—thus violating the most basic principles of the concept of sustainability—but also because the general tendency to attribute more significance to the doctrine of "loss of profit" ends up reducing the meaning of ecological catastrophes to a matter of economic calculations. By way of expropriating the intrinsic value of forms of life other-than-human, confining all possible manifestations of nonanthropological alterity to the category of (owned or unowned) property, the current legal order functions as a mechanism that conceals the most important ethical and cultural implications of environmental crimes.

LEGAL ANIMISM

By positing nature as the subject of rights, the lawsuit filed in the constitutional court of Ecuador was meant as a critique of the existing legal regime and its inherent anthropocentric/capitalist logic of cost-benefit calculations. Under contemporary conditions, when global natural resources tend toward scarcity, large-scale extraction activities expand, and climatic chaos ensues, the contemporary political consequences of this juridical gesture are doubtless meaningful, though the idea itself is not completely new. Perhaps the oldest and oddest of the possible examples one might mention in this respect is the fact that in medieval courts it was not unusual to bring criminal proceedings against animals, for example accusing rats of wanton destruction of crops. Similar cases have been recorded well into the twentieth century.[24] Because jurisprudential procedures of this kind today sound bizarre, and to a large extent rightly so, the concept of nature as a rights-holder tends to be criticized as being at best a naive idea that belongs to a premodern form of social contract in which the ontological distinction between things and persons was not yet clearly demarcated. At worst, the notion is condemned by a persisting colonial thinking as a facet of "primitive" belief. What its enlightened critics usually do not take into account is that modern law is essentially "animist," populated by right-holder entities whose personhood is a product of legal fiction. Corporations, for example, are defined as "fictitious persons," and under various international statutes and national constitutional provisions have their own specific rights in a manner similar to human citizens—the ultimate fetish of capital made real by law.

"We are inclined to suppose the rightlessness of rightless 'things' to be a decree of Nature, not a legal convention acting in support of some status quo," concludes law professor Christopher Stone in his groundbreaking manifesto "Should Trees Have Standing? Towards Legal Rights for Natural Objects," arguably the first in-depth modern legal study on the possibilities of endowing natural elements with rights.[25] As with the lawsuit filed against BP in the constitutional court of Ecuador, Stone prepared this text as a form of intervention into a specific legal dispute, *Sierra Club v. Morton*, which at the time of his writing, in late 1971, was about to enter the US Supreme Court. The renowned environmental advocacy organization Sierra Club brought an injunction against Walt Disney Enterprises to try to block a development project in the Mineral King Valley, an important wilderness area in the Sierra Nevada Mountains in California. In the first round of hearings at lower federal courts, judges rejected the suit based on the opinion that, insofar as Sierra Club had no property holdings in the area and therefore would not suffer any direct injury from the project, the organization had no "standing" to bring the case to court. Although the Supreme Court upheld this decision, Justice William Douglas, referring directly to Stone's text, presented a dissident opinion contesting the idea that it was impossible to make a legal demand on behalf of the environment per se:

> The critical question of "standing" would be simplified and also put neatly in focus if we fashioned a federal rule that allowed environmental issues to be litigated before federal agencies or federal courts in the name of the inanimate object about to be despoiled, defaced, or invaded by roads and bulldozers and where injury is the subject of public outrage. Contemporary public concern for protecting nature's ecological equilibrium should lead to the conferral of standing upon environmental objects to sue for their own preservation.[26]

Therefore, Justice Douglas further stated, the suit would be more properly named *Mineral King v. Morton*. Animated by law, natural elements could assume a degree of personhood in a similar manner to the way other nonhuman entities such as corporations are endowed with legal personality:

> Inanimate objects are sometimes parties in litigation. A ship has a legal personality, a fiction found useful for maritime purposes. The corporation sole—a creature of ecclesiastical law—is an acceptable adversary and large fortunes ride on its cases. The ordinary corporation is a "person" for purposes of the adjudicatory processes, whether it represents proprietary, spiritual, aesthetic, or charitable causes. So it should be as respects valleys, alpine meadows, rivers, lakes, estuaries, beaches, ridges, groves of trees, swampland, or even air that feels the destructive pressures of modern technology and modern life.[27]

Justice Douglas's dissent set a jurisprudential milestone for subsequent claims and *Should Trees Have Standing?* became an influential reference in theoretical debates on environmental law. Drawing on historical relations between law, ethics, and politics, Stone shows how the modern system of rights has been progressively enlarged throughout history, demonstrating that elements once considered alien to the arena of rightness were made rights-holders following paradigmatic ethical-political transformations. He situates the question of the rightlessness of nature in relation to the jurisprudential history of the objectification of humans, as for example in the case of the legal status of women and slaves, similarly reduced to the category of property until very recently. Writing in the early 1970s, when the ecological agenda was at its height, Stone's arguments that Western-modern systems of law should consider nature as something more than just a collection of things to be possessed and mastered by humans were part of a larger set of concerns and debates that occupied diverse fields of knowledge and political forums. Yet reading his text one can grasp how the possibility of attributing rights to the environment was an idea still considered unreasonable and discredited. Such conservative reactions were not only comprehensible, Stone argued, but symptomatic. Likewise in the historical disputes concerning whether slaves should or should not have rights, whenever there was a movement to amplify the system of rights beyond those already recognized, there were fierce oppositional views, for such proposals were seen as ridiculous in themselves as much as they were considered to threaten the established power structures that they helped to maintain.

NATURAL HEGEMONY

The philosophical and juridical debates that evolved from this discussion were key elements in the development of the articles that established the rights of nature in the Ecuadorian constitution of 2008.[28] But even more crucial was the local historical-political context from which the law emerged. Ecuador, like the rest of Latin America, is a country whose history has been determined by the heritage of colonialism. An extremely unequal distribution of land and social relations dominated by racist ideologies conditioned every domain of the country's political and economic structures, engendering a social order that was sustained by a double and entangled form of violence: the exhaustive exploitation of natural resources and the exclusion of native culture from political representation. The new constitution established in 2008 was designed to break up that regime, reframing the role of the environment within the economy and opening up spaces of political participation for Ecuador's large Indigenous population.

The foundations of this new constitution can be traced back to the landmark Indigenous uprising of 1990, when thousands marched from the Sierra and Amazonia to the capital, Quito, demanding territorial rights and proposing the reconstitution of the state as a "plurinational" polity; i.e., formed by multiple cultures/nationalities. In parallel, ecological issues became increasingly important within Ecuadorian politics. The maintenance of a colonial-style economy, largely conditioned by the demands of the global market for natural commodities—cacao, bananas, and, since the mid-'60s, petroleum—and dependent on the subordination of Indians as a cheap labor force, generated a highly exclusionary, polarized regime. On the one side were concentrated patterns of primitive accumulation; on the other, spoliation and expropriation of Indigenous lands, plunder of common resources, and endemic poverty. This context pushed territorial and environmental issues to the center of political struggles. Furthermore, although territorially small, Ecuador is probably one of the most socio-biodiverse countries on Earth and has been the stage of one of the most devastating cases of petroleum-led environmental destruction ever recorded in history, the so-called Amazon Chernobyl mentioned above.

At the confluence between Indigenous culture and modern environmentalism—the former characterized by carrying a powerful land ethics grounded on nonutilitarian relations to nature, and the latter informed by scientific visions of a living biosphere—the politics of the rights of nature were gradually forged in Ecuador. "When we approved the rights of nature in the Constitution, this process implied a reflection on what exactly nature is," explained ecologist Esperanza Martínez:

> For science it is one thing, for the Indigenous people another, for law another one, and for capitalism yet another.

> For capitalism nature is environment: a place where one extracts resources within certain limits. The Indigenous people have a distinct notion: nature is not only the ecosystem, but also spiritual beings. In the case of biological sciences, it depends from which scientific paradigm you depart: are human beings included inside nature or not? So we are not speaking about the same thing. When we decided to adopt the term *Pachamama*, it was foremost an act of acknowledging the wisdom of those who are so closely tied to the earth. It was also a critical act in relation to the classical notions of environment and nature.[29]

Considering the "plurinational" constituency of Ecuador, and therefore the diverse cosmologies nurtured by the different Indigenous cultures and nationalities, it was necessary to take into account multiple forms of conceiving of and relating to nature—that is, to acknowledge by law the existence of, quite literally, different natural worlds. "It was an act of openness, of opening to diversity," recounted Martínez, which required the introduction of a concept broad enough to allow for a certain mediation between the conflictive conceptions of nature that coexist within the geographical borders of the Ecuadorian state. *Pachamama*—usually interpreted as *Madre Tierra*, "Mother Earth," a mythical deity entity that is omnipresent in Andean Indigenous cultures—was the chosen concept to guarantee that Amerindian cosmologies were politically represented within constitutional law. "If modernity has adopted a single paradigm, one single rationality, one sole model of nature," Martínez concluded, "what we are saying is that there is not only one but many, as many as there are cultures."[30]

By radicalizing the "animist" condition of modern law, extending the notion of universal rights toward nature, the Ecuadorian Constitution could be interpreted as a pragmatic attempt to critically respond to the limits and flaws of what philosopher Bruno Latour has called "the modern constitution"—i.e., the foundational law of modernity, which is instituted by the sectioning of reality in two great and separated poles, namely the world of objects and the world of subjects, nature and culture.[31] While defining a common ground between humans and nonhumans, the rights of nature disrupt this border regime, projecting a form of social contract that seeks to break with the hierarchies and systems of domination between humans and nature upon which Western politics and culture are based.

There are of course many unsolved questions that emerge together with the idea that nature should be granted rights that had previously only been attributed to humans. The most important of these, perhaps, concerns the way that, while it attempts to break with the divisions established by "modern constitutionalism," the law seems simultaneously to be reinforcing them. By granting exclusive rights to nonhuman natural elements it risks drawing even stronger borders between one domain and the other, thus contributing to the reaffirmation of the validity of nature as an epistemic category when the reality on the ground constantly points to its obsolescence. And this at a moment when, given the historically unprecedented anthropogenic impact over the whole planet, nature, once and for all, is definitely "over." But it is also true that nature has been declared over many times before. And yet we have never felt its presence so strongly, bursting over shores, flooding cities, disrupting agricultural cycles, and triggering large-scale human catastrophes.

The concept of the rights of nature is above all a tactical tool, a political instrument, as perhaps all laws are, either in the hands of those who oppress or of those who resist oppression and domination. On which side does it stand in the current order of things? The Ecuadorian Constitution must be interpreted as part of a larger historical process of reformulating the state apparatus that, together with the new Bolivian constitution of 2009, represents what Brazilian jurist Carlos Marés de Souza has called "the second moment" of Latin America's "new constitutionalism."[32] The first moment emerged in the context of the constitutional reforms of the late 1980s/early '90s amid the paradoxical conjuncture between the process of transition toward democracy after the long period of military dictatorships that ruled most of the continent during the Cold War, and the subsequent consolidation of the neoliberal order. During this period there were significant advancements in relation to the rights of the so-called minority groups, chiefly Indigenous and Afro-descendant peoples. This led to the introduction of a set of legal measures aimed at including linguistic, ethnic, and cultural diversity as part of the reformulation of the national polity, a movement that followed a broader cultural turn in the definition of civil rights. In many different ways, the "exclusive rights" that were implemented by these constitutional reforms came as a form of reparation; that is to say, they functioned as a legal instrument designed to guarantee the transformation of a social order that had been built upon structural cultural violence.

This politicization of culture was fundamental to breaking with the homogenizing authoritarianism inherited from the former regimes and for responding to the demands of marginalized groups to whom access to legal and political representation has been historically denied. Nevertheless, as multiculturalism gradually became a useful engine within the managerial logics upon which neoliberal forms of power were formulated and deployed in the following decades, the agenda of political transformations that was put forward at this crucial moment, which implied a radical and broad reconfiguration of the whole state apparatus, came to be absorbed into the localized and specific demands of particular communities: merely culture, no longer politics.

Whereas cultural homogenization is certainly intrinsic to modernity's foundational law, alterity is fundamental to modernity, even if only in the form of a domesticated or repressed imaginary or as a subjected position that functions to legitimize hierarchies of power. Rather than cultural difference, what escapes the framework of modern constitutionalism is a "different nature"—that is to say, dissident political ecologies that implicitly challenge the objectified idea of nature upon which the multicultural universalism of modernity was constructed. As anthropologist Eduardo Viveiros de Castro has insisted on different occasions, modernity is fundamentally "mononatural."[33] This is neither an innocent nor a merely symbolic construction. Thus when considering the global hegemony of modernity's constitutional laws, we must take into account the history, the methods, and the means by which mononaturalism was enforced on the ground and the latent violence that was implicit in that process. Not in the sense, or not only in the sense, of forms of epistemic violence that attempt to subjugate or eradicate alterity in forms of knowledge and practice, but in terms of the proper material, geophysical dimension of that violent process through which nature was confined to the position of an object of human mastery and possession. Alongside cultural domination, the enforcement of a natural order has been one of the most powerful instruments of power and domination in modern/colonial history. Subjective and objective violence—violence against cultures and violence against nature, ethnocides and ecocides—have often come hand in hand. The "new constitutionalism" of the '90s created legal codes that should limit and restrain the former; the recent new constitutionalism represented by the Ecuadorian constitution addresses the latter.

The colonial and postcolonial histories of the process through which modernity's "constitutional law" became globally hegemonic have been criticized primarily because of their intrinsic cultural violence; i.e., because the diffusion of modernization has been largely guided by the destructive intent of modeling and homogenizing other cultural formations according to occidental paradigms of civilization. The emerging practices around the concept of nonhuman rights might allow us to frame that process from a different and complementary perspective, one according to which it is not only the question of cultural hegemony, but also that of the imposition of a "natural hegemony," that is politically crucial. While the modern constitution has been increasingly tolerant in relation to the cultural pole, it has remained firmly grounded in the perpetuation of mononaturalism. Perhaps the most relevant question therefore is not so much how culturally tolerant the enlightened, Western-based category of universal humanity can be, but precisely the opposite: how modernity, together with its laws and politics, might accommodate "different natures" besides and beyond the monolithic version upon which it was constituted. This is arguably the crucial political and ethical imperative whose time has come.

August 15, 2013. Working in my office in the old colonial center of Quito, rereading this text and making final adjustments, it feels necessary to add some extra closing remarks. Though short, and somewhat misplaced, these final notes are so relevant to the current political situation here (and elsewhere in the world, I believe) that, if not left to the bottom margins, the entire text would probably have to be changed completely. I recall Eduardo Galeano's famous axiom, "In the outskirts of the world the system reveals its true face," hoping that this marginal conclusion can shed new light on what was written above.[34]

From 2008 to 2013, the situation in Ecuador has significantly changed. The revolutionary hopes that were attached to the new constitution—the product of what has been named the "citizenship revolution" or "twenty-first-century socialism"—have dissipated under the resurgence of an aggressive and authoritarian neoextractivist politics. Fueled by the demands of the Chinese market, much of the Ecuadorian Amazon is being opened up for large-scale oil-drilling and mining operations. In parallel, peasant and Indigenous leaders and environmental activists who are fighting under banners such as "rights to water" are being constantly harassed by an increasing policing apparatus. Despite the hopes raised by the constitution of a radically new "civilizational pact," old schemes of power are being reinforced: human-rights violations and violence against nature operating entangled within a single political engine. Perhaps nothing is more expressive of this process than the executive legal decree that was launched today by the government to halt the Yasuni-ITT project, the utopian initiative of leaving large reserves of oil underneath the Amazonian soil. The current conflicts that ensued reveal much about the coming into being of a new logic of geo-power—one that will attempt to enclose large tracts of the seabed, plunder the earth's deep crust, appropriate lands that are surging below melting glaciers, and colonize the last realms of tropical forests. It is, however, within this context that the law of the rights of nature has proved to be a powerful instrument of political action, within and outside formal courts of law, all the way down to the streets of Quito.

First published in Forensic Architecture, ed., *Forensis: The Architecture of Public Truth* (Berlin: Sternberg Press, 2014), 553–71.

1 Information drawn from the forensic engineering and risk-analysis report published in March 2011 by an independent research unit based at the University of California. See Deepwater Horizon Study Group, *Final Report on the Investigation of the Macondo Well Blowout* (Berkeley: University of California Press, 2011).

2 By early June, clean-up operations involved approximately 47,000 people and 7,000 vessels. See Jonathan L. Ramseur, Congressional Research Service, R42942, "*Deepwater Horizon* Oil Spill: Recent Activities and Ongoing Developments," 2013. Available online at http://www.fas.org/sgp/crs/misc/R42942.pdf (accessed May 3, 2024).

3 See Rocky Kistner, "The Macondo Monkey on BP's Back," *Huffington Post*, September 30, 2011. Available online at www.huffpost.com/entry/the-macondo-monkey-on-bps_b_988262 (accessed May 3, 2024).

4 For a detailed account of the controversial claims over the leak-flow rate see Julia Whitty, "The BP Cover-Up," *Mother Jones*, September/October 2010. Available online at www.mother-jones.com/special-reports/2010/09/bp-oceans (accessed May 1, 2024).

5 See Gianluigi Zangari, "Risk of Global Climate Change by BP Oil Spill," Frascati National Laboratories and Italian National Institute of Nuclear Physics, June 12, 2010. Available online at https://documents.pub/document/institute-of-atmospheric-sciences-and-climate-risk-of-atmospheric-sciences-and.html?page=2 (accessed May 3, 2024).

6 Paul Virilio, *The Original Accident*, 2005 (Eng. trans. Julie Rose, Cambridge, UK, and Malden, MA: Polity, 2007), 10–11.

7 See David Bond, "The Science of Catastrophe: Making Sense of the BP Oil Spill," *Anthropology Now* 3, no. 1 (April 2011): 36–46.

8 See Paul Quinlan, "Less Toxic Dispersants Lose Out in BP Oil Spill Cleanup," *New York Times*, May 13, 2010.

9 See David Biello, "Is Using Dispersants on the BP Gulf Oil Spill Fighting Pollution with Pollution," *Scientific American*, June 18, 2010.

10 Whitty, "The BP Cover Up."

11 Esperanza Martínez, in an interview with the author, Quito, November 2011.

12 Lawsuit on Behalf of the Rights of Nature under the Principle of Universal Jurisdiction, filed November 26, 2010, in the Constitutional Court of Ecuador, at 1. Emphases are mine.

13 Ibid., 2.

14 Ibid.

15 Wolfgang Sachs, "Environment," in Sachs, ed., *The Development Dictionary: A Guide to Knowledge as Power* (New York: Zed Books, 2010), 25.

16 See Zhendi Wang and Scott A. Stout, *Oil Spill Environmental Forensics: Fingerprinting and Source Identification* (Burlington, MA: Academic Press, 2007). See also Brian L. Murphy and Robert D. Morrison, eds., *Introduction to Environmental Forensics* (Burlington, MA: Academic Press, 2007).

17 Under US environmental law, the methodology for establishing ecological harm is defined by "Natural Resource Damage Assessments"

(NRDA). One year after the disaster, Suzanne Goldenberg reported that more than 100 NRDA were being conducted. See Goldenberg, "Deepwater Horizon aftermath: how much is a dolphin worth?," *Guardian*, April 19, 2012.

18 See Dramé Ibrahima, "Recovering Damage to the Environment per se following an Oil Spill: The Shadows and Lights of the Civil Liability and Fund Conventions of 1992," *Review of European Community & International Environmental Law* 14, no. 1 (April 2005): 63–72.

19 See ibid., and Michael Mason, "Transnational Compensation for Oil Pollution Damage: Examining Changing Spatialities of Environmental Liability," *LSE Research Papers in Environmental and Spatial Analysis* no. 69 (2002). Available online at http://eprints.lse.ac.uk/570/ (accessed May 3, 2024). US legal provisions to "make the environment and public whole" are defined by the Code of Federal Regulations, 15 CFR § 990.30. Another more recent example that attempts to deal with "pure ecological damage" as opposed to traditional forms of injury is the EU Environmental Liability Directive, Council Directive 2004/35/CE, 2004 O.J. (L143).

20 See James Liszka, "Lessons from the *Exxon Valdez* Oil Spill: A Case Study in Retributive and Corrective Justice for Harm to the Environment," *Ethics & the Environment* 15, no, 2 (Fall 2010): 1–30.

21 See Whitty, "The BP Cover-Up."

22 Dominic Rushe, "BP compensation fund for Gulf oil spill victims at risk of running out," *Guardian*, July 2, 2013.

23 David Uhlmann, quoted in Goldenberg, "Deepwater Horizon aftermath."

24 Jeffrey Kastner, "Animals on Trial," *Cabinet* no. 4, *Animals* (Fall 2001).

25 Christopher Stone, "Should Trees Have Standing?" Towards Legal Rights for Natural Objects" was originally published in 1972 in *Southern California Law Review* 45. The quotation is from the version that later appeared as the main article of the book *Should Trees Have Standing? Law, Morality, and the Environment* (New York and Oxford: Oxford University Press, 2010), 2.

26 Justice William Douglas, in *Sierra Club vs. Morton*, 405 US 727 (1972). Available online at http://supreme.justia.com/cases/federal/us/405/727/case.html (accessed May 3, 2024).

27 Ibid.

28 A reconstruction of the debates around the Rights of Nature in the Constitution of Ecuador is included in two books edited by Alberto Acosta and Martínez: *Derechos de la naturaleza: El futuro es ahora* (Quito: Abya-Yala, 2009) and *La naturaleza con derechos: De la filosofía a la política* (Quito: Abya-Yala, 2011). On the influence of Stone's legal analysis, see especially the text of lawyer Mario Melo, "Los derechos de la naturaleza en la nueva constitución ecuatoriana," included in *Derechos de la Naturaleza*.

29 Martínez, interview with the author.

30 Ibid.

31 Bruno Latour, *We Have Never Been Modern* (Cambridge, MA: Harvard University Press, 1993).

32 Carlos Marés de Souza, in an interview with the author, Curitiba, Brazil, April 2012.

33 Eduardo Viveiros de Castro, "Os pronomes cosmológicos e o perspectivismo amerindio," *Mana* 2, no. 2 (1996): 115–44. See also Latour, "Facing Gaia: Six Lectures

on the Political Theology of Nature," Gifford Lectures, February 18–28, 2013. Recordings available online at https://www.ed.ac.uk/arts-humanities-soc-sci/news-events/lectures/gifford-lectures/archive/series-2012-2013/bruno-latour (accessed May 3, 2024).

34 Eduardo Galeano, *Days and Nights of Love and War* (London: Pluto Press, 2000), 170.

EDUARDO GUDYNAS

ENVIRONMENTAL AND ECOLOGICAL FORMS OF JUSTICE

The recognition of nature as a subject of rights, and possible changes in politics and environmental management from a biocentric perspective, are intimately linked to questions of justice, a complex concept that touches on different fields and meanings. It is often invoked when noncompliance with legal norms of environmental protection is denounced. In this regard, the justice embodied in judicial power is targeted as a system that guarantees compliance with rights and obligations. It also takes on another meaning, when, for example, the injustice of allowing the destruction of a forest is raised, beyond the fact that this action is not prevented by a legal norm. Likewise, these and other meanings superimpose the idea of justice on the moral terrain, in which the virtue of being just is also defended. This approach in turn enables one to reach the field of ethics. Perspectives on justice, then, are diverse.

These types of questions are examined in the essay that follows, beginning with the classic positions in environmental justice. Then I shall move toward the reformulation of justice in order to be able to attend to the rights of nature and discuss the possible foundations of this radical change.[1]

ENVIRONMENTAL JUSTICE

Calls for environmental justice appear in very diverse cases, for example, lawsuits over polluted waters that affect a city neighborhood, or demands to protect an endemic species. At the same time, we find situations to which people react by realizing that there are environmental problems that express an injustice by violating their rights, as often happens in the face of mining or oil extraction. Finally, there are those who hope for an expansive justice that will attend to the new rights of nature.

An initial approach to this problematic might restrict itself to considering environmental justice by focusing on norms (constitutions, laws, decrees, etc.) and judicial practices.[2] Various environmentalist organizations working on environmental law are active in this field. A second, broader approach is to analyze the connections between ethical perspectives and justice. Values recognized in the environment in turn generate moral considerations and rights, and from these we reach the field of justice.

The predominant position understands justice as a matter between humans (and is therefore part of the anthropocentric perspective). This may follow at least three pathways in environmental matters. On the one hand, we can rely on corrective or punitive justice, which imposes punishments for environmental crimes (for example, penalizing whoever dumps toxic substances into sources of drinking water). Other measures involve a compensatory form of justice, where monetary compensation is made for environmental damages (paying victims of pollution, for example), although this may also apply to those demanding an environmental action (compensating owners of a ranch, say, who wish to leave it uncultivated as a conservation measure). Finally, distributive justice aims at a just distribution of environmental benefits and damages, assigning rights and obligations.

The close relationship between justice, environment, and rights was approached in an innovative way in the United States some decades ago with the so-called environmental-justice movement. This was a reaction to "environmental racism," and attacked the coincidences between poverty, marginalization, racial segregation, and poor environmental quality. Its practices emerged from sites where the poorest communities or racial minorities inhabited contaminated zones, or worked in areas that were either high risk or had low environmental quality. For these reasons, "environmental racism" became a point of reference. Strictly speaking, this movement expresses a superimposition of classic questions of social justice and environmental aspects.[3] The ways of confronting this problematic have focused above all on distributive justice.

In the United States, the Environmental Protection Agency (EPA) is tasked with implementing a strategy for environmental justice. In its mission, this agency maintains that all people should live in sustainable and clean communities, and that no segment of the population, independent of race, color, national origin, or income, should suffer disproportionately from adverse environmental or sanitary conditions. In addition, the strategy declares that people must have "every opportunity for public participation in decision-making," and that informed and actively involved communities are a "necessary and integral" part of measures for protecting the environment.[4] This kind of strategy aims at assuring citizens' rights in the environmental sphere, but also at enabling access to information, participation, legal assistance, and reparation in case of damage or disturbance.

In recent years this topic has achieved greater visibility in Latin America, especially in Brazil. Indeed, it is invoked there by a broad array of academics, social organizations (many of them environmentalist), and in the recent past by trade unions as well. The Brazilian initiatives have maintained contact with the movement for environmental justice in the United States since at least 1998; shortly thereafter, a collection of classic texts by the Anglo-Saxon wing of this movement was published in Portuguese, along with Brazilian case studies.[5] The Rede Brasileira de Justiça Ambiental (Brazilian network for environmental justice) was soon formed and a similar network also exists in Chile, the Red de Acción por la Justicia Ambiental y Social (Action network for environmental and social justice).

Taking the Brazilian case as an example, the emphasis is on situations in which disproportionate harm from environmental hazards is imposed on populations less endowed with financial, political, and informational resources.[6] The Rede Brasileira de Justiça Ambiental defines environmental justice as the "just treatment and full involvement of all social groups, independent of origin or income, in decisions about the occupation, access to, and use of the natural resources in their territories." Among various points, the group demands the rights of populations to equitable environmental protection against socioterritorial discrimination and environmental inequality, and on the basis of these postulates proposes a radical alteration of the patterns of production and consumption, although this point is not elaborated in detail.[7]

The demand for environmental justice from any of these aspects has several positive aspects, such as enhancing environmental awareness, linking social conditions to their ecological contexts, reinforcing civic recognition, the scaffolding of rights and the judicial system, opening the doors for some form of social regulation of the state and market, and enabling struggle over urgent concrete situations.

But it is also true that it confronts several limitations. Some are practical, such as the weakness of the judicial system in almost every country, where processes are slow and costly; others are in design and structure, which may have very little breadth in terms of other cultures or alternative conceptions of citizenship. It must be warned that this type of environmental justice unfolds within the classical conceptions of citizenship and rights, and therefore in almost all cases remains within a conventional anthropocentric ethic. In the major reflections in Brazil, for example, there is no substantive discussion on intrinsic values, such as can be gleaned from the contributions of Henri Acselrad and his colleagues.[8] The emphases point to problems such as the asymmetries of power that lead to environmental injustices among the poorest groups or racial minorities, remaining specifically within the field of political, social, and economic rights. But intrinsic values or the rights of nature, for example, are not explored.

It is also common to arrive at environmental justice through environmental conflicts. A widely disseminated vision in Latin America, promoted by, for example, Joan Martínez Allier, understands these protests as expressing environmental conflicts of a distributive nature.[9] In his view, the emphasis is on the conflicts over environmental resources and services, commercialized or not, and their social, spatial, and temporal patterns. Since these conflicts express unequal modes of distribution, their distributional emphasis leaves these positions within the realm of environmental justice. The problem is that this theoretical approach does not offer substantive space for a specific ethic of value, and beyond all the calls for the environmental struggles of the poor, the porosity toward cultural plurality is narrow.[10] In other words, although ecological price corrections may be made use of, this position would express an ecologically corrected distributive justice that nonetheless remains anthropocentric.

A JUSTICE AMONG HUMANS

In this way, among both those who reject environmental justice and many of those who promote it, the premise that justice is restricted to the community of human beings is almost always accepted. They are the moral agents able to articulate their preferences and ideals and to aspire to reciprocity and cooperation within an impartial system, and from that standpoint they are the ones to build justice.

In this discussion, one cannot avoid reference to the US philosopher John Rawls, surely the most influential theorist in contemporary discussions of justice.[11] His perspective, nowadays subscribed to by many, maintains that justice is specific to humans as citizens within a nation-state, dealing with the distributional inequalities that affect them, and defending positions such as neutrality of values, in which the procedures and emphases center on how to combat a poor distribution of goods (or prejudices). Rights are only for humans, and goods are delimited from an anthropocentric perspective, and could include environmental components.

Those who defend Rawlsian positions are not insensitive, and they understand that one may be compassionate toward plants and animals or attend to nature when damage affects people or their belongings, but do not see these problems as expressions of injustice. They may agree that it is morally incorrect to bring a species to extinction, but its extermination would not be a case of "injustice" toward this species. Note that this liberal (in the philosophical sense) position can generate environmental management of an administrative type and on an anthropocentric basis, with a form of justice expressed in the reaction of defense of natural resources to the extent that those resources are the property of people or affect individuals' health or quality of life.

The justice generated by this position can defend the quality of life of humans, or the environment as a function of humans, removed from the recognition of the rights of nature. What is just or unjust in environmental terms may be resolved in relation to the rights of humans, or the implications for humans. It is a form of justice corresponding to an object-nature, and therefore its perspective is anthropocentric. Its conventional expression is the inclusion of the environment in human rights of the third generation.

But this perspective also encloses a practical problem: the increasing diffusion and acceptance of ideas of justice reduced to fit economic frameworks. Indeed, in several countries use is already being made of vouchers and other forms of monetary compensation in order to deal with the justice system, especially in the social field (in situations of extreme poverty), and in some circumstances on the environmental terrain as well. In this way, justice too becomes a commodity, presuming that everything can be economically compensated. But the destruction of nature cannot be justified through appeal to economic compensation, nor does this generate real solutions for damaged ecosystems and endangered species. Distributive justice among humans is not a genuine solution to environmental problems. Analogously, it offers no real solutions in a multicultural context where other cultures define their community of moral and political agents in a broader way, integrating the nonhuman.

For these reasons, another type of justice is necessary that breaks with anthropocentrism and complements environmental justice in order to reinforce it but at the same time to go beyond it, incorporating intrinsic values and the rights of nature.

A FORM OF JUSTICE FOR NATURE: ECOLOGICAL JUSTICE

As we know, at present in Latin America, even the framework of rights to a healthy environment has insufficient and precarious coverage. Even where there have been advances, there exists a tendency to fall into a tangle of economic compensations for environmental damage. The limitations of this approach were pointed out in the preceding section, since it compensates people without necessarily restoring nature.

The recognition of values specific to nature obliges us to promote another perspective that we term *ecological justice*. This is part of recognizing nature from within its own values. It is a necessary and inevitable consequence of the recognition of the sequence that begins with intrinsic values and continues with the rights of nature. Meanwhile, the label of *environmental justice* should be reserved for the type of justice based on the rights to a healthy environment and quality of life, resting on the classical conceptions of human rights.

The transition toward ecological justice is necessary, since the destruction of plants and animals is a matter not only for compassion but for justice as well; the disappearance of ecosystems not only generates economic problems but also encloses questions of justice, as does much of the environmental problematic.

This distinction between two forms of justice, the one environmental and the other ecological, is recent. In this survey, the contributions of Nicholas Low and Brendon Gleeson, Brian Baxter, and David Schlosberg stand out.[12] Low and Gleeson, for example, declare that the former form of justice must focus on the distribution of environmental space among people, while ecological justice should tackle the relationships between humans and the rest of the natural world. These authors defend two basic points of departure: 1) all living beings have the right to enjoy their development as such, to complete their own lives; 2) all forms of life are interdependent and depend in turn on physical support. Only in a limited way have these ideas penetrated Latin American debates, since the prevalent perspective is that of environmental justice. As seen above, ecological justice is not a central topic in discussions among many networks in civil society, at least for now.

The idea of ecological justice is not opposed to environmental justice, but complements it, encompassing it in order to go beyond it. As such, it is a field under construction; its foundational sources are diverse. The arguments are varied, in some cases timid but in others more radical, expressing different intentions, not always connected, even contradictory at times, to go beyond the outlooks on justice and the environment that are specific to the cultural tradition of modernity.

THE MULTIPLE FOUNDATIONS OF ECOLOGICAL JUSTICE

One set of arguments in favor of a form of justice capable of going beyond the field of the human may be derived even from Rawlsian ideas. Within this complex, one must begin with those that continue to emphasize the human being but expand their temporal horizon in the name of a commitment to future generations. They maintain that this is necessary in that current wastage and environmental destruction are limiting our descendants' options to reach an adequate quality of life or enjoy biological diversity. This component is recognized in the Ecuadorian constitution. (In Article 395, among the "environmental principles," an environmentally sustainable development is postulated that "assures the satisfaction of the needs of present and future generations.")

Next, other approaches go further, puncturing the classic ideas of the community of justice. Indeed, one of the most common criticisms of the idea of ecological justice insists that decisions about justice or injustice can only be expressed by conscious agents articulating their preferences on a scale of values or within moral frameworks. This type of questioning is Rawlsian, concluding that ecological justice cannot exist since an intelligibility of this sort on the part of nonhuman living beings is impossible, and that nature is not a moral agent. Plants and animals cannot express their ideas or feelings of value, nor can they publicly debate their moral preferences.

But even from a Rawlsian perspective, this exclusion can be rebutted. In fact, this position accepts that individuals who by their life circumstances or disabilities are not conscious moral agents (e.g., fetuses, those affected by mental limitations, etc.) may be included in the realm of justice. Approaches of this type are defended by Derek Bell, for example.[13] The key point is that just as these amplifications can be made, the same can apply to other living beings.

This tendency considers in some cases that the separation between humans and superior species of mammals and birds is blurred in the light of current knowledge of the latter's cognitive and effective attributes. The assumed specificities that make the human species unique are in reality matters of degree. Following Amartya Sen, Martha Nussbaum argues that there are problems of asymmetry in justice involving animals, which possess moral status and should be included in questions of justice.[14] This already implies a transition beyond the classic Rawlsian positions. There also exist redefinitions of concepts of self-consciousness and moral agency that assert their presence in other living beings.

Within the same tendency, others defend animal well-being in particular, or even animal rights, such as avoiding their suffering, guaranteeing them adequate life conditions, etc. Good examples are the contributions of Peter Singer and, more recently, in Spanish, Jorge Riechmann.[15] Following this path, animal rights can be interpreted as a subset of the rights of nature. Or the point of departure would be to consider them as ends in themselves, and therefore with a moral status.[16]

Some take feminist ecology as a starting point, in particular the promoters of an *ethics of care*.[17] In Latin America, this has many similarities to the positions of Leonardo Boff, although there are no explicit links between them.[18] Beyond differences, these contributions center on sensitivity and empathy as the driving forces of justice, complemented by aesthetic and affective visions. In this way, conventional utilitarianism is rejected, and reciprocity is not demanded as a key factor in contractual relations.

Independently, the alternative regimes of justice proposed by Luc Boltanski—grounded on a very strong affective bond with what surrounds us, where nothing is expected in exchange nor is reciprocity aspired to—provide additional arguments for ecological justice.[19] These positions resemble ecological anthropology's calls to attend to the role of the gift as a nonmercantile action that includes environmental aspects, and of which there are many examples in Andean space.

Other tendencies broaden the ideas of distributive justice into other dimensions, as lucidly accomplished by Nancy Fraser.[20] Her approach recognizes that justice unfolds in several dimensions: one redistributive, another focusing on recognition, and another pointing toward representation. In Fraser's opinion, each of these dimensions corresponds to different types of injustice, each with its own specificity, and it is impossible to reduce them to a single aspect. Fraser's proposal does not engage the environmental question, but as Schlosberg sets forth, it contains many possibilities.[21] For example, it facilitates tackling other cultural expressions, including the incorporation of demands for the rights of nature wielded by ecological groups, Indigenous organizations, and peasant communities, as well as offering concrete paths toward adding an ecological dimension to this aggregate.

Although not directly linked to the previous positions, other authors also explore a multidimensional form of justice. In this regard, Michael Walzer maintains that there exist "spheres of justice" in which the criteria for one cannot necessarily be transferred or extrapolated to others.[22] This position has limitations from an environmental perspective, especially when it has to engage with intrinsic values,[23] but it provides important lessons by warning that (for example) Rawlsian approaches to equitable distribution can function on some social and economic levels, but not necessarily in the environmental field, since other conditions require attending to, such as the survival of species and the protection of ecosystems.

Other authors understand that while nonhumans are not moral agents, they are on the other hand recipients of value and moral judgments by humans and are therefore subjects of justice.[24] Following this position, the community of justice cannot be restricted solely to those who express values and morals, but it must also incorporate their recipients. In this case, the criterion of belonging rests in the quality of being recipients of human beings' actions, valorizations, and even interests. The problem with this path is that it could leave the biocentric perspective in order to return to an anthropocentrism founded on Rawlsian distributive justice. This is what happens with Baxter, who maintains that every living being should receive a just portion of environmental resources, whether at the individual or the collective level, and that humans are just one more part of this whole.[25] In this way, a species has the right to use a "share" of resources, and this would be achieved by large-scale distributive justice, both human and nonhuman, but it is not necessarily protected by its intrinsic values.

Finally, another source of argumentation resides in the mandates derived from the recognition of intrinsic values. Included here are the different expressions of biocentrism, beginning with Aldo Leopold and followed by deep ecologists.[26] These positions recognize intrinsic values in species and ecosystems, and in several cases defend identification and empathy with nature as a form of being in it. For example, Arne Naess's deep ecology defends "personal" realization but one that goes beyond the individual self, based on an identification with nonhuman surroundings.[27] Likewise, let us recall the positions of the broadened communities specific to Pachamama, expressing Indigenous people's perspectives based on relational networks that integrate into an equal hierarchy different living beings and other components of the environment.[28]

Broadened forms of relationality in turn modify the traditional concepts of civil rights. By incorporating nature as a subject of rights, the notion of citizenship must be adjusted, and this has an impact on the application of justice. Both civil rights and the rights of nature must be defended.[29]

CRITIQUES OF ECOLOGICAL JUSTICE

In the preceding essay, some criticisms have been presented of the idea of ecological justice in particular, or of nature as the subject of rights in general. This analysis must be complemented with other details. Some maintain, for example, that ecological justice would impair the desired impartiality of justice, in that a group would be imposing its values and morals on the rest of society. The response to these warnings recalls that ecological justice is not a matter of imposing a few values but of broadening values as a whole; nor are the means that must be taken or the prohibited or sanctionable actions predetermined—rather, a political discussion is opened to work through all this. Neither decisions nor results are predetermined, though of course the debate will be different if contents and subjects are changed, and this is precisely one of the advantages of ecological justice.

Nor is invoking a violation of impartiality in order to reject ecological justice very realistic in Latin America. The current problem is that in many cases, the state shows partiality toward practices with high environmental impact, promoted by private businesses and even the governments themselves. There are repeated denunciations and a long list of cases where the state has abandoned all impartiality by promoting environmentally negative undertakings, or has denied or minimized their effects, whether through weak controls, defective applications of the environmental norm, or economic subsidies and supports of different kinds, explicit or hidden, for these kinds of undertakings. (This has occurred in conservative and progressive governments alike.)

On the other hand, the Ecuadorian case has become particularly relevant, since, as was noted earlier, the polis accepted a new social contract that recognizes the rights of nature. In this case the citizen majority approved a constitutional text that includes another vision of the environment. This does not imply overlooking or refusing those who do not believe in nature as the subject of rights, but it compels consideration of these rights along with others in debates about justice and its administration. Also, since the rights of nature act in parallel and empower the classic visions of environmental justice and human rights to a healthy environment, compromises may be reached in conservation and development from very different ethical, religious, and moral points of origin, which may be biocentric in some cases while continuing to be anthropocentric in others. Moreover, with this broadening of discussions about rights, what occurs in reality is a more radical democratization of environmental policies.

THE NECESSITY OF JUSTICE FOR NATURE

The recognition of values specific to nature engenders a chain of consequences in several dimensions. It immediately compels the postulation of nature as the subject of rights, and consequently makes it necessary to rethink justice in order to accommodate this new situation.

An analogous situation occurs in the inverse sense, since in many cases harsh controversies are postulated in the field of justice when confronted by, for example, calls for ensuring social justice in exchange for sacrificing nature. This problem is becoming increasingly common in Latin American countries, where, for instance, high-impact extractivism is justified as a supposed necessity for ensuring the financing of antipoverty plans, especially the payment of vouchers and monthly monetary assistance. Moreover, justifying environmental damage on the basis of vague calls for social justice only serves to forestall serious discussion of developmental strategies, of how to distribute economic surpluses, and of the insistence on conceiving justice as mere monthly monetary assistance to the poorest. All of this conceals the fact that these environmental impacts almost always hurt the poorest and most marginal and subaltern groups, and that they are in reality mortgaging alternatives to future development.

In reality, complete justice is only possible if it is achieved in both the social and the environmental realms. Ecological justice is not against justice among humans, but is a necessary ingredient for it.

First published as “8. Justicias ambiental y ecológica” in Gudynas, *Derechos de la Naturaleza: Ética biocéntrica y políticas ambientales* (Lima: Programa Democracia y Transformación Global and Red Peruana por una Globalización con Equidad, Rio de Janeiro: CooperAcción, and Montevideo: Centro Latino Americano de Ecología Social, 2014), 135–46.

1 Parts of this essay are based on my article “La senda biocéntrica: valores intrínsecos, derechos de la Naturaleza y justicia ecológica,” *Tábula Rasa* (Bogotá) no. 13 (July–December 2010): 45–71.
2 For a contribution of this type see Enrique Leff, ed., *Justicia ambiental. Construcción y defensa de los nuevos derechos ambientales, culturales y colectivos en América Latina*, Foros y Debates Ambientales (Mexico City: Programa de las Naciones Unidas para el Medio Ambiente, 2001).
3 See, for example, the recent contributions in David Schlosberg, *Environmental Justice and the New Pluralism: The Challenge of Difference for Environmentalism* (New York: Oxford University Press, 1999), Kristin Shrader-Frechette, *Environmental Justice: Creating Equality, Reclaiming Democracy* (New York: Oxford University Press, 2002), and the essays in Ronald Sandler and Phaedra C. Pezzullo, eds., *Environmental Justice and Environmentalism: The Social Justice Challenge to the Environmental Movement* (Cambridge, MA: The MIT Press, 2007).
4 Office of Environmental Justice, United States Environmental Protection Agency, “Environmental Justice Implementation Plan” (Washington, DC: Environmental Protection Agency, 1996), 2.
5 Henri Acselrad, Selene Herculano, and José Augusto Pádua, eds., *Justiça ambiental e cidadania* (Rio de Janeiro: Relume Dumará, 2004).
6 See Acselrad, Cecília Campello do Amaral Mello, and Gustavo das Neves Bezerra, *O què e justiça ambiental* (Rio de Janeiro: Garamond, 2008).
7 Based on the declaration of principles of the Rede Brasileira de Justiça Ambiental, available online at https://rbja.org (accessed December 30, 2023).
8 Acselrad, Herculano, and Pádua, eds., *Justiça ambiental e cidadania*, and Acselrad, Mello, and Bezerra, *O què e justiça ambiental*.
9 See Joan Martínez Alier, *El ecologismo de los pobres. Conflictos ambientales y lenguajes de valoración* (Callao, PE: Espiritrompa, 2010).
10 See critiques similar to these in Jorge Riechmann, *Todos los animales somos hermanos. Ensayos sobre el lugar de los animales en las sociedades industrializadas* (Madrid: Catarata, 2005), and a different conceptual framework on conflicts in my essay “Conflictos y extractivismos: conceptos, contenidos y dinámicas,” *Decursos. Revista en Ciencias Sociales* (CESU, Centro de Estudios Superiores Universitarios, Universidad Mayor de San Simón, Cochabamba) no. 27–28 (2014): 79–115.
11 See, for example, John Rawls, *A Theory of Justice* (Cambridge, MA: Belknap Press, 1971).
12 See Brendan Gleeson and Nicholas Low, *Justice, Society and Nature: An Exploration of Political Ecology* (London and New York: Routledge, 1998); Brian Baxter, *A Theory of Ecological Justice* (Abingdon: Routledge, 2005); and Schlosberg, *Defining Environmental Justice: Theories, Movements, and Nature* (Oxford: Oxford University Press, 2007).

13 See Derek R. Bell, "Political Liberalism and Ecological Justice," *Analyse & Kritik* 28 (2006): 206–22.

14 Martha C. Nussbaum, *Frontiers of Justice: Disability, Nationality, Species Membership* (Cambridge, MA: Harvard University Press, 2006).

15 See, e.g., Peter Singer, *Animal Liberation: A New Ethics for Our Treatment of Animals* (New York: Random House, 1975), and Riechmann, *Todos los animales somos hermanos*.

16 See Riechmann, *Todos los animales somos hermanos*.

17 See, e.g., Virginia Held, *The Ethics of Care: Personal, Political, and Global* (Oxford: Oxford University Press, 2006), and Nel Noddings, *O cuidado. Uma abordagem feminina à ética e à educação moral* (São Leopoldo, BR: Editora Unisinos, 2002).

18 See, e.g., Leonardo Boff, *El cuidado esencial. Ética de lo humano, compasión por la tierra* (Madrid: Trotta, 2002).

19 See Luc Boltanski, *El amor y la justicia como competencias. Tres ensayos de sociología de la acción* (Buenos Aires: Amorrortu, 2000).

20 See, e.g., Nancy Fraser, *Escalas de justicia* (Barcelona: Herder, 2008).

21 Schlosberg, *Defining Environmental Justice*.

22 Michael Walzer, *Las esferas de la justicia. Una defensa del pluralismo y la igualdad* (Mexico City: Fondo de Cultura Económica, 1993).

23 See, e.g., Baxter, *A Theory of Ecological Justice*.

24 See ibid.

25 Ibid.

26 See my *Derechos de la Naturaleza. Ética biocéntrica y políticas ambientales* (Lima: Programa Democracia y Transformación Global and Red Peruana por una Globalización con Equidad, Rio de Janeiro: CooperAcción, and Montevideo: Centro Latino Americano de Ecología Social, 2014), chapter 2.

27 See Alicia Irene Bugallo, *La filosofía ambiental en Arne Naess. Influencias de Spinoza y James* (Río Cuarto, AR: Ediciones del ICALA, 2011).

28 See my *Derechos de la Naturaleza*, chapter 6, and Nina Pacari, "Naturaleza y territorio desde la mirada de los pueblos indígenas," in Alberto Acosta and Esperanza Martínez, eds., *Derechos de la Naturaleza* (Quito: Abya Yala, 2009), 31–37.

29 See my *Derechos de la Naturaleza*, chapter 9.

ARTURO ESCOBAR

AUTONOMOUS DESIGN AND THE POLITICS OF RELATIONALITY AND THE COMMUNAL

La tierra manda, el pueblo ordena, y el gobierno obedece. Construyendo autonomía/The earth commands, the people order, and the government obeys. Constructing autonomy.
—Zapatista slogan

Cambiar el mundo no viene de arriba ni de afuera/Changing the world does not come from above or from outside.
—Tramas y mingas para el Buen Vivir, Popayán, Colombia, June 2013

In fact, the key to autonomy is that a living system finds its way into the next moment by acting appropriately out of its own resources.
—Francisco Varela, *Ethical Know-How: Action, Wisdom, and Cognition*, 1999

AUTONOMY IN THE SOCIAL AND CULTURAL DOMAIN

Ever since the irruption of the Zapatistas and their cry of *Ya Basta!* (Enough is enough!), the struggle for autonomy has raged in Latin America, principally among Indigenous peoples but also among other rural and urban groups. "*Que se vayan todos, que no quede ninguno!*" (Let them all go away, let not one remain!), shouted the Argentinean unemployed to all the politicians and economic elites in whose representations, the protesters claimed, nobody could ever be trusted again after the economic collapse of 2001. Similar calls have been heard since, for instance, among the Indignados movement of southern Europe and the Occupy protesters in the

United States. In Latin America the call for autonomy involves not only a critique of formal democracy but an attempt to construct an altogether different form of rule anchored in people's lives, a struggle for liberation and for a new type of society in harmony with other peoples and cultures.[1]

The Mexican development critic Gustavo Esteva has provided the following useful distinction from the perspective of the tenacious resistance to development, modernity, and globalization by Indigenous and peasant communities in southern Mexico. He distinguishes among three situations in terms of the norms that regulate the social life of a collectivity:

- Ontonomy: when norms are established through traditional cultural practices; they are endogenous and place specific and are modified historically through embedded collective processes.
- Heteronomy: when norms are established by others (via expert knowledge and institutions); they are considered universal, impersonal, and standardized and are changed through rational deliberation and political negotiation.
- Autonomy: when the conditions exist for changing the norms from within, or the ability to change traditions traditionally. It might involve the defense of some practices, the transformation of others, and the veritable invention of new practices.[2]

"Changing traditions traditionally" could be a description of autopoiesis; its correlate, "changing the way we change," designates the conditions required to preserve it, that is, to shift back from heteronomy to autonomy and ontonomy, from allopoiesis to autopoiesis (for instance, from heteronomous developmentalism to life projects). So understood, *autonomía* (autonomy) describes situations in which communities relate to each other and to others (say, the state) through structural coupling while preserving the community's autopoiesis. It tends to occur in communities that continue to have a place-based (not place-bound), relational foundation to their existence, such as Indigenous and peasant communities, but it could apply to many other communities worldwide, including those in cities who are struggling to organize alternative life projects.[3]

The crucial elements for maintaining a mode of existence that is both relational and communal include particular types of relations among persons, relations to the Earth and to the supernatural world, forms of economy, food production, and of nurturing plants and animals, healing practices, and forms of deliberation and decision making. The concept of *territory*, as utilized by some social movements, is a shorthand for the system of relations whose continuous reenactment re-creates the community in question. In the context of the long historical resistance of Indigenous and Afro-descendant peoples in countries like Colombia, *autonomía* is a cultural, ecological, and political process. It involves autonomous forms of existence and decision making. Its political dimension is incontrovertibly articulated by Indigenous organizations in Colombia during the past two decades: "When we fail to have our own proposals we end up negotiating those of others. When this happens we are no longer ourselves: we are them; we become part of the system of global organized crime."[4] The statement also points at the continuous slippage between autonomy and heteronomy, particularly in social movements' relations to the state. There is no absolute autonomy in practice; rather, *autonomía* functions as a theoretical and political horizon guiding political practice.

Autonomía in these cases involves the ontological condition of being communal. The Zapatista put it well in their remarkable Sixth Declaration from the Lacandon Jungle in 2005: "[our] method of autonomous government was not simply invented by the Ejército Zapatista de Liberación Nacional (EZLN); it comes from several centuries of Indigenous resistance and from the Zapatistas' own experience. It is the self-governance of the communities."[5] In describing the autonomous movements in Oaxaca during the same period, Esteva similarly writes, "It is a social movement that comes from afar, from very Oaxacan traditions of social struggle, but it is strictly contemporary in its nature and perspectives and view of the world. It owes its radical character to its natural condition: it is at the level of the earth, close to the roots.... It composes its own music. It invents its own paths when there are none.... It brings to the world a fresh and joyful wind of radical change."[6] *Autonomía* is thus exercised within a long historical background, which has led some researchers to argue that, particularly in cases of Indigenous-popular insurrection such as those that have taken place in southern Mexico, Bolivia, and Ecuador over the past two decades, it would be more proper to speak of societies in movement rather than of social movements.[7] We can go farther and speak of "worlds in movement."[8] These societies/worlds in movement are moments in the exercise of cultural and political autonomy—indeed, of ontological autonomy.[9]

This characterization of *autonomía* is a response to the *current conjuncture of destruction of communal worlds by neoliberal globalization*. Interestingly, the aim of autonomous movements is not so much to change the world as to create new worlds (community, region, nation) *desde abajo y a la izquierda* (from the bottom and to the left), as the Zapatistas like to put it. *Autonomía* is not achieved by "capturing the state" but by taking back from the state key areas of social life that it has colonized. Its purpose is to create spheres of action that are autonomous from the state and new institutional arrangements to this end (such as the well-known *juntas de buen gobierno*, or "councils of good government," in Zapatista territories). At its best, *autonomía* seeks to establish new foundations for social life. Zapatista autonomy, for instance, involves the transformation of the procurement of key social functions, particularly in the following domains: eating, learning, healing, dwelling, exchanging, moving, owning (collective ownership of land), and working.[10] While it would be impossible to analyze here how the practices in each of these domains have been transformed along the axis "heteronomy-autonomy," making them more autonomous, in all likelihood this experience constitutes the best example of design for autonomy.[11]

Autonomía often has a decided territorial and place-based dimension. It stems from, and re/constructs, territories of resistance and difference, as the cases of Black and Indigenous movements in many parts of the Americas show; however, this applies to rural, urban, forest, and other kinds of territories in different ways. In the case of the well-known movements of the unemployed in

Buenos Aires after the crisis of 2001, the exercise of autonomy included both a critique of capitalism and the creation of new forms of life (from day care centers and urban gardens to free clinics, the restructuring of public schools, and the recovery and self-management of abandoned factories); in other words, it involved the creation of noncapitalist spaces and other forms of territoriality. New practices began to emerge, such as workplace democracy and horizontality in the self-managed factories and communitarian values rather than market values in the communities. The goal of the movements was to produce in different ways and to create nonexploitative labor relations, not so dependent on capital and the state, over an entire range of activities involving production and social reproduction. In urban movements one can see the interplay among territorial organizing, collective identities, and the creation of new forms of life that is often at the core of autonomy.[12]

The place-based dimension of *autonomía* often entails the primacy of decision-making by women, who are historically more likely than men to resist heteronomous pressures on their territories and resources and to defend collective ways of being.[13] There is often, in *autonomía*-oriented movements, the drive to re/generate people's spaces, their cultures and communities, and to reclaim the commons. These processes involve epistemic disobedience and foster cognitive justice.[14] Some say that *autonomía* is another name for people's dignity and for conviviality;[15] at its best, *autonomía is a theory and practice of interexistence and interbeing, a design for the pluriverse*.

It is important to remark, however, that the capacity of communities to create and maintain their autonomy depends on their transversal skillful coordination of efforts at many levels, from the local and regional to the transnational. For autonomy to take root, "there has to obtain the conjunction of a local regime of autonomy, understood as the basis for the self-government of social life, and a planetary network open to the collaborative interconnection of living entities."[16] As they free themselves from the state form, autonomous collectives tend to self-organize as a plurality of worlds through intercultural planetary networks. As the salience of the *Planes de Vida* and life projects of communities reveals, control over a basic level of production is indispensable for an effective translocal politics of articulation. For Jérôme Baschet, this basic production infrastructure is a sine qua non for liberated spaces to grow and go beyond their determination by capital, the dominant economy, and the law of value.

The Colombian anthropologist Astrid Ulloa similarly sees territorial autonomy as a multiscalar process.[17] We have already cited her work with Indigenous groups in the Sierra Nevada de Santa Marta, in the Colombian northwest. Based on the strategies of these groups, she suggests the notion of Indigenous relational autonomy, stemming from the confrontation between Indigenous groups and local and translocal actors. Anchored in the ontology of the circulation of life, Indigenous groups develop strategies in their dealings with diverse actors, from the direct local intermediaries of extractive operations and regional megadevelopment projects to transnational legal regimes that not infrequently act as mechanisms of symbolic appropriation, given the neoliberal understanding of nature and forms of eco-governmentality they often deploy through, say, carbon markets and Reducing Emissions from Deforestation and Forest Degradation (REDD) schemes. In so doing, as she proposes, the Arhuaco, Kogui, Kankuamo, and Wiwa peoples engage in a complex interepistemic and interontological geopolitics aimed at creating alternative territorialities that might result, to the greatest extent possible, in an effective articulation of territory, culture, and identity for the defense of their lifeworlds.[18]

THE REALIZATION OF THE COMMUNAL: NONLIBERAL FORMS OF POLITICS AND SOCIAL ORGANIZATION

Let us consider an important concept of the mobilization of the Nasa people of Colombia, the Minga social y comunitaria (Social and communal collective work). "The word [*la palabra*] without action is empty. Action without the word is blind. The word and the action outside the spirit of the community are death."[19] Notions of community are making a comeback in diverse epistemic-political spaces, including Indigenous, Afro-descendant, and peasant mobilizations, particularly in Mexico, Bolivia, Colombia, Ecuador, and Peru; this rekindled interest in things communal is also present in some urban struggles throughout the continent. The communal has also become an important concern for decolonial feminism. It is also found in some transition-related approaches, for instance those that speak of commoning and community economies.[20] Talk of community in Latin America may take a number of forms: *comunalidad* (communality), the communal, the popular-communal, struggles for the common, communitism (community activism), and so forth. Here I will use "the communal" or "communal logics" to encompass this range of concepts.

The historical background of this "return of the communal," if we are allowed to put it in these terms, is very complex; for the case of Latin America, it includes the emergence of Indigenous movements after 1992, the political turn to the left and the rise of progressive regimes after 1998, and the particularities of the Indigenous-popular insurrections in countries like Bolivia and Ecuador. A recounting of this context is beyond our scope here, as is a discussion of the many critiques raised against communal notions—from charges of romanticism and going back to the past to warnings about the repressive character of communities.[21]

Communal thought is perhaps most developed in Mexico, based on the experiences of social movements in Oaxaca and Chiapas. For Esteva, *la comunalidad* (the condition of being communal) "constitutes the core of the horizon of intelligibility of Meso-American cultures. . . . It is the condition that inspires communalitarian existence, that which makes transparent the act of living; it is a central category in personal and communitarian life, its most fundamental *vivencia*, or experience."[22] As the Oaxacan activist Arturo Guerrero puts it,

> *Comunalidad*, or Communality, is a neologism that names a mode of being and living among the peoples of the Sierra Norte of Oaxaca, plus other regions in this state located in southeastern Mexico. . . . It expresses a stubborn resistance to all forms of development that have arrived in the area, which has had to accept diverse accommodations as well as a contemporary type of life that incorporates what arrives from afar without allowing it to destroy or dissolve what is its own—*lo propio*. . . . Communality is the verbal predicate of the We. It names its action and not its ontology. Incarnated

> verbs such as eat, speak, learn are collectively created in a specific place. It only exists in its execution.... We open ourselves to all beings and forces, because even if the We manifests itself in the actions of concrete women, men, and children, yet in that same movement, all that is visible and invisible below and on the Land also participates, following the principle of "complementarity" among all that is different. The communal is not a set of things, but an *integral* fluidity.[23]

The Mexican sociologist Raquel Gutiérrez Aguilar has recently proposed the concept of *entramados comunitarios* (communitarian entanglements) as opposed to "coalitions of transnational corporations," two contrasting modes of the organization of the social. By "communitarian entanglements" she means "the multiplicity of human worlds that populate and engender the world under diverse norms of respect, collaboration, dignity, love, and reciprocity, that are not completely subjected to the logic of capital accumulation even if often under attack and overwhelmed by it." As she further explains, "such community entanglements... are found under diverse formats and designs.... They include the diverse and immensely varied collective human configurations, some long-standing, others younger, that confer meaning and 'furnish' what in classical political philosophy is known as 'socionatural space.'"[24] Gutiérrez Aguilar's distinction also aims to make visible "the *gigantic and global confrontation* between diverse and plural communitarian entanglements, with a greater or lesser degree of relationality and internal cohesion, on the one hand; and, on the other, the most powerful transnational corporations and coalitions among them, which saturate the global space with their police and armed bands, their allegedly 'expert' discourses and images, and their rigidly hierarchical rules and institutions."[25]

It is important to emphasize, however, to return to Guerrero, that communality can be understood only in its relation with the noncommunal exterior: "This is the 'outside spiral': it begins with an external 'imposition,' which unleashes, or not, an internal 'resistance' and develops into an 'adaptation.' This result is *lo propio*—what is one's own—and the We."[26] In other words, the communal does not refer to an ontological condition that preexists a social group's interactions with its surrounding worlds *but is the very product of such interactions*. Said otherwise, the "We" is never produced in isolation but is always coproduced through an interplay among heteronomy, autonomy, and ontonomy. At the same time, it is clear that communitarian entanglements involve a type of human relation centered on *lo común* (the common), always attempting to overflow their determination by capital.

The massive mobilizations and popular insurrections that took place in Bolivia during the years before the election of the country's first Indigenous president, Evo Morales, in 2006 have been another fertile ground for the theorization of *autonomía* and the political. The literature is already vast and cannot be summarized here;[27] only a few contributions of particular relevance for this chapter's purposes will be presented, based on the work of Indigenous and non-Indigenous intellectuals. In her important work on liberalism and modernity in Bolivia from Indigenous perspectives, the Bolivian scholar Silvia Rivera Cusicanqui interprets Indigenous struggles, starting with the famous rebellion of Túpac Amaru and Túpac Katari in 1780–81, in terms of the tension between liberal and communal forms of life and social organization.[28] The tension between these forms, as she states, has shaped much of Bolivia's history, as they are interwoven "in a chain of relations of colonial domination."[29] It remains so today, as shown by the intense insurrections of 2000–2005, before Morales's election, when the collective memory of the events of 1781, including the dismemberment of Katari and the exhibition of his lifeless body parts in different public spaces in La Paz, yielded a desire "for the reunification of the fragmented body politic of Indigenous society."[30]

Rivera Cusicanqui gestures at a crucial dimension of politics in relation to communal groups, namely, their nonlinear conception of time and history and yet their strict contemporaneity. It is against this background that El Alto, the largely Aymara city near La Paz that grew to close to a million people in less than three decades, heavily populated by peasant migrants expelled by the neoliberal reforms of the 1980s (largely on the advice of Jeffrey Sachs, which was adopted by the military ruler of the time), became, for sociologist Félix Patzi Paco, a school for communal thought. For this Aymara intellectual, the transformation pursued by these movements took place "from the perspective of their own philosophy and their own economic and political practices."[31] Similarly, writing about the insurrections against neoliberal reforms in 2000–2005, Pablo Mamani Ramírez speaks of an "Indigenous-popular world" in movement, stemming from a society different from liberalism,[32] and Gutiérrez Aguilar writes about the fracture of the liberal paradigm effected by the communal popular forms.[33] As she concludes, the insurrections demonstrated "the possibility of transforming social reality in a profound way in order to preserve, transforming them, collective and long-standing lifeworlds and to produce novel and fruitful forms of government, association, and self-regulation. In some fashion, the central ideas of this path can be synthesized in the triad: dignity, autonomy, cooperation."[34]

These interpretations unveiled the existence of a Bolivian society "characterized by noncapitalist and nonliberal social relations, labor forms, and forms of organization."[35] The main features of nonstatist and nonliberal regulation include deliberative assemblies for decision-making, horizontality in organizations, and rotation of assignments. The struggles created forms of self-organization aimed at the construction of nonstate forms of power. These forms appeared as *micro-gobiernos barriales* (neighborhood microgovernments) or *anti-poderes dispersos*, that is, diffuse and quasi-microbial, intermittent forms of power.[36] The struggles (a) aimed to reorganize society on the basis of local and regional autonomies; (b) set in movement noncapitalist and nonliberal forms of organization, particularly in urban areas; (c) introduced self-managed forms of the economy, organized on communal principles, even if articulated with the market; and (d) engaged with the state, but only to dismantle its colonial rationality. The objective was not to control the state but "*organizarse como los poderes de una sociedad otra*" (to become organized on the basis of the powers of an other society).[37]

Emerging from this interpretation is a fundamental question, that of "*being able to stabilize in time* a mode of regulation outside of, *against, and beyond* the social order imposed by capitalist production and the liberal state."[38] Patzi Paco's concept of the communal system spells out this hypothesis: "Our point of departure for the analysis of communal systems is doubtlessly the Indigenous societies.

In contradistinction to modern societies, Indigenous societies have not reproduced the patterns of differentiation nor the separation among domains (political, economic, cultural, etc.); they thus function as a single system that relates to both internal and external environments [*entorno*]. . . . The communal system thus presents itself as opposed to the liberal system. The communal system can appropriate the liberal environment without this implying the transformation of the system [and vice versa]."[39] One can relate this conceptualization to the theory of autopoiesis and autonomy.[40] In the communal economy, as practiced by urban and rural Indigenous groups, natural resources, land, and the means of labor are collectively owned, although privately distributed and utilized. The entire system is controlled by the collectivity. The political dimension is just as important as the economic dimension; power is not anchored in the individual but in the collectivity. In the communal form of politics, "social sovereignty is not delegated; it is exercised directly" through various forms of authority, service, assembly, and so on; in short, the representative "*manda porque obedece*," or "rules through obedience," which is also a main Zapatista principle.[41]

The proposal of the communal system implies three basic points: (1) the steady decentering of the capitalist economy and the expansion of communal enterprises and noncapitalist forms of economy; (2) the decentering of representative democracy in favor of communal forms of democracy, or *comunalocracia*;[42] and (3) the establishment of mechanisms for genuine interculturality.[43] Patzi Paco is emphatic in stating that the communal system is not predicated on excluding any group. It utilizes the knowledge and technological advances of liberal society but subordinates them to the communal logic; in the process, the communal system itself becomes more competitive and fairer. The proposal is a call not for a new hegemony but for an end to the hegemony of any system, for taking leave of the universals of modernity and moving into the pluriverse of interculturality. To achieve this goal may perhaps require a refounding of the societies of the continent based on other principles of sociability. Patzi Paco's conceptualization of the communal system offers persuasive principles for autonomy-oriented redesign.

To sum up: in lieu of state-driven development based on imputed needs and market-based solutions, *autonomía* builds on ways of learning, healing, dwelling, producing, and so forth that are freer from heteronomous commands and regulation. This is crucial for design projects intended to strengthen autonomy. Thus, *autonomía* means living, to the greatest extent possible, beyond the logic of the state and capital by relying on, and creating, nonliberal, nonstate, and noncapitalist forms of being, doing, and knowing. Yet it also requires organization, which tends to be horizontal in that power is not delegated, nor does it operate on the basis of representation; rather, it fosters alternative forms of power through types of autonomous organization such as communal assemblies and the rotation of obligations. *Autonomía* is anticapitalist but not necessarily socialist. If anything, it can be described in terms of radical democracy, cultural self-determination, and self-rule. In linking design and democracy, design theorist Gui Bonsiepe actually defines democracy as the reduction of heteronomy, that is, of domination by external forces, and as the process by which dominated citizens transform themselves into subjects, opening spaces for self-determination and autonomous projects.[44]

This does not mean autarky or isolation; on the contrary, *autonomía* requires dialogue with other peoples, albeit under conditions of greater epistemic and social equality. Moreover, it requires alliances with other sectors or groups in struggle—strategies of localization and interweaving not intended to insert "the local" into "the global," following conventional views, but a type of place-based globalism that connects autonomous movements with each other.[45] These alliances are seen in terms of "walking the word" (*caminar la palabra*), a concept developed by the Colombian Minga social y comunitaria to point at the need to come into visibility, make demands on society, and collectively weave knowledges, resistances, and strategies with other movements.

One final caveat about the notions of community and the communal: as the Buenos Aires militant research collective Colectivo Situaciones put it, rather than being a preconstituted entity or an "unproblematic fullness," the community "is the name given to a particular organizational and political code, a singular social technology"; in resisting being rendered an anachronism by the modern, the community summons "actualized collective energies"; as such, and "against all common sense, the community produces dispersion," and this dispersion could become central to the invention of amplified nonstatist modes of cooperation.[46] The appeal to community itself is thus not anachronistic, as moderns often dismissively reply; on the contrary, "the community summons actualized collective energies. . . . Communal doings and their openness to internal contradictions and ambivalence are a reflection of the radical contemporaneity of communities with respect to other modes of organization and cooperation"—including, one might say, standard modern forms that are by now more anachronistic.[47]

To speak of "communities in resistance" does not imply an essentialist or homogenizing vision of the community, as some critics adduce. It means understanding how, despite communities' fracture and fragmentation, communal actions might reveal "transition paths, beyond the dualism between modernity and postmodernity, universalism and communitarianism. . . . [They reveal] collective biographies of microrevolutions for self-determination."[48] That said, it is important to investigate exactly how, in the midst of the conflicts and heterogeneity that inevitably shape these communities because of their subaltern condition, there appear in them new forms of life, solidarity, and militancy. While the internal diversity of the communities might generate strife and disorganization under the pressure of intense repression and displacement, it often also yields types of intercultural diversity capable of broadening the processes by which they endow their worlds with meaning. These dynamics usually escape the attention of researchers too intent on finding antagonist oppositions in the midst of communities in struggle (which of course are also there). To go beyond this habitual research behavior requires a different epistemic positioning, one in which the researcher genuinely sees herself or himself as part of the collective action she or he is studying, as well as the willingness to interact with it. Far from disappearing, in many communities in struggle the collective dimension is woven out of plurality and disagreement.[49]

There is perhaps no clearer example of the openness of the communal at present than the decolonial and communitarian feminisms that are emerging in some popular and ethnic communities.

For the Aymara intellectual/activist Julieta Paredes, communitarian feminism is a strategy for pursuing the twin goals of depatriarchalization (in relation to both autochthonous and modern patriarchies) and decolonization (in relation to liberal, modernizing, and capitalistic hegemonies, including individualizing Western feminisms). In this framework, the community is seen as "the inclusive principle for the caring of life."[50] The community "is another manner of understanding and organizing society and of living otherwise.... It is an alternative proposal to that of individualistic societies."[51] This is why it implies an entire *tejido* (weave) of complementarities, reciprocities, and forms of autonomy and interculturality that include, for rural communities, relations to urban communities and transnational groups and, of course, the entire range of nonhumans. The community links together body, space, memory, and movement within a dynamic cyclic vision; this is the complex process that anchors the *Vivir Bien*, or collective well-being.

In all of these experiences, the community is thus understood in deeply historical, open, and nonessentialist terms. If anything, there is an emphasis on the creation of new spaces for the communal. It follows that the realization of the communal is always an open-ended historical process. In the language of Humberto Maturana and Francisco Varela, as social systems, communities are third-order autopoietic entities whose "operational closure" (often coded in terms of "the defense of our culture" by locals) is maintained throughout the communities' relation to their "environment" (the sociopolitical and ecological context, broadly speaking); through this always ongoing form of relating (structural coupling), communities may undergo structural changes of various types (e.g., by adopting the use of information and communication technologies or novel market practices); however, the basic system of relations has to be maintained for the community to preserve its autopoiesis, that is, its capacity for self creation.[52] *Autonomy* is the name given to this process.

Needless to say, communities' exercise of autonomy takes place today under astonishingly inimical conditions. The ongoing war on things communal also means that communal-based transition initiatives and territorial struggles indeed prefigure potential worlds to come (like the musics they sometimes create), yet they have to realize themselves within incredibly hostile environments that relentlessly undermine their efforts. This is the political ontology—held in place by capitalism, corporate coalitions, expert institutions, repressive and police states, and dualist rationalities—within which autonomous initiatives have to struggle. They surely cannot flourish in isolation, but perhaps the strategies of interweaving being tried out across myriad tiny islands of attempted autonomy might result in the renovated continents, however small for the time being, imagined by transition activists and designers. Can autonomous design contribute to this pluriversal realization of the communal? This long historical, political, and theoretical background on *autonomía* and the communal has been necessary to convey the importance of placing autonomy within the scope of design, on the one hand, and of constructing the communal as a design space for ontologically oriented design, on the other. It was also a way to convey what the Latin American struggles' specific contributions to the *pensamiento* (thought) of transition might be.

AN OUTLINE OF AUTONOMOUS DESIGN

The remainder of this chapter will lay down additional elements for thinking about the relations among autonomy, design, and the realization of the communal. This will be done in three parts. The first identifies some principles for autonomous design, drawing on a particular experience in Colombia in the late 1990s; the second extends these lessons based on the chapter's discussion of autonomy and the communal. The third, finally, sketches a transition imagination exercise for a particular region in the Colombian southwest. Autonomous design—as a design praxis with communities that has the goal of contributing to their realization as the kinds of entities they are—stems from the following presuppositions:[53]

1. *Every community practices the design of itself*: its organizations, its social relations, its practices, its relation to the environment. If for most of history communities practiced a sort of "natural design" independent of expert knowledge (ontonomy, spontaneous coping), contemporary

situations involve design based on both detached and embodied forms of reflection.

2. Every design activity must start with the strong presupposition that *people are practitioners of their own knowledge* and from there must examine how people themselves understand their reality. This epistemological, ethical, and political principle is at the basis of both autonomy and autonomous design. (Conventional development planning is intended to get people to practice somebody else's knowledge, namely, the experts'!)
3. What the community designs, in the first instance, is *an inquiring or learning system about itself*. As designers, we may become coresearchers with the community, but it is the latter that investigates its own reality in the codesign process.
4. Every design process involves a *statement of problems and possibilities* that enables the designer and the group to generate agreements about objectives and to decide among alternative courses of action (concerning the contamination of the river, the impact of large-scale mining, a particular food-production project, landlessness, the struggle to defend place and culture, discrimination against women, availability of water, etc.). The result should be a series of scenarios and possible paths for the transformation of practices or the creation of new ones.[54]
5. This exercise may take the form of building a model of *the system that generates the problem of communal concern*. Given this model, the question that every autonomous design project must face is: what can we do about it? The answer will depend on how complex the model of reality is. The concrete result is the design of a series of tasks, organizational practices, and criteria by which to assess the performance of the inquiry and design task.[55]

In building the model for the particular concern, it is important to recognize that *problem statements always imply solution statements*; problems never stand as neutral statements about reality;

the entire process is political since any construction entails choices that affect people in particular ways. Problem statements are by the same token necessarily partial. The group's perception of the problem is continuously evolving as the conceptualization of it becomes more complex in light of new thinking, new information, more involved experimentation, and the like. The more complex the conceptualization of the system that produces the problem, the sharper the sense of purpose and of what needs to be done. Problem statements need to address the question, "Why do we/I see this as a problem?," and to follow each "because..." with another "why" until participants' values are made explicit. The design process also needs to broach the questions, What/who needs to change? Why is this change not happening now? What consequences would follow if such changes were to happen? And these inquiries must be repeated at various scales, including the household, community, and regional (e.g., river basin) levels and beyond.[56]

A problem statement is thus the *expression of a concern* that the group has about people's condition (ideally shared by the designer). In the last instance, what the autonomous design process wants to accomplish is to make not only the community but also the larger society more sensitive and responsive to the newly articulated concerns of the collectivity. This can be seen in terms of generating, out of the breakdowns that the systems' exercise unveils, a range of possibilities for disclosing new spaces for the exercise of community autonomy as the group deals with the problems at hand. It should be apparent by now that, according to this perspective, the ideal situation for autonomous design obtains when the client, the designer, the decision maker, and the guarantor of the system are the same entity, namely, the community and its organizations.[57]

The workshop's systems methodology might seem a bit dated now, yet it is useful as a starting point for understanding autonomous design practices. In this particular instantiation in the Colombian Pacific, the workshop contributed to the creation of concepts and scenarios that, eventually, resulted in a framework developed by Afro-descendant movements (to some extent in conversation with Indigenous activists) and provided the basis for a sophisticated political ecology by the movement, which I have analyzed at length elsewhere.[58] Some of the key notions included that of the Colombian Ecuadorian Pacific as a "region-territory of ethnic groups," the conceptualization of the territory as the space for the "life projects of the communities," a framework for the conservation of biodiversity based on the defense of territory and culture (very different from the established frameworks designed by conservation biologists and economists), and a set of guiding principles for the region's own vision of development and perspective on the future.[59] *Autonomía* became central to the entire process. Figure 1 is a representation of the process.[60]

We can recognize in this model the pillars of a design imagination centered on autonomy and the realization of the communal. *Autonomía* involves the articulation of the life project of the communities, centered on the *Vivir Bien* (the well-being of all, humans and nature), with the political project of the social movement, centered on the defense of the region-territory. (Notably, the notion of *Vivir Bien* in this framing is very similar to that of *Buen Vivir* that became well known in the 2000s.) While the life project is grounded in the long-standing relational ontology of the river communities (referred to as *cosmovision* during those years), the political project is based on the work of ethnoterritorial organizations, requiring the effective appropriation of the territories and guided by the communities' own vision of the future. Would it be too far-fetched to suggest that this particular social movement was pursuing a strategy of autonomous, ontologically oriented design? In this and similar cases, one could argue that a codesign process is at play in which communities, activists, and some outside participants (including expert designers) engage in a collaborative exercise, with planner, designer, decision maker, and guarantor coinciding to a great extent with the communities and the movement.

It is noteworthy that this experience was based on the organizational principles agreed on by the Proceso de Comunidades Negras (PCN; Black communities process) since 1993 (and which remain in force to this day, even if in an enriched form that nevertheless maintains their basic structure). These principles include the affirmation of identity (the right to be Black); the right to the territory (as the space for the exercise of being); autonomy (as the right to the conditions for the exercise of identity); the right

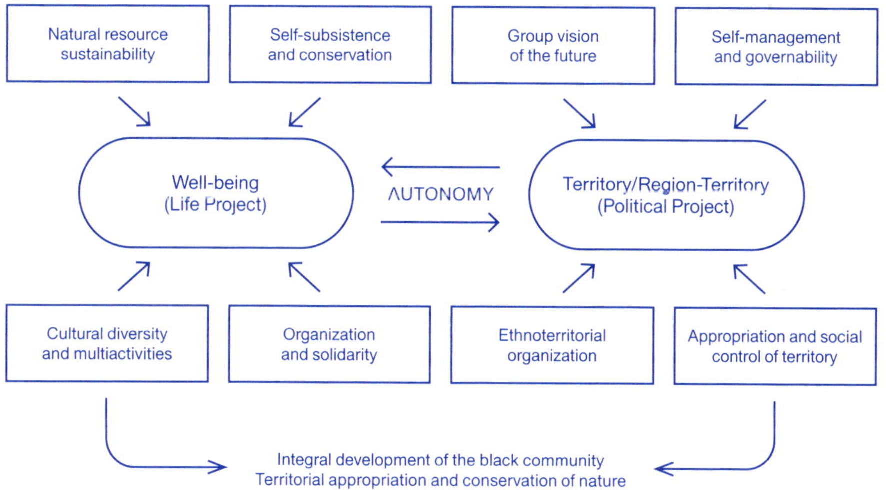

1 Basis for a culturally and ecologically sustainable development and perspective of the future. Redrawn based on diagram from the Proceso de Comunidades Negras (PCN).

to their own vision of the future, including the communities' right to choose their own model of development and of the economy according to their cosmovision; and the right to historical reparations.[61] These principles anchor not only the internal decision-making of the organization but its relation to the state and to other actors. In cases such as this, *it is of crucial importance for designers to develop a profound understanding of the political project of the movement* (not necessarily to share it in its entirety but to apprehend it fully) and to be willing to submit all codesign activities to the same principles. This is a sine qua non for working with political (say, ethnoterritorial) organizations under the rubric of autonomous design.

First published in Escobar, *Designs for the Pluriverse: Radical Interdependence, Autonomy, and the Making of Worlds* (Durham, NC: Duke University Press, 2018), 172–88.

1 See Gustavo Esteva, "The Hour of Autonomy," *Latin America and Caribbean Ethnic Studies* 10, no. 1 (2015): 134–45.
2 I heard Esteva make this distinction in a lecture in the mid-2000s. A version of it is found in ibid. This entire issue of the *Latin American and Caribbean Ethnic Studies* journal is devoted to Indigenous autonomy in Latin America.
3 These features of *autonomía* emerge from discussions by and about social movements particularly in southern Mexico (Chiapas, Oaxaca), southwestern Colombia (Black and Indigenous movements), and parts of South America, especially Bolivia and Ecuador. There are resonances with themes in contemporary theory (e.g. Gilles Deleuze and Félix Guattari, *A Thousand Plateaus*, 1980 [Eng. trans. Brian Massumi, Minneapolis: University of Minnesota Press, 1987]) and with anarchist thought.
4 From a document signed by the main Indigenous organizations of Colombia (Organizaciones Indígenas de Colombia), 2004. Available online on the website of Minga Informativa de los Movimientos Sociales, http://www.movimientos.org/es /show_text.php3%3Fkey%3D3282 (accessed August 5, 2023).
5 Subcomandante Marcos and the Zapatistas, *The Other Campaign/La Otra Campaña* (San Francisco: City Lights, 2006), 77–78.
6 Esteva, "The 'Other Campaign' and the Left: Reclaiming an Alternative" (Oaxaca: unpublished ms, 2006), 36–38.
7 Raúl Zibechi, *Dispersar el poder: Los movimientos como poderes anti-estatales* (Buenos Aires: Tinta Limón, 2006).
8 See Arturo Escobar, *Sentipensar con la tierra. Nuevas lecturas sobre desarrollo, territorio y diferencia* (Medellín: UNAULA, 2014).
9 In fact, social movements can be considered autopoietic units; established theories see movements as allopoietic, that is, produced by and referring to another logic, whether capital, the state, nationalism, or what have you. See Escobar, "Imagining a Postdevelopment Era? Critical Thought, Development, and Social Movements," *Social Text* 31–32 (1992): 20–56.
10 See Esteva, "Nuevas formas de la revolución" (Oaxaca: unpublished ms, 2013), and Jérôme Baschet, *Adiós al capitalismo: Autonomía, sociedad del buen vivir y multiplicidad de mundos* (Buenos Aires: Futuro Anterior Ediciones, 2014).
11 According to Baschet, rebellious autonomy is the general principle of both Zapatista organizing and their actions aimed at the reconstruction of life beyond capitalism. See Baschet, *Adiós al capitalismo*.
12 See Elizabeth Mason-Deese, "The Unemployed in Movement: Struggles for a Common Territory in Argentina's Urban Peripheries," PhD diss., University of North Carolina, Chapel Hill, 2015, and Matrina Sitrin and Dario Azzelini, *They Can't Represent Us! Reinventing Democracy from Greece to Occupy* (London: Verso, 2014). The case of the *piqueteros* in Argentina is one of the best-known cases of autonomous politics in urban Latin America. See Mason-Deese, "The Unemployed in Movement."
13 See, e.g., Wendy Harcourt and Escobar, eds., *Women and the Politics of Place* (Bloomfield, CT: Kumarian Press, 2005); Janet M.

Conway, *Edges of Global Justice: The World Social Forum and Its "Others"* (London: Routledge, 2013).

14 See Boaventura de Sousa Santos, *Epistemologies of the South: Justice against Epistemicide* (Boulder: Paradigm, 2014).

15 See Esteva, "Celebration of Zapatismo," *Humboldt Journal of Social Relations* 29, no. 1 (2005): 127–67, and Esteva, "The 'Other Campaign' and the Left."

16 Baschet, *Adiós al capitalismo*, 72.

17 See Astrid Ulloa, "Reconfiguraciones conceptuales, políticas y territoriales en las demandas de autonomía de los pueblos indígenas en Colombia," *Tabula Rasa* 13 (2010): 73–92; Ulloa, "The Politics of Autonomy of Indigenous Peoples of the Sierra Nevada de Santa Marta, Colombia: A Process of Relational Indigenous Autonomy," *Latin American and Caribbean Ethnic Studies* 6, no. 1 (2011): 79–107; and Ulloa, "Los territorios indígenas en Colombia: De escenarios de apropiación transnacional a territorialidades alternativas," *Scripta Nova: Revista Electrónica de Geografía y Ciencias Sociales* 16, no. 65 (2012): n.p.

18 I should make it clear that I am discussing here the Latin American perspectives on *autonomía*. There are many other sources for the concept, including in anarchism and Italian autonomous Marxism. The alter-globalization movements of the late 1990s and early 2000s did much to bring the question of autonomy into discussion. See, for instance, Conway, *Edges of Global Justice*; Michal Osterweil, "The Italian Anomaly: Place and History in the Global Justice Movement," in Cristina Fominaya and Laurence Cox, eds., *The European Social Movement Experience: Rethinking "New Social Movements," Historicizing the Alterglobalization Movement and Understanding the New Wave of Protest* (London: Routledge, 2013), 33–46; and Andrej Grubačić and Denis O'Hearn, *Living at the Edges of Capitalism: Adventures in Exile and Mutual Aid* (Oakland: University of California Press, 2016).

19 This principle is repeated frequently in Nasa writings, particularly those by the Asociación de Cabildos Indígenas del Norte del Cauca (ACIN). See, for example, the call for the "Tercer Congreso Zonal de Çxhab Wala Kiwe—ACIN," June 15–21, 2017, available online at https://nasaacin.org/3o-congreso-acin-cxhab-wala-kiwe/ (accessed August 9, 2023).

20 J. K. Gibson-Graham, Jenny Cameron, and Stephen Healy, *Take Back the Economy: An Ethical Guide for Transforming Our Communities* (Minneapolis: University of Minnesota Press, 2013).

21 For detailed accounts of both the context of these critiques and the responses to them see Escobar, "Latin America at a Crossroads: Alternative Modernizations, Post-liberalism, or Post-development?," *Cultural Studies* 24, no. 1 (2010): 1–65, and Escobar, *Sentipensar con la tierra*.

22 Esteva, "La noción de comunalidad" (Oaxaca: unpublished ms, n.d.), 1.

23 Arturo Guerrero Osorio, "Comunalidad," in Ashish Kothari, Ariel Salleh, Escobar, et al., eds., *Pluriverse: A Postdevelopment Dictionary* (New Delhi: Tulika Books and Authorsupfront, 2019), 130–31. The term *comunalidad* was coined at the end of the 1970s by two Oaxacan thinkers, Floriberto Díaz Gómez and Jaime Martínez Luna. Esteva also introduces *comunalitario*, or "communalitarian," different from the well-established *comunitario* (communitarian). This neologism is helpful in establishing some distance from

the association of the communal with what is often described as "communitarian violence" in South Asia.

24 Raquel Gutiérrez Aguilar, "Pistas reflexivas para orientarnos en una turbulenta época de peligro," in Gutiérrez, Natalia Sierra, Pablo Dávalos, et al., *Palabras para tejernos, resistir y transformar en la época que estamos viviendo* (Oaxaca: Pez en el Árbol, 2012), 12.

25 Ibid., 13.

26 Guerrero, "Comunalidad," 132.

27 Escobar, "Latin America at a Crossroads."

28 See Silvia Rivera Cusicanqui, "Democracia liberal y democracia de ayllu: El caso del Norte Potosí, Bolivia," in Carlos Toranzo Roca, ed., *El difícil camino hacia la democracia* (La Paz: ildis, 1990), 9–51, and Rivera Cusicanqui, *Hambre de huelga: Ch'ixinakax Utxiwa y otros textos* (Querétaro, MX: La Mirada Salvaje, 2014).

29 Rivera Cusicanqui, "Democracia liberal," 20.

30 Rivera Cusicanqui, *Hambre de huelga*, 9.

31 Félix Patzi Paco, *Sistema comunal: Una propuesta alternativa al sistema liberal. Una discusión teórica para salir de la colonialidad y del liberalismo* (La Paz: Comunidad de Estudios Alternativos, 2004), 187–88.

32 Pablo Mamani Ramírez, *Geopolíticas indígenas* (El Alto, BO: CADES, 2005).

33 Gutiérrez Aguilar, *Los ritmos del Pachakuti: Movilización y levantamiento indígena-popular en Bolivia* (Buenos Aires: Tinta Limón, 2008).

34 Ibid., 351.

35 Zibechi, *Dispersar el poder*, 52.

36 Mamani Ramírez, *Geopolíticas indígenas*.

37 Zibechi, *Dispersar el poder*, 75.

38 Gutiérrez Aguilar, *Los ritmos del Pachakuti*, 46. There has been a clear fallback into statist and developmentalist positions in Bolivia in recent years, certainly at the level of the state.

39 Patzi Paco, *Sistema comunal*, 171–72.

40 Patzi Paco's conceptual framework includes a distinction between system and environment reminiscent of Humberto Maturana and Francisco Varela's.

41 Patzi Paco, *Sistema comunal*, 176.

42 Guerrero, "Comunalidad."

43 Patzi Paco, *Sistema comunal*, 190.

44 Gui C. Bonsiepe, "Design and Democracy," paper presented at the Metropolitan University of Technology, Santiago de Chile, June 2005.

45 See Osterweil, "Place-Based Globalism: Locating Women in the Alternative Globalization Movement," in Harcourt and Escobar, eds., *Women and the Politics of Place*, 174–89.

46 Colectivo Situaciones, "Epílogo: Notas sobre la noción de 'comunidad' a propósito de Dispersar el poder," in Zibechi, *Dispersar el poder*, 212, 215.

47 Ibid., 213, 215.

48 Patricia Botero, "Presentación: Antología de los pueblos en resistencia," in *Resistencias: Relatos del sentipensamiento que caminan la palabra* (Manizales, CO: Universidad de Manizales, 2015), 17–19.

49 See Botero and Alicia Itatí Perdomo, eds., *La utopía no está adelante: Generaciones, resistencias e institucionalidades emergentes* (Manizales, CO: Consejo Latinoamericano de Ciencias Sociales/Centro de Estudios Avanzados en Niñez y Juventud, 2013).

50 Julieta Paredes, *Hilando fino desde el feminismo comunitario* (La Paz: Communauté Mujeres Creando Comunidad, with the support of the Deutscher Entwicklungsdienst, 2012), 27.

51 Ibid., 31.

52 See Maturana and Varela, *Autopoiesis and Cognition: The Realization of the Living* (Boston: Reidel, 1980).

53 This part is based on a weeklong workshop on ecological river-basin design that I designed and implemented in 1998 in the port city of Buenaventura, in the Colombian Pacific, together with the activists of the Proceso de Comunidades Negras (PCN; Process of Black communities), under the explicit rubric of autonomous design. The participants were leaders of grassroots river organizations and activists in the social movement of the Black communities. The background to the exercise was the need for river communities to develop their own *plan de ordenamiento territorial* (territorial action plan), mandated by the government. The workshop workbooks are available, although they were never published. See PCN (Proceso de Comunidades Negras) and Escobar, "Taller de capacitación sobre diseño de sistemas de ríos" (Buenaventura, CO: PCN, 1998). The workshop followed my own version of a systems approach, significantly influenced by Charles West Churchman and Leonard Joy.

54 For Victor Papanek, "the most important ability that a designer can bring to his work is the ability to recognize, isolate, define, and solve problems." See Papanek, *Design for the Real World* (Chicago: Academy Chicago, 1984), 151. Today everybody agrees that design goes beyond problem-solving, and that inquiring into problems needs to be participatory. To be fair, Papanek advocated for "integrated, comprehensive, anticipatory design" (322), arguing against narrow problem definitions and planning.

55 The further one departs from established Cartesian methodologies, the more engaging the discussions leading to what I have called a model (surely not the best term) become. By "engaging" I mean an intense, open-ended conversation that brings forth, and at its best challenges, the cultural background of the collectivity. This type of engaging conversation is well-known in community assemblies or social movements' political meetings, which often go on for hours, seemingly without a concrete agenda. Planners miss this dynamic altogether with their fixed routines, or they consider it a waste of time.

56 These questions stem from Joy's systems approach to food and nutrition planning (from class notes, University of California, Berkeley, summers of 1978 and 1979). See Joy, ed., *Food and Nutrition Planning: The State of the Art* (Guildford: IPC Science and Technology Press, 1978).

57 See Churchman, *The Design of Inquiring Systems* (New York: Basic Books, 1971).

58 See Escobar, *Territories of Difference: Place, Movements, Life, Redes* (Durham, NC: Duke University Press, 2008).

59 Ibid.

60 See PCN, *Fortalecimiento de las dinámicas organizativas del Proceso de Comunidades Negras del Pacífico Sur Colombiano, en torno al ejercicio de los Derechos étnicos, culturales y territoriales. Proyecto PCN-Solsticio. Segundo informe técnico trimestral, Septiembre—Noviembre 2000*. (Cali, CO: PCN, 2000), 5, and PCN, *Construyendo Buen Vivir en las comunidades negras del río Yurumanguí y en Pílamo, Cauca* (Cali, CO: PCN/Solsticio, 2004), 38.

61 See Escobar, *Territories of Difference*, 221–27.

EDUARDO KOHN AND DANIEL STEEGMANN MANGRANÉ

THE THINKING FOREST

Tendrils of Sylvan Thinking

EDUARDO KOHN

The claim I made in my book *How Forests Think*—namely that forests think—is a psychedelic one. From the Greek roots *psyche* and *dēlos*, psychedelics are literally mind manifesting. I was saying this without really knowing that I was saying it. Since then, my work has become more explicitly psychedelic, in the sense that the collaborative research methodology we use with the communities with which I am now working includes psychedelics. When I work on documents with the community, when I try to help them express in writing what they want to say about their relationship to the forest and its many beings—plants, animals, and spirits—and what the forest spirits themselves want to express in these documents, we do so through many means. This is part of my current work to capacitate the kind of thinking that forests express, in the many valences in which it can become manifest.

Part of this collaborative work is with the Sápara Nation, whose home is in the South Central portion of the Amazon region of what today is Ecuador. One of these collaborative projects involves the Sápara political leader and healer Manari Ushigua. With him I've been thinking a lot about sylvan thinking and its relation to aesthetics and ethics—the ways in which sylvan thinking helps us see how aesthetics and ethics are linked. Sylvan thinking, the kind of thinking that forests do, the kind of thinking that conjoins us to the rest of life, is essentially an imagistic form of thought. I've become convinced that sylvan thinking is a good thing—that we should struggle to hold open spaces for its flourishing. But this leads to the question: what makes a good image?

A few months ago Manari was in Montreal to take part in a research project bringing Indigenous leaders and academics together to discuss issues of environmental and territorial conflict.

As part of his visit Lisa Stevenson and I organized a workshop where we screened an experimental film by a really amazing Quebecois artist named Philippe Léonard. The film was about the 2016 Fort McMurray fire in Alberta. He was there during the fire and made a powerful experimental film, one that to my mind went directly into sylvan thought. No context, no explanation, it just took you directly into what it feels like to be a forest under assault. You could sense the spirit-being of the forest as well as the obdurate agencies of the forces of its destruction. I wanted to hear Manari's thoughts about this film because I felt that Philippe, through his film work, was making an ecopolitical argument at the level of spirit.

I was excited to hear Manari's comments on the film because I took it to be an instance of image manifesting spirit in the context of environmental destruction. But I realized just how naive this was of me. Manari's reaction to the film, as he told us, was the sense of becoming literally disembodied: he felt one arm flying off across the room in one direction, another in the other direction, his head floating above, his legs dispersing. Since then I've been reading one of Charles Sanders Peirce's lectures on pragmatism in which he discusses the relationship between aesthetics and ethics (you might say between art and politics). Peirce thinks of ethics as nested within aesthetics. For him, a "good" image is one in which parts come together to form a larger emergent whole. What became interesting for me was that this film that Philippe Léonard showed was clearly a powerful and a good film artistically, and it also certainly had an important—albeit dark—political message. But Manari was saying that it was a bad film. It was a bad film in the sense that it was doing bad spirit work, because it was conjuring images/spirits that dissipated the self, ones that were not allowing for the manifestation of a larger self. They were literally breaking Manari into pieces. This is the opposite of the kind of integration that he is seeking in his spirit ecopolitics. I was naively content with seeing the animacy in the film. Manari was concerned with the ethics within this animacy and he insisted that as important as this film was, it should not stand alone. It requires that we find an aesthetic resolution and in practical terms should, in any screening, be followed by another film that would do this. Aesthetics, for him, carries an ethical content. For him, art work and spirit work are the same thing. The thinking that is happening in the spirit world is imagistic thinking.

This question of the "good" image guides my current work. I've argued that forests think, but now I want to argue that such thinking is good. A thinking forest is a good thing. Of course, forests are good. But I'm explicitly saying that a *thinking* forest is good. It's the kind of thinking that they do, and the kinds of thoughts this kind of thinking can generate, that make them good. The important attribute of what a forest is is its thinking quality, and that's what's good. So the question is how do you hold open spaces for that and how do you allow this "sylvan" kind of thinking to flourish? How do you hold open spaces for it in a very practical sense? Part of this involves supporting the kinds of struggles that people like the Sápara face in keeping extraction out of the Amazon. But, at another level, how do you allow sylvan thinking to continue to move through us? This does not necessarily involve literal forests (nor does it necessarily involve psychedelics). It could involve listening to our dreams, being in touch with all these forms of sylvan thought that we as symbolic creatures need also to be able to ignore. So at one level there is an ethics to allowing sylvan thinking to flourish, but at another there still remains the question of how to think ethically both within and around a sylvan mode.

DANIEL STEEGMANN MANGRANÉ

This is something that's concerned me lately. We need to overcome the paradigm we live in, with its clear-cut division between mind and body, culture and nature. The modern paradigm is clearly exhausted: we are all facing this ecological crisis, but we are not going to overcome it if we don't start thinking differently, no?

EK

Absolutely. This is exactly where the project is going. Ecuador is a hub for another kind of ecopolitical thought. I'm working with a diverse group of people there—artists, activists, lawyers, shamans, and political leaders—who are all in some sense saying that regrounding thought is the way to go. This is a radically different way to think. One of the communities I work with, Sarayaku, calls this change a "*tiam*." *Tiam* is like a flip, a mind flip.

Sylvan thinking ultimately is a form of regrounding into something more capacious. Ayahuasca can access this ground. You can see the thinking forest and you can see your connection to it. Ayahuasca is impractical. When you take it you become detached from your everyday practical life. It dissipates the self. But that in and of itself is not the end but rather a means to something else. This breakup can allow for a semiotic regrounding in all of these other forms of communication that are embedded ecologically. But the problem is that you still have to do the everyday work of maintaining those embedded relations. Ayahuasca is neither necessary nor sufficient for the creation of this broader relational or ecological self, but it can be an "ally" in this attempt. But I worry about the current Western fascination with psychedelics. Our quest for it might lead us to forget that we already have what we are looking for. So is it a symptom of the illness or part of the cure?

DSM
I would like for you to talk a bit about your relationship with the rainforest itself. It's clear reading *How Forests Think* that you are very interested in biology. I wanted myself to be a biologist before becoming an artist.... Did you want to be a biologist before becoming an anthropologist?

EK
Yeah, my fascination with the forest began as a discovery, somewhat by chance. When I was living in Ecuador in the 1980s a friend invited me to a field station in the Amazon and I was mesmerized by this dense, quiet, cool, living network. Then I realized that this was what I really wanted to understand. I have always been interested in biology and have studied it a fair bit. I think certain limitations in my form of thinking, and inabilities as a youth to overcome those limitations, have kept me from being a biologist. But that's probably for the better because as an anthropologist I have more freedom to think and to think more freely about the kind of thinking that biological life manifests. I see scholarship as art. I see myself as an artist and I think what artists can do is very similar to what anthropologists and scientists can do. Claude Lévi-Strauss says something wonderful in *Tristes Tropiques* [1955]. He says that the only "true" disciplines

are anthropology, art—well, he says music—and mathematics. All of these allow you to derive the properties of the world from yourself and from your engagement with the world.

What I like so much about the possibility of poetic engagement that science, math and art and anthropology can give us is that they allow a tremendous freedom in the sense that they encourage us to dispel any kind of supposition that we have about what we're engaging with. I think this is somehow harder to do in some of the natural sciences—although I think good scientists always do it. It's somewhat easier perhaps in mathematics and in anthropology and in art in the sense that you're less constrained. The social sciences need to understand complex social structures. We know just how constrained by them we are and our work is to critique them. But anthropology says: "Go have an engagement with the world and allow that engagement to radically disrupt your thinking," right? And I think art does something like that too.

DSM
What I like about working with art is that it can reconfigure your relationship to reality. The important thing is not what happens inside the show but what happens when you face reality after leaving the exhibition. Art seems to operate in a state of openness.

EK
Art is constantly breaking the rules. It's a form of thinking aimed at freeing us from the constraints society places on thought. Of course artists are steeped in their representational traditions and a lot of artistic energy is directed toward reflecting on these traditions. But its freedom from societal constraints lies in its ability to resonate with the world. That's why I think artists can be at the avant-garde. My work resonates with artists because they are already thinking with and like forests. They have the pulse of the planet because they know how to dream. They are the psychic avant-garde. They resonate with what's going on and they know what the problems are.

DSM
It's funny because the title of the show [at Marres, Huis voor Hedendaagse Cultuur, Maastricht, in 2018] is *Dreaming Awake*.

EK
Exactly! That's what art can do because it's unconstrained in that sense. I mean obviously it is constrained. There's good art and there's bad art. Art has its traditions and standards just like any other endeavor. It requires skill and talent and there are power structures to contend with and all that stuff. But nonetheless it privileges resonance and this sylvan mode of thought with which it always operates opens it to sylvan kinds of thoughts.

DSM
I got to know about Peirce while I was doing the *Phasmides* film [2012]. Talking to a friend, I realized just how connected biology and semiotics are. Since then I've been thinking about the differences between a fragile and a strong sign. A swastika or a Christian cross is a strong sign in the sense that its meanings are highly historically determined. But you are thinking of signs that are much more fragile.

EK
Yeah, I think we are trying to tap into emergent meanings—before they are stabilized. Dreaming traffics in just these sorts of fragile signs. I had a powerful dream a couple of nights ago. What it means is still not fully clear to me. I try to create the conditions to continue to dream with it—often through other imagistic registers like drawing.

DSM
[Laughing] That's a nice dream!
I would like to quote two passages in your book. In one you write:

> A more complete understanding of representation, which can account for the ways in which that exceptionally human kind of semiosis grows out of and is constantly in interplay with other kinds of more widely distributed representational modalities, can show us a more productive and analytically robust way out of this persistent dualism.

And the other reads:

> The Amazon's many layers of life amplify and make apparent these greater-than-human webs of semiosis. Allowing its forests to think their ways through us can help us appreciate how we too are always, in some way or another, embedded in such webs and how we might do conceptual work with this fact.

EK
Yeah, the idea here is not to do away with the human, but to understand that what we take the human to be as a particularly special emergent possibility is actually grounded in this larger field of semiotic processes. Any appropriate way of actually thinking new thoughts is always part of this process of immersion into these other forms. When we analytically ignore that, we get into trouble. Forests are wonderful reminders of that larger form of thinking. Because they do so much sylvan thinking, they literally come to manifest sylvan thinking. Forests are literally psychedelic—they are mind manifesting. Sylvan thinking exists wherever there is life—human and otherwise—but in a tropical forest you can't but notice it. Such forests make it obvious.

DSM
I was talking with a specialist about the Brazilian Mata Atlântica, the coastal rainforest that is one of the most diverse ecosystems ever known and it's endangered because it is on the coast where everyone lives. To my surprise he told me that this is a super-resilient rainforest. It's very difficult to reforest because of its dense web of interactions and interdependencies, but at the same time there are so many forms of life that it's very difficult to kill it. You can obviously cut it down and then it's gone, but even with all the pressure of millions of people living around, it's still relatively healthy. But what really stuck to my mind is the resilient strength of these "million forms of life," even as they are also all individually quite fragile.

EK
This is an interesting thing about what we can call "weak semiosis." There's a sense that the symbol is strong and in fact that's why

you can still use words that no longer refer to anything real. They're dead but they're still alive in a certain way. But on the other hand, you've got this other form of thinking whose individual threads are much weaker. Dreams are like this. Oh, you had a dream? If you manage to remember it you are still left with figuring out what it means (if anything). But if you pay enough attention to the weak signs that make up a dream, maybe you can understand something. A forest is sort of like this. It has many layers and at some point it achieves a sort of solidity. It is a dense tangle of tendrils of thought that, collectively and in their mutual resonances, becomes quite strong.

DSM
One of the things that really struck me about your book is how it makes clear the ways we create significance through a web of signs that do not belong completely to us.

EK
Yes! This gets at my quibble with how in the academic world we usually think about reference. On the one hand there is a world of signs, and on the other the world of things to which signs refer in a contingent and arbitrary fashion. But in this other way of understanding an animate lifeworld that I'm proposing, humans are interacting with a whole set of entities that themselves are communicating, signifying, representing. And part of the game is to get that, to enter into that world which is not made by us. And the way to enter it is to listen. It's the listening—not the speaking—that's important. You have to learn to open yourself to something that is not part of your world or way of thinking.

DSM
This also raises a complex question. We start to realize how thought is not only something that happens in our head but is always part of a larger ecology of thoughts, as you describe so beautifully. But how, when we are inside this network of thoughts, can we represent this network to ourselves?

EK

One of the insights of Amazonian animism is that we are not the only thinkers; we are not the only persons. We are not alone. One of the insights of the Peircian semiotic framework is to insist that all thought is a semiotic process and indeed the self is also a semiotic process. Selves are not thinkers; they are the outcome of thinking. They are fleeting stabilized waypoints in the flow of thoughts. I don't want to completely dissolve the self. I think the self is something real. But the self is the product of semiotic interpretation. It's not the other way around. So it's not like in the Saussurean framework where you have the famous diagram of the talking heads from which thoughts flow. It's not that way. The thoughts think the thinker. When you and I come together in this wonderful conversation, something new emerges in the sense that you are affected by what I say, I am affected by what you say, and that new thing—a new self, a new us—has emergent properties. You say something and I'm not a machine that is simply transmitting what you say. If I do, I'm not thinking, right? But when you say something and I digest it and I resay it in the way that I understand it and thus reinterpret what you're saying in a way that allows you to understand it anew, then all of a sudden there appears an emergent mind, at all sorts of levels. A psychedelic moment is the moment of seeing that thought in its emergence, seeing that process and celebrating that as what life is. It's the moment when the emergence becomes manifest to us. A shamanistic aesthetic is after this: it looks for the places where disparate things come together and emerge as part of a larger whole. And this matches Peirce's definition of the aesthetic: our appreciation of what happens when different things come together as parts of a larger emergent whole. Our sense of being a living self is the feeling of this happening.

DSM

Regarding this idea of thinking together, or thinking with, or being thought through, my experience as an artist is that art objects are somehow alive in and of themselves. They are independent and have their own agendas and they always end up being something that I didn't plan them to be. They resist my attempts to frame them analytically.

EK

This is something that I've struggled with a lot. The semiotic framework that I have adopted allows me to say that thought is much larger than the human. It allows me to consider selves in relation to all sorts of dynamics. It allows me to understand something real about the presence of the dead and spirits who all have a certain kind of selfhood. But in my own work, I've been very careful to limit this. Partially I do this intellectually because I think that the way that agency has been handled academically is to basically extend it to everything. Just because this table in front of me resists my ability to push it or to represent it well in a drawing doesn't give it the kind of agency that I'm talking about semiotically. So this all works very neatly in my theory and I can argue it very cleanly. But it doesn't really fit ethnographically or phenomenologically. In fact, many Amazonians wouldn't necessarily agree with what I am saying. Amazonian animists will consider many kinds of things to be animate that I, speaking semiotically, would consider inanimate. This is a problem. But I reject the anthropological solution. Since Franz Boas and Émile Durkheim, anthropologists have insisted that their task is, as Bronisław Malinowski put it, to grasp the natives' point of view. That is, we are trying to understand the historically and culturally constructed systems of meaning that people use to make sense of the world. You don't ask whether a spirit is real, you ask how people understand it to be real. Those are two very different questions. So the proper title of my book in that vein would have been "How the Runa *Think* Forests Think," not "How Forests Think."

I want to do something else. I want to be able to say that I can think with forests, and those thoughts will not always match up with the thoughts of other humans—because if they did, they would be thoughts fully framed by human culture. In my work it's very important to try to arrive at a certain conceptual closure. Thinking semiotically about sylvan thinking helps me understand things in a new way. I can speak in very precise terms about the semiotic properties of things and I can work with those properties. But I don't have everything fully worked out. I haven't worked out the animacy problem. Although I can tell you why I think so much of the science-and-technology-studies literature is wrong about agency, I also have to confront my very real psychedelic experiences involving communication with stones! I'm still trying to understand the emergent aesthetic position from which this will all make sense.

DSM
I'm talking to stones all day long!

EK
That's the thing about being an artist, right? You don't let our limited thoughts about the world stop you from thinking with the world. I always joke with my students that anthropology, in its ability to resonate with the world, is a flaky science. Well, art is an even flakier science and that's good! It's empirical. It's a product of engagement, actual engagement with the world.

DSM
But what about what we could call new subjectivities? In the West we've done so much work to disenchant the world, to remove agency and subjectivity from the world, but now things like Gaia are coming back to haunt us with their own agendas. I'm thinking here of climate change but also of things like the algorithms that are basically ruling our lives, as well as artificial intelligence, which is growing exponentially. . . . What about these "abiotic selves"?

EK
Let me first address this from within this semiotic framework; there's also a way to think about this shamanistically. But just in terms of the semiotic tradition, a lot of what we try to delimit as a unit of thought is actually more accurately a distributed kind of thought. The British anthropologist Gregory Bateson, for example, in his book *Steps to an Ecology of Mind* [1972], would always emphasize that when you think of a human thinker as the thinker you're missing the relevant environment. He asks us to consider a blind man walking with a stick, tapping to see where the path is. Bateson asks us to consider where along the stick does the man end and the world begin? His answer is that it's the wrong question.

DSM
It's the wrong question?

EK
He says it's the wrong way to approach the question because you can't locate this in any one particular place along the stick.

DSM
Right, it just doesn't end.

EK
Bateson also makes a very important distinction between what he calls "pleroma" and "creatura." "Pleroma" refers to entities that affect and are affected. And "creatura" are those entities for which differences can make a difference. Think of a rock; I'm not talking now in shamanic terms but in semiotic terms. A rock can make a difference for us. We can use it. We can be affected by its properties. But differences don't make a difference to it. They don't affect its internal dynamics or how it acts in the world. Take an organism. Why does it have a skin? Why does it have digestion? Why does it have a head? Why does it have eyes? All of those things—they are the product of differences making a difference that make it what it is. So even when you are talking about a distributed kind of a mind, there are places where mindlike ways of thinking are appropriate. Think of computers. They are mindlike things. But they're mindlike things within highly distributed networks that include humans and corporations and capitalism and all these kinds of things. In this sense they are like the blind man's stick, which only becomes a cane in his hands. We still don't know exactly where mind begins and ends, but without that man, there would be no mind there. Computers are still sort of like that. Canes and computers both are and are not like the distributed networks of a forest ecology. The difference between my iPhone and a plant is that even though the iPhone has algorithms that increasingly addict me and destroy my ability to think, nonetheless it was designed in a very traditional way by someone. It was built. iPhones are completely dependent on that. But a plant is not designed. Its design emerged. It's the thinking that thought it.

DSM
You're ambitious enough to propose a new cultural paradigm that advocates an open world. This comes with its own political beliefs.

In this regard, I would like to quote a passage from your book again:

> The property that most interests me here is hierarchy. The life of signs is characterized by a host of unidirectional and nested logical properties—properties that are consummately hierarchical. And yet, in the hopeful politics we seek to cultivate, we privilege heterarchy over hierarchy, the rhizomatic over the arborescent, and we celebrate the fact that such horizontal processes—lateral gene transfer, symbiosis, commensalism, and the like—can be found in the nonhuman living world. I believe this is the wrong way to ground politics.

EK
If you don't recognize how some of the world's dynamics are nested then you won't be able to understand or work with its properties. Semiosis is nested; you cannot have symbolic thinking—human thinking—without having the larger, sylvan thinking out of which it emerges. Yet you can have sylvan thinking without symbolic thinking. This is neither good nor bad. It just is. There are many kinds of dynamics like that in the world. So my point is that although we strive for a certain kind of equality politically, naming everything as equal in the world is not the way to get there. Yes, we must fight racialized, gendered, and species-specific forms of supremacy but not by appealing to a flat ontology.

That said, I think one very interesting thing about psychedelic politics is that it can disrupt the kinds of hierarchies that get stabilized when we get too caught up in our human nests. Psychedelics break the hold that language has on our sense of who we humans are as selves. They allow us to reground in a denser ecology of thought. In this sense states are similar to human minds. They too are ecologically nested in deeply democratic processes that involve humans as well as nonhumans. But they too can get too caught up in their "nests." Pyschedelic politics can disrupt this as well.

So for example, I have been working with Indigenous activists from the Amazon. Amazonians have never fully been incorporated into the nation-states in which they now live, their forests have never been fully domesticated, and somehow these two facts are related. A sylvan psychedelic politics, one that grows out of the thinking that forests do, can provide a way to disrupt and then reground these "higher" forms, whether they are individual human minds or nation-states, thanks to the relatively more entropic, self-organizing, or anarchic thinking that they encourage. In this sense dreaming is also anarchic. It's autopoetic. I'm not saying we should do away with our human symbolic capacities. And I'm not saying we should do away with the state. I don't think we can and I don't think we should. But I do think that we can ecologize the state, which would mean to radically democratize it. The goal would be to reconnect us with the playful diversity of all the things that make us as a new kind of us. Top-down organization of any kind can kill this potential. Totalitarian regimes at the political level are what depression is at the psychological level. Both are instances in which a higher-order mind becomes disconnected from the larger ecology of thoughts that make it mind. They are instances of mind becoming matter. The goal of a psychedelic politics is to break this with the hopes that these selves can reemerge in a way that is more connected with the sylvan thinking that holds them.

DSM
That's the political question per antonomasia. Who is included when we say "we"?

EK
Within an Amazonian ontology of predation, as the Brazilian anthropologist Eduardo Viveiros de Castro describes it so well, the self, the I, is also formed in relation. It's not always clear who is who. Jaguars are not always the predators.

DSM
The perspectival idea that what you are depends on the point of view that frames you is dear to me. Who you are is the product of how you are seen by others. Subjects and objects are not fixed and you are at the mercy of the dynamics of possession and transformation. If you translate this idea to the art world, you have to dispense with the opposition between artworks and viewers.

And I think this is a more interesting way of thinking about art and how it operates in the world.

EK
The artist's goal is to inculcate—I'm thinking about this a lot in terms of an exhibit that I'm developing in Quito which seeks to capacitate the animacy inherent to a collection of pre-Hispanic ceramic figures. José Gualinga, a brilliant Amazonian philosopher from Sarayaku, likes to say that if the greatest European contribution to the world is the machine, the enduring Amerindian one is the development of a series of technologies for recognizing the selves in the world. Animism encourages a sense of undying curiosity. You never know where the animacy is and you don't know whether it's good or bad. A shamanic ethics involves the inculcation of techniques that can help us try to figure out these things. My goal in creating this exhibit is to create for a viewer the sense that things that seem to be objects may not be objects, that you can learn to listen for the possibility that there may be a person in those figures and that they might—anachronistically—have something to say for our times.

DSM
Are you thinking of yourself as an anthropologist when you're doing this exhibition?

EK
Well, first of all, I think anthropology is art. Let me be clear. And—I'm going to make a claim about what art is and maybe you can correct me—I think art is a particular vehicle, a set of techniques, for engaging with poiesis—not a poiesis that we can impose but a poiesis that we can capture, that can capture us—one that we can channel. As an anthropologist I do have my biases. I expand the borders of the human, but I stay with human problems. I'm not a mycologist. I'm ultimately thinking about the role humans play within something larger. I celebrate the anthropological method—ethnography—a method that involves techniques for immersing oneself in a world that can change how we think. Now of course I expand that because I think dreaming, psychedelics, drawing, film,

all of these can be ethnographic techniques that open up how you can listen. I often find myself much more interested in talking to artists or biologists than I do to anthropologists. And these conversations, and those dreams and visions, do radically change what it means for me to be an anthropologist and a human, but I'm still an anthropologist (and a human). . . .

First published in Luiza Mello and Valentijn Byvanck, *Dreaming Awake*, exh. cat. (Maastricht: Marres, 2018).

GLADYZ TZUL TZUL

INDIGENOUS SYSTEMS OF GOVERNMENT

The Organization of the Reproduction of Life

WHY THINK INDIGENEITY IN COMMUNAL TERMS?

For more than 500 years, Indigenous men and women, through their communities, have defended their territories, recovered the assets of which others have wished to rob them, and halted the neoliberal projects that sought to change them into small landowners in order to eradicate the *amaq'* (structures of communal government and Indigenous communal lands in Guatemala) and the ayllus (the name given to Indigenous communal landholdings).

The fight has been constant, and even as Indigenous people resist expropriation, they create legal and political strategies for controlling the concrete means of their daily lives, organize massive communal and religious fiestas, carry out ceremonies asking for rain or expressing gratitude for the harvest, and self-regulate. On several occasions they have also dismissed their authorities when they do not implement the decisions of the assemblies.

The ability to disrupt and sabotage projects of domination involves strength and power, but this ability is nurtured by a communal framework of men, women, and children who produce government and defend territory. In this text I will call this a *system of Indigenous communal government*.

I understand systems of Indigenous communal government to be the plural connections of men and women creating historico-social relationships that take on shape, force, and content in a concrete space: communal territories that produce governmental structures in order to share, defend, and recover the material means for the reproduction of the life of humans and domesticated and wild animals.

These forms of Indigenous communal government produce and control the concrete means for the reproduction of daily life through at least three political forms: *k'ax k'ol* (communal labor), which gives life to the concrete means of life; *kinship ties*, a powerful and sometimes contradictory strategy employed to defend the territory's communal property and to organize its use; and the *assembly as a communal form of deliberation* for resolving daily problems and issues of state aggression, or for discussing how and in what form what is produced on communal lands can be redistributed.

When I speak of concrete means for the reproduction of life, I refer to the land and everything it contains, namely water, roads, forests, cemeteries, schools, sacred places, rituals, fiestas—in sum, the concrete and symbolic wealth that the communities produce and govern by means of a series of strategies marked within a concrete space and a specific time structured on the basis of each unit of reproduction.

To clarify, I would specify that the unit of reproduction is the space where daily life is carried out; that is, the dwellings inhabited by nuclear and/or extended families who benefit from a water supply, make use of the roads, nourish themselves on the mushrooms that grow in the forest, and other things.

In this text, Indigenous communality does not appear as an essence that must be maintained and cared for lest it be contaminated by external forces, nor is it an archaic form of the past. I propose that we think of Indigenous communality as the functioning of the strategies of men and women who daily manage, self-regulate, and defend their land.

These communal ties are not exempt from contradictions and political hierarchies; on the contrary, they are aggressively encroached upon and surrounded by them. Even in these conditions, however, they are able to fight an autonomous struggle to control their means of daily life.

Nor is Indigenous communality an archaic or superseded form of the past; quite the contrary, Indigenous communality functions as a political strategy that, despite hierarchical textures (such as kinship), have the ability to actualize and recompose themselves and structure their authority.

To illustrate what I have said, I will present some examples. If we observe that in Guatemala as in other areas of Latin America, a sequence of processes of the enclosure and expropriation of land is taking place, the defense and recovery of these territories are spearheaded by communal structures. Those structures adapt the most varied methods, ranging from struggles in the courtroom, where they present orders of protection demanding that their property be respected, to open mobilization against extractive enterprises and their security forces, private or state. We have seen how systems of communal governance (or Indigenous communities) display temporary control over the courts[1]—not that they find solutions in the courts, but it is a path they consistently travel in order to demand equality from the state in terms of rights. At the same time, they are watchful lest the state—the same state from which they demand equality—interfere in the management of their daily lives. These debates in the statist structures of justice are always nourished by the open struggles waged from the concrete spaces of the communities to hinder and limit extractive enterprises—open-air mining projects or hydroelectric enterprises produced against the will of the communities. It is in this context that we may understand the political fire and overflowing creativity of the youth, women, and even entire families who organize blockades, art festivals, vigils, and communal meals, and who may even decide to move to the places where they wish to establish these projects. The Q'anjob'al communities of Barillas, for example, in the Guatemalan department of Huehuetenango, organized to prevent the functioning of the machinery intended for the construction of a hydroelectric plant, funded by Spanish capital. Innumerable actions such as this are taking place throughout Abya Yala.

On several occasions, these communitarian mobilizations have made use of memories emanating from the long history of rebellions and uprisings organized by past generations. In this sense, communitarianism, according to the form it takes, is actualized and functions as a singular form of governance. And depending on the level of aggression that has had to be confronted, these forms of communal governance may consolidate at the moment of response, or on other occasions may only be able to structure the struggle. In this way, a singular form of Indigenous politics emerges from Indigenous communality, responding to historical density, the internal textures of hierarchies, and the situated way in which aggressions are countered.

In this way, a series of strategies and practices emerge from systems of Indigenous communal government to organize and energize the forms of halting and/or disrupting capitalist state domination, whether locally, municipally, or in other local practices of state power. Systems of Indigenous communal government understand how to read the ways in which the exploitative colonial paradigm continues to operate. Regarding the strategies of the signifying practices of the politics of an Indigenous community and the local composition of state power, Mario Rufer reflects, "Colonial tutelage remains in force, reproducing in new garments the idea that there are peoples who cannot be complete citizens because in some way they are 'minors' ('our Indians,' 'our communities'), and in this sense are worthy of legal protection but at the same time unworthy of sovereignty."[2]

Communities know how to interpret this ambivalence and therefore have ways of deciding when to make use of the law and when not to. How is this knowledge produced? How are these strategies produced? I now wish to explain the three political forms: *k'ax k'ol*, kinship ties, and the assembly as a form of political deliberation. All three are endowed with elements for understanding the strengths and antagonisms of Indigenous systems of government.

K'AX K'OL: COMMUNAL POWER LIES IN SERVICE

In the K'iche' language, *k'ax k'ol* designates a labor strategy.[3] Etymologically, *k'ax* means "pain" and *k'ol* "labor" or "service." There are two ways of designating labor in this language: *k'ol* and *cha'k*. Both verbs refer to labor but *k'ol* alludes to communal labor, namely, an activity in which men, women, and children work to produce communal well-being—cleaning the cemetery, for example. *Cha'k*, on the other hand, means paid labor, namely, what everyone has to carry out for the family's daily sustenance; for example, being a seamstress, weaver, trader, carpenter, migrant, etc.

A free English interpretation of *k'ax k'ol* would be labor that requires exertion, or obligatory community service. Perhaps the

English words that come closest to defining it are "service" or "communal labor." I will here use the three terms interchangeably to designate the obligation we all have to work for the sustenance of communal life. Thus, communal labor is the social relation—labor power—that enables the production of what we need to live and produce our lives, and that must be standardized, organized, and regulated. Here we have a direct connection to the assemblies.

Let's continue. Service is not remunerated; it is the obligatory labor that we all have to do in order to sustain a life in common. Taking charge of maintaining roads, participating in a march, handling legal matters, taking the minutes in assemblies, shopping for the fiesta, organizing dances for the fiesta, worshiping in Catholic or Protestant faiths, reforesting the woods, digging graves for the dead—all this is *k'ax k'ol*, in such a way that one of the standards for communal equilibrium is that nobody live off the service of others.

If we think on the basis of the notion of *k'ax k'ol*, domestic society (which organizes the world of reproduction) and political society (which organizes public life) are not fully separate (if indeed the distinction is worth making). Rather, in the communal world the one sustains the other, and at the same time, they mutually nourish each other. That's why it is possible to go to university and get a profession, to work as a rural schoolteacher, or to go into business, and this is possible—even with all the difficulties—because our daily material conditions are guaranteed; in communal settings, personal trajectories are possible and achievable. That is to say, personal trajectories in work, business, and the professions develop support from a common foundation in which daily life unfolds more freely.

In a conversation in La Paz with Silvia Rivera Cusicanqui, we spoke about the connections between work, community, and individual trajectories. Silvia said that life in communal forms isn't normalized; plurality is possible in communal settings, and the individual's fullness comes out. I think communal settings provide a grounding from which to sustain one's intimate, personal life, and communal work is a general condition in which we all provide ourselves with this strength in order to make our personal lives.

Indigenous communal government, then, is the political organization that guarantees the reproduction of life in the communities, where *k'ax k'ol* is the fundamental ground on which these systems of communal governance rest and are produced, and where the full participation of everyone is involved. I propose a classification of communal labor in order to take account of how all of us work or could work, and that Indigenous communality is not limited to an identity but to the capacity that all of us possess:

1. *Decision-producing service*. This includes varied strategies and forms of labor involving collective meeting and deliberation. It is a communal labor, with regular cycles that depend on the rhythm of communal relations; for that work to be carried out, the proper day and time must be determined.

2. *Coordination-producing service*. Strategies for building external alliances with other communities, for coordinating struggles or fiestas at a national or international level. Here, business people and travelers play a fundamental role, placing their knowledge and experience at the service of the systems of governance and transmitting and fostering information. Also, those who participate in Indigenous conferences or meetings support these coordination efforts.

3. *Work at running the fiesta*. Planning dates, organizing rituals, estimating the best days in terms of the weather, calculating the food, managing the musical groups and other sources of amusement.

4. *Communal work to restrain grief*. That is, the entire labor force mobilizing to organize funerals and burials, making arrangements to recover bodies from morgues and organizing the repatriation of migrants who die in the United States.

Here are at least four forms communal labor takes. My intention in analyzing them separately is to explain one of the mottos used most often in the Indigenous communities of the southern altiplano of Guatemala: "The power of the people is in service," because service collectively constructs the conditions for material self-determination. Herein resides the power of the great Indigenous rebellions.

PATRILOCAL NETWORKS AND THE FORMS OF WOMEN'S LIBERATION

Patrilocal networks are a legal and political strategy constructed by systems of communal governance in order to halt the expropriation of lands and to prevent the consolidation of the national project to convert communal lands into small properties. In this long Indigenous struggle against colonialism, kinship alliances work to occupy and delimit communal lands, thereby succeeding in halting expropriation, invasion, and privatization.

I don't presume to say that in all systems of communal governance, patrilocal alliances are the only—or the strongest—structure for defending communal lands; rather, I propose to think of them as a more or less present feature. When I speak of patrilocal alliances, what am I referring to? I allude to the communal networks of extended families that organize life collectively. These are the affective and political ties that structure a particular form of maintaining and inheriting communal land, where patrilineal surnames are those that appear and act as protagonists in the communal struggle for the defense of the land.

I use the term "communal network" to endow patrilineal alliances with content in two senses: first, to note that both men and women live in political and affective relationships; that is, they do not exist for themselves alone, without the support and management of resources for organizing daily life. In this regard, the units of reproduction—namely, houses inhabited by nuclear families and extended families (grandparents, aunts and uncles, cousins, sons- and daughters-in-law)—are where life is organized and connected.

The other aspect to which I would like to call attention when considering patrilineal networks is that through them, land ownership has been preserved, and this presents us with a range of forces and antagonisms for those of us with the experience of living as women in communal lands.

Issues such as the communal inheritance of land are major matters to elucidate. Inasmuch as those who inherit land are men and women, we need to ask the question: how does one inherit communally? And with this question and its subsequent response, several questions unfold, but I will restrict myself to the forms in which communal power schemas are expressed.

I paraphrase Gayatri Spivak when she says that the structures of communal power are based in large part on the family—or family networks, as I prefer to call them—where women have more than marginal structural roles, as Spivak says, citing Partha Chatterjee's quotation of Victor Turner, "who suggests that the resurgence of communal modes of power often generates ways to fight feudal structures: 'resistance or revolt often takes on the form of . . . *communitas*.'"[4]

Spivak's argument is relevant in that it enables awareness of the internal textures of hierarchies, and suggests that even despite these, revolts draw their strength from the communal form. Another relevant issue involves understanding that the communal mode of power is intimately bound up with dynamics of defense and territorial regulation. In other words, women's struggles must be understood not only from women's individual liberal horizon; they are constituted according to concrete limits, although this does not mean that women do not develop flexible and broad strategies of liberation.

In this way, patrilocal networks appear as a legal and political strategy that declares communal land to be their property in such a way as to restrict private property, but at the same time, these networks shape internal textures for the communal use of land.

Immersed in these singular conditions and power relations, Indigenous women have produced conditions of possibility for living in a communal network and thereby having a concrete and affective ground for resolving difficulties ourselves. I refer to the multiple obstacles in the kinship ties that organize property and communal use. There is a political game regarding use and property that re-creates hierarchies and has a central node: land ownership and the transmission of the inheritance of communal lands through the structure of patriarchal surnames.

That is, communal lands can be used by all of us because we perform service, but they are communally inherited by the male descendants of patriterritorial surnames. This does not mean that women, even when entrapped in the limits of patrilineal networks, do not mock and efface this hereditary form, even as it serves as a strategy to defend ourselves from private property.

One of the difficulties of systems of governance, then, is that while they have certainly thought of paths to inherit communal lands through the paternal surname, they have not taken it upon themselves to create strategies of communal land inheritance to pass to their daughters or sisters without the mediating structure of marriage. Women can be widows, wives, single, or can decide to enter into matrimonial contracts with men who do not belong to the patrilineal enclave, but clearly these decisions endow women with or deprive them of strength in communal lands.

How has this been resolved in the past? Communal systems of government have devised a strategy, and it is through communal labor. Namely, men and women coming from other communities who do not belong to the kinship alliances gradually bring about their incorporation as they carry out *k'ax k'ol*.

THE ASSEMBLY AS THE POLITICAL FORM OF DECISION-MAKING

Communal authority is produced in the assembly. How does this come about? Authorities are elected annually by the technique of delegating representatives in charge of organizing *k'ax k'ol*, collecting monetary contributions to repair belongings, and organizing and maintaining communal spaces such as houses and social meeting halls. The assembly annually nominates a leadership team in charge of organizing the ways and times needed for making decisions on matters that are the responsibility of all. In this way, through self-convocated meetings, communities discuss the forms of organizing daily life according to the changes and problems to be confronted. The assemblies are a structure of authority and take on the form inherited from particular times in specific historical moments, but they have the capability of adapting themselves to daily contingency in order to halt plunder.

In this sense, decision-making capacity is delegated not to an abstract entity but to women and men who practice it in the assemblies. Communal authorities—embodied in men and women—are the ones who carry out the decisions of the assemblies, but are also subject to communal decision.

What is decided on? The most varied and multiple matters, from "With what species of tree shall we reforest?" to "How do we train ourselves to carry out an Indigenous uprising in order to block the law intended to declare our communal lands as wastelands?" Also: "How are we going to support struggles in other territories that are being invaded?" To do this, exhaustive information must be gathered in order to arrive at decisions, and prior experience is vital, along with arriving at the time and arrangements that are more or less convenient for everyone. If the matter is urgent, assemblies are organized on an extraordinary basis, in order to define shifts, times for leaving and staying, in the places where the marches or reforestations will be carried out, to name a few activities.

The communal assembly does not allow the expropriation of command, precisely because command is subject to the collective entity. Again, here is where we find the strength of the great uprisings that have succeeded in halting state aggression and in structuring a way of living where the basic conditions for the reproduction of life are not subject to the world of value.

BY WAY OF CONCLUSION

The three strategies I have named allow me to think about the strength, form, and content of systems of governance. Sometimes one is more prominent or at least more visible than another, but it is clear that the three forces are interwoven and more or less in balance.

Working to produce water, roads, forests, fiestas—in sum, working communally to maintain and make use of concrete wealth—is what enables Indigenous communal life. It is not an identity or an ancestral necessity that allows the processes of communal inclusion, but rather *k'ax k'ol*. We find ourselves, then, with service as the powerful force that produces communal life, in such a way that whatever kinship tends to close off, *k'ax k'ol* erodes and opens. Here we are faced with the interconnections and dynamism of communal politics.

And women do not find themselves entrapped in the roles of victims or heroines; rather, they move forward in their struggles on the basis of *k'ax k'ol*. The goal of this essay is to give content to what it initially suggested: Indigeneity is political. There are analytical mechanisms that seek to think Indigeneity only in the light of ethnicity, culture, Otherness, or custom; my interest is to supply ideas that help us to understand that from Indigeneity flow historical strategies and political structures. This is where enduring struggles have been forged, struggles that have succeeded in breaking colonial domination and that in many territories continue to supply the horizon of life.

First published as "Sistemas de Gobierno Comunal Indígena: La organización de la reproducción de la vida," in Ana Silvia Monzón, ed., *Antología de pensamiento crítico guatemalteco contemporáneo* (Buenos Aires: CLACSO, 2019), 71–80.

1 For example, when communities turn to accompanying their authorities to demand that their boundaries be respected, their lands not invaded, or that political prisoners be freed. To attend such hearings, real pilgrimages take place in which places to sleep and ways to obtain food are obtained through communal ties. In short, people activate familial ties in the city so as to support those in the tribunals by not abandoning their precincts.

2 Mario Rufer, "Estado, violencia y condición poscolonial: breves consideraciones desde México," *Trazos y contextos: Miradas oblicuas y crítica social* (Mexico City: Universidad Autónoma Metropolitana, 2021), 18.

3 K'iche' is a language of the Maya family that is spoken and written in a large area of the Western altiplano region of Guatemala.

4 Gayatri Chakravorty Spivak, "Subaltern Studies: Deconstructing Historiography," in *In Other Worlds: Essays in Cultural Politics* (New York: Routledge, 1988), 219.

GASTÓN GORDILLO

THE METROPOLIS

The Infrastructure of the Anthropocene

In May 2003, I drove for the first time along the foot of the Argentine Andes in the province of Salta, a region where the mountains meet the lowlands of the Gran Chaco region. As I was heading north toward the town of Las Lajitas amid a landscape dominated by agricultural fields, the roads were in terrible shape. The pavement was riddled with potholes and cracks, which forced me to drive my car carefully, zigzagging to avoid bumps or collisions. Adding to the sense of decaying infrastructure, decrepit railroads lay silently parallel to the roads—the result of privatization schemes that had terminated the passenger trains a decade earlier. The roads I was driving on were the main if precarious infrastructure that allowed for mobility to and from Las Lajitas. What struck me the most about the roads, however, was not their disrepair but the heavy traffic of trucks moving along them. The trucks advanced slowly, carefully sensing the damaged asphalt, but their motion was persistent and was part of a long caravan of vehicles that, I soon learned, was converging on a cluster of large silos that had emerged in the 1990s south of Las Lajitas. The silos poured streams of soybeans into the trucks, which then turned around and drove to the southeast for over 1,300 kilometers to the shores of the Paraná River around the city of Rosario, the third largest in the country and soon to be called Argentina's "soy capital." There, crushing factories turned the soybeans into oil and meal, which were loaded onto ships that headed downstream on the Paraná toward the ocean and then ports across the world, mostly in Asia and Europe.

In the seemingly mundane motion of those trucks navigating roads in bad shape, in retrospect I was witnessing, first, the materialization of the record-breaking soybean harvest of 2003 and the beginning of the "Argentine soy boom," which in a few years turned this country into the third-largest producer of soy in the world after the United States and Brazil. More significantly, Argentina became the first global exporter of soy meal and oil, made possible by the existence of an enormous cluster of crushing factories around Rosario. In those early days of the boom, however, the flow of trucks and the large silos around Las Lajitas surprised me. I was not aware that agribusinesses were so active in that area of Salta, a northern province seemingly removed from the agricultural core of central Argentina. In the 1970s, Las Lajitas was a sleepy village surrounded by forests and with little infrastructure except for the railways. The region was inhabited by campesinos and rural workers: gauchos, or cowboys of mestizo background who raised cattle "in the bush." By 2003, a good part of that physical and social geography was clearly gone. While forests remained in the foothills and in the interior of the Chaco, around Las Lajitas they had been replaced by corporate, mechanized farms operating through fields, roads, and silos to respond to the global commodity-boom triggered by China's rapid urbanization. Amid the growing demand for land to plant soy, deforestation north and east of Las Lajitas was then accelerating. In my early weeks in the region, I was struck not only by the traffic of trucks on the road but also by the flipside of their momentum: the bulldozers crushing forests and myriad life forms in order to create even more soy fields.[1]

In the following thirteen years (2003–16), I returned to Las Lajitas multiple times and witnessed how the combined effect of deforestation, eviction of campesino residents, and the expansion of infrastructures generated a more urbanized terrain.[2] The first infrastructures to be upgraded were the roads, the main medium that allows for the extraction of the soybeans away from Salta and toward Rosario, the gateway to the ocean. On my second visit, in 2004, I began encountering crews and heavy equipment replacing the damaged asphalt. By 2007, the regional roads had been turned into smooth, rapid channels of connectivity. All gas stations were renovated and now offer Wi-Fi to a growing volume of vehicles. At the entrance of Las Lajitas, modern buildings house the companies that sell inputs to farmers, from Monsanto herbicides to John Deere tractors. A few meters away, the upscale Las Lajitas Hotel attracts mostly male and white managers, engineers, or Cargill agents who meet to discuss how to plant, harvest, store, sell, and transport soybeans and other crops to Rosario. Most of them are Argentines from other parts of the country, primarily "the south" (Córdoba, Rosario, or Buenos Aires). These men are often on their phones, tablets, and laptops, attentive to the pulses coming from across the world that may affect their strategies and profit margins. Some of them manage farms whose fields look "rural." But the farms are mechanized, are sprayed with herbicides, and are part of the networks of infrastructures that expanded with deforestation.

As I witnessed the expansion of urbanized infrastructures produced by corporate elites who are newcomers to this area of Argentina, it was apparent that Las Lajitas's dramatic transformation did not involve the growth of a self-referential, bounded object. On the contrary, those assemblages of roads, machinery, vehicles, and silos were all designed to facilitate the transportation of soybeans and their industrialized derivatives to densely populated urban centers across the world. In investigating how the soy supply chains process and move their products from northern Argentina to a planetary network of urban centers, I gradually learned to view the infrastructures around Las Lajitas as an entry point, a path into something much larger, planetary in scale, and therefore very difficult to apprehend: what Timothy Morton would call a "hyperobject" but that I prefer to think of as a vast assemblage, a wildly diverse, differentiated, yet interconnected constellation of nodes and conduits repetitively moving objects, bodies, energy, and information across national borders and continents.[3] In this chapter I draw from the case of the soy supply chains to analyze this plurality in motion as "the metropolis."

The most important thinkers of the metropolis have been radical theorists and activists, chiefly Antonio Negri, Alberto Toscano, and the collective The Invisible Committee.[4] Negri conceptualizes the metropolis as "the general structure that capitalism assumes in its imperial phase" but also as the urban spatiality that has replaced the factory as the main site of struggle and emancipation.[5] Influenced by Negri, the writings by Toscano and The Invisible Committee in turn dissect the infrastructures of the metropolis to reveal the power of logistics in global capitalism. In the words of Toscano, "The metropolis has the intensification and expansion of supply lines as its precondition, and logistics becomes its primary concern, its foremost product, and the basic determinant of its power."[6]

In what follows, I build from these ideas to conceptualize the metropolis in more depth and along more materialist and ethnographic lines, especially through an examination of how the soy supply chains connect Argentina to urban constellations worldwide. In particular, I draw from Henri Lefebvre and the literatures on infrastructures, logistics, and planetary urbanization to decompose the metropolis into its constitutive multiplicities and fractal differentiations.[7] The metropolis, in this regard, is not a figure of the One: a totalizing object, an extended version of "the city," or a homogenous whole; rather, it is motion guided by capitalist-imperial patterns of accumulation and the infrastructures that make it possible; it is a rhizomatic multiplicity of encounters and friction; it is a heuristic, concrete abstraction to name the materiality of the continuum that moves matter from one continent to another in order to reproduce and expand the metabolism of the largest urban agglomerations on Earth. The metropolis's infrastructures are therefore "the secret weapon of the most powerful people in the world."[8] And logistics becomes "capital's art of war."[9]

Conceptualizing the metropolis along these lines is an ambitious and complex task, chiefly because it evokes the controversial idea of totality. With the rise of poststructuralist, posthuman, and actor-network sensibilities, the idea of totality has come to symbolize to many, as Toscano and Jeff Kinkle put it, "totalitarianism at worst, or paranoid criticism at best," a concept "corrupted by the metaphysical desire for coherence and the hubris of intellectual mastery."[10] Yet understanding the daunting scale of an urbanized and interconnected world in the era of climate change cannot be done without a critical understanding of totality in the most immanent sense of the term. As Theodor Adorno insisted, "Totality is not an affirmative but rather a critical category," which seeks to identify "the whole of social relations" not to reify them but to negate them.[11] And the main source of a nontotalizing understanding of the urban totality is Lefebvre, who wrote that "the urban phenomenon can only be comprehended as a totality, but its totality cannot be grasped. It escapes us. It is always elsewhere."[12] The metropolis, likewise, is an ungraspable, nonrepresentable totality that escapes our capacity to make sense of it. And what makes it partly opaque to thought is both its planetary scale and that it is permanently redefined by practices woven locally and regionally.

There are therefore countless dimensions of human and nonhuman life in both urban and nonurban spaces that escape the reach and control of the metropolis. Each node in urbanized planetary networks is also a place with a distinct history, texture, and political flavor, and is irreducible to this transnationality, even in highly globalized urban centers such as London, Dubai, or Shanghai. This is more so the case in regions with strong nonurban traditions. Most residents of Las Lajitas still celebrate the gauchos as icons of the region and view the agribusiness elites as arrogant newcomers with little love for the land. And in this and other towns, large multitudes go in pilgrimage to shrines in the forest and in places in ruins to venerate dark-skinned virgins believed to be miraculous.[13] The regional experiences, in this regard, are shaped by memories, relations, and dispositions that far exceed the rhythms of for-export agribusiness. Yet the latter's impact on this region forced me to interrogate the ways in which this place is also part of an intricate, physical planetary formation. This connectivity has global implications at an even larger scale. Industrial agriculture is one of the main sources of global warming and environmental ruination in the world, due to its obliteration of tropical forests, its reliance on toxic pesticides, and its subordination to the demands of animal farming—for most soybeans in South America are grown not for direct human consumption but to create animal feed. And this is why the question of the metropolis is also that of the Anthropocene, which I view, rather than as a generic "era of humans," as the era of climate change and severe environmental disruptions produced by capitalism.[14]

In the first section, I argue why a conceptual shift away from "globalization" and toward the metropolis can help us better appreciate both the global continuum and the differences that structure supply chains. I subsequently examine the metropolis not from its most densely populated and northern nodes of agglomeration but *from its edges* in the "global South": in our case, in areas in which urbanizing infrastructures encounter and destroy forests inhabited by campesinos who actively defend them from the advance of bulldozers. In the last sections, I show how the violence of the supply chains reveals the imperial hierarchies that structure the metropolis, the salience of national differences in it, as well as the struggles and nonhuman forces that interrupt its rhythms. These interruptions bring to light that the metropolis is also a generative space of collective encounters that offers the potential to reinvent the commons.

FROM GLOBALIZATION TO THE METROPOLIS

In the 1990s, scholars tried to make sense of the new world order created by the collapse of communism, the planetary expansion of a triumphant capitalism, and the rise of new communication technologies as a process of "globalization." As noted by Stuart Rockefeller, this early work engaged the question of globalization largely through a disembodied notion of "flows,"[15] emblematic in the influential work of Arjun Appadurai on "ethnoscapes"[16] and of Manuel Castells on "the information age."[17] The work of Saskia Sassen on "global cities" was important for thinking globalization along more spatial lines: as a process that redefined the nature of urbanization by creating tightly interconnected and powerful urban nodes that structure financial transactions worldwide, such as New York, London, and Tokyo.[18] Yet as argued by Ananya Roy,[19] the focus on "global cities" re-created a Euro-American gaze that overlooked the urbanizing networks of "the global South," which can no longer be dismissed as irrelevant to globalization.[20] Neil Brenner adds that "global cities" are insufficient to understand planetary urbanization, which also mobilizes "vast grids of accumulation and spatial regulation that cascade along intercontinental transportation corridors" as well as "large-scale infrastructural, telecommunications, and energy networks."[21]

The more recent research on infrastructures and logistics, in this regard, has been important to redefine these questions in more materialist terms, especially through the analysis of the assemblages, objects, and movements that make global capitalism possible.[22] And the literature on "urban metabolism," in turn, brought into focus how urban nodes attract through their infrastructures (and expel as waste) flows of water, energy, food, and myriad commodities.[23]

Yet amid the literatures tackling the infrastructures of globalization, the work on "planetary urbanization" led by Brenner, Christian Schmid, and Andy Merrifield has been the most ambitious effort to conceptualize the urban form of the metropolis, even if without using this concept. Profoundly influenced by Lefebvre, these authors argue that the globalized nature of urbanization in the twenty-first century forces us to move beyond the idea of "the city" as a localized, bounded object and examine, rather, the complex ways in which urban processes are, in the words of Merrifield, "decoupling from their traditional city forms" and becoming "formless" and "inevitably polycentric," "stretched and torn up, sprawled and ghettoized."[24] Brenner and Schmid,[25] in particular, write that urban forms can no longer be restricted to their agglomeration in urban centers, for they also include what they call "extended urbanization": infrastructures and networks that extend into zones of extraction, such as those transformed by mining in the Andes.[26] This spatial extension demands for Brenner an "urban theory without outside," which dissolves the urban/rural divide and looks at the differentiated, varied ways in which even the ocean and the mountains are traversed by infrastructures.[27]

This body of work has been criticized by postcolonial and feminist scholars for being too "totalizing" and for failing to account for the everyday, difference, locality, and struggles in the making of the urban.[28] Some of these critics raise important questions that help us think of the limits of the metropolis, as we shall see. Yet as Kanishka Goonewardena argues, the use of the term "totalizing" to dismiss the work on planetary urbanization reproduces a problematic equation between the philosophical idea of totality and totalitarianism and determinism.[29] Goonewardena shows that there is a long tradition not only in Marxism but also in radical feminism and postcolonial thinking that views "the social totality" relationally: that is, not to downplay differences but as a political injunction to "think holistically" and understand differences amid wider relations of domination. Understood in this sense, the totality posed by planetary urbanization is crucial for thinking holistically about the Anthropocene. In proposing to name this fraught constellation of urban agglomerations and extensions "the metropolis," this chapter nonetheless goes beyond Merrifield, Schmid, and Brenner and engages the theorizations on this concept by Negri, Toscano, and The Invisible Committee. And what this latter set of authors adds to these debates is their insistence on the imperial dimensions of global capitalism and the metropolis, something I examine in the next section. The Invisible Committee introduces an important concept into the discussion: that the motion of these infrastructural networks defines the metropolis as "a global continuum."[30] The question of this continuum is that of what Lefebvre called "the urban form," the main concept through which he sought to move past the often fetishized ideology of "the city."[31] This urban form is much more than a "system," a term Lefebvre disliked because it "implies fulfillment and closure." The urban form "cannot achieve closure" because it is "made manifest as movement."[32] Further, as Lefebvre put it, "the urban is most forcefully evoked by the constellations of light at night, especially when flying over a city."[33] These constellations capture that the urban form combines the agglomerations and extensions identified by Brenner and Schmid.[34] Yet they also show a continuum of *textured urban materials* (asphalt, cement, steel, glass, plastic, timber) through which people, vehicles, and energy move, congregate, and disperse. If the nocturnal view of the urban terrain from an airplane is zoomed out much farther to what satellite images show, we see a form that is vaster: for instance, the European peninsula crisscrossed by high-density clusters of light and rhizomatic lines forming a nebulous continuum across borders. To paraphrase Lefebvre, it is in those images that the urban forms of the metropolis as a continuum are most forcefully evoked.

But this image of continuous urban connectivity is, at the same time, misleading. In order to avoid falling into what Donna Haraway would call the "God trick" of looking at satellite images from above and from nowhere, it is crucial to examine the uneven textures of the metropolis at the ground level.[35] This quickly reveals that the metropolis's continuum is highly selective, differentiated, and never wholly continuous. Its infrastructures are closely policed and made violently *discontinuous* by the efforts by states and corporations to restrain the mobility of undesired and racialized populations and the flow of illicit goods. The rise to power of Donald Trump in the United States, the Brexit vote in the United Kingdom, and the surge of xenophobic nationalisms confirm that borders and nation-states are far from withering away. But it is precisely the perceived erosion of borders by transnational and subaltern mobilities that is creating calls for the militarization of borders and the rise of infrastructures of separation. The metropolis, in this regard, is a decentered structuration of motion and acceleration and also a structuration of containment, stasis, and slowness by closely surveilled walls and fences.

When Brian Larkin wrote that "infrastructures are matter that enable the movement of other matter" he was referring, in this regard, to the policed and uncoded motion of the metropolis's continuum.[36] In its extremely diverse expressions, this continuum requires technical standardizations, made possible by infrastructures such as the strips of asphalt that allow roads to smooth out rugged terrain;[37] the power grids and pipelines that channel electricity, gas, and oil across borders;[38] or the submarine cables through which Internet data flow from one continent to the next.[39] This continuum is also made possible by logistical standardizations, such as the use of what Hannah Appel calls "modular" ways of outsourcing services[40] and the use of same-size containers that are transportable by ship, train, and truck.[41] Currents of illegal objects are part of the continuum, which includes narco-trafficking organizations moving drugs in disguise through public infrastructures and through clandestine infrastructures of their own making, such as the myriad tunnels underneath the US-Mexico border fence.[42]

Yet the metropolis's continuum, for all its seemingly unstoppable momentum, is a networked connectivity that can easily break down locally and regionally as soon as its components cannot be properly assembled. In September 2017, the severe disruptions that several hurricanes inflicted on the infrastructures of Houston, Florida, Puerto Rico, and several Caribbean islands are a clear example of the power of climate change to disrupt key nodes of the metropolis and their patterns of motion by preventing airplanes, airports, ships, ports, roads, grids, and people from being assembled in a particular way. The idea of assemblage is crucial in this regard to make sense of the intrinsic multiplicity of the metropolis; this is why the concept has long been used in the literature on infrastructures to highlight that the latter are not homogeneous entities but associations of different materialities and technologies.[43] As Manuel DeLanda put it, assemblages have no essence and do not constitute a totality because they are the temporary aggregation of a multiplicity.[44] The metropolis, in short, is not an already formed object but something that demands for its existence the articulation of countless materials.

These assemblages bring together not only infrastructures and human bodies but also the terrain of the planet in its solid, liquid, and atmospheric expressions.[45] The ocean and the atmosphere, in particular, have been turned into spaces of global connectivity without which the metropolis would not exist. The continuum of roads, railroads, or pipelines operating on land continue by other means at the most important and long-range nodes of planetary transport: ports and airports. Departing from and converging on those nodes, the metropolis's continuum is re-created by the thousands of ships and airplanes that at any given time are moving across the ocean and the sky. This continuum is also made possible by the numerous satellites circling the earth and the flows of data traversing the metropolis via submarine cables, computers, and ATMs, like the ATMs in Las Lajitas through which I get cash from my bank account in Canada.

Yet this connectivity is not only discontinuous but also fragmented, disjointed, and extremely diverse in the uneven urban densities that define it. Some of the most powerful nodes of the metropolis are certainly what Sassen called "global cities" that structure financial flows worldwide.[46] In the case of the soy supply chains, this includes the prominent influence of Chicago, the home of the Chicago Board of Trade, where the future prices of soy and grains are set. Yet the nodes of power of the metropolis have become so globalized that the old distinction between "core" and "peripheral" nations articulated by Immanuel Wallerstein has been spatially dispersed and made more fractal, nebulous.[47] Urban nodes based in South America and Asia such as São Paulo, Buenos Aires, Rosario, Shanghai, or Beijing all exert various degrees of influence on the configuration of the soy supply chains.

Often within these highly urbanized nodes of power, the metropolis is also made up of large zones of urban poverty served by poor or derelict infrastructures, embodied in the shantytowns of many megacities the world over.

As one moves away from the metropolis's most dense nodes of agglomeration, the spatially extended urbanization created by infrastructures usually becomes more capillary, dispersed, and rural. In areas of the world heavily populated by small farmers living in poverty, such as many areas of Latin America, Africa, and South Asia, the rhythms of the metropolis encounter and are constrained by local rhythms of a different nature, which while affected by and entangled with urban forms are not urban in origin, for they draw from an affective, tactile engagement with agrarian, forested, and mountainous terrains. Brenner's idea of an "urban theory without outside" highlights that these places are part of our globalized present but at the cost of silencing that much of what happens in those places is not reducible to urbanization, as critics point out.[48] In his response to critics, Brenner in fact concedes that he is happy to abandon the idea of urban theory "without outside" for the sake of conceptual precision.[49]

This is a crucial point. The metropolis is a nontotalizing totality, first, because there are myriad places "outside" of it. While the metropolis has been created by a capitalist system that affects the totality of the planet, its urban textures form a configuration of a scale smaller than the world. That is, if by "outside" of the urban we mean nonurban textures and rhythms, the world is full of outsides of the metropolis, which are nonetheless "the outsides of a world without outside": that is, they are areas that may be beyond the direct reach of grids and infrastructures, for instance in the still-forested heart of the Argentine Chaco, yet are not beyond the reach of capital. This is clear in the worldwide impact of climate change and in the destructive effects of agribusiness in the forests of South America.

For all its power, the metropolis therefore becomes thinner and more diffused in those areas of the world it does not fully regiment. The latter, however, are neither a pristine "nature" nor a rurality disconnected from the metropolis, for "to miss the city's relation to nature and the country is in fact to miss much of what the city is."[50]

And authors such as Ashley Carse[51] and Stephanie Wakefield and Bruce Braun have shown that "nature" is regularly mobilized by infrastructures.[52] The salience of the ocean in supply chains is a clear example of this inseparability. But rather than "nature" or "rural" places, what the rhythms of the metropolis disrupt are non-urban terrains, such as the forests of Argentina.

ON THE EDGE OF THE METROPOLIS

The infrastructures and urbanized networks that agribusinesses have recently created around Las Lajitas, in this regard, can be said to be "on the edge of the metropolis" because they tensely coexist with forested areas inhabited by campesino families and still largely defined by rural and peasant rhythms. Ninety kilometers east of Las Lajitas, the urban forms brought about by industrial farming fade away and encounter an edge or threshold in the physical sense of the term. At kilometer ninety, the gravel road originating in Las Lajitas comes to an end; it is replaced by a dirt road in terrible shape. By then, the fields have been replaced by forests. The dirt road takes you through heavily forested terrain that is beyond the reach of electricity grids or cell-phone coverage. Here, residents have lived *en el monte*, "in the bush," raising cattle in an artisanal way for generations. They identify as gauchos, the mestizo cowboys who settled the area once the Argentine Army defeated and displaced local Indigenous people in the early 1900s.[53] Their subaltern rurality, in this regard, is not primordial but was created by the territorial expansion of the state. Yet while they are not "outside" of the nation their distance from the infrastructures around Las Lajitas places them on an edge of sorts. This is a porous and elastic edge, clear in that most residents have cell phones for when they go to Las Lajitas and that their land is affected and threatened by the soy frontier. This is why this area where fields give way to campesino forests is more than an edge; it is a zone of what Anna Tsing calls "friction."[54]

Around Las Lajitas, the expulsion of campesino residents by the expansion of industrial agriculture began in the late 1970s and 1980s, when agricultural firms based in other parts of Argentina began investing in the area, attracted by tax benefits. Until then,

Las Lajitas was a village with a strong gaucho identity that had coalesced in the 1940s around a train station. For decades, the main infrastructure was the railways, inaugurated in the 1930s, as well as dirt roads that became impassable after heavy rains. The cattle raised in this region were taken west on caravans to slaughterhouses in cities in Salta and Jujuy and in Chile. While connected to regional urban centers through the railroads, Las Lajitas was surrounded by forests and for decades maintained a low-paced rhythm. The first wave of evictions and deforestation in the 1980s and subsequently the "soy boom" destroyed this spatial configuration, turning Las Lajitas into an agribusiness hub that expanded its reach toward the Chaco. Led by companies from other areas of Argentina, this was a textbook example of "land grabs" that took advantage that many residents lacked titles to the land despite having lived there for generations. Despite multiple resistances that I briefly describe later, many evictions prevailed because they were legitimized by courts and officials and violently enforced by the police or armed civilians.

The destruction of these expropriated forests subsequently follows a methodical plan, which begins with what Kregg Hetherington calls a foundational infrastructural practice: the surveying of the land.[55] First, engineers reduce the complex, dense texture of the forested terrain to a Euclidian geometry of straight lines and angles. The Salta government then draws from these surveys to authorize *el desmonte*, the bulldozing of forests, which is subcontracted to companies that specialize in this task. In addition to bulldozers and drivers, these companies assemble RVs where the drivers will reside, tanks with fuel, and the main tool of destruction: an extremely heavy chain that is more than one hundred meters long and whose steel knots are twenty centimeters thick.

In order to obliterate a relatively large section of forest, bulldozers operate in pairs connected through the chain, forming an assemblage whose action is called *cadenear*, or "chaining." As the bulldozers pull in unison parallel to each other, the chain steadily crushes everything on its path, not only trees but also large amounts of wildlife and even cattle. Once the forest has been flattened, the debris is set ablaze. Manual workers then dig, remove, axe, drag, pile up, and reburn all stumps and remains, further disintegrating them in order to smooth out the terrain and turn it into agricultural fields.

The fields then become the medium of a mechanized production that demands very few workers, for the planting of seeds, the spraying of pesticides, and the harvesting is done by combines, some of which are automated and controlled by satellite. The use of agrichemicals is crucial to the whole process, for making the fields grow soy demands spraying them with herbicides aimed at killing all plants growing there except the one genetically modified by Monsanto or Syngenta to resist their toxicity. The devastation created by agribusiness in the Chaco region, in short, does not stop with deforestation and continues every farming season, killing small animals, contaminating water supplies, and negatively impacting the health of residents in nearby areas, something that has been widely documented by scientists, activists, and ethnographers in many parts of Argentina.[56] Yet this is a destruction that has been silenced by the corporations and state officials who celebrate the value of the commodity, soybeans, that once produced in those fields is inserted in supply chains. At harvest time in April and May, caravans of trucks intensify their march from Salta to Rosario, where the soy boom created a real estate boom that dramatically expanded the city vertically through the construction of skyscrapers and horizontally through the industrialization of the shores of the Paraná. Between 2003 and 2006, the giants that dominate the global market of food and logistics as well as Argentine corporations built around Rosario the largest concentration of crushing factories and private ports in the world, a process that evicted thousands of working-class residents from the shores of the river.[57] Today, the twenty factories and nineteen ports around Rosario generate half of the soy meal in the world. The largest factories can process 1,000 trucks per day. But at harvest time the volume of trucks is such that thousands of them clutter the surroundings of the factories, often waiting in line for days on end before dropping off their load. Further confirming the transnational nature of this infrastructural zone, the factories around Rosario attract not only the soybeans produced in Argentina but also those that descend the Paraguay and Paraná Rivers on barges from Paraguay, Bolivia, and southern Brazil.

The urbanization generated by the soy boom in this part of Argentina also expanded into the vast farming region west of Rosario: the Pampas, or prairies, that in the twentieth century became the agricultural and cattle-raising heart of the nation. Settled by European immigrants, these prairies were for decades devoted to producing beef and wheat for domestic consumption as well as exports, but in the 1990s they became the epicenter of soy farming in Argentina.[58] Valeria Hernández has shown that this expansion of industrial agriculture has transformed the towns of these prairies into "agro-cities":[59] modernized urban centers that are surrounded by mechanized farms and infrastructures serving agribusiness (roads, silos, gas stations, banks) and based on urban patterns of consumption.[60] A similar transformation has been noted in soy country in Brazil, which has created what Gustavo Oliveira and Susanna Hecht call "a new agroindustrial urbanism."[61] In Las Lajitas, this agroindustrial urbanism is less developed, for its growth has been limited and unequal. Most of the streets remain unpaved, and the local public hospital and schools are underfunded. In working-class neighborhoods, dark-skinned people live in poverty, depending on odd jobs and social programs. It is largely around the Las Lajitas Hotel that one can see the local expression of an agroindustrial urbanism defined by modern buildings serving corporate actors attentive to the global market.

Once it is industrially processed around Rosario as feed and oil (or simply as beans), the soy originally produced around Las Lajitas is promptly loaded on ships at the ports next to the factories. Ports mark a threshold in the planet's terrain, for their presence indicates the end of landmass's solidity and the beginning of a world of water that requires a different means of transport: ships. While the Paraná is a wide river that has long been a medium of trade, it needs to be regularly dredged by heavy machinery to make it navigable for freighters. This is part of an ambitious infrastructural project called "La Hidrovía" (The water highway), which was first implemented in the 1990s by the four countries of the basin (Argentina, Uruguay, Paraguay, and Brazil) and is now the largest dredge operation in the world.[62] This infrastructural maintenance cannot be slowed down, because the sediments carried by the Paraná permanently fill back what was dredged.

The motion-oriented infrastructures of the metropolis, in short, have to permanently struggle against the nonhuman motion of water and sediment flowing toward the Atlantic.

Once the ships are in the ocean, their trajectory depends on multiple factors and market fluctuations. Minor changes in the price of soy or shifting political conditions across the world may make the corporations controlling the supply chains change the trajectory of ships mid-course, redirecting them to slightly more profitable destinations.[63] A good percentage of the ships that leave the ports on the Paraná River dock in the ports of southeastern China, which serve one of the most urbanized, industrialized, and heavily populated zones of the metropolis. The soy meal is then distributed to thousands of farms to feed half of the world's population of pigs. The Chinese working class, whose increasingly pork-based diet is partly produced with soy from South America, then assemble the phones, computers, and countless other stuff that are then shipped in containers, through those same ports, to the world. In Vancouver, where I live, one can regularly see container ships arrive from China and head to the port, the largest in Canada and yet another logistical node moving matter through the metropolis.

THE UNITED REPUBLIC OF SOY: ZONES OF IMPERIAL EXTRACTION

In August 2016, to my initial surprise, I came across in the streets of Las Lajitas a group of about ten men of East Asian appearance getting off two vans. What gave them away as foreigners, in addition to their facial features, was that they were more formally dressed than locals. All wore dark dress pants and white shirts, and some wore jackets. A few Argentine men were with them. I initially assumed they were from one of the largest state-run companies from China, China National Cereals, Oils, and Foodstuffs Corporation (COFCO). This company had recently arrived in Argentina by purchasing Nidera (a domestic leader in seeds and trade) and the agribusiness branch of Noble, a Hong Kong–based logistical giant that controlled many facilities in the country including a silo near Las Lajitas. A few days earlier, I had seen COFCO flags at this local silo formerly owned by Noble, publicly signaling the arrival of Chinese capital in the region. Later that afternoon, however, I met those men again at the Las Lajitas Hotel and learned they were not from COFCO but from another major Chinese corporation, China Machinery Engineering Corporation (CMEC). Following a deal signed between China and Argentina, CMEC was in the process of investing billions of US dollars to repair and upgrade the decaying railroads connecting northern Argentina to the crushing factories and ports around Rosario. This has been a long-standing demand by the local business elites, for the high cost of transporting beans and grains by truck to Rosario makes farms in Salta less competitive than those in central Argentina. The Chinese team visiting Las Lajitas was made up of engineers who were inspecting, guided by Argentine officials, the state of decay of the regional railways in preparation for their radical makeover. While they had dinner a few meters away from me, half of them were on their smartphones, communicating with urban centers located thousands of kilometers away and taking advantage of the metropolis's continuum.

While the Chinese investment in this region of Argentina is still relatively thin, the presence of those engineers as well as the recent acquisition of silos near Las Lajitas by COFCO signal that some of the regional infrastructures of the supply chains are owned and repaired by actors and capital based in Asian nodes of the metropolis. And it was not a coincidence that the Chinese delegation stayed at Las Lajitas Hotel, sharing the restaurant with this hotel's regular clientele: a few North American and European visitors but mostly Argentine white men who stay for a few days, usually coming from Salta, Córdoba, Santa Fe, or Buenos Aires to oversee farms administered or serviced by their companies or to sell farmers seeds and herbicides and buy their crops. The hotel and the modern buildings that surround it, in this regard, have become crucial political nodes that encapsulate the new hierarchies that exist in Las Lajitas. While most locals descend from gauchos who once raised cattle in the forest, the power in the town council, the hospital, the agrotechnical school is held by white men and women who came from other parts of Argentina. This elite regularly holds meetings at "the hotel," which is viewed by residents as the embodiment of the new relations of power that connect Las

Lajitas with the world. And these relations are not only capitalist but also respond to spatial hierarchies of an imperial nature.

In their book *Empire*, published in 2000, Michael Hardt and Antonio Negri argued that the globalizing expansion of capitalism that followed the collapse of communism marked the weakening of the sovereignty of nation-states, the withering away of nationally based imperialisms, and the rise of the planetary and decentered system of capitalist sovereignty "without outside" that they called "Empire."[64] This provocative argument created a flurry of debate and many authors were quick to criticize the idea of "Empire" as a totalizing abstraction unable to explain the heterogeneous complexity of the world and the ongoing power of nation-states and nationally based imperialisms.[65] Tsing, in particular, argued that the idea of an Empire "without outside" reduces global capitalism to a "unitary, homogeneous object" and misses that supply chains often depend on differences and local specificities that are not based on capitalist relations.[66] The phenomenal rise of China and the parallel popularity of xenophobic nationalisms, further, reveal that the world is more disjointed than *Empire* could have foreseen. Yet despite the limits of their analysis, what Hardt and Negri bring to light, and is worth redefining along more materialist lines, is that capitalist globalization has an imperial dynamic that is more decentered than the ones that defined classical, nationally based imperialisms. Rather than speaking of "Empire," however, I prefer to outline the imperial dimensions of the supply chains, that is, that they are structured by long-range spatial hierarchies that capture resources for powerful nodes in the world at the expense of zones of sacrifice.

That urban nodes based on different continents such as Rosario, Shanghai, and Chicago profit from the devastation of the forests of South America suggest that while nation-states and imperialism are indeed part of these networks, they are subsumed to the imperial constellation that exceeds them and constrains them: the metropolis. Tsing wrote that studying how supply chains operate on the ground provides "an antidote" to the abstractions of Hardt and Negri's Empire.[67] But it could be argued that the local texture and trajectories of supply chains—especially those of planetary reach—can help us better appreciate the imperial nature of globalization, whose extended forms of urbanization could be described as imperial apparatuses of capture. These apparatuses are akin to "a hostile environment" that in some parts of the world acquires more physical density than in others.[68] This density creates what I call "zones of imperial extraction": areas of the planet with a high density of infrastructures created to satisfy, as Lewis Mumford once put it, "the voracious mouth of the metropolis."[69]

These zones are discontinuous from each other and are scattered around the world: the Tar Sands in Canada; the open-pit mines in the Andes, Appalachia, or Papua New Guinea; or the vast nodes of oil extraction in the Gulf of Africa, the Gulf of Mexico, or western Amazonia. Andrew Barry highlighted the technical standardization of these areas by calling them "technological zones."[70] But these zones foster a distinctly imperial type of territoriality, for while they are constrained by their host nation-states they are more subsumed than other zones to the power of global capital. Stuart Elden articulated a theory of territory that conceives it not as a "space" but as political-legal technologies (such as laws, surveys, census, map-making) for the control of terrain.[71] In the case of zones of imperial extraction, the main legal-territorial technologies for their expansion have been the free-trade and neoliberal deregulations and privatizations of the past decades, which have maximized capital's power to control terrain by overruling the rights of local people and governments. But a more physical set of territorial weapons have been used to create these zones: infrastructures and logistics.

The roads, silos, and farms around Las Lajitas, in this regard, are part of one of the largest zones of imperial extraction on Earth: lowland South America. This zone expanded rapidly across the continent after Argentina authorized Monsanto in 1996 to sell in the country its genetically modified soy seeds together with its star herbicide Roundup, one of the first in the world to do so. Using Argentina as a beachhead, in a few years Monsanto's seeds and pesticides expanded toward, and took over, much of lowland South America.[72] Enjoying the support of both the neoliberal and center-left governments of the continent, this zone of extraction was the child of the chemical companies that dominate the global market of GM seeds and pesticides: Monsanto, Syngenta, Bayer,

Dow, DuPont, and BASF, a group that is going through a rapid process of further concentration. The recent fusion between Dow and DuPont, the purchase of Monsanto by Bayer, and the purchase of Syngenta by ChemChina signal the further tightening of this oligopoly and the growing presence of the Chinese state among the myriad actors that profit from this zone.

This agribusiness zone in the heart of South America forms a vast archipelago that extends north from central Argentina all the way to Amazonia in Brazil and is the main source of soy on the planet, generating almost 60 percent of the global output. In a well-known advertising campaign, the Swiss-based agrichemical giant Syngenta (now controlled by ChemChina) captured the existence of this transnational zone with an imaginary map of "the United Republic of Soy" covering much of Argentina, Uruguay, Brazil, Paraguay, and lowland Bolivia, with the heading "soy knows no borders." This vast zone is geographically and culturally very diverse and certain subregions cultivate specific modalities of soy farming.[73] Yet the soy farms around Las Lajitas operate in ways that are similar to those in Brazil or central Argentina, for they all plant genetically modified seeds, use large amounts of pesticides, and rely on mechanized operations that hire very few workers.[74]

Whereas in zones of imperial extraction that involve open-pit mining the production of ore is usually carried out by multinational corporations, in "the United Republic of Soy" the farming of soybeans is largely in the hands of Argentine, Brazilian, or Uruguayan companies and farmers. Yet the action of these domestic actors is constrained by one of the most powerful corporate conglomerates in the world: the trading and logistical companies that control the supply chains of soybeans and grains and that sell farmers their inputs (seeds, pesticides, insurance, technical assistance) and buy their crops. The most powerful of these highly globalized companies are ADM (Archer Daniel Midlands), Bunge, Cargill, and Dreyfus, a group known collectively as "ABCD." These companies played a decisive role in helping create this zone of extraction in the 1990s, when neoliberal deregulations allowed them to take over much of the crushing, trade, and logistics sector in South America, eroding the prior prominence of domestic companies and the state on this market.[75] The more recent arrival in the supply chains of companies based in Asia, led now by COFCO, are further decentering these globalized forms of power.

As noted by Tsing, global traders need to engage in a work of cultural translation in order to work with the diverse social actors they interact with at different stages of the supply chains.[76] In other words, the assembling of supply chains is not a mechanical, top-down process but demands a sensibility to local specificities and cultural differences. In South America, several ethnographers including myself have noted this sensibility among the local agents of corporations such as Cargill or Bunge. These agents try to gain the trust of potential clients by adapting to their practices and cultural rhythms and by contributing to local institutions through donations and the organization of social events.[77] Yet these localized adaptations, as Valdemar João Wesz Jr. and Amalia Leguizamón show, do not change the fact that these global traders operate in Brazil and Argentina in very similar ways.[78] They all impose oligopolic practices that leave farmers and midsize farming companies with little space for maneuver. This is why most farmers and administrators I talked to tend to resent the globalized conglomerates that control the market.

Understanding how these trading companies operate across the multiplicity of the planetary terrain is key to appreciating the imperial nature of their strategies. All of these companies share a strategy of vertical integration through which they control the different stages of the chain in dozens of different countries. Once they purchase crops from farmers, the ABCD group as well as other companies mobilize their vast networks of ports, ships, and silos to arrange for their storage, industrial processing, and transport to their destination in Asian, European, or African nodes of the metropolis. Throughout the process, these companies are guided by a planetary field of vision that is clearly articulated, for instance, on the website of Noble Group, the logistics corporation based in Hong Kong. In 2013, when it owned a silo near Las Lajitas before it sold it to COFCO, Noble was number 73 in the Fortune 500 ranking and stood out among the logistics giants because it was the first to be based in Asia. For Noble, the soy it stored in its facility near Las Lajitas is just one of the many commodities—from iron ore to natural gas and oil—it transports and processes across the world.

The company's website boasted about the planetary and hierarchical trajectories mobilized by silos such as those in Argentina: "We connect low cost producing countries with high demand growth markets." The website also highlighted that the company's geographical "diversification," by which it meant spatial multiplicity and planetary reach, is "crucial" to overcome "times of uncertainty": "As an example, we can source agricultural products at any time from Argentina or the Ukraine, iron ore products from India, Mongolia, or Brazil, coal products from Colombia, Australia, or Indonesia."[79] This is what logistics corporations call "agility."[80] This "agile" spatiality is imperial because of its capacity to *move* at any time *any* commodity almost anywhere in the world.

As Deborah Cowen shows, this means that global logistics prioritize *the security of global currents* over national forms of control, therefore pushing for imperial, transnational forms of oversight. "Rather than territorial borders at the edge of the national space guaranteeing the sovereignty of the nation-state, a new cartography of security aims to protect global networks of circulation."[81] And while borders do not disappear, they "are superseded by transnational networks, flows, and urban nodes."[82] In the case of the soy supply chains, the destruction created in forested areas of South America by the practices and infrastructures that Bunge, Cargill, or COFCO mobilize to extract and transport soy meal across the world respond to so many transnational actors that it cannot be said to respond to "US imperialism" or "Chinese nationalism." To be clear, imperialist and nationalist agendas standing in tension with each other are part and parcel of the disjointed currents of the metropolis. The US government, for instance, permanently lobbies South American nations to favor Monsanto and other US-based corporations. And the state-run companies from China active in the soy supply chains, such as COFCO and ChemChina, follow (unlike their private competitors) the nationalist agenda of the Chinese state and its goal to guarantee food self-sufficiency.[83] Yet lowland South America has been transformed into a zone of imperial extraction not by particular nation-states but by a more transnational, dispersed, and nebulous constellation of actors and logistical operations that seek to capture cheap animal feed, oil, and beans regardless of the environmental and social cost.

These corporations are so entangled with each other that they often cooperate rather than compete. China's investment in the repair of the railways between Las Lajitas and Rosario, after all, will benefit not just Chinese consumers but also the whole commodity chain.

As Aihwa Ong has argued, local actors and nation-states can definitively have an impact in shaping the spatial patterns of globalization.[84] Likewise, zones of imperial extraction are profoundly affected by national differences. The explosive rise of an infrastructural cluster of factories and ports around Rosario in 2003–06, for instance, was the product of the efforts by the center-left administration of President Néstor Kirchner (2003–07) to create a crushing industry on Argentine soil. In particular, his government created a differential tax policy on soy exports, which slightly lowered taxes on soy meal and oil in relation to unprocessed soybeans. This lower tax on industrialized soy succeeded in attracting heavy investment in crushing factories by global conglomerates and in rapidly turning Argentina into the world's lead exporter of soy meal and oil. Brazil, in contrast, exports most of its much larger soy output as beans for the reverse reason: its taxes are lower for nonindustrialized exports.[85] This territorial policy by the Brazilian state means that soybeans harvested in Amazonia are crushed in factories in Asia, confirming how national differences affect the composition of the supply chains.

Revealing the agency of domestic actors in these chains, some Argentine corporations have become crucial players in the subordination of this area of South America to the metropolis. In Las Lajitas, for instance, despite the strong presence of Cargill and Bunge in the area, the main buyer of local crops is the Argentine corporation AGD, Aceitera General Deheza. And the main local seller of GM seeds is Don Mario, an Argentine company that holds the right to produce Monsanto technology in seeds. Further, these companies have developed a transnational strategy that has made them expand operations to neighboring countries such as Brazil and work closely with the corporations that dominate the global market.[86] Something similar occurs in Brazil, where the most powerful agribusiness in the country, Amaggi, has built a port near Rosario. In short, the infrastructures of these apparatuses of capture in South America are also produced and controlled by corporations that are based on the continent.[87]

This overlap within Argentina of different technologies of territorial control by national companies, transnational corporations, and the state often leads to conflicts. In March 2016, for instance, Monsanto began to set up checkpoints at the gates of factories and ports around Rosario to subject the trucks arriving there to the type of controls usually carried out by officials or the police. In these checkpoints, Monsanto agents tested the beans looking for traces of their Intacta soybeans (genetically modified to resist insects) and billed farmers if they had not paid royalties.[88] Notably, they did so in collaboration with the companies that own the crushing facilities. This corporate appropriation of the right to police, however, proved too controversial and the outcry from farmers and officials forced Monsanto to back down. Yet the backstory was Monsanto's efforts to force a change in the Argentine legislation on seeds, which in contrast to North American or Brazilian law does not allow corporations to claim GM seeds as private property subject to royalties. Ironically, while Argentina is a crucial node in this zone of imperial extraction, it is (at least for now) one of the last bastions of the principle that seeds should be part of the commons: owned by none, usable by all. And this key difference within "the United Republic of Soy" exists because of the social movements that have fought hard to keep seeds unpatented.

In the first decade of the century, the administrations of Néstor and Cristina Kirchner encouraged the soy boom as a source of national progress, development, and social justice. And they redistributed part of the wealth generated by the export taxes on soy, which funded the social programs that sustained their popularity among the poor, at least during the early boom years. When they were in power, Cristina Kirchner and other center-left presidents in South America such as Luis Inácio Lula da Silva in Brazil thereby celebrated the expansion of agribusiness as a source of national pride defined against "US imperialism." But their complicity with the high levels of dispossession, violence, and destruction in the Indigenous and mestizo heart of the continent has made of former presidents Kirchner and Lula what could be called "anti-imperialists for Empire": that is, leaders who may have put limits to US imperialism in South America but did so by embracing the extractive rhythms of the metropolis.

The openly right-wing government that has ruled Argentina since late 2015 has embraced extractivism further by immediately lowering taxes on soy exports and identifying the interests of agribusiness as official policy. Yet the negative legacy of two decades of deforestation, evictions, violence, and poisoning has also generated vibrant social movements committed to confronting the powerful agribusiness lobby.

INTERRUPTING THE SOY SUPPLY CHAINS

The fact that the different stages of the supply chain require the reproduction of localized assemblages means that protests against agribusinesses in Argentina have attempted to interrupt and disarticulate the assemblages themselves. In forested areas of the Chaco, the attempts at interruption start as soon as the bulldozers seek to move forth against the forest and people's homes. Struggles against evictions are fought on multiple fronts, involving courts, the media, and protests on the streets of nearby towns. But most people who fight deforestation do so with their own bodies, placing themselves in the path of bulldozers to demand that the drivers halt their advance. In this low-intensity warfare, residents have even burnt bulldozers to defend their homes, and many people I know remain on their land because of their determination to stay put. But others have been jailed, wounded, or killed, and twelve people have been murdered in the past few years for defending their land on the soy frontiers of northern Argentina.[89]

Road blockades have also proliferated in protests against agribusiness. And this has included blockages to interrupt the construction of infrastructural nodes by some of the most powerful companies in the world. The most notable defeat suffered by Monsanto in Argentina, for instance, was its failure to construct what was hailed as its largest factory in Latin America: a high-end facility in the town of Malvinas Argentinas, province of Córdoba, where the company intended to produce GM corn seeds. Because of well-founded fears that the factory would spread toxic dust into their homes, residents organized and blockaded the gate and interrupted construction for three years (2013–16). Despite the hostility of the media and attacks by the police and thugs, the determination of local men and women to defend the blockade generated a strong national and international solidarity campaign that eventually forced Monsanto to abandon the project.[90]

In the agroindustrial ring around Rosario, the unionized workers who operate the crushing factories have regularly asserted their power at this bottleneck of the supply chain by blockading access to the factories and ports during strikes, thereby shutting down the flow of beans, oil, and soy meal out of the country.[91] These protests are a reminder that infrastructures require laborers who have the power to shut down the continuum. As Toscano argues,[92] in a neoliberal era in which most workers have been offshored and downsized, the workers in charge of logistical nodes are therefore empowered by their partial control of infrastructures.[93] And around Rosario, the capacity of workers to shut down the ports has made them among the best paid in Argentina.

Not all of the interruptions slowing down or disrupting the supply chains are the product of human political actions. In Las Lajitas, a recurring concern among farm administrators and engineers is the growing number of “super weeds” that are becoming resistant to herbicides. These nonhuman life forms have become so resilient and expansive that they threaten to overgrow the soy fields and interrupt the farms’ productive rhythms. In the Paraná River, conjunctural events generated by the becoming of terrain such as the formation of a sandbank can create massive traffic jams involving dozens of 30,000-ton freighters. In March 2014, for instance, a ship stranded on the Paraná near Rosario and immobilized eighty vessels, shutting down exports for ten days.[94] These are contingent interruptions generated by encounters between human-made objects and the shifting depth of a river; yet they confirm that for all its power the metropolis is an aggregation of multiplicities that always already involve, and are affected by, the textures of terrain.

CONCLUSIONS: THE AGE OF THE METROPOLIS

The idea of the Anthropocene has been important to discuss the disruptive impact of climate change because it highlights that the world is an interconnected totality. This is a totality in the immanent sense of the term, for no part of the planet and none of its life forms are immune from a warming atmosphere. The Anthropocene is another way of admitting, in other words, that "there's no outside" of the dislocations produced by climate change. In this chapter, I sought to contribute to debates about the Anthropocene by highlighting that the atmosphere is warming not just because of capitalism but because of its incarnation in the metropolis, "a flow of beings and things, a current that runs through fiber-optic networks, high-speed train lines, and video surveillance cameras, making sure that this world keeps running straight to its ruins."[95]

The trajectories of the soy supply chains affecting Argentina remind us that the current configuration of the metropolis was produced by China's phenomenal urbanization and its transformation into the planetary factory for global capital, and by the resulting suction of voluminous flows of resources to build up such an expansion of the urban form. The Chinese state has since adopted an even more assertive role in the expansion of the metropolis by becoming the largest builder of infrastructures the world over: from the "New Silk Road" across central Asia to massive investment in mines, ports, railways, and dams in South Asia, Africa, Oceania, and Latin America. The repair of the railways around Las Lajitas is just one sample of this planetary presence of Chinese capital. While the nationalist agenda behind China's infrastructural drive has generated friction with other global powers, it is not at all incompatible with the imperial apparatuses of capture of global capitalism. On the contrary, China is "not outside" of these apparatuses. The huge industrial zones that exist in southeastern China to produce stuff for the metropolis under conditions of intense exploitation, after all, are also zones of imperial extraction. These are zones in which the most powerful capitalists on Earth give the Chinese state the mandate to provide them with a cheap, docile, and industrious workforce. The metropolis may have today a Chinese engine, but its apparatuses of capture remain rhizomatic and imperial. In its nonrepresentable multiplicity and vastness, the metropolis is certainly not reducible to its destructive and exploitative dimensions. Infrastructures that provide for clean water, renewable energy, sanitation, health care, education, or public transit are central to the production of collective and egalitarian forms of well-being. And as Lefebvre, Negri, Merrifield, and David Harvey insist, the metropolis is a space of encounters full of egalitarian and revolutionary potential: a networked terrain that in putting millions of people in close proximity with each other, and in helping them communicate across borders, can create transformative solidarities.[96] In the words of Hardt and Negri, "the metropolis is a factory for the production of the commons."[97]

In a metropolis structured by capitalist forms of acceleration, the defense of the commons has emerged with particular force in the grassroots attempts to interrupt the assembling of infrastructures that reproduce this acceleration. The politics of interruption through which people in Argentina try to disrupt the advance of bulldozers captures, in this regard, the politics of many social movements all over the world. Well-known examples are the shutdown by 25,000 protesters of the port of Oakland, California, in November 2011, following the police repression of Occupy Oakland; the rise of blockades by Indigenous people to interrupt the expansion of pipelines and other fossil-fuel-related infrastructures in North America, which Naomi Klein calls "blockadia"; or the blockade of oil refineries in France in 2015 by workers opposing the erosion of their labor rights.[98] Joshua Clover has argued that the growing popularity of blockades all over the world means that the dominant anticapitalist struggle is no longer the strike but "riots" over circulation.[99] The Invisible Committee, further, argue that insurrections should involve the blockage and sabotage of the infrastructures of the metropolis and the derailing of its restless currents of matter. In their words, "Power is logistic. Block everything!"[100]

Toscano has nonetheless criticized this overemphasis on interruption and blockade in the radical left.[101] He argues that any attempt to move beyond capitalism will have to confront the hard fact that the supply chains cannot be just shut down, first because the livelihood of millions of people depend on their ongoing mobility. To call for a shutdown of the metropolis, he wrote, is "to make something of a fetish out of rupture."[102] The infrastructure and logistical operations of the metropolis, Toscano argues, should rather be reconfigured toward new ends: to satisfy collective uses and needs rather than private profits. In turn, Jasper Bernes criticized Toscano by counterarguing that exploitation and disruption are so constitutive to supply chains, so central to their design as imperial weapons, that their appropriation for egalitarian ends is unfeasible. But he adds that blockades are not enough. What is needed, rather, is a "delinking from the planetary factory" through which local communes organize to cut off their dependence on supply chains.[103] This act of withdrawal is "a matter of survival," Bernes writes.

These debates are multifaceted and unanswerable in the abstract. But struggles against the extractivist drive of the metropolis usually involve a combination of interruptions, delinking, and positive appropriations and creations. Activists all over Argentina confront agribusiness not only through blockades but also through efforts to create different agrarian places by cutting off links with the supply chains of GM seeds, agrichemicals, and feed. They embrace an organic, socially conscious agriculture, *la agroecología*, to grow food for people living in regional towns and cities. These efforts withdraw from some supply chains but without withdrawing from the metropolis; rather, they embrace its collective potential and exploit the connectivities of its continuum in order to forge links with social movements based elsewhere in the world.

A political interrogation of the metropolis as the fast-paced infrastructure of the Anthropocene, however, confronts us with the bleak future it hints at, confirming what Hetherington and Jeremy Campbell argue about infrastructures: that they are "always about the future, or different futures."[104] Amid the worsening patterns of class polarization, racism, atmospheric disruptions, and ambient toxicity that define this early phase of the Anthropocene, it is not surprising that in films and popular culture the future evokes a catastrophe in which the grand infrastructures and skyscrapers of the metropolis are reduced to huge piles of rubble. Any collective attempt to create a future not defined by a planetary ruination, in this regard, has to face that the undoing of imperial and capitalist relations will require the reinvention of the metropolis as a collective constellation that does not depend for its existence on the creation of sacrifice zones.

First published in Kregg Hetherington, ed., *Infrastructure, Environment, and Life in the Anthropocene* (Durham, NC: Duke University Press, 2019), 66–94.

1 See Gastón Gordillo, “Longing for Elsewhere: Guaraní Reterritorializations,” *Comparative Studies in Society and History* 53, no. 4 (2011): 855–81. In 2003–07 alone, over a million hectares of forests were destroyed in the province of Salta. See Andrés Leake, Omar Enríque López, and María Cecilia Leake, *La Deforestación del Chaco Salteño* (Salta, AR: SMA Ediciones, 2016).

2 I first carried out the fieldwork that led to my book *Rubble: The Afterlife of Destruction* (Durham, NC: Duke University Press, 2014) in 2003–07. I subsequently returned in 2010, 2011, 2013, and 2016.

3 Timothy Morton, *Hyperobjects: Philosophy and Ecology after the End of the World* (Minneapolis: University of Minnesota Press, 2013).

4 See Antonio Negri, *From the Factory to the Metropolis* (Cambridge, UK: Polity, 2018); Alberto Toscano, “Factory, Territory, Metropolis, Empire,” *Angelaki: Journal of the Theoretical Humanities* 9, no. 2 (2004): 197–216; Toscano, “Logistics and Opposition,” *Mute* 3, no. 2 (August 9, 2011): 30–41; Toscano, “Lineaments of the Logistical State,” *Viewpoint Magazine*, September 28, 2014; The Invisible Committee, *The Coming Insurrection*, 2007 (Eng. trans. Los Angeles: Semiotext(e), 2009); and The Invisible Committee, *To Our Friends*, 2014 (Eng. trans. Robert Hurley, Los Angeles: Semiotext(e), 2015).

5 Toscano, “Factory, Territory, Metropolis, Empire,” 210.

6 Toscano, “Logistics and Opposition,” 6.

7 See Henri Lefebvre, *The Urban Revolution* (Minneapolis: University of Minnesota Press, 2003), and Lefebvre, *Rhythmanalysis* (New York: Continuum, 1992).

8 Keller Esterling, *Extrastatecraft: The Power of Infrastructure Space* (London: Verso, 2014), 15.

9 Jasper Bernes, “Logistics, Counterlogistics and the Communist Prospect,” *Endnotes* 3 (September 2013): 203.

10 Toscano and Jeff Kinkle, *Cartographies of the Absolute* (Alresford, UK: Zero Books, 2015), 24.

11 Theodor Adorno, quoted in Dimitri Vouros, “Hegel, ‘Totality,’ and ‘Abstract Universalism’ in the Philosophy of Theodor Adorno,” *Parrhesia* 21 (2014): 179.

12 Lefebvre, *The Urban Revolution*, 186.

13 See Gordillo, *Rubble*.

14 See Jason W. Moore, *Capitalism in the Web of Life: Ecology and the Accumulation of Capital* (London: Verso, 2015).

15 See Stuart Alexander Rockefeller, “Flow,” *Current Anthropology* 52, no. 4 (August 2011): 557–78.

16 See Arjun Appadurai, *Modernity at Large: Cultural Dimensions of Globalization* (Minneapolis: University of Minnesota Press, 1996).

17 See Manuel Castells, *The Information Age: Economy, Society, and Culture*, vol. 1, *The Rise of the Network Society*, 1996 (2nd ed. Malden, MA, Oxford, and Chichester: Wiley-Blackwell, 2011).

18 See Saskia Sassen, *The Global City: New York, London, Tokyo* (Princeton, NJ: Princeton University Press, 2001).

19 Ananya Roy, “The 21st Century Metropolis: New Geographies of Theory,” *Regional Studies* 43, no. 6 (2009): 819–30.

20 See Roy and Aihwa Ong, eds., *Worlding Cities: Asian Experiments and the Art of Being Global* (Malden, MA, and Oxford: Wiley-Blackwell,

2011), and AbdouMaliq Simone, *For the City Yet to Come: Changing African Life in Four Cities* (Durham, NC: Duke University Press, 2004).

21 Neil Brenner, "Theses on Urbanization," *Public Culture* 25, no. 1 (Winter 2013): 88.

22 See Andrew Barry, *Material Politics: Disputes along the Pipeline* (London: Wiley-Blackwell, 2013); Deborah Cowen, *The Deadly Life of Logistics: Mapping Violence in Global Trade* (Minneapolis: University of Minnesota Press, 2014); Esterling, *Extrastatecraft*; and Brian Larkin, "The Politics and Poetics of Infrastructure." *Annual Review of Anthropology* 42, no. 1 (2013): 327–43.

23 See Matthew Gandy, "Rethinking Urban Metabolism: Water, Space and the Modern City," *City* 8, no. 3 (2004): 363–79; Christopher Kennedy, John Cuddihy, and Joshua Engel-Yan, "The Changing Metabolism of Cities," *Journal of Industrial Ecology* 11, no. 2 (April 2007): 43–59; and Nikhil Anand, *Hydraulic City: Leaks and the Infrastructures of Citizenship in Mumbai* (Durham, NC: Duke University Press, 2017).

24 Andy Merrifield, *The Politics of the Encounter: Urban Theory and Protest under Planetary Urbanization* (Athens: University of Georgia Press, 2013), xx.

25 Brenner and Christian Schmid, "The 'Urban Age' in Question," in Brenner, ed., *Implosions/Explosions: Towards a Study of Planetary Urbanization* (Berlin: Jovis, 2014), 310–37.

26 See Martín Arboleda, "Spaces of Extraction, Metropolitan Explosions: Planetary Urbanization and the Commodity Boom in Latin America," *International Journal of Urban and Regional Research* 40, no. 1 (January 2016): 96–112.

27 Brenner, "Introduction: Urban Theory without an Outside," in Brenner, ed., *Implosions/Explosions*, 14–30.

28 See Kate D. Derickson, "Urban Geography I: Locating Urban Theory in the 'Urban Age,'" *Progress in Human Geography* 39, no. 5 (October 2015): 647–57; Michelle Buckley and Kendra Strauss, "With, against and beyond Lefebvre: Planetary Urbanization and Epistemic Plurality," *Environment and Planning D: Society and Space* 34, no. 4 (August 2016): 617–36; and Linda Peake, "The Twenty-First Century Quest for Feminism and the Global Urban," *International Journal of Urban and Regional Research* 40, no. 1 (January 2016): 219–27.

29 Kanishka Goonewardena, "Planetary Urbanization and Totality," *Environment and Planning D: Society and Space* 36, no. 3 (June 2018): 456–73.

30 The Invisible Committee, *The Coming Insurrection*, 52.

31 Lefebvre, *The Urban Revolution*.

32 Ibid., 173–74.

33 Ibid., 118.

34 Brenner and Schmid, "The 'Urban Age' in Question."

35 Donna Haraway, "Situated Knowledge: The Science Question in Feminism and the Privilege of Partial Perspective," *Feminist Studies* 14, no. 3 (Autumn 1988): 575–99.

36 Larkin, "The Politics and Poetics of Infrastructure," 329.

37 See Richard Kernaghan, "Furrows and Walls, or the Legal Topography of a Frontier Road in Peru," *Mobilities* 7, no. 4 (2012): 501–20, and Penny Harvey and Hannah Knox, *Roads: An Anthropology of Infrastructure and Expertise* (Ithaca, NY: Cornell University Press, 2015).

38 See Jane Bennett, *Vibrant Matter: A Political Ecology of Things* (Durham, NC: Duke University Press, 2010), and Barry, *Material Politics*.

39 See Andrew Blum, *Tubes: Journey to the Center of the Internet* (New York: Ecco, 2012).

40 See Hannah Appel, "Offshore Work: Oil, Modularity, and the How of Capitalism in Equatorial Guinea," *American Ethnologist* 39, no. 4 (November 2012): 692–709.

41 See Cowen, *The Deadly Life of Logistics*.

42 See Shaylih Muehlmann, *When I Wear My Alligator Boots: Narco-Culture in the US-Mexican Borderlands* (Berkeley: University of California Press, 2013).

43 See Stephen J. Collier and Ong, "Global Assemblages, Anthropological Problems," in Ong and Collier, eds., *Global Assemblages: Technology, Politics, and Ethics as Anthropological Problems* (Malden, MA: Blackwell, 2005), 3–21.

44 Manuel DeLanda, *A New Philosophy of Society: Assemblage Theory and Social Complexity* (New York: Continuum, 2006).

45 See Gordillo, "Terrain as Insurgent Weapon: An Affective Geometry of Warfare in the Mountains of Afghanistan," *Political Geography* 64 (May 2018): 53–62.

46 Sassen, *The Global City*.

47 See Immanuel Wallerstein, *The Modern World System* (New York: Academic Press, 1974).

48 See Buckley and Strauss, "With, against and beyond Lefebvre."

49 Brenner, "Debating Planetary Urbanization: For an Engaged Pluralism," *Environment and Planning D: Society and Space* 36, no. 3 (June 2018): 570–90.

50 William Cronon, *Nature's Metropolis: Chicago and the Great West* (New York: Norton, 1991), 19. See Raymond Williams, *The Country and the City* (New York: Oxford University Press, 1973).

51 See Ashley Carse, "Nature as Infrastructure: Making and Managing the Panama Canal Watershed," *Social Studies of Science* 42, no. 4 (August 2012): 539–63, and Carse, *Beyond the Big Ditch: Politics, Ecology, and Infrastructure at the Panama Canal* (Cambridge, MA: The MIT Press, 2014).

52 See Stephanie Wakefield and Bruce Braun, "Infrastructure, Profanation, and the Sacred Figure of the Human," in Kregg Hetherington, ed., *Infrastructure, Environment, and Life in the Anthropocene* (Durham, NC: Duke University Press, 2019).

53 Gordillo, *Rubble*.

54 Anna Lowenhaupt Tsing, *Friction: An Ethnography of Global Connection* (Princeton, NJ: Princeton University Press, 2005).

55 See Hetherington, "Waiting for the Surveyor: Development Promises and the Temporality of Infrastructure," *Journal of Latin American and Caribbean Anthropology* 19, no. 2 (July 2014): 195–211.

56 See Jorge Eduardo Rulli, *Pueblos fumigados: Los efectos de los plaguicidas en las regiones sojeras* (Buenos Aires: Del Nuevo Extremo, 2009); Pablo Lapegna, "Genetically Modified Soybeans, Agrochemical Exposure, and Everyday Forms of Peasant Collaboration in Argentina," *Journal of Peasant Studies* 43, no. 2 (March 2016): 517–36; Lapegna, *Soybeans and Power: Genetically Modified Crops, Environmental Politics, and Social Movements in Argentina* (Oxford: Oxford University Press, 2016); and Medardo Avila Vazquez et al., "Association between Cancer and Environmental Exposure to Glyphosate," *International Journal of Clinical Medicine* 8, no. 2 (February 2017): 73.

57 See "Rosario, auge y caída de la Argentina dorada," *El País*, October 8, 2015.

58 See Carla Gras and Valeria Hernández, "El modelo agribusiness y sus traducciones territoriales," in Gras and Hernández, eds., *El agro como negocio: Producción, sociedad, territorios en la globalización* (Buenos Aires: Biblos, 2013), 49–66; Gras and Hernández, "Los pilares del modelo agribusiness y sus estilos empresariales," in Gras and Hernández, eds., *El agro como negocio*, 17–46; and Gras and Hernández, *Radiografía del nuevo campo argentino: Del terrateniente al empresario transnacional* (Buenos Aires: Siglo XXI, 2016).

59 Hernandez, "La ruralidad globalizada y el paradigma de los agronegocios en la pampa gringa," in Gras and Hernández, eds., *La Argentina rural: De la agricultura familiar a los agronegocios* (Buenos Aires: Biblos, 2009).

60 See also Hernández, María Florencia Fossa Riglos, and María E. Muzi, "Agrociudades pampeanas: Usos del territorio," in Gras and Hernández, eds., *El agro como negocio*, 123–49, and Hernández, Fossa Riglos, and Muzi, "Figuras socioproductivas de la ruralidad globalizada," in Gras and Hernández, eds., *El agro como negocio*, 151–69.

61 Gustavo Oliveira and Susanna Hecht, "Sacred Groves, Sacrifice Zones and Soy Production: Globalization, Intensification and Neo-Nature in South America," *Journal of Peasant Studies* 43, no. 2 (March 2016): 251–85.

62 See Héctor Huergo, "La hidrovía abona la inversión," *Clarín*, September 2, 2006.

63 See Oliveira and Mindi Schneider, "The Politics of Flexing Soybeans: China, Brazil, and Global Agroindustrial Restructuring," *Journal of Peasant Studies* 43, no. 1 (January 2016): 173.

64 See Michael Hardt and Antonio Negri, *Empire* (Cambridge, MA: Harvard University Press, 2000).

65 See Timothy Brennan, "The Empire's New Clothes," *Critical Inquiry* 29, no. 2 (Winter 2003): 337–67; Stephen Reyna, "Empire: A Dazzling Performance according to a Simpleton," *Anthropological Theory* 2, no. 4 (December 2002): 489–97; and Ong, "Powers of Sovereignty: State, People, Wealth, Life," *Focaal: Journal of Global and Historical Anthropology* 64 (November 2012): 24–35.

66 Tsing, "Empire's Salvage Heart: Why Diversity Matters in the Global Political Economy," *Focaal: Journal of Global and Historical Anthropology* no. 64 (November 2012): 37–38. See also Tsing, *The Mushroom at the End of the World: On the Possibility of Life in Capitalist Ruins* (Princeton, NJ: Princeton University Press, 2015), 65.

67 See Tsing, "Empire's Salvage Heart," and Tsing, *The Mushroom at the End of the World*.

68 See Tiqqun, *This Is Not a Program*, trans. Joshua David Jordan (Los Angeles: Semiotext(e), 2011), 171.

69 Lewis Mumford, *The City in History: Its Origins, Its Transformations, and Its Prospects*, 1961 (reprint ed. New York: Harcourt Brace, 1989), 539.

70 Barry, "Technological Zones," *European Journal of Social Theory* 9, no. 2 (May 2006): 239–53.

71 See Stuart Elden, "Land, Terrain, Territory," *Progress in Human Geography* 34, no. 6 (December 2010): 799–817, and Elden, *The Birth of Territory* (Chicago: University of Chicago Press, 2013).

72 See Marie-Monique Robin, *The World according to Monsanto: Pollution, Corruption, and the Control of the World's Food Supply*, 2008 (Eng. trans. George Holoch, New York: New Press, 2010).

73 See Mateo Mier y Terán Giménez

Cacho, "Soybean Agrifood Systems Dynamics and the Diversity of Farming Styles on the Agricultural Frontier in Mato Grosso, Brazil," *Journal of Peasant Studies* 43, no. 2 (March 2016): 419–41.

74 See Oliveira and Schneider, "The Politics of Flexing Soybean."

75 See Oliveira and Hecht, "Sacred Groves," 257.

76 See Tsing, *The Mushroom at the End of the World*.

77 See Valdemar João Wesz Jr., "Strategies and Hybrid Dynamics of Soy Transnational Companies in the Southern Cone," *Journal of Peasant Studies* 43, no. 2 (March 2016): 286–312; Amalia Leguizamón, "Disappearing Nature? Agribusiness, Biotechnology and Distance in Argentine Soybean Production," *Journal of Peasant Studies* 43, no. 2 (March 2016): 313–30; and Oliveira and Hecht, "Sacred Groves."

78 Wesz, "Strategies and Hybrid Dynamics," and Leguizamón, "Disappearing Nature?"

79 Noble Group, "About Noble Resources." Available online at http://www.thisisnoble.com (accessed December 2013).

80 Bernes, "Logistics, Counterlogistics and the Communist Prospect," 172–201.

81 Cowen, *The Deadly Life of Logistics*, 69.

82 Ibid.

83 "ChemChina Deal for Syngenta Reflects Drive to Meet Food Needs," *New York Times*, February 3, 2016.

84 Ong, *Flexible Citizenship: The Cultural Logics of Transnationality* (Durham, NC: Duke University Press, 1999).

85 Wesz, "Strategies and Hybrid Dynamics," 292.

86 See Clara Craviotti, "Which Territorial Embeddedness? Territorial Relationships of Recently Internationalized Firms of the Soybean Chain," *Journal of Peasant Studies* 43, no. 2 (March 2016): 331–47.

87 See "Inversión: Grupo brasileño construirá una terminal portuaria en Rosario," *Infocampo*, June 18, 2016. Available online at www.infocampo.com.ar/inversion-grupo-brasileno-construira-una-terminal-portuaria-en-rosario/ (accessed April 2024).

88 See "Monsanto: Crece la pelea por el pago de regalías," *La Nación*, May 11, 2016.

89 See Darío Aranda, *Tierra arrasada* (Buenos Aires: Sudamericana, 2015).

90 See "Chau Monsanto: Triunfo de los vecinos en Malvinas Argentinas," *La Vaca*, August 3, 2016.

91 See "Dani, el aceitoso," *Revista Crisis* 27 (November 4, 2016).

92 Toscano, "Logistics and Opposition."

93 See Timothy Mitchell, *Carbon Democracy: Political Power in the Age of Oil* (London: Verso, 2011), and Cowen, *The Deadly Life of Logistics*.

94 See "Un barco varado en el Paraná impide que se exporte soja," *La Nación*, March 20, 2014.

95 See The Invisible Committee, *The Coming Insurrection*, 58–59.

96 Lefebvre, *The Urban Revolution*; Negri, *From the Factory to the Metropolis*; Merrifield, *The Politics of the Encounter*; and David Harvey, *Rebel Cities: From the Right to the City to the Urban Revolution* (London: Verso, 2012).

97 Hardt and Negri, *Commonwealth* (Cambridge, MA: Harvard University Press, 2009), 250.

98 See Naomi Klein, *This Changes Everything: Capitalism vs. the Climate* (New York: Simon and Schuster, 2015).

99 See Joshua Clover, *Riot. Strike. Riot: The New Era of Uprisings* (Brooklyn: Verso, 2016).

100 The Invisible Committee, *To Our Friends*, 81.
101 Toscano, "Logistics and Opposition," and Toscano, "Lineaments of the Logistical State."
102 Toscano, "Logistics and Opposition," 2.
103 Bernes, "Logistics, Counterlogistics and the Communist Prospect," 11.
104 Hetherington and Jeremy M. Campbell, "Nature, Infrastructure, and the State: Rethinking Development In Latin America," *Journal of Latin American and Caribbean Anthropology* 19, no. 2 (July 2014): 193.

OF DREAMS AND THE EARTH

From the northeast to the eastern border of Minas Gerais, where the Doce River flows through the Krenak Indigenous reserve, and in the Amazon, where Brazil meets Peru and Bolivia along the Upper Negro River, in all of these places our Indigenous families are experiencing moments of great tension in their political relations with the Brazilian state.

This tension is nothing new, but it has worsened because of recent political changes that have severely affected Indigenous communities. For decades now, we have been pressing the government to honor its constitutional pledge to protect the rights of our peoples over the ancestral territories that the prevailing legal framework calls "homelands."

I don't know how familiar the reader is with the language used to describe the relationship between the Indigenous peoples and their ancestral lands, or the functions the Brazilian state has ascribed to these territories down through its history. The truth is, since colonial times, the issue of what to do with the remaining tribal pockets that survived the first tragic encounters between the European conquistadores and the tribes inhabiting what we now reductively call "Indigenous homelands" led to a terribly wrongheaded relationship between the state and these communities.

Of course, we have since ceased to be a colony and become the Brazilian state, and our tribes have made it into the twenty-first century, when most predictions said we could never survive the occupation of our territories, at least not in any autonomously organized and self-sustaining form. It may not have materialized, but that gloomy prognosis was not unwarranted, given the way the state machine works to undermine our communities and subsume us into wider Brazilian society.

The political bind our surviving communities have found themselves in is to have to fight for the last remaining swaths of land where nature still prospers, where our needs for food and a home can be met, and where our societies, however small, can make their own way in the world, without excessive reliance on the state.

The Doce River, which we, the Krenak Nation, call Watu—our grandfather—is a person, not a resource, as the economists like to call him. He is not something you can own or appropriate; he is part of our construction as a collective society that dwells in a specific place into which we have been gradually corralled by the government, forcing us to live and breed in bubbles subject to increasingly crippling external pressure.[1]

Taking the Krenak as an example in speaking about the relationship between the Brazilian state and the Indigenous societies came to us as an inspiration, a relatable way to reach those who perhaps don't know how Brazil actually treats these communities—an estimated 900,000 individuals across 250 tribes, a population far smaller than found in many Brazilian cities.

What lies at the root of the nation's incapacity to embrace its Indigenous peoples—who have seen ways of life they have maintained for thousands of years relentlessly and inhumanely attacked by ferocious colonial attitudes that endure to this very day—is the idea that the tribes ought to be contributing to the success of modernity's nature-depleting project, just like everybody else. Watu, the river that has nurtured and nourished our life along a 600-kilometer stretch of the Doce River, from Minas Gerais to Espírito Santo, finds itself today sunk under toxic mud from a burst tailings dam that orphaned our tribes and plunged the river into a coma. It's been a year and a half since this crime—accident just doesn't cut it—sent our lives into a tailspin and conjured a real end-of-the-world scenario.

Here, the intention is to broach the impact that we, humanity, are having on the living organism that is the earth, which some cultures still consider to be our mother and provider, not only in terms of subsistence and sustenance but also transcendentally, as the wellspring of all meaning in our existence. In many places today we have become so uprooted from our homelands that we cross oceans as if we were merely popping across the street. We give little thought to making continent-size journeys. While technological advancements mean we can travel easily between places, moving around the planet in vast numbers and at incredible speeds, this has also stripped our journeys of real significance.

We feel today as though we've been cut adrift in a cosmos devoid of meaning and shorn of any shared ethic. It weighs upon our lives, and at every turn we are reminded of the consequences of these choices. If we could only heed some vision beyond the blindness that has descended upon the world, perhaps it would open our minds to some form of cooperation among peoples, not so we can save others but so we can save ourselves. For the last thirty years, I have built an extensive contact network so as to spread the word to other peoples and governments about the realities we are facing in Brazil, and my goal was always to trigger bonds of solidarity with other Indigenous peoples.

What I have learned over these decades is that we all need to wake up, because whereas before it was just us, the Indigenous peoples, who were facing a loss of meaning in our lives, today everyone is at risk, without exception. As our planet Earth teeters on the verge of collapse beneath our impossible weight, none of us can ignore this reality. As the Yanomami Chief Davi Kopenawa says, the world today believes that everything is merchandise, and it projects upon those goods the full range of its experience. Everything must be product, resource, commodity. This tragedy has been avoided in some places, where politics—political power, political choice—has provided temporary safe havens for communities that, though already drained of any real sense of shared space, are still protected, for want of a better word, by an apparatus that is becoming increasingly dependent on sucking the life out of our forests, rivers, and mountains. An apparatus, in short, that puts us in a disastrous position where it seems the only way we can survive is at the expense of all other forms of life.

The conclusion or understanding that we are living in a time that can only be identified as the Anthropocene ought to set alarm bells ringing in our minds. If we're leaving such a deep mark on the earth that it defines an epoch—a mark that runs so deep it may well outlast us, given how quickly we are depleting the resources that allowed us to prosper and feel at home—it's because we are once again faced with that same bind. We are excluding all local forms of organization that are not integrated into the world of merchandise, thus threatening with extinction all other ways of life—at least those we used to recognize as such, to which we ascribed some co-responsibility and respect for shared spaces and fellow beings, not just this single humanity, an abstraction we've allowed ourselves to create to the exclusion of all other creatures. This humanity refuses to recognize that the river, now in a coma, is also our grandfather; that the mountains mined in Africa or South America and transformed into merchandise elsewhere are also the grandfather, grandmother, mother, brother of some other constellation of human beings that want to go on sharing the communal home we call Earth.

The name "Krenak" is formed by two words: *kre*, which means "head," and *nak*, which means "land." Krenak is the legacy of our forebears, the memory of our origins, which we identify as our "headland," as a humanity that cannot understand itself without this connection, this deep-set communion with the earth. Not the earth in the sense of a property, but as the place we share and from which we, the Krenak, feel increasingly disconnected. I'm speaking of the earth as this place that has always been sacred to us, but which our neighbors are ashamed to admit could ever be seen in such terms. When we say that our river is sacred, the response is always the same: "That's just their folklore." When we say that the mountain is telling us it is going to rain and that today will be a prosperous day, they say: "Mountains are just mountains. They don't tell us anything."

When we depersonalize the river, the mountain, when we strip them of their meaning—an attribute we hold to be the preserve of the human being—we relegate these places to the level of mere resources for industry and extractivism. The result of our divorce from our integrations and interactions with Mother Earth is that she has left us orphans—not just those termed, to a greater or lesser degree, Indigenous peoples, Natives, Amerindians, but everyone. I hope that the creative encounters we still manage to muster can nourish our practice, our action, and give us the courage to step back from this negation of life and toward a commitment to it, wherever it and we may be. I want to see us move beyond our incapacities to extend that vision to places beyond those we call home, and not only the vision, but forms of sociability and organization from which a large part of this human community is excluded; those who burn up the earth's energies just to feed their demand for merchandise, comfort, and consumption.

How can we find a point of contact between these two worlds, which share the same origin but have drifted so far apart that today we have, at one extreme, those who need a river in order to live, and at the other, those who consume rivers as mere resources? Regarding this notion of "resource" attributed to a mountain, a river, a forest, where can we possibly find this intersection between our opposing visions that might save us from this state of blindness toward one another?

When I said I was going to speak of dreams and the earth, what I had in mind was a place and a practice that is perceived in so many different cultures, by so many different peoples, not merely as part of the daily experience of sleeping and dreaming, but as the disciplined exercise of deriving guidance for our actions in the waking world from the dreams that visit us in our slumber.

For some people, to dream is to step outside of reality, relinquish the practical meaning of life. For others, however, there is no meaning to life unless informed by dreams, the place we go in search of songs, cures, inspiration, and even solutions to practical problems that befuddle and elude us in the daytime, but which are laid out in all their possibilities in the realm of dream. I was very content this afternoon when more than one colleague mentioned the institution of dreaming not as mere oneiric experience but as a discipline related to our formation, to our cosmovision, to the traditions of different peoples who approach dreams as a path toward learning, self-knowledge, and awareness of life, and the application of that knowledge in our interaction with the world and other people.

First published in Krenak, *Ideas to Postpone the End of the World*, trans. Anthony Doyle (Toronto: House of Anansi Press, 2020).

1 According to the National Indian Foundation, FUNAI, Brazil currently has 117 million hectares of Traditionally Occupied Indigenous Territories, with a further million in the process of regularization, and somewhere in the region of 90,000 hectares of Indigenous reserves (often lands donated for this purpose). These territories are constantly encroached upon and often violently invaded by illegal logging, mining, and wildlife extraction operations. —Trans.

NETWORKS OF ECOTERRITORIAL FEMINISMS IN LATIN AMERICA

Patriarchy does to our bodies what capitalist extractivist economies do to our land.
—Thirteenth Latin American and Caribbean Feminist Gathering, Peru, 2015

There is a great deal of historical evidence for the empirical and epistemic ties between gender and ecology, between feminism and the environment.[1] This is the case because it is those women in charge of care-giving and social reproduction who are the first to detect the sociosanitary impacts linked to developmental models, risks that are usually minimized or go unrecorded in official statistics. In Latin America, ecoterritorial feminisms also emerge from the condemnation of pollution and the defense of living conditions in the face of its impacts on health, air, and environment. But they add to this critique a strongly decolonizing touch, which brings to the fore the relationship between sociosanitary impacts and developmentalist neoextractivism.

This essay analyzes some of the major topics addressed by women's struggles in the Latin American region, and their connection to environmentalist and antiextractivist organizations. To emphasize the connection to the ecoterritorial turn of these struggles, as well as the mobilizations of those affected by socioenvironmental conditions, I adopt the label "ecoterritorial feminisms," which I developed in a previous text.[2]

My point of departure is the fact that in the last twenty years the manifestations of communitarian, rural, and popular feminisms have continually multiplied in the context of expanded neoextractivisms. They involve popular feminisms that emerge on the social, ethnic, and geographical margins: Indigenous, peasant, Africa-descended women, poor and/or vulnerable women from rural and urban áreas, which leave silence behind, mobilize themselves in the public sphere, and re-create relationships of solidarity and new forms of collective self-management in the face of the impacts of industrial and/or extractivist projects, whether already in place or in the process of expansion.

In their origins, many of these women-led ecoterritorial struggles do not explicitly recognize themselves as feminist, since feminisms seem to be associated with the urban middle classes. This initial reticence shows that, far from resulting from an automatic labeling, the reappropriation of feminism and the critique of patriarchy form part of a cultural and collective process of construction. It is through the recursive dynamic of struggles, largely through an intergenerational dialogue—and a back-and-forth between collective, public involvement and the return to a privacy traversed by the ties of patriarchal oppression—that women begin to name this type of domination and the struggle begins to be resignified as feminist and antipatriarchal.

Nevertheless, the processes of reappropriation and the interconnection of networks are not the same for all women. Many women from Indigenous communities, as Francesca Gargallo points out, live "a transition toward feminism whenever they can make feminism not into something that distances them from their people's history and culture, but rather a theoretical instrument for living well."[3] By affirming difference, a space of variable geometry is constructed that unites anticolonial expressions (confrontations with big mining and petroleum corporations, in charge of large dams, which are always associated with national governments) with the critique of patriarchy and of previously naturalized domestic and familial oppression, or that silences patriarchal violence. All this strengthens action in defense of land and territory, which broadens languages of valorization and leads to democratization in the heterogeneous feminist space of struggle.

On the other hand, ecofeminist theory originated by emphasizing the patriarchal character of a double domination by men, on the levels of both interpersonal relations and the relationship to nature. This double domination rests on a dualist, binary paradigm separating the human from the nonhuman, man from woman, public from private, reason from emotion, modern from nonmodern, and whose correlative is an identitarian, monocultural logic that dismisses and devalues those who are different.

Regarding the woman/nature relationship, ecofeminism's point of departure is the inversion of the stigma, positively resignifying identification through different paths—in the first place, through a different connection with the body and with nature. This interconnection has contributed to the development of other valorizing languages, based on relational approaches or ontologies emphasizing interdependence, complementarity, and caring—in short, ecodependency, especially within the framework of the present environmental and civilizational crisis. Second, together with a feminist economy, ecofeminisms placed on the public agenda the importance of the invisibilized and unrecognized reproductive labor carried out by women. This caring work, necessary for sustaining life, has been traditionally devalued, as has the task of sustaining nature and maintaining its cycles, now threatened by the commodifying dynamic of capital. Consequently, ecofeminism emphasizes that just as there exists an ecological debt and an ecological trace, there also exists a caring debt and caring trace, associated with the sexual division of labor: caring work falls to women, particularly poor women.[4]

Within ecofeminism, there exist different currents ranging from identitarian ecofeminism, which naturalizes/biologizes the tie between women and nature, to constructivist ecofeminism, which conceives of it as a sociohistorical construction linked to the sexual division of labor. In Latin America, voices have arisen, like that of the Brazilian ecofeminist theologian Ivonne Gevara, that emphasize an ecofeminist epistemology based on relationality, interdependence, and emotionality in knowledge. In the Global South, the impact of the theories and experience in struggle of Vandana Shiva is particularly notable, since she was the one who identified ecofeminism with the defense of diversity, questioning not only productive but mental monocultures. As will be shown, ecoterritorial feminisms are distinguished by their anticolonial dimension and their defense of diversity, making it evident that the sustainability of life and the planet is founded on another link, at once material and spiritual, to the body and nature, within the framework of an epistemology of emotions and affects.

Not all these voices explicitly recognize themselves as ecofeminist.[5] Even so, there is an observable tendency toward the emergence of spaces of reexistence that oppose capital to life, animated by a relational ecofeminist perspective. In these social and intergenerational crossings, women who had been relegated to the domestic sphere, with few interclass contacts of a nonhierarchical nature, continue to forge important changes in subjectivity through collective mobilization, and to find their own voice. *Environmental Impact and Environmental Justice*, *Water for the Territories*, *Territoriality and Care*, *Health and Nature*, *Access to the Land*, and *Food Sovereignty* are among the major issues addressed by the region's movements.

In summary, through the experiential convergence between public and private, an unprecedented space of interchange between collectives of women of differing ages and social sectors has opened up: on the one hand, between groups of poor women from rural backgrounds, peasants and Indigenous women, or from small localities, groups of women affected by the environment, marginalized both socially and ethnically; and on the other hand, between activists and specialized NGOs, collectives of critical professions advocating for the broadening of women's rights. This has resulted in the broadening of the feminist and ecological dimension of struggle, and beyond that, in the elaboration of a new feminist epistemology on the basis of the inclusion of new issues and perspectives on the land, and the development of a more horizontal vision of presently existing feminisms, beyond ethnic and class differences or the distances between urban and rural.

ENVIRONMENTALLY AFFECTED PEOPLE AND SACRIFICE ZONES

The popular environmentalisms of the South and the movement for environmental justice that began in the United States during the 1960s have, as is known, a common root: namely, the coincidence between the map of poverty and the map of pollution. Degraded lands, spills of polluting effluents in urban areas, the construction of low-income neighborhoods on landfill, leaks of industrial chemicals, leaks of minerals and hydrocarbons that destroy the soil and pollute the water, open-air garbage dumps, the alteration and destruction of flora and fauna, the die-off of animals, desertification, pitted fields—these are some of the damages that extractive industry, petroleum, and mining activities have left behind over time. Thus one of the first issues addressed by women in the territories is the impacts of traditional industrial and extractive activity on health and daily life. Social resistance is born with the denaturalization of the pollution of air, water, and land, giving names to the poor quality of life and its consequences on bodies: "environmentally affected people," "environmental justice," "environmental suffering," "sacrifice zones."

In Latin America, women play a central role in numerous examples of the denaturalization of environmental damage. These include the case of Quintero-Puchuncaví, near Valparaíso, Chile, the most polluted industrial site in the country, where the first copper-smelting plant and refinery of the Empresa Nacional de Minería (National mining enterprise), known as ENAMI, began operations in 1964. Here, women created the group Mujeres de Zonas de Sacrificio en Resistencia de Puchuncaví-Quintero (Women of sacrifice zones resisting Puchuncavi-Quintero). "We women living in sacrifice zones," states the group's cofounder, Carolina Orellano, "have been rendered dramatically invisible where symptoms of pollution and the scale of abuses are concerned. . . . Possibly because we have naturalized not only the violence of patriarchy but also the alienation imposed by extractivist neoliberalism, with women and the earth being objects of abuse. It is urgent that this expression of violence motivate struggle on a national level as well."[6]

Another case is the pollution hub in Argentina's Cuenca Matanza-Riachuelo basin, which covers sixty-four square kilometers and traverses fourteen municipalities, as well as the city of Buenos Aires. In this genuine sacrifice zone, the difficulty of struggling with naturalization and with verbalizing a situation of extreme environmental suffering reappears. The short film *Las Mujeres del Río* (The women of the river, 2018), directed by Soledad Fernández Bouzo, shows women from different affected neighborhoods demanding to be heard and to have their rights recognized: "We need to be seen, and we need to be considered and to have channels able to exercise our rights. Our goal is to ensure that young people can have their voices heard, that garbage not be naturalized, that air not be naturalized, that sewage water not be naturalized, that being in a dirty school and neighborhood not be naturalized, that there are alternatives and they can be the protagonists of creating an alternative."[7]

Situations such as Puchuncaví-Quinteros and Matanza-Riachuelo show how "epidemiological profiles" have developed in urban space that illustrate not only the unsustainability of large cities but the different impact of pollution on health in terms of social class.[8] The new extractivisms of the twenty-first century have introduced other elements into this situation of social and environmental injustice. A paradigmatic example is the sociosanitary impacts of glyphosate, the herbicide marketed by Monsanto/Bayer and associated with transgenic soy, a monoculture that has spread through Argentina, Brazil, Paraguay, and Bolivia. In Argentina, for example, transgenic soy occupies around 25 million hectares.[9] The impacts of the soybean model on health were placed on the public agenda by the women of the Barrio Ituzaingó Anexo, in the southeast of the city of Córdoba, Argentina. In 2000, a group of mothers began to mobilize, concerned by an increase in spontaneous abortions and pathologies such as cancer. Against the lack of environmental and epidemiological controls, the women organized to carry out their own survey: "a map of the illnesses and deaths in the barrio, going from house to house through our territory, speaking with the neighbors. We started noticing what agrotoxics were, because up until then we didn't know what they were, what they were used for, who produced them and for what."[10] In 2012, these accusations led to the first penal judgment in relation to glyphosate fumigation in Argentina, and gave national and international visibility to a problem that struck at the core of the Argentine economic model.[11]

In sum, in the cases of both old and new extractivisms, the path from the detection of damage to the leap to public mobilization is long, from the denunciation that seeks visibility to the demand for rights, reparation, and compensation.[12] The naturalization of environmental suffering and pollution often constitutes the first and only reality. The women-led collectives of those affected have to construct, almost from zero, a language able to name what they are suffering from.

WATER FOR LIFE

The monopolization and appropriation of water within the framework of extractivist expansion have lately worsened, whether through overconsumption and intensified use or through privatization and pollution. Open-pit mining, the most-resisted extractive activity in Latin America, not only leaves behind huge environmental liabilities, utilizes polluting substances, and uses up a great deal of energy but also consumes millions of gallons of water, competing for it with traditional and sustainable economic activities as well as with domestic and residential water uses. The result has been an increase in preexisting environmental, social, and ethnic inequalities.

An emblematic case of the defense of water against mining is that of Máxima Acuña, known as "the lady of the blue lagoon," who opposed the Conga project in Cajamarca, Peru. The Conga megamining project had caused the drying up and destruction of four interconnected lagoons in the area, permanently affecting the whole system, a situation also affected by the dwindling of glaciers.[13] Acuña, a subsistence farmer living 13,000 feet above sea level, became famous for her fight against the Yanacocha mine. "Have you heard that they're selling the lagoons?," she asked herself while lifting a heavy rock on the day the company tried to evict her, "or that they're selling the rivers, and the spring's being sold and we're being kept away from it?"[14] Finally, thanks to the struggle of communities and popular pressure, the Conga project was suspended in 2011.

In the region as a whole, the Red Latinoamericana de Mujeres Defensoras de Derechos Sociales y Ambientales (Latin American network of women in defense of social and environmental rights) includes organizations with a long history of struggle against extractivism, including CENSAT Agua Viva (Colombia), Acción Ecológica (Ecuador), Colectivo CASA (Bolivia), Mujeres ecofeministas de El Salvador, GRUFIDES (Peru), and others.[15] As for Chile in particular, it is suffering hydrological stress linked to the mining of copper and more recently lithium. Indeed, the expansion of the lithium frontier, another mining system dependent on water, has aggravated the dispute over this scarce resource in arid ecosystems in the face of the impacts of the climate crisis. The new constitution that was rejected in 2022 proposed the deprivatization of waters and the inclusion of the "Rights of Nature." The anthropologist Francisca Fernández Droguett, a member of the Movimiento por el Agua y los Territorios (MAT; Movement for water and the territories), speaks of the "hydropolitics of waste" in the face of the intensification of extractivisms and resistances.[16]

What has happened in the last few decades is that the serpentine South American rivers have become aquatic routes of extractivism, whose function is to mobilize and transport raw materials—minerals, metals, soy, petroleum, palm, and other commodities exported to the world from Latin America—out of the territories. The expansion of megadams and the transformation of rivers into hydro-highways is another illustration of privatization through diversion, since these processes leave communities, farming groups, Indigenous peoples, and small localities without water.

The Movimiento de Afectados por las Represas (Movement of those affected by dams; known as MAB after its Portuguese name, Movimento dos Atingidos por Barragens) emerged in the 1990s in Brazil, where it was associated with the struggle against the Belo Monte megadam. As in so many cases, the movement involves a demand for environmental justice related to water. Under the slogan "The water belongs to the people, not Belo Monte," it was women who told the story of resistance and the loss of community ties as a result of the displacements forced by the dams, which they recorded in narratives woven into the fabric patchwork images called *arpilleras*.[17]

One of the most resonant cases in defense of rivers and women's rights is that of Berta Cáceres, the activist and founder of the Consejo Cívico de Organizaciones Populares e Indígenas de Honduras (COPINH; Civic council of popular and Indigenous organizations of Honduras), who received the 2015 Goldman Environmental Prize and was murdered shortly thereafter, in March 2016, by repressive forces in Honduras because of her opposition to a hydroelectric megaproject: "When we began the struggle against Agua Zarca, I knew how hard it would be but I knew we would win; the river told me," Cáceres told the BBC in 2015. "We will continue not only as the Lenca people but with other organizations with the hope of changing the situation in our country. The only path left to us is struggle." She also said, "Right now, we are more than 400,000 Lencas. Our people go back thousands of years in Honduras and Eastern El Salvador. . . . We consider ourselves guardians of nature, the earth, and especially the rivers."[18] COPINH's struggle has been taken up by one of Cáceres's daughters, who is reinforcing the messages of water and the connection between women, Indigenous peoples, and nature.

Also worth mentioning here are the women in the foothills of Orinoquía and Amazonía—in the eastern cordillera of Colombia, the region of the Caribbean and the Magdalena River Basin—who are defending the rivers against the expansion of the petroleum frontier.[19] But any listing would be incomplete. What is most noteworthy is how a language of valuation opposed to the dominant territoriality is repeated and amplified in defense of rivers, hydrologic basins, glaciers—an entire feminist political ecology of water that marks the interconnection between water, life, biodiversity, and nature.

In summary, the messages of water are multiple. Doubtless the process of appropriation and privatization of water will intensify on the basis of its listing on Wall Street, beginning in 2020. As Ecuador's Acción Ecológica, mostly composed of women, declares: "Every click on the stock exchange computers will have an effect on the water found in nature and on the land where communities live; it is a direct attack on the rights of rural and urban populations to have access to water for their survival."[20]

BODIES-TERRITORIES AND OTHER POSSIBLE FEMINISMS

To the political ecology of water and its multiple voices must be added the praxis of and reflection on the "body-territory" and the "body as territory" of recent years, an inflection specific to the communitarian feminism of Central American Indigenous women. The perspective was introduced by Lorena Cabnal of the Red Ancestral de Sanadoras del Feminismo Comunitario (Ancestral network of healers of communitarian feminism), or Tzk'at (in the K'iche' Maya language), for whom the notion of body-territory connects different types of violence: patriarchal, colonial, and extractivist. Cabnal expresses the idea powerfully:

> You see how the compañeros zealously defend the land, but look at what happens to the women. Right here they're raping girls and women. White and mestizo men aren't doing that. Indigenous men are. . . . what happened? From this comes our first watchword: "my body, my first territory of defense." In 2007, the struggle against mining awoke with greater force and we began to call the Indigenous government to account, saying: as Indigenous people we aren't coherent, we defend the territory-earth but we don't defend the territory-body. Because we women are defending this territory but within it we're getting killed.[21]

Communitarian, decolonizing feminism counts on healing "as a personal and political act," a process that seeks women's spiritual recovery on the basis of reconnecting with bodies and nature and that returns to ancestral knowledge, questioning both the patriarchy in its various modalities and neoliberal and extractivist capitalism.[22]

We have here a praxis and a political epistemology that posit other possible feminisms and other possible modernities. In the words of Aura Lolita Chávez, a member of Guatemala's Consejo de Pueblos K'iche' por la Defensa de la Vida, Madre Naturaleza, Tierra y Territorio (CPK; Council of K'iche' peoples for the defense of life, mother nature, land, and territory), "Other feminist worlds are possible where patriarchies don't exist, neither ancestral, Western, nor business patriarchies, where it is possible to dream that transnational enterprises, like hydroelectric mines and petroleum and other monocultures, have left and are no longer here."[23]

This feminism also has diverse expressions. The differences may be explained in part by the way in which the role of patriarchy in the pre-Hispanic world, and its relation to Western patriarchy, are problematized. While some associate patriarchy only with colonial history (that is, with the West), others call attention to its "refunctionalization" (Cabnal) or its "colonial connection" (Julieta Paredes, Asamblea Feminista, Bolivia) within the framework of present rural Indigenous communities.[24] The most disruptive communitarian feminisms, however, are those that denounce the idea of refunctionalization, emphasizing the triple violence—patriarchal, colonial, and extractivist—on territories/bodies. The power of this anticolonial narrative has been such as to challenge academic-militant feminist activism. Collectives have emerged to create workshops and develop critical tools to map the body-territory, including, for example, the Colectivo Miradas Críticas del Territorio desde el Feminismo (Collective of critical views of the territory from a feminist perspective; Ecuador, Mexico, Spain, Brazil, Uruguay, Peru), the Colectivo de Geografía Crítica de Ecuador (Ecuadoran collective of critical geography), and others. "In the mapping of the body there appear the wounds, marks, particular memories, places, spaces, ways of knowing and doing as part of the registers of bodies through which we are able to tell our personal histories in different territories," declares the Colectivo Miradas Críticas del Territorio desde el Feminismo.[25] It is worth transcribing some of the testimonies collected by the collective's workshops, the first from a Kichwa woman from the Ecuadoran sierra: "I made myself: I'm a lagoon, a plain, the birds, the rainbow. I represent myself as part of nature, of the territory, that I myself am a part of. All of us human beings are in harmony." Another, from a Colombian woman affected by a dam: "When I had a thrombosis, I heard the dam in my body."[26]

The possibility of representing the corporeality of conflicts illustrates the productiveness of the critical cartography of bodies/territories, in a context of sisterhood and mutual support among women of different occupations, ethnicities, and social classes. From this space, ties are woven with urban feminisms that are more centered on struggles against femicides and the expansion of rights. Such ties are neither immediate nor obvious, because ecoterritorial feminisms, more centered on the idea of the sustainability of life, tend to differ from the urban feminisms that demand abortion rights. But the narrative of the corporeality of struggle and the marks of violence offers a possibility of connection between communitarian ecoterritorial feminisms and urban feminisms, since denouncing violence is also the trigger for the demand for the autonomy of the body.

ACCESS TO LAND AND FOOD SOVEREIGNTY

Disputes over land and territory empower the role of women in the rural environment. Historically, women have played a crucial role in the production of food and the transmission of ancestral knowledge, a strategic role based on the sexual division of labor. It follows that for years now, as Claudia Korol writes, "The women of the Coordinadora Latinoamericana de Organizaciones del Campo (CLOC) and La Vía Campesina Internacional have been advocating for a rural feminism, whose axes are the cultivation of native seeds, the struggle for food sovereignty and thoroughgoing land reform, and against patriarchal violence."[27]

The difficulty women faced in planting their flag and challenging gender stereotypes in the early years of La Via Campesina (LVC) appears clearly in the words of Francisca Pancha Rodríguez, leader of the Asociación Nacional de Mujeres Rurales e Indígenas (ANAMURI; National association of rural and Indigenous women), Chile:

> The process of discussion and debate on food sovereignty enabled the recognition and valorization of our activities as subsistence farmers, namely, that we women have been fundamental to the development of agriculture and continue to be key to the production and preparation of food. So our space within LVC is not decorative; we have gender parity because we demand and invoke the right to be equal. If we're equal in field labor, we should also be equal in directing the movement. Because of this, we women always hold an assembly before the conference, in which we speak of our issues, look at our actions, and make proposals to the movement.[28]

Central American countries such as Costa Rica and Nicaragua also harbor numerous agro-ecological feminisms that challenge gender stereotypes. Here, Indigenous and Afro-descended women of all ages seek to carry out healing work in the aftermath of war, drought, and the consequences of climate change.[29]

Another type of woman-led organized experiment is the cooperative La Verdecita, an agro-ecological farm in operation since 2002 in the province of Santa Fe, Argentina. This experiment emerged in a territory besieged by agribusiness, and many of its members are migrants from other provinces and neighboring countries. Alongside it, the Escuela Vocacional Agroecológica (Agro-ecological vocational school) has embraced the mission of providing informal training, open to the community, since 2010.[30]

In sum, the denunciation of corporatist and environmental violence—whose brutal face is agribusiness, which advances through deforestation, bulldozers, forced evictions, and the commodification of the earth—has as its flip side a regenerative discourse and praxis through the creation of spaces of reexistence characterized by the cultivation of seeds and the land, the valorization of ancestral, agro-ecological knowledge, and the recovery of medicinal plants, all in the name of food sovereignty. Added to these practices is the denunciation of machismo and the patriarchal model, not only from outside the community but also from within it, in tune with a model of inhabiting territory that includes the democratization of gender in terms of access to the land.

"THE RIVER TOLD ME"

We live on a planet of ontological plurality, founded on the idea of a multiplicity of worlds and pluriverses, of interculturalism, of respect for other ways of understanding culture and organizing life. Latin American ecoterritorial feminisms add to already existing ecofeminisms an ecofeministically oriented praxis and narrative centered on the defense of water, on bodies-territories, on food sovereignty, on environmental justice as social and gender justice. Associated with spirituality and emotion, they illustrate an interconnection with the Earth and the whole of life, opposed to the dominant one.

The COVID-19 pandemic and the multiple other crises we are experiencing have shown how necessary it is to transform the relationship between society and nature, superseding the dualist anthropocentric paradigm, a conception and linkage at the origin of the models of misdevelopment that support the idea of unlimited growth and the expansion of the frontiers of commodification. Additionally, the pandemic has made visible the importance of care in its multiple dimensions. More than ever, the public-health crisis and its economic repercussions have made evident the unsustainability of the present organization of care and the tasks of communitarian self-management that fall to women, especially to poor women. Even before the pandemic, "women devoted three times as much time than men to unremunerated care work, a situation aggravated by the growing demand for care and the reduction of service provision caused by the measures of confinement and social distancing adopted to curb the health crisis."[31] Thus, more than ever, any discussion of care as a human right necessarily involves its democratization and universalization.

It is worth noting the epochal shift in social activism, now leaning toward women and youth. The rise of progressive governments with the commodities boom (2000) and the questioning of neoliberalism had Indigenous peoples as its major protagonists and bearers of a new emancipatory narrative (*buen vivir* [a dignified life], the rights of nature, autonomy, the plurinational state). The end of the progressive cycle in 2015, the worsening socioecological crisis, and the ascent of new authoritarian right-wing movements reflect the prominence of women, illustrated by the various ecoterritorial and urban feminisms, with an important presence of young women of different social and ethnic origins. Expressed otherwise, between 2000 and 2015 we went from the "Indianist moment" to the "feminist moment," a tendency that accompanies the narrative of *buen vivir* and the rights of nature and adds to it the ecofeminist language of the defense of the body as territory against all types of violence, of the defense of water for life, of the belief that women are the guardians of the earth. This passage may be exemplified by new political leaders such as Francia Márquez, now vice-president of Colombia. This Afro/ecofeminist leader illustrates an intersectional struggle, bringing together antiracism, the defense of the land (her struggle began twenty years ago against mining in the Colombian Pacific region), and the defense of diversity and democracy.

Such activism must also be seen in the light of the increase in gendered violence. I have written elsewhere about how, alongside the expansion of territorial and socioenvironmental conflict, violence against defenders of the environment has worsened.[32] In a context of marked wage asymmetries, gender inequalities have been accentuated and the traditional role of women has been fortified. Sexual exploitation and violence against women (trafficking, prostitution, femicide) have increased, as well as the chains of violence—physical and sexual—against female defenders of the environment. Following the mining and petroleum route, in the heat of the masculinization of territories and the twenty-first century's expansion of extractivisms, it is possible to see how trafficking, prostitution, and physical violence against women are also expanding, unveiling the revival of a matrix of patriarchal domination. The murder of Berta Cáceres and those that followed show that this femicide was not an isolated incident. In an article of 2019, the scholar Mina Navarro provides a shocking list of murdered female defenders of territory in Mexico.[33]

In summary, at a moment when we are rethinking our ties to nature within the framework of the environmental and civilizational crisis, the ecoterritorial feminisms of the South point toward a profound renewal. They contribute to a reformulation of the ties between the human and the nonhuman, the questioning of the false sense of autonomy or of exteriority with regard to nature, and an understanding of the negative role of androcentrism at the basis of our modern conception of the world, of science and technology at the service of corporations, of the belief in unlimited growth. In the embodied struggle for the defense of the earth and the land, women feel themselves to be, and to live as, "guardians of nature," but far from falling into a kind of essentialist ecofeminism, this belief articulates a decolonizing narrative that questions capitalism and the patriarchy. At the same time, it forges an epistemology of affects and emotions by placing on the agenda the problem of health and our interrelation with nature—our spiritual and material contact with other, nonhuman sentient beings, with water, hills and mountains, seeds and plants. This interconnection and spirituality are reflected in the brief, concise phrase of the great Honduran activist Berta Cáceres: "The river told me." The phrase appears tossed off, but testifies to a situated political praxis and the horizon of a relational ecofeminist epistemology that insists on reminding us that we are part of an interconnected whole indistinctly named "Pacha," "Mother Earth," "Nature."

1 See Alicia H. Puleo, "Ecofeminismo: la perspectiva de género en la conciencia ecologista," *Claves del ecologismo social* (Madrid: Libros en Acción, Ecologistas en Acción, 2009), 169–72.

2 Maristella Svampa, "Feminismos ecoterritoriales en América Latina. Entre la violencia patriarcal y extractivista y la interconexión con la naturaleza," *Documento de trabajo* no. 59 (Madrid: Fundación Carolina, 2021). Available online at https://www.fundacioncarolina.es/feminismos-ecoterritoriales-en-america-latina-entre-la-violencia-patriarcal-y-extractivista-y-la-interconexion-con-la-naturaleza/ (accessed April 16, 2024).

3 Francesca Gargallo Celentani, *Feminismos desde Abya Yala. Ideas y proposiciones de las Mujeres de 607 pueblos en nuestra América* (Bogotá: Ediciones Desde Abajo, 2015), 148.

4 Yayo Herrero, "Apuntes introductorios sobre el Ecofeminismo," *Boletín de recursos de información* no. 43 (June 2015). Available online at https://publicaciones.hegoa.ehu.eus/uploads/pdfs/278/Boletin_nº43.pdf?1488539850 (accessed April 16, 2024).

5 The social scientist Marian Solá Álvarez uses the concept of "ecofeminist praxis" in her analysis of the leading role of women in the struggle against mining in La Rioja, Argentina, linked to the defense of water and their identification with the peak of the Sierra de Famatina. See Solá Álvarez, "El conflicto socioambiental en torno a la minería a gran escala en la provincia de La Rioja, Argentina. Territorios en disputa y praxis ecofeministas," PhD diss., Universidad de Buenos Aires, 2021.

6 See Paola Bolados García and Alejandra Sánchez Cuevas, "Una ecología política feminista en construcción: El caso de las 'Mujeres de zonas de sacrificio en resistencia,' Región de Valparaíso, Chile," in *Psicoperspectivas. Individuo y Sociedad* 16, no. 2 (2017): 33–42.

7 Soledad Fernández Bouzo, *Mujeres del río*, video documentary (Buenos Aires: Instituto de Investigaciones Gino Germani, 2018). Available online at https://www.youtube.com/watch?v=6IlwE5ZocIw (accessed April 18, 2024).

8 See Jaime Breilh, "La epidemiología crítica: una nueva forma de mirar la salud en el espacio urbano," *Salud Colectiva* (Buenos Aires) 6, no. 1 (January–April, 2010): 83–109. Available online at https://www.scielosp.org/pdf/scol/2010.v6n1/83-101/es (accessed April 18, 2024).

9 At least 12 million people live in zones where more than 500 million liters of agrotoxics are dumped annually and where the levels of (no longer potential) exposure have risen to forty to eighty liters/kilos per person per year. See Svampa and Enrique Viale, *El colapso ecológico ya llegó. Una brújula para salir del (mal)desarrollo* (Buenos Aires: Editorial Siglo XXI, 2020), 118.

10 Testimony collected in Mauricio Berger and Cecilia Carrizo Sineiro, *Afectados Ambientales. Aportes conceptuales y prácticas para la lucha por el reconocimiento y la garantía de derechos* (Córdoba, AR: Ediciones Ciencia y Democracia, 2019), 15–16.

11 Ibid.

12 Ibid., 125.

13 See Nelly Luna Amancio, "Sed y conflicto en los andes peruanos," *Ojo Público*, December 18, 2016. Available online at https://ojo-publico.com/348/sed-y-conflicto-en-los-andes-peruanos (accessed April 22, 2024).

14 Ibid.

15 See Red Latinoamericana de Mujeres Defensoras de Derechos Sociales y Ambientales, "Memoria Anual 2020." Available online at https://www.redlatinoamericanademujeres.org/

(accessed April 22, 2024).

16 See Observatorio Plurinacional De Aguas, "Francisca Fernández Droguett: hidropolítica del despojo: hacia una intensificación del extractivismo y de las resistencias," April 6, 2021. Available online at https://oplas.org/sitio/2021/04/06/francisca-fernandez-droguett-hidropolitica-del-despojo-hacia-una-intensificacion-del-extractivismo-y-de-las-resistencias/ (accessed April 22, 2024).

17 See Marina Ertzogue and Monise Busquets, "'El agua es de la gente, no de Belo Monte.' Represas y pérdida de redes de sociabilidad entre las poblaciones afectadas, representadas en arpilleras amazónicas," *Tabula Rasa* no. 30 (January–June 2019): 109–31.

18 Berta Cáceres, quoted in Alejandra Martins, "Honduras: matan a Berta Cáceres, la activista que le torció la mano al Banco Mundial y a China," *BBC News Mundo*, April 24, 2015, rev. March 3, 2016. Available online at www.bbc.com/mundo/noticias/2015/04/150423_honduras_berta_caceres_am (accessed April 22, 2024).

19 See Tatiana Roa Avendaño and Luisa María Navas, eds. the poem reads, *Extractivismo, conflictos y resistencias* (Bogotá: Censat-Agua Viva, Amigos de la Tierra de Colombia, 2014). Available online at http://extractivismo.com/wp-content/uploads/2016/07/RoaNavasExtractivismoConflictosResistencias.pdf (accessed April 23, 2024).

20 Acción Ecológica, "El agua en la bolsa de valores y en los bolsillos de los especuladores y empresarios," December 30, 2020. Available online at www.accionecologica.org/el-agua-en-la-bolsa-de-valores-y-en-los-bolsillos-de-los-especuladores-y-empresarios/ (accessed April 22, 2024). See also the campaign we undertook in Argentina with various collectives, including the Pacto Ecosocial e Intercultural del Sur and Escritoras No Hay Cultura Sin Mundo, on *Los Mensajes del Agua* (The messages of water); available online at https://docs.google.com/forms/d/e/1FAIpQLS-cYRoKwqRB7Yr-zPWbadiRnErF-sk-MTJfHqEsrEv_0Dot13Lg/viewform?vc=0&c=0&w=1&flr=0&fbclid=IwAR2ywcXN8yGyrX7Cw8t-8wHJRTPnQM6eptrVAboiHV1LS-2MOzwQ4bKOldMI4 (accessed April 22, 2024).

21 Lorena Cabnal, quoted in Florencia Goldsman, "Lorena Cabnal: 'Recupero la alegría sin perder la indignación, como un acto emancipatorio y vital,'" *Pikara*, November 13, 2019. Available online at https://www.pikaramagazine.com/2019/11/lorenacabnal-recupero-la-alegria-sin-perder-la-indignacion-como-un-acto-emancipatorio-y-vital/ (accessed April 22, 2024).

22 Cabnal, "Las niñas no se tocan," *DW*, October 26, 2020. Available online at https://decolonial.hypotheses.org/2147 (accessed April 22, 2024).

23 Aura Lolita Chávez, quoted in Anayeli García Martínez, "Lolita Chávez: 'Me llaman loca, guerrillera y bruja' por defender la tierra," *Prensa Comunitaria*, October 13, 2017. Available online at https://prensacomunitaria.org/2017/10/lolita-chavez-me-llaman-loca-guerrillera-y-bruja-por-defender-la-tierra1/ (accessed April 23, 2024).

24 See Julieta Paredes, "Hilando fino. Desde el feminismo comunitario" (Lima: Lesbianas Independientes Feministas y Socialistas, 2008). Available online at http://mujeresdelmundobabel.org/files/2013/11/Julieta-Paredes-Hilando-Fino-desde-el-Fem-Comunitario.pdf (accessed April 23, 2024). Differences are also

to be found in the academic world. Whereas Rita Segato sees a low-intensity patriarchy in the communitarian forms of pre-Hispanic Indigenous villages, Maria Lugones argues that patriarchy did not exist in the pre-Hispanic Indigenous world. See Segato, "Género y colonialidad: en busca de claves de lectura y de un vocabulario estratégico descolonial," in Karina Bidaseca and Vanesa Vazquez Laba, eds., *Feminismos y poscolonialidad: Descolonizando el feminismo desde y en América Latina* (Buenos Aires: Ediciones Godot, 2016), 11–40.

25 Colectivo Miradas Críticas del Territorio desde el Feminismo, "Mapeando el Cuerpo-Territorio. Guía metodológica para mujeres que defienden sus territorios" (Quito: Colectivo Miradas Críticas del Territorio desde el Feminismo, 2017). Available online at https://miradascriticasdelterritoriodesdeelfeminismo.files.wordpress.com/2017/11/mapeando-el-cuerpo-territorio.pdf (accessed April 23, 2024).

26 Ibid.

27 Claudia Korol, "Feminismos populares: Las brujas necesarias en los tiempos de cólera," in Korol, ed., *Feminismos populares. Pedagogías y Políticas* (Buenos Aires: América Libre-El Colectivo, 2016). Available online at https://www.comisionporlamemoria.org/wp-content/uploads/sites/21/2018/03/Korol-Feminismos-populares.pdf (accessed April 23, 2024).

28 Francisca Rodríguez, in Eneko Calle García, "'Somos las guardianas de la tierra,'" *Revista Pueblos*, August 3, 2017. Available online at https://viacampesina.org/es/8960-2/ (accessed April 23, 2024).

29 See *La Agroecológa* no. 4 (April 2020). Formerly available at http://agroecologa.org/wp-content/uploads/2021/04/WEB-LAAGRO4.pdf.

30 See Silvia Papuccio de Vidal, "La experiencia del Colectivo de Mujeres de La Verdecita en Argentina," *Leisa: Revista de Agroecología* 36, no. 1 (March 2020). Available online at https://leisa-al.org/web/revista/volumen-36-numero-01/la-experiencia-del-colectivo-de-mujeres-de-la-verdecita-en-argentina/ (accessed April 23, 2024).

31 CEPAL (Comisión Económica para América Latina y el Caribe) and ONU Mujeres (UN Women), "Cuidados en América Latina y el Caribe en tiempos de COVID-19. Hacia sistemas integrales para fortalecer la respuesta y la recuperación," United Nations, Brief v 1.1, August 19, 2020. Available online at https://repositorio.cepal.org/bitstream/handle/11362/45916/190829_es.pdf (accessed April 23, 2024).

32 See Svampa, *Neo-extractivism in Latin America, Socio-environmental Conflicts, the Territorial Turn, and New Political Narratives,* Cambridge Elements series (Cambridge: Cambridge University Press, 2019). In 2021, Global Witness recorded 200 murders of women defending the earth and the environment, an average of almost four people a week. Mexico is the country with the most recorded murders, with fifty-four in 2021. More than three-quarters of the recorded attacks occurred in Latin America. In Brazil, Peru, and Venezuela, 78 percent of the attacks occurred in Amazonia (Global Witness, 2022).

33 Mina Lorena Navarro Trujillo, "Mujeres en defensa de la vida contra la violencia extractivista en México," *Política y Cultura* no. 51 (Universidad Autónoma Metropolitana México, 2019): 11–29. Available online at https://www.redalyc.org/journal/267/26760772002/ (accessed April 23, 2024).

CONTRIBUTORS

CARLA ACEVEDO-YATES is a writer, researcher, and the Marilyn and Larry Fields Curator at the MCA Chicago.

JENS ANDERMANN is a professor of Spanish and Portuguese at New York University and the author of *Entranced Earth: Art, Extractivism, and the End of Landscape* (Northwestern, 2023).

LISA BLACKMORE is a curator, writer, and the director of entre—ríos, a confluence of projects that explore continuities between bodies of water, human bodies, and land.

HELENA CHÁVEZ MAC GREGOR is a philosopher, curator, and researcher at the Universidad Nacional Autónoma de México's Instituto de Investigaciones Estéticas.

PATRICIO DEL REAL is an architectural historian who teaches in the Department of History of Art and Architecture, Harvard University.

ARTURO ESCOBAR is professor of anthropology emeritus at the University of North Carolina at Chapel Hill, and coauthor of *Relationality: An Emergent Politics of Life beyond the Human* (Bloomsbury, 2024).

JOSÉ LUIS FALCONI is an assistant professor of art and human rights at the University of Connecticut and the president of the NGO Cultural Agents, Inc.

ARNAUD GERSPACHER is an art historian, a scholar of animal studies, and an adjunct assistant professor in the art department at City College, City University of New York.

JULIETA GONZÁLEZ is Head of Exhibitions at the Wexner Center for the Arts, Columbus, Ohio, and curator-at-large of modern and contemporary art at the Museu de Arte de São Paulo.

GASTÓN GORDILLO is a professor in the Department of Anthropology at the University of British Columbia and the author of *Rubble: The Afterlife of Destruction* (Duke, 2014), and other books.

EDUARDO GUDYNAS is a senior researcher at the Centro Latino Americano de Ecología Social (CLAES), Montevideo.

EDUARDO KOHN is an associate professor of anthropology at McGill University, Montreal.

DAVI KOPENAWA, from the Yanomami people, is a shaman, a spokesperson for the Yanomami, and one of the most respected Indigenous leaders in Brazil.

AILTON KRENAK, from the Krenak people, was born in Minas Gerais, Brazil, and is a leading environmental activist and campaigner for Indigenous rights.

MIGUEL A. LÓPEZ is a writer, researcher, former codirector and chief curator of TEOR/ética in Costa Rica, and, currently, curator of the 2024 Toronto Biennial of Art.

CARLA MACCHIAVELLO is an art historian and educator from Santiago, Chile. She is an associate professor of art history at the Borough of Manhattan Community College, City University of New York.

CAMILA MARAMBIO is a curator, writer, and somatic-care worker. They were founding director of Ensayos (2011–24) and are currently curator of new perspectives at Para la Naturaleza, Puerto Rico.

JOANNA PAGE is a professor of Latin American studies at the University of Cambridge and the director of the university's Centre for Research in the Arts, Social Sciences and Humanities.

CONTRIBUTORS

MARA POLGOVSKY EZCURRA is a filmmaker and senior lecturer in contemporary art at Birkbeck, University of London.

VÍCTOR MANUEL RODRÍGUEZ-SARMIENTO is a Colombian artist, independent curator, professor, and researcher in visual and cultural studies.

FLORESMILO SIMBAÑA is a political analyst and a former leader of the Confederación de Nacionalidades Indígenas del Ecuador (CONAIE).

IRENE V. SMALL is an associate professor in the Department of Art and Archaeology and is affiliated with the programs in Media and Modernity and Latin American Studies, Princeton University.

GRACIELA SPERANZA is a writer and critic. She was a professor of Argentine literature at the Universidad de Buenos Aires and is a professor in the Departamento de Arte, Universidad Torcuato Di Tella.

DANIEL STEEGMANN MANGRANÉ is a multidisciplinary artist based in Rio de Janeiro, Brazil, whose works explore such themes as nature, perception, and technology.

MARISTELLA SVAMPA is a sociologist, writer, and activist, and a senior researcher at the Consejo Nacional de Investigaciones Científicas y Técnicas (CONICET) in Argentina.

PAULO TAVARES is an architect, author, and educator.

GLADYS TZUL TZUL is a Maya K'iche' activist, public intellectual, sociologist, and visual artist.

PATRICIA VIEIRA is a research professor at the Centro de Estudos Sociais of the Universidade de Coimbra.

EDUARDO VIVEIROS DE CASTRO is a Brazilian anthropologist and a professor at the Museu Nacional, Universidade Federal do Rio de Janeiro.

INDEX

Italic type indicates an image.

PHOTOGRAPH CREDITS

In reproducing the images contained in this publication, the Museum obtained the permission of the rights holders whenever possible. If the Museum could not locate the rights holders, notwithstanding good-faith efforts, it requests that any contact information concerning such rights holders be forwarded so that they may be contacted for future editions.

Cover: © Ximena Garrido-Lecca, photograph by Ramiro Chaves. • P. 22: © Cisneros Institute, The Museum of Modern Art (MoMA), New York, photograph by María del Carmen Carrión. • P. 23: © Maria Thereza Alves. • P. 24: © and courtesy Ursula Biemann and Paulo Tavares. • P. 25: © Cisneros Institute, MoMA, New York, photograph by Natanael Guzmán. • P. 27: © BY-SA 4.0 by Aerocene Community, photograph by Florencia Montoya. • P. 28: © Carolina Caycedo and MoMA, New York, digital image © 2024 Imaging and Visual Resources Department, MoMA, New York, photograph by Jonathan Dorado. • P. 41: © Denilson Baniwa; courtesy A Gentil Carioca, Rio de Janeiro and São Paulo; photograph by DuHarte Fotografia. • P. 46: © Juan Downey. • P. 51: © Ximena Garrido-Lecca, photograph by Ramiro Chaves. • Pp. 66–67: © Familia Benedit, digital image © 2024 Imaging and Visual Resources Department, MoMA, New York, photograph by Robert Gerhardt. • P. 72: © Archivo Roth, photograph by Pedro Roth. • Pp. 75, 76: © Marta Minujín Archive. P. 76: photograph by Ignacio Lasparra. • Pp. 89, 92–93, 97: © Eduardo Kac, courtesy Henrique Faria Fine Art, New York. Pp. 92–93: photograph by Belisário Franca. • Pp. 108, 115, 120–21: © 2024 Cecilia Vicuña/Artists Rights Society (ARS), New York, courtesy the artist and Lehmann Maupin, New York, Hong Kong, Seoul, and London. P. 108: photograph by César Paternosto. Pp. 120–21: photograph by Christian Chierego. • P. 138: © and courtesy Bartolina Xixa. • Pp. 142–43: courtesy Studio Dan Lie, photograph by Edouard Fraipoint. • P. 146: © Seba Calfuqueo, documented by Sebastián Melo in Bosque Pewen, Fundación Mar Adentro, Chile. • Pp. 158–59: © and courtesy Adrián Villar Rojas, photograph by Carla Barbero. • Pp. 164–65: © Hulda Guzmán, courtesy Hulda Guzmán, Alexander Berggruen, New York, and Stephen Friedman Gallery, London. • P. 168: © Eduardo Navarro, courtesy Eduardo Navarro and Proyectos Ultravioleta, Guatemala City. • P. 191: courtesy Espace Frans Krajcberg, Centre d'art contemporain Art & Nature, Paris. Pp. 202–3: © Miguel Angel Rojas, photograph by Clemencia Echeverri. • Pp. 210–11: © 2024 José Alejandro Restrepo, digital image © 2024 Imaging and Visual Resources Department, MoMA, New York. • Pp. 214–15: © María Elvira Escallón, courtesy María Elvira Escallón and Galería Espacio Continuo, Bogotá. • Pp. 221, 228–29: © Vivian Suter, courtesy Vivian Suter; Gaga, Mexico City, Guadalajara, and Los Angeles; Gladstone Gallery, New York; Karma International, Zurich; and Proyectos Ultravioleta, Guatemala City. P. 221: photograph by Sebastian Lendenmann. Pp. 228–29: photograph by Margo Porres. • P. 247: © Allora & Calzadilla, courtesy Allora & Calzadilla and Galerie Chantal Crousel, Paris; Kurimanzutto, Mexico City; and Lisson Gallery, London et al. Photograph © Jack Hems. • Pp. 252–53: © Allora & Calzadilla, courtesy Allora & Calzadilla and Dia Art Foundation, New York, photograph © Myritza Castillo/Dia Art Foundation, New York. • P. 263: © Regina José Galindo, courtesy Prometeo Gallery di Ida Pisani, Milan, photograph by Bernard Huet. • P. 269: © María Evelia Marmolejo, photograph by Fabio Arango. • P. 270: © Beatriz Santiago Muñoz. • Pp. 286, 289, 300: courtesy Archivo Histórico José Vial Armstrong, Pontificia Universidad Católica de Valparaíso. • P. 308: courtesy Ursula Biemann. • Pp. 314–15: © Christy Gast. • P. 324: © Ines Linke and Louise Ganz (thislandyourland). • P. 338: © Ala Plástica–Proyecto Junco: Especies Emergentes 1995, photograph by Alejandro Meitin. • P. 352: © and courtesy Mujeres Creando. • P. 355: © Margarita Azurdia, courtesy Colección Milagro de Amor. • P. 363: © and courtesy Valentina Desideri and Denise Ferreira da Silva, digital image designed by Arely Amaut. • Pp. 376–77: © 2024 Fernando Palma Rodríguez, all rights reserved; courtesy the artist and Gaga, Mexico City, Guadalajara, and Los Angeles; photograph by Gaga.

This publication is a project of the Patricia Phelps de Cisneros Institute for the Study of Art from Latin America, The Museum of Modern Art, New York.

Produced by the Department of Publications, The Museum of Modern Art, New York

Michelle Kuo, Chief Curator at Large and Publisher
Curtis R. Scott, Associate Publisher
Joseph Mohan, Production Director
Hannah Kim, Business and Marketing Director

Edited by David Frankel
Design by Chad Kloepfer with Willis Kingery
Production by Matthew Pimm
Index by Alan Dean
Proofread by A. F. Dunlap Smith
Printed and bound by Brizzolis, S.A., Madrid

The essays by Miguel A. López, Helena Chávez Mac Gregor, and Graciela Speranza were translated from the Spanish by Jen/Eleana Hofer. The essays by Eduardo Gudynas, Camila Marambio, Víctor Manuel Rodríguez-Sarmiento, Floresmilo Simbaña, Maristella Svampa, and Gladys Tzul Tzul were translated from the Spanish by Christopher Winks.

This book is typeset in Union and Rhetorik Serif. The paper is 100 gsm Munken Lynx.

Published by The Museum of Modern Art
11 West 53 Street
New York, NY 10019-5497
www.moma.org

Library of Congress Control Number: 2024946300
ISBN: 978-1-63345-148-3

Distributed by Duke University Press, Durham, North Carolina
www.dukeupress.edu

Cover: Ximena Garrido-Lecca. *Insurgencias botánicas: Phaseolus Lunatus* (Botanical insurgencies: Phaseolus lunatus; detail). 2017. Installation, dimensions variable. Collection the artist. The colors of this photograph have been deliberately shifted in the printing process, which includes fluorescent inks; see pp. 50–51 for the original image.

Printed in Spain